Gmelin Handbook of Inorganic Chemistry

8th Edition

Volumes published on "Rare Earth Elements" (Syst. No. 39)

A 1 Introductory Review. History. Occurrence – 1938

A 2 Scandium: History. Occurrence – 1973

A 3 Y, La, and the Lanthanoids: Cosmochemistry – 1980

A 4 Y, La, and the Lanthanoids: Crystal Chemistry – 1979

A 5 Y, La, and the Lanthanoids: Geochemistry – Total Earth. Magmatic Cycle – 1981

* **A 7** Y, La, and the Lanthanoids: Minerals (Excluding Silicates) – 1984 (present volume)

° **B 1** Sc, Y, La, and the Lanthanoids: History, Position in the Periodic System. Separation from Raw Materials – 1976

B 2 Sc, Y, La, and the Lanthanoids: Separation of the Rare Earth Elements from One Another. Preparation of the Pure Metals. Uses. Toxicology – 1976

B 3 Sc, Y, La, and the Lanthanoids: Physical Properties of the Metals – 1974

° **B 4** Sc, Y, La, and the Lanthanoids: Nuclear, Atomic, and Molecular Properties – 1976

B 5 Sc, Y, La, and the Lanthanoids: Preparation, Enrichment, and Separation of Isotopes (Sc, Y, La to Sm) – 1978

B 6 Sc, Y, La, and the Lanthanoids: Preparation, Enrichment, and Separation of Isotopes (Eu to Lu). Detection and Determination of Isotopes. Chemical Reactions of the Elements – 1978

B 7 Sc, Y, La, and the Lanthanoids: Reactions of the Ions in Solution. Electrochemical Behavior – 1979

C 1 Sc, Y, La, and the Lanthanoids: Hydrides. Oxides – 1974

C 2 Sc, Y, La, and the Lanthanoids: Compounds with H + O, Alkali Metals + O, N, as well as the Related Alkali Double Salts – 1974

C 3 Sc, Y, La, and the Lanthanoids: Fluorides, Fluoride Oxides, and the Related Alkali Double Salts – 1976

* **C 4a** Sc, Y, La, and the Lanthanoids: Chlorides. Comparative Data – 1982

* **C 4b** Sc, Y, La, and the Lanthanoids: Data on Individual Chlorides – 1982

C 5 Sc, Y, La, and the Lanthanoids: Oxide Chlorides, Hydroxide Chlorides, Salts of Oxoacids of Chlorine, and Alkali Chlorometallates – 1977

C 6 Sc, Y, La, and the Lanthanoids: Bromides. Iodides, Corresponding Basic Halides. Salts of Halogen Oxoacids and Alkali Double Salts – 1978

* **C 7** Sc, Y, La, and the Lanthanoids: Sulfides, Oxide Sulfides, Alkali Thiometallates – 1983

° **C 8** Sc, Y, La, and the Lanthanoids: Sulfide Halides. Sulfates, Salts of Other Sulfuric Acids and Corresponding Alkali Double Compounds – 1981

* **D 1** Sc, Y, La, and the Lanthanoids: Coordination Compounds 1 – 1980

* **D 2** Sc, Y, La, and the Lanthanoids: Coordination Compounds 2 – 1982

* **D 3** Sc, Y, La, and the Lanthanoids: Coordination Compounds 3 – 1981

* **D 6** Sc, Y, La, and the Lanthanoids: Ion Exchange and Solvent Extraction Reactions. Organometallic Compounds – 1983

* Completely or ° in part in English

Gmelin Handbook of Inorganic Chemistry

8th Edition

Gmelin Handbuch der Anorganischen Chemie

Achte, völlig neu bearbeitete Auflage

Prepared and issued by
Gmelin-Institut für Anorganische Chemie der Max-Planck-Gesellschaft zur Förderung der Wissenschaften
Director: Ekkehard Fluck

Founded by
Leopold Gmelin

8th Edition
8th Edition begun under the auspices of the Deutsche Chemische Gesellschaft by R. J. Meyer

Continued by
E. H. E. Pietsch and A. Kotowski, and by Margot Becke-Goehring

Springer-Verlag Berlin · Heidelberg · New York · Tokyo 1984

Gmelin Handbook of Inorganic Chemistry

8th Edition

Sc, Y, La–Lu Rare Earth Elements

A 7

Y, La, and the Lanthanoids

Minerals (Excluding Silicates)

With 13 illustrations

AUTHORS Isa Kubach, Peter Schubert

EDITORS Isa Kubach, Wolfgang Töpper

CHIEF EDITOR Isa Kubach

System Number 39

Springer-Verlag Berlin · Heidelberg · New York · Tokyo 1984

LITERATURE CLOSING DATE: 1982
IN SOME CASES MORE RECENT DATA HAVE BEEN CONSIDERED

Library of Congress Catalog Card Number: Agr 25-1383

ISBN 3-540-93489-8 Springer-Verlag, Berlin · Heidelberg · New York · Tokyo
ISBN 0-387-93489-8 Springer-Verlag, New York · Heidelberg · Berlin · Tokyo

Typesetting, printing, and bookbinding: LN-Druck Lübeck

Preface

The present volume Rare Earth Elements A 7 treats the RE minerals with the exception of the RE silicates which together with the RE deposits, will follow as Rare Earth Elements A 8, in 1984. The minerals with RE in their formulas are classified according to their anions. Occurrence and paragenesis, chemistry, crystal forms and crystal structures, as well as physical properties and chemical behavior of each mineral are described in detail. RE minerals which have been treated in earlier Gmelin volumes are included with references to these volumes.

The carbonates of the RE minerals comprise the largest portion of this volume. The bastnaesite group is the most important within the carbonates due to their mutual cationic (RE, Ca, Ba) and anionic (F, OH) diadochic replacements as well as their syntactic intergrowths and polycrystals.

The geochemistry of the rare earth elements will be completed by volume A 6 which will contain the sedimentary and metamorphic cycle of RE as well as their contents and behavior in the hydrosphere and atmosphere.

Frankfurt am Main
November 1983

Isa Kubach

Table of Contents

Page

10 Minerals . . . 1

10.1 Halogenides . . . 1

10.1.1 Yttrofluorite . . . 1
Occurrence and Paragenesis. Alteration . . . 1
Chemistry . . . 2
Crystal Forms and Structure . . . 3
Physical Properties . . . 4
Chemical Behavior . . . 5

10.1.2 Cerfluorite . . . 5

10.1.3 Tveitite . . . 7

10.1.4 Gagarinite . . . 8
Occurrence and Paragenesis. Alteration . . . 8
Chemistry . . . 8
Crystal Forms and Structure . . . 11
Physical Properties . . . 11
Chemical Behavior and Differential Thermal Analysis . . . 13

10.1.5 Tysonite . . . 13
Occurrence and Paragenesis. Alteration . . . 13
Chemistry . . . 16
Crystal Forms and Structure . . . 17
Physical Properties . . . 19
Chemical Behavior and Differential Thermal Analysis . . . 21

10.1.6 Chukhrovite . . . 21

10.2 Oxidic Compounds . . . 23

10.2.1 Simple Oxides . . . 23
Cerianite . . . 23
RE-bearing Uraninite . . . 25
History and Occurrence . . . 25
Chemistry . . . 25
Crystal Forms and Structure . . . 26
Physical Properties . . . 27
Mozambikite . . . 27
"Thucholite" . . . 28
Hibonite . . . 28
Occurrence and Paragenesis . . . 28
Chemistry . . . 29
Crystal Forms and Structure . . . 30

	Page
Optical Properties	30
Other Physical Properties	30
Chemical Behavior	31
10.2.2 Complex Oxides with Ti (and sometimes Nb)	31
Davidite	31
Occurrence and Paragenesis	32
Chemistry	34
Crystal Forms and Structure	36
Optical and Other Physical Properties	38
Chemical Behavior and Differential Thermal Analysis	41
Loveringite	41
Unnamed Mineral	42
Knopite	43
Yttrocrasite	44
Brannerite	45
Paragenesis, Inclusions, and Alteration	46
Occurrence and Paragenesis	46
Inclusions and Alteration	48
Chemistry	50
Crystal Forms and Crystal Structure	52
Optical Properties	54
Other Physical Properties	56
Chemical Behavior	58
Loparite	59
Dysanalyte	59
Latrappite	59
Murataite	59
Zirconolite	60
10.2.3 Complex Oxides with Nb, Ta$\pm$Ti	60
Koppite	60
Brocenite	60
Fergusonite	62
Formanite	62
Ashanite	62
Yttrotantalite	62
Samarskite	62
Fersmite	62
10.2.4 Complex Oxides with Nb, Ta, and Ti	63
Yttrobetafite	63
Schetelegite	63
Yttromicrolite	63
Yttropyrochlore	63

	Page
Ceriopyrochlore	64
Pisekite	64
Euxenite	64
Tanteuxenite	64
Polycrase	64
Kobeite	65
Blomstrandine	65
Eschynite (Aeschynite)	65
Polymignite	65
Vigezzite	65
Sinicite	66
10.3 Hydroxides	66
10.3.1 Yttrotungstite	66
10.3.2 Cerotungstite	68
10.4 Borates	68
Braitschite	68
10.5 Carbonates	69
10.5.1 Anhydrous Carbonates	69
Sahamalite	69
Carbocernaite	71
Occurrence and Paragenesis	71
Chemistry	71
Crystal Forms and Structure	72
Physical Properties	74
Chemical Behavior and Differential Thermal Analysis	74
Burbankite	75
Occurrence and Paragenesis	75
Chemistry	77
Crystal Forms and Crystal Structure	80
Optical Properties	82
Other Physical Properties	84
Chemical Behavior and Differential Thermal Analysis	85
Khanneshite	85
Ewaldite	87
Occurrence and Paragenesis	87
Chemistry	87
Crystal Forms and Crystal Structure	88
Physical Properties	88

Page

10.5.2 Anhydrous Carbonates with Additional Anions 89

Review 89
RE-Ca Series 89
Subseries of the RE-Ca Series with Y for Ce and OH for F 92
RE-Ba Series 92

Bastnaesite Group 93
Name and Varieties 93
Occurrence and Paragenesis 94
Review 94
Magmatic Rocks 95
Granites and Granite Pegmatites 95
Alkali Syenites and Related Rocks 96
Carbonatites 96
Metasomatic Rocks 97
Hydrothermal Formations 98
Sedimentary and Metamorphic Rocks 99
Inclusions and Admixtures 102
Alteration 103
Replacement of Other Minerals by Bastnaesite 103
Replacement of Bastnaesite by Other Minerals 106
Chemistry 107
Crystal Forms and Intergrowths 117
Crystal Structure 122
Optical Properties 125
IR Spectra and Luminescence 131
Other Physical Properties 132
Chemical Behavior and Thermal Analyses 134

Parisite 137
Occurrence, Paragenesis, and Alteration 138
Chemistry 142
Crystal Forms and Crystal Structure 146
Morphology 146
Crystal Structure 148
Optical Properties 150
Other Physical Properties 153
Chemical Behavior and Thermal Analyses 153

Roentgenite 154

Synchysite 156
Name, Synonyms, and Varieties 156
Occurrence, Paragenesis, and Alteration 158
Occurrence and Paragenesis 158
Common Synchysite 158
Synchysite-(Nd) and Synchysite-(Y) 159
Alteration and Inclusions 160
Chemistry 162
Crystal Forms and Crystal Structure 167

	Page
Morphology	167
Crystal Structure	168
Physical Properties	171
Chemical Behavior and Thermal Analyses	174
Cordylite	175
Synonyms, Occurrence, and Paragenesis	175
Chemistry	176
Crystal Forms and Crystal Structure	177
Physical Properties	179
Chemical Behavior	180
Huanghoite	181
Name, Occurrence, and Paragenesis	181
Chemistry	181
Crystal Forms and Crystal Structure	183
Physical Properties	183
Chemical and Thermal Behavior	184
"Barium-Parisite" ("Cebaite")	185
Zhonghuacerite	186
Ambatoarinite	187
Unnamed Orthorhombic Y-Ca-RE Carbonate	188
10.5.3 Hydrous Carbonates	188
Calkinsite	189
Lanthanite and Lanthanite-(Nd)	191
Name and Synonyms	191
Occurrence and Paragenesis	191
Chemistry	192
Crystal Forms and Crystal Structure	193
Physical Properties	196
Chemical Behavior and Differential Thermal Analysis	197
Donnayite	198
Mackelveyite	199
Name, Occurrence, and Paragenesis	199
Chemistry	199
Crystal Forms and Crystal Structure	200
Optical and Other Physical Properties	203
Tengerite	203
Definition and Related Minerals	203
Occurrence and Paragenesis	204
Chemistry	205
Crystal Forms and Crystal Structure	207
Physical Properties	208
Chemical Behavior and Thermoanalysis	209
Lokkaite	209

Page

Unnamed Y-Ca Carbonates with H_2O 210

Monoclinic RE Carbonate with H_2O 212

10.5.4 Hydrous Carbonates with Additional Anions 212

Ancylite and Calcio-ancylite 212
Name and Varieties 212
Occurrence, Paragenesis, and Alteration 213
Occurrence and Paragenesis 213
Inclusions 216
Pseudomorphs 216
Alteration 217
Chemistry 218
Crystal Forms and Crystal Structure 221
Optical and Other Physical Properties 224
Chemical Behavior and Differential Thermal Analysis 228

Schuilingite 229

Bijvoetite 231

Lepersonnite 232

10.6 Sulfates 233

"Texasite" 233

10.7 Phosphates 233

10.7.1 Phosinaite 233

10.7.2 Unnamed Ce Phosphate 234

10.7.3 Monazite 235

10.7.4 Xenotime 235

10.7.5 Cheralithe 235

10.7.6 Vitusite 235

10.7.7 Churchite (Weinschenkite) 236

10.7.8 Rhabdophane 237

10.7.9 Ningyoite 237

10.7.10 Smirnovskite 237

10.7.11 Lermontovite 237

10.7.12 Unnamed Minerals U 2 and U 3 237

10.7.13 Cer- and Yttriumapatite, Saamite, Belovite 238

10.7.14 Florencite 238

10.7.15 Petersite 238

Page

10.8 Arsenates 238

10.8.1 Chernovite 238

10.8.2 Retzian and Retzian-(Nd) 241

10.8.3 Agardite 242

10.8.4 Kemmlitzite 243

10.8.5 Goudeyite 244

10.8.6 Unnamed Arsenate 245

10.9 Vanadates 245

10.9.1 Wakefieldite 245

10.9.2 Kusuite 245

Table of Conversion Factors 247

10 Minerals

10.1 Halogenides

10.1.1 Yttrofluorite $(Ca,Y)F_{2\ to\ 2.17}$

10.1.1.1 Occurrence and Paragenesis. Alteration

Yttrofluorite was first described in 1911 by Vogt [1], see "Seltenerdelemente" A 1, 1938, p. 56, from a pegmatite in Hundholmen, Tysfjord, northern Norway. A RE-fluorite of this locality (described as yttrofluorite by Vogt [1]) is later shown to be cerfluorite [2, pp. 250/1], see pp. 5/6. Further occurrences of yttrofluorite in Norway are Eivolden, Drag district, Tysfjord, where this mineral is associated with quartz, microcline, plagioclase, kainosite, and ? lombardite in a pegmatite [2, pp. 245, 247], Jennyhaugen in the Drag district, and Høydalen, Tørdal, Telemark, southern Norway; in the last two occurrences yttrofluorite is intergrown with tysonite [3], see also p. 18.

Russian deposits of yttrofluorite are: the Ukrainian Crystalline Shield, where it occurs in form of lenses and as crystals between block quartz and microcline at the rims of pegmatites within biotite-granite massifs [4, p. 106]; Ploskaya Mountain, Kola Peninsula, where yttrofluorite forms isometric grains together with monazite, zircon, and galena in a very late association of RE minerals in an amazonite pegmatite [5]; Eastern Siberia, where yttrofluorite occurs in metasomatic granitoids of the subalkalic plumasite, and in the alkalic agpaite types together with zircon, malacon, ilmenite, and, in places, chevkinite, and sometimes tysonite (fluocerite) [6]. An occurrence in Central Asia has yttrofluorite in form of irregular grains in alaskitic granites, paragneisses, and contact metamorphic, altered schists [7]. For yttrofluorite in metasomatic biotite-sericite-chlorite rocks, USSR, see [8].

Yttrofluorite from Suishoyama, Kawamatamachi, Fukushima Prefecture, Japan, occurs as irregularly shaped, fine grained agglomerates in spessartine, which is imbedded in potash feldspars of a pegmatite [9]. Yttrofluorite was also found in the area of the Little Spitzkopje, Namibia, southern Africa [10], and in a weathered carbonatite deposit from Salpeterkop, Republic of South Africa [26].

In the White Cloud pegmatite, South Platte district, Jefferson County, Colorado, "yttrofluorite"[1)] occurs as pseudomorphs after fluorite within the outer to intermediate zone of the pegmatite [12]. For yttrofluorite together with thalenite just outside the quartz core of this pegmatite, see [11, 13]. The description of these deposits can be found in "Rare Earth Elements" A 8, 1984, Chapter 11.7.2.3.3.2.

Inclusions in yttrofluorite from the Ukrainian Shield are RE carbonates, such as bastnaesite, parisite, synchisite [4, pp. 106/7]; yttrofluorite from the White Cloud pegmatite contains inclusions of bastnaesite, with some hematite and thalenite [11], and yttrofluorite from Central Asia, inclusions of yttroparisite, xenotime, monazite, and quartz [7].

Yttrofluorite from the Little Spitzkopje, Namibia, is altered to stiepelmannite [10], and yttrofluorite from the White Cloud pegmatite to thalenite and other minerals [11], e.g., synchysite-(Y) (doverite) [12]. Exsolution intergrowth of tysonite (fluocerite) + RE-fluorite is observed in yttrofluorite from Eastern Siberia [6].

[1)] Called yttrian fluorite by [11].

10.1.1.2 Chemistry

Analyses and Formulas

For a compilation of analyses and the $(RE,Y)_2O_3$ content, mean and variances of 9 samples, see "Seltenerdelemente" A 4, 1979, pp. 35, 7.

The name yttrofluorite has been used for minerals with various contents of ΣRE+Y: Adams and Sharp separate yttrian fluorite with 9.45% ΣRE+Y from yttrofluorite with 23.49% ΣRE+Y [11], see footnote p. 1. Gurov et al. have named yttrofluorite samples with 4.97, 6.07, and 7.87% ΣRE+Y; the formulas of the last two samples are given as $Ca_{0.95}(RE+Y)_{0.05}F_{2.04}$, and $Ca_{0.94}(RE+Y)_{0.06}F_{2.12}$, respectively [4, p. 108]. Further fluorites classified as yttrofluorites have ΣRE+Y in %: 18.0 [2, p. 251] and 18.83 (mean of three samples); the formula of the last minerals is $Ca_{1-n}(Y+RE)_nF_{2+n}$ [5, pp. 54/5]. Yttrofluorite with ΣRE+Y=18.84% has the formula $(Ca_{0.8107}(RE+Y)_{0.1613}Mg_{0.0158}Fe,Al,Mn)_{1.00}F_{2.1524}$ [9]. Two pure yttrofluorites contain 33.84 and 40.15 $\Sigma(RE+Y)_2O_3$ [6].

As can be seen from these formulas, Ca is replaced by Y and RE, see also [14], [5, p. 56], and under structure p. 3. A trace element is Th, occurring in traces in uranium-free yttrofluorite from Eivolden, Tysfjord [2, p. 249], and with 0.15 and 0.16% Th in two samples from Central Asia [7].

Distribution of Rare Earth Elements

Per definition, yttrofluorite has Y-earth+Y>Ce-earth [15], see e.g., Y-earth>Ce-earth in yttrofluorite from the Ukrainian Shield [4, p. 111], and Y-earth+Y/Ce-earth=1.85 and 1.89 in two samples of pure yttrofluorite from Eastern Siberia [6]. Yttrofluorites from Kola Peninsula are selectively yttrian with 62.6% Y (mean of three analyses) and 63.5% Y (one analysis) of ΣRE+Y; distribution of single RE in these yttrofluorites are as follows: RE_1 (La to Nd)=38.2 and 29.2%, RE_2 (Sm to Ho)=18.6 and 22.9%, and RE_3 (Er to Lu)=43.2 and 47.9% [5, p. 56]. For distribution of RE±Y in yttrofluorites from different localities see the following table:

Locality	$\Sigma(RE\pm Y)=100$							
	Y	La	Ce	Pr	Nd	Sm	Eu	Gd
Hundholmen, northern Norway	—	4.3	13	3.7	9	4.7	0.3	10
Suishoyama, Japan	—	—	2.9	—	3.4	5.9	—	9.8
Eivolden, northern Norway	63.3	2.1	6.9	0.9	5.4	2.6	—	5.4
Ukrainian Shield	56.6	4.5	9.6	1.6	6.6	2.5	—	3.8
Central Asia	54.8	4.4	11.0	1.0	4.7	2.5	0.1	5.1
White Cloud, Colorado	45.5	7.2	12.2	1.8	7.7	4.1	1.4	5.0

Locality	$\Sigma(RE\pm Y)=100$							Ref.
	Tb	Dy	Ho	Er	Tm	Yb	Lu	
Hundholmen, northern Norway	2	13	3	13	1.9	19	3.3[1)]	[16]
Suishoyama, Japan	3.2	22.6	3.2	24.4	3.9	20.6	—[2)]	[9]
Eivolden, northern Norway	—	5.9	—	4.3	—	3.3	—[3)]	[2, p. 250]
Ukrainian Shield	0.7	3.6	0.9	4.3	0.8	3.9	0.8[4)]	[4, p. 110]
Central Asia	0.9	5.2	1.0	2.5	0.4	5.1	1.0[5)]	[7]
White Cloud, Colorado	1.4	4.5	0.9	4.1	0.5	3.6	0.5[6)]	[11]

[1] After Vogt [1] the yttrofluorite from Hundholmen contains 17.35% Y-earth and 1.68% Ce-earth. An analysis of Zambonini [17] gives 10.59% YF_3 and 0.95% CeF_3, corresponding to Y = 6.45% and Ce = 0.68%. – [2] At.%; Y-content not given. – [3] Sum of sesquioxides (exception CeO) = 100.1; Y_2O_3 = 11.02%, RE_2O_3 = 6.98% of total analysis. – [4] Mean of 3 determinations. ΣRE+Y in these fluorites see p. 2. – [5] ΣCe(La to Nd) = 21.1%, ΣY(Sm to Lu+Y) = 78.6%. – [6] Calculated from wt% as elements; ΣRE+Y of this fluorite see p. 2. Distribution in % of La through Lu in at.% see [18].

10.1.1.3 Crystal Forms and Structure

Yttrofluorite is cubic [1, 19]. Crystals are rare, they sometimes show tetrahedral habit [4, p. 106]. The faces {100} and {111} are concavely bent at yttrofluorite from Spitzkopje, Namibia [10]. The cleavage is imperfect after {111} [1, 6, 19], imperfect after {100}, sometimes passing into parting [7], or there is no cleavage [5, p. 52].

For lattice parameters of yttrofluorite from the different occurrences compared with synthetic material, see the following table:

Locality	a in Å	Ref.
Hundholmen, northern Norway	5.50	[2, p. 247], [14]
	5.492	[20, p. 276]
Eivolden, northern Norway	~5.51(8)	[2, p. 247]
Jennyhaugen, northern Norway	~5.48(9)	[2, p. 247]
Tørdal, southern Norway	~5.48(2)	[2, p. 247]
Kola Peninsula, USSR	5.494 ± 0.0005	[5, p. 52]
Eastern Siberia, USSR	5.4982	[6]
	5.5141	[6]
Central Asia, USSR	5.446	[7]
Spitzkopje, Namibia	5.493 ± 0.005	[21]
Suishoyama, Japan	5.500	[9]
White Cloud, Colorado	5.51	[11]
Synthetic CaF_2 + 20 mol% YF_3	5.500	
+ 50 mol% YF_3	5.511	[20, p. 276]
Synthetic CaF_2 + 20 mol% YF_3	5.500	
+ 40 mol% YF_3	5.537	[22]

Yttrofluorites with only 6.07 and 7.87% ΣRE+Y have a = 5.47 and 5.48 Å, respectively [4, p. 108]. Space group $Fm3m - O_h^5$ (No. 225), Z = 4 [27].

The lattice parameter of $CaF_2-(RE,Y)F_3$ mixed crystals increases in direct proportion to their $(RE,Y)F_3$ content [6], see also [11]. The F atoms that are present in excess, due to substitution of CaF_2 by YF_3, are placed in interstitial positions of the fluorite lattice [22, 23], [4, p. 107], see however [24]. The fluorite lattice is stable up to 50% YF_3, with 90 and 95% YF_3 the lattice has lost its uniformity [20, p. 272/4]. Different stability ranges of mol% YF_3 and CeF_3 in synthetic crystals see [6].

10.1.1.4 Physical Properties

Optical Properties

Luster is glassy on cleavage faces, fatty on fracture faces [1], or dimly fat [5, p. 52]. Yttrofluorite is transparent to translucent [1], semitransparent [5, p. 52], but also opaque [12]. The color is yellowish to yellowish brown and light yellowish green [1], yellow-brown [10], light brown [11], dark violet, due to inclusions of bastnaesite, to pale green and colorless [7], weak creamy or yellowish green [5, p. 52], cream or pink [12], and white to colorless [2, p. 247]. Crystals and powder are gray [9]. Isotropic [1], [5, p. 52]. The refractive index of analyzed samples is given as:

Locality	n_{Na}	Ref.
Hundholmen, northern Norway	1.4572 and 1.4522	[1]
	1.4452 and 1.4483	[17]
Eivolden, northern Norway	1.460	[2, p. 247]
Kola, USSR	1.468 ± 0.001	[5, p. 52]
Ukrainian Shield, USSR	1.4370 to 1.445	[4, p. 107]
East Siberia, USSR	1.46	[6]
Central Asia, USSR	1.443 and 1.452	[7]
Suishoyama, Japan	1.454	[9]
White Cloud, Colorado	1.463 to 1.464	[11]

There exists a distinct correlation between refractive index and content of ΣRE + Y in yttrofluorites; n_{Na} increases with increasing content of Y and RE [4, pp. 107, 111], [11], [5, p. 52], [7]. Synthetic yttrofluorites, $(Ca, Y)F_{2\ to\ 3}$, with 5 to 13% YF_3 have n = 1.442 to 1.450, and excellent optical properties [25].

For fluorescence and absorption spectra of natural and synthetic yttrofluorite, see [20, pp. 248/72]. Yttrofluorite from Kola shows no luminescence in ultraviolet light and in cathodic rays, but intense thermoluminescence [5, p. 52].

Other Properties

Hardness H_M = 4.5 [1, 6]. Yttrofluorite is brittle, its fracture is uneven [1] and hackly [5, p. 52]. Density increases with increasing content of ΣRE + Y. With some exceptions this can be seen from the following table:

ΣRE + Y	D g/cm³	Locality	Ref.
0.0n	3.18	Optical fluorite	[5, p. 52]
4.97	3.25 to 3.32	Ukrainian Shield	[4, p. 108]
6.07	3.32 to 3.42	Ukrainian Shield	[4, p. 108]
7.13	3.319	Hundholmen, Norway	[17]
7.87	3.42 to 3.45	Ukrainian Shield	[4, p. 108]
15.1	3.557	Hundholmen, Norway	[1]
16.7	3.50	Central Asia, USSR	[7]
17.5	3.41	Central Asia, USSR	[7]
18.84	3.51	Suishoyama, Japan	[9]
19.55	3.61 ± 0.01	Kola, USSR	[5, p. 52]
23.49	3.698	White Cloud, Colorado	[11]

This is also shown in the specimen from Spitzkopje, Namibia: the core of yttrofluorite has D = 3.355 g/cm^3, the rim of pure CaF_2, D = 3.190 g/cm^3 [10]. D(calc.) = 3.62 g/cm^3 [9].

10.1.1.5 Chemical Behavior

Yttrofluorite is easily soluble in sulfuric acid, hardly in hydrochloric acid; the powder is soluble in hot 3N HCl after heating for 6 to 7 h [7].

References for 10.1.1:

[1] T. Vogt (Centralbl. Mineral. **1911** 373/7). – [2] T. L. Sverdrup (Norsk Geol. Tidsskr. **48** [1968] 245/52). – [3] T. L. Sverdrup, P. C. Saebö, K. Ö. Bryn (Norsk Geol. Tidsskr. **45** [1965] 177/88, 180/1). – [4] E. P. Gurov, E. P. Gurova, L. G. Loginova, Yu. A. Lavitskaya (in: P. M. Tatarinov et al., Mineraly i Paragenezisy Mineralov Metasomaticheskikh i Metamorficheskikh Gornykh Porod, Nauka, Leningrad 1975, pp. 106/14). – [5] A. M. Vetoshkina, V. V. Gordienko, N. A. Elina, L. I. Polezhaeva (Mineral. Zh. **2** No. 4 [1980] 51/8).

[6] V. V. Arkhangel'skaya (Dokl. Akad. Nauk SSSR **195** [1970] 1411/4; Dokl. Earth Sci. Sect. **195** [1970] 142/5). – [7] V. N. Zuev, A. V. Kosterin (Tr. Inst. Mineral. Geokhim. Kristallokhim. Redk. Elem. No. 4 [1960] 136/8). – [8] D. A. Mineev (in: L. N. Ovchinnikov, N. A. Solodov, Mestorozhdeniya Litofil'nykh Redkikh Metallov, Nedra, Moscow 1980, pp. 178/83). – [9] K. Uetani, Y. Ogimura, A. Kato, K. Nagashima (Bull. Chem. Soc. Japan **41** [1968] 603/5). – [10] P. Ramdohr (Neues Jahrb. Mineral. Geol. Paläontol. Abhandl. Beilagebd. A **76** [1941] 1/35, 15/6, 34).

[11] J. W. Adams, W. N. Sharp (U.S. Geol. Surv. Profess. Paper No. 800-C [1972] 63/9, 63, 65/7). – [12] C. V. Haynes (Geol. Soc. Am. Bull. **76** [1965] 441/61, 447/9). – [13] J. W. Adams, W. N. Sharp (Can. Mineralogist **10** [1971] 907). – [14] V. M. Goldschmidt, L. Thomassen (Videnskapsselskap. Skrifter Akad. Oslo I Mat. Naturv. Klasse **1923** No. 2, pp. 1/48, 41). – [15] C. Palache, H. Berman, C. Frondel (Dana's System of Mineralogy, 7th Ed., Vol. 2, Wiley, New York 1951, pp. 32/3).

[16] E. I. Semenov, R. L. Barinskii (Geokhimiya **1958** 314/33; Geochemistry [USSR] **1958** 398/419, 406). – [17] F. Zambonini (Riv. Mineral. Crist. Ital. **45/46** [1915] 1/185, 150/2). – [18] M. Fleischer (Indian Mineralogist **10** [1969/71] 36/9, Table 1). – [19] E. S. Larsen, H. Berman (U.S. Geol. Surv. Bull. No. 848 [1934] 1/266, 48). – [20] N. Chatterjee (Z. Krist. **102** [1940] 245/84).

[21] H. Haberlandt (Sitz.-Ber. Akad. Wiss. Wien Math. Naturw. Klasse IIa **147** [1938] 137/50, 138/9). – [22] J. Short, R. Roy (J. Phys. Chem. **67** [1963] 1860/1). – [23] V. M. Goldschmidt (Videnskapsselskap. Skrifter Akad. Oslo I Mat. Naturv. Klasse **1926** No. 2, pp. 1/117, 89). – [24] W. Gettman, O. Greis (J. Solid State Chem. **26** [1978] 255/63, 255/6). – [25] Anonymous (Ti Chih K'o Hsueh **1978** 88/91 [Chinese with English Abstract]; Mineral. Abstr. **29** [1978] No. 78-5193).

[26] E. A. Viljoen, J. Reis, J. Turner, M. L. Veitch (Rept. Natl. Inst. Met. [S. Africa] No. 1829 [1976] 1/6, 2). – [27] H. Strunz (Rend. Soc. Mineral. Ital. **14** [1958] 326/8).

10.1.2 Cerfluorite $(Ca,Ce)F_{2 \text{ to } 2.17}$

The synonym yttrocerite [1] is a name that should be discarded as misleading [2]. The first description of cerfluorite was in 1915 (under the name yttrocerite) of a sample from Finbo, near Falun, Sweden, by Gahn and Berzelius [1]. Further occurrences of this mineral are Khibina, Kola Peninsula, USSR, and Colorado; see "Seltenerdelemente" A 1, 1938, p. 57. But Deer et al. consider its natural occurrence to be doubtful [3]. Sverdrup has identified a RE-containing fluorite from Hundholmen, Tysfjord, northern Norway, as cerfluorite; the author thinks it to be possible that this pegmatite contains cerfluorite besides yttrofluorite (see p. 1) [4, pp. 250/1]. Gurov et al. describe cerfluorite from pegmatitic veinlets in granites of the Ukrainian Crystalline Shield, where it forms irregular grains near the rims of the thin veins and in the contact zone of the surrounding rocks; this cerfluorite has been replaced by violet fluorite and alteration products of red color with allanite and hematite [5, 6]. For cerfluorite in deposits near Jamestown, NW of Boulder, Colorado, see "Rare Earth Elements", A 8, 1984, Chapter 11.7.2.3.3.2.

Analyses of cerfluorite from Hundholmen yield 33.74% RE_2O_3 and 6.49% Y_2O_3 [4, p. 251], those from the Ukrainian Shield 3.91% ΣCe-group and 1.92% ΣY-group + Y [5, 6]. The distribution of RE in these cerfluorites, given by the same authors, is compiled in the following table:

Locality	Y_2O_3	La_2O_3	CeO	Pr_2O_3	Sm_2O_3	Eu_2O_3	Gd_2O_3
Hundholmen	16.1	15.8	36.5	3.0	3.3	—	3.2
Ukraine	16.9	12.7	32.4	3.7	2.7	—	2.7

Locality	Tb_2O_3	Dy_2O_3	Ho_2O_3	Er_2O_3	Tm_2O_3	Yb_2O_3	Lu_2O_3
Hundholmen	0.4	3.0	0.5	1.9	0.3	1.7	—
Ukraine	—	4.1	1.0	5.3	3.7	2.1	0.4

These data show a strong preponderance of Ce-earth elements with a ratio (La to Sm):(Gd to Yb) + Y = 2.7:1 for cerfluorite from Hundholmen, and (La to Sm):(Gd to Lu) + Y = 1.8:1 for cerfluorite from the Ukraine. This corresponds with the data of cerfluorite of Falun, Sweden, as given by Gahn and Berzelius who calculate Ce-earth:Y-earth ≈ 2:1 [1].

Cerfluorite from Hundholmen contains traces of Th, but no U [4, p. 251].

Cerfluorite sometimes forms cubic crystals [5, 6]. It is cubic, space group $Fm3m-O_h^5$ (No. 225) with a = 5.47 Å, and Z = 4 (Finbo) [7, 8]; a = 5.46 Å (Ukraine) and a ≈ 5.50 Å (Hundholmen) [4, p. 248].

The color varies from dark violet to grayish red, and may also be white or gray [1] and yellow-brown [4, p. 247]. Fresh samples are pale yellow, altered samples red [5, 6], see above. Refractive index of the analyzed Ukrainian samples n = 1.440 [5, 6].

Density D = 3.447 g/cm^3 [1], 3.32 g/cm^3 [5, 6].

References for 10.1.2:

[1] J. G. Gahn, J. Berzelius (Afhandl. Fys. Kemi Mineral. **4** [1815] 148/216, 151/64). – [2] E. N. Goddard, J. J. Glass (Am. Mineralogist **25** [1940] 381/404, 400, footnote). – [3] W. A. Deer, R. A. Howie, J. Zussman (Rock-Forming Minerals, Vol. 5, Longmans Green, London 1962, pp. 1/371, 348). – [4] T. L. Sverdrup (Norsk Geol. Tidsskr. **48** [1968] 245/52). – [5] E. P. Gurov, E. P. Gurova, E. Ya. Marchenko (Dopov. Akad. Nauk Ukr. RSR B **1971** 297/9).

[6] E. P. Gurov, E. P. Gurova, L. G. Loginova, Yu. A. Lavitskaya (in: P. M. Tatarinov et al., Mineraly i Paragenezisy Mineralov Metasomaticheskikh i Metamorficheskikh Gornykh Porod, Nauka, Leningrad 1975, pp. 106/14, 108/9). – [7] H. Strunz (Rend. Soc. Mineral. Ital. **14** [1958] 326/8). – [8] H. Strunz, C. Tennyson (Mineralogische Tabellen, 6th Ed., Geest and Portig, Leipzig 1977, pp. 1/621, 157).

10.1.3 Tveitite $Ca_{1-x}(Y,RE)_x(F,O)_{2+x}$

Occurrence, Paragenesis, and Chemistry

Tveitite first was found at Høydalen, north of Tveit farm, Tørdal, Telemark, southern Norway, in a pegmatite dike together with quartz, amazonite, microcline, muscovite, beryl, and monazite [1]; and afterwards in the pegmatite from Barringer Hill, Texas, as small inclusions in yttrian fluorite; it is there associated with gadolinite, allanite, and albite [2]. Tveitite from Norway contains small inclusions of a gadolinite-like mineral. On the surface and in cracks it is altered to kainosite [1].

Based on one cation per formula unit (as in fluorite), the chemical analysis of the Norwegian tveitite, with 8.06% RE and 22.2% Y, leads to $Ca_{0.7}Y_{0.25}RE_{0.05}(F,O)_{2.25\ \text{to}\ 2.30}$ [1]. Bevan et al. interpret $Ca_{13+d}Y_{6-d}F_{44-d}$, $Ca_{14}Y_5F_{43}$ or $Ca_{13}Y_6F_{42}O$ as possible formulas [3]. RE distribution calculated from the chemical analysis of the Norwegian tveitite in % (without Y): La 15.9, Ce 47.9, Pr 6.9, Nd 11.2, Sm 1.5, Eu 0.5, Gd 1.5, Tb 0.4, Dy 3.2, Ho 0.6, Er 2.1, Tm 1.6, Yb 4.3, Lu 2.4; $(Y+\Sigma Y)/\Sigma Ce=3.5$ [1]. A predominance of Y and the heavier lanthanides over the lighter cerium group elements, with $(Y+\Sigma Y)/\Sigma Ce\approx 4$ to 5 is also observed in tveitite from Texas [2].

Crystal Form and Structure

Tveitite shows complex polysynthetic twinning [1]. Cleavage {111} is good [2]. The unit cell is rhombohedral with hexagonal parameters a = 16.6920(9), c = 9.6664(8) Å, Z = 57, space group is $R\bar{3}$-C_{3i}^2 (No. 148) or R3-C_3^4 (No. 146) [3]. Indexed on a pseudocubic, monoclinic pseudocell a′ = 3.924, b′ = 3.893, c′ = 5.525 Å, β = 90.26°, V = 84.38 Å^3 [1]. The structure is built up of discrete M_6X_{37} groups, which consist of an octahedral arrangement of cation-centered square antiprisms sharing corners around a threefold axis to enclose a filled cubooctahedron of anions [3]. Tveitite is a member of the homologue series M_mF_{2m+5}, see [4], with m = 19 [3]. For a refinement of the structure in the space group $R\bar{3}$, see [5].

Physical Properties

Luster is resinous to greasy [2] or greasy; the mineral is transparent [1]. The color is white to cream-yellow [1, 2]. The streak is white [2]. Tveitite is optically negative, $n_x=1.476$, $n_y=1.479$, $n_z=1.481$, $2V_x=34°$ [1] or $n_x=1.476(2)$, $n_y=1.479(2)$, $n_z=1.481(2)$, 2V = 35° [2]. Faint [1] or strong [2] yellow-orange fluorescence on short wave UV light has been observed.

Hardness $H_M=3.5$ to 4. Fracture is subconchoidal to uneven [2]. D (meas.) = 3.789(5) g/cm^3 [2], 3.94 g/cm^3; D(calc.) = 3.97 to 4.02 g/cm^3 [1]. Low magnetic susceptibility [1].

Differential Thermoanalysis

DTA studies of tveitite show a sharp endothermic reaction at ~670°C; material heated at 800°C for about 2 h gave a diffraction pattern corresponding to that of fluorite [1].

References for 10.1.3:

[1] S. Bergstøl, B. B. Jensen, H. Neumann (Lithos **10** [1977] 81/7). – [2] W. W. Crook III (Mineral. Rec. **9** [1978] 387). – [3] D. J. M. Bevan, O. Greis, J. Strähle (Acta Cryst. A **36** [1980] 889/90). – [4] W. Gettman, O. Greis (J. Solid State Chem. **26** [1978] 255/63, 261). – [5] D. J. M. Bevan, J. Stähle, O. Greis (J. Solid State Chem. **44** [1982] 75/81).

10.1.4 **Gagarinite** $NaCaYF_6$

10.1.4.1 Occurrence and Paragenesis. Alteration

Gagarinite often occurs in metasomatic riebeckite-microcline-albite granitoids with pyrochlore and zircon, and in metasomatic riebeckite-aegirine rocks with Fe-polylithionite, astrophyllite, zircon, Ce-Pb-pyrochlore and cryolite [1]. Thus it first was found in intensively albitized granites, pegmatites, and albitites in Kazakhstan, associated with riebeckite, aegirine, zircon, pyrochlore, and bastnaesite [2, 3]. In a granitic massif in Tuva gagarinite was found in albitized rocks in subalkalic quartz syenites in close association with riebeckite, aegirine, cyrtolite, fergusonite, polylithionite, siderite, sphalerite (cleiophane), and fluorite [4]; particularly in the Erzinsk massif, eastern Tuva, it is spatially associated with cryolite and thomsenolite [5], and also zinnwaldite and accessory malacon and pyrochlore [6]. In the northwestern Tarbagatai Mountains, eastern Kazakhstan, gagarinite occurs (in three different varieties) in metasomatites of the exocontact, in pegmatites, and in parts of the albitized apical zone of the massif [7]; cf. "Seltenerdelemente" A 4, 1979, p. 49. For gagarinite in metasomatically altered rocks above granite bodies in the deposit Verkhne Epinskoe, eastern Kazakhstan, see "Rare Earth Elements" A 8, 1984, Chapter 11.7.6.2.1. In fenites and quartz-albite metasomatites of the intrusive Nygvaganskii complex, Far East, gagarinite is closely associated with hematite and late hornquartz, which is partly replaced by aegirine and riebeckite [8]. In Eastern Siberia gagarinite occurs in two generations in metasomatic alkali granites together with amphibole [9, 10], and in biotite-riebeckite metasomatites; it is associated with pyrochlore, zircon, pachnolite, tysonite (fluocerite), and is intimately intergrown with cryolite [11]. In occurrences of Transbaikalia gagarinite occurs in riebeckite and aegirine metasomatites with Ce-pyrochlore, cryolite, and zircon [12].

Outside the USSR gagarinite has been found in Norway in the elpidite granite near Lake Gjerdingen, Nordmarka, Oslo region [13], where it occurs together with other fluorides in cavities in ekerite [14, 15]; and in the riebeckite granite at Washington Pass, Okanogan County, Washington [16].

Gagarinite from metasomatites in Eastern Siberia has fine-dispersed inclusions of hematite and minute inclusions of probably aegirine or arfvedsonite [9, 10]. For inclusions of tysonite (fluocerite) and Fe oxides in East Siberian gagarinite, see [11]. For inclusions in gagarinite from Erzinsk massif see p. 11.

Gagarinite from Kazakhstan is often altered and easily replaced by yttrofluorite, "tengerite", bastnaesite, or synchisite [3], see especially for "tengerite" [2] and p. 210. Bastnaesite as an alteration product of gagarinite is shown to be Y-bastnaesite with Y > Ce [17, 18]; see also [7]. For distribution of RE in bastnaesite as pseudomorphs after gagarinite see p. 115. East Siberian gagarinite too is altered to RE carbonates of the bastnaesite type [11], cf. [12].

10.1.4.2 Chemistry

For a compilation of analyses and Y_2O_3 and $(RE,Y)_2O_3$ contents, mean and variance, of 3 and 19 analyses, respectively, see "Seltenerdelemente" A 4, 1979, pp. 21 and 5.

From these analyses the **formulas of gagarinite** from different localities are given as follows:

Kazakhstan	$Na_2Ca_2Y_3(F,Cl,OH)_{15}\cdot H_2O$ or	[3]
	$Na_2Ca_2(Y,Al)_3(F,Cl,OH)_{15}$	[2]
Tuva	$Na_{1.0}RE_{1.3}Ca_{0.5}F_6$	[4]
Erzinsk massif, Tuva	$Na_{0.822}(Ca_{0.731}Fe_{0.081}Ti_{0.052}RE_{1.133})F_6$	[6]

Far East	$Na_{0.87}(Ca_{1.0}RE_{1.04})F_6$	[8]
Eastern Siberia	$Na_{0.95}(Ca_{0.66}Mg_{0.11}Al_{0.05}Th_{0.03}RE_{1.2})(F_{5.44}OH_{0.46})$	[9, 10]
	$Na_{0.88}(Ca_{0.91}RE_{1.11})F_{5.96}$	[11], cf. [12]
Transbaikalia	$Na_{0.89}(Ca_{0.91}RE_{1.12})F_{6.0}$	[12]

Tarbagatai Mountains. The analyses of 6 samples are combined to the three types:

1. $Na_{1.0}(Na_{0.04}Ca_{0.92}RE_{0.98}Al_{0.04}Fe_{0.02})F_{6.0}$;
2. $Na_{1.0}(Na_{0.13}Ca_{0.44}RE_{1.22}Al_{0.18}Fe_{0.03})F_{6.0}$;
3. $Na_{0.8}(Ca_{0.81}RE_{1.09}Al_{0.03}Fe_{0.01})F_{6.0}$ [7].

These types are characterized by their RE and other cations as: 1. Ce-gagarinite with RE = Na > Ca, 2. Na-RE-gagarinite with RE > Na > Ca, and 3. Y-gagarinite with RE > Na = Ca [7]. (For different physical properties of these gagarinites, see pp. 11 and 12.)

In gagarinite from the Erzinsk massif, eastern Tuva, with 1.89% ThO_2, Th possibly replaces RE; Ti and Fe are perhaps due to inclusions; trace elements in this mineral are Sr, Al, Mg, Mn [5, 6]. Two of the six samples from Tarbagatai Mountains contain 0.94 and 0.50% ThO_2 [7], see for one of these samples also [3]. Gagarinite from the Far East has 0.13% ThO_2 [8]. In East Siberian samples Ti, Zr, Mn, Nb, Pb, Sn, Be, Sc, and P are found as trace elements [10], and in gagarinite from Kazakhstan there are Be, Ti, Mg, Pb, Sn [2].

The **distribution of RE ± Y** in gagarinite varies. Most gagarinites show predominance of Y-earth, but there are some that have predominance of Ce-earth, as it is shown by the following table:

Locality	$\Sigma Ce^{1)}$	$\Sigma Y^{2)}$	$\Sigma Ce/\Sigma Y$	Ref.
Kazakhstan				
sample 1	48.8	51.2	0.95	
sample 2	$34.6^{+)}$	$65.4^{++)}$	0.53	[3]
	39.2	60.8	0.64	[12]
	$50.6^{+)}$	$49.4^{++)}$	1.02	
	$43.7^{+)}$	$56.4^{++)}$	0.77	[2]
Tarbagatai Mountains,	37.8	62.4	0.61	
Eastern Kazakhstan	35.6	64.4	0.55	[7, 17]
Tuva	45.3	54.7	0.83	[4]
	$33.3^{*)}$	66.7	0.50	[5, 6]
Eastern Siberia	63.1	37.0	1.71	[11]
Transbaikalia	45.1	36.9	1.22	[12]

$^{1)}$ ΣCe = La to Gd. – $^{2)}$ ΣY = Tb to Lu. – $^{+)}$ Without Eu. -- $^{++)}$ Without Lu. – $^{*)}$ Only Nd, Sm, Gd.

For analyses with Y included in ΣRE the following ratios result: two generations (I and II) of gagarinites from Eastern Siberia $\Sigma Ce/(Y + \Sigma Y(\text{without Lu})) = 0.89$(I) and 0.84(II) [10]. Gagarinite from the Far East $\Sigma Ce/(Y + \Sigma Y) = 1.61$ [8]; cf. for both localities "Seltenerdelemente" A 4, 1979, p. 49. The distribution of the single RE in gagarinites from three localities with a Dy maximum are shown in **Fig. 1**, p. 10. For calculation of RE distribution in at.% in gagarinites from Kazakhstan and Tuva, see [20].

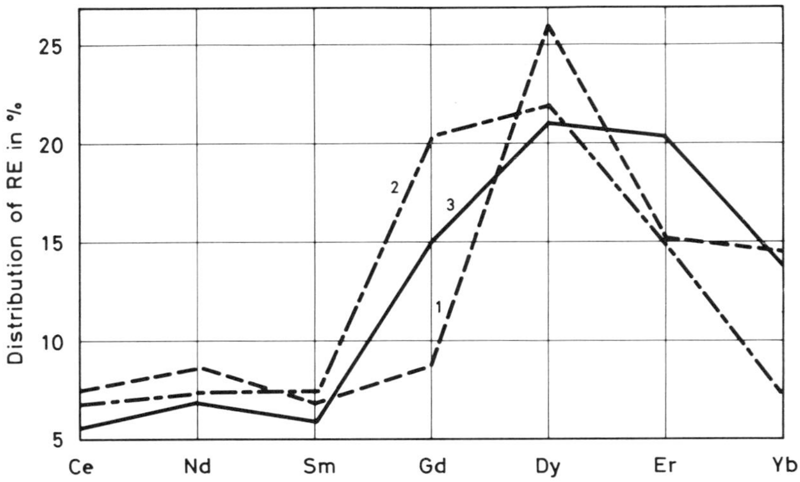

Fig. 1 from [6]. Distribution of RE in 3 gagarinites ($\Sigma RE_2O_3 = 100$): 1 from Kazakhstan with data from [3], 2 and 3 from Tuva with data from [4] and [6], respectively.

References for 10.1.4.1 and 10.1.4.2:

[1] Yu. L. Kapustin (in: L. N. Ovchinnikov, N. A. Solodov, Mestorozhdeniya Litofil'nykh Redkikh Metallov, Nedra, Moscow 1980, pp. 210/25, 210/1, 216/7, 224). – [2] A. V. Stepanov (Tr. Kaz. Issled. Inst. Mineral'n. Syr'ya No. 5 [1961] 147/61, 147, 149/51). – [3] A. V. Stepanov, E. A. Severov (Dokl. Akad. Nauk SSSR **141** [1961] 954/7; Dokl. Earth Sci. Sect. **139/141** [1961] 1290/3). – [4] N. A. Akelin, M. E. Kazakova (Dokl. Akad. Nauk SSSR **149** [1963] 672/4; Dokl. Earth Sci. Sect. **149** [1963] 111/3). – [5] V. S. Kudrin, M. A. Kudrina (Tr. Mineral. Muzeya Akad. Nauk SSSR No. 18 [1968] 80/97, 82, 92/4).

[6] V. S. Kudrin, M. A. Kudrina, T. I. Stolyarova (Geol. Mestorozhd. Redk. Elem. Vses. Nauchn. Issled. Inst. Mineral'n. Syr'ya No. 30 [1966] 131/9, 132, 134/8). – [7] D. A. Mineev (Geokhimiya Apogranitov i Redkozemel'nykh Metasomatitov Severo-Zapadnogo Tarbagataya, Nauka, Moscow 1968, pp. 1/183, 93/6). – [8] P. G. Nedashkovskii, E. I. Lagovskaya, V. S. Demchenko (in: I. N. Govorov, P. G. Nedashkovskii, Geokhimicheskie Tsikly Dal'nego Vostoka, Dalnevost. Geol. Inst., Vladivostok 1969, pp. 29/36, 31, 33/4). – [9] V. V. Arkhangel'-skaya (Sb. Kratk. Soobshch. Mineral. Geokhim. No. 1 [1968/70] 108/12 from Ref. Zh. Geol. **1970** No. 10 V 216). – [10] V. V. Arkhangel'skaya, T. I. Stolyarova (Tr. Mineral. Muzeya Akad. Nauk SSSR No. 21 [1972] 174/8).

[11] N. E. Zalashkova, D. I. Agapova, N. B. Yusupov (Mineral. Issled. **2** [1972] 69/78, 70/6). – [12] N. E. Zalashkova, D. A. Mineev, N. B. Yusupov (in: D. A. Mineev, Osobennosti Raspredeleniya Redkikh Elementov v Izverzhennykh Gornykh Porodakh, Inst. Mineral. Akad. Nauk, Moscow 1970, pp. 16/32, 24/5, 27). – [13] P. C. Saebø (Norsk Geol. Tidsskr. **46** [1966] 260/1). – [14] G. Raade (Mineral. Rec. **3** No. 1 [1972] 7/11, 8). – [15] G. Raade, J. Haug (Mineral. Rec. **11** No. 2 [1980] 83/91, 83/4).

[16] B. Cannon (from [15]). – [17] D. A. Mineev (Geokhimiya **1963** 1082/100; Geochemistry [USSR] **1963** 1129/49, 1143, 1145). – [18] D. A. Mineev, T. I. Lavrishcheva, A. V. Bykova (Zap. Vses. Mineral. Obshchestva **99** [1970] 328/32, 328). – [19] A. A. Voronkov, N. G. Shumyatskaya, Yu. A. Pyatenko (Zh. Strukt. Khim. **3** [1962] 691/6; J. Struct. Chem. [USSR] **3** [1962] 665/9, 668). – [20] M. Fleischer (Indian Mineralogist **10** [1969] 36/9, 37).

10.1.4.3 Crystal Forms and Structure

Hexagonal. Forms observed at crystals from Norway are $\{0001\}$, $\{10\bar{1}0\}$, $\{11\bar{2}0\}$, and rhombohedral faces. The habit is short- or long-prismatic and tabular [1]. Gagarinite often forms irregular grains [2, 3], but also hexagonal prismatic crystals [4, 5]. Hexagonal prismatic crystals occur in pegmatites and xenomorphic gagarinite in albitites [6]; the first generation of gagarinite from Eastern Siberia forms irregular, xenomorphic grains, the second generation idiomorphic hexagonal prismatic crystals, sometimes of sheaf-like form [7]. Gagarinite from metasomatites in Kazakhstan occurs as idiomorphic crystals, in metasomatites from Transbaikalia it forms rounded, irregular grains [8]. Gagarinite has imperfect [2, 6] to medium prismatic cleavage [4], and cracks semiparallel (0001) [1].

Inclusions of an unkown, needle-shaped mineral in gagarinite from the Erzinsk massif, Eastern Tuva, are regularly orientated after the faces of the trigonal prism or the dipyramid [3], they cut each other under 60° [9]. Parallel intergrowth of gagarinite and tysonite (fluocerite) is observed by [5].

Lattice parameters of gagarinite: $a = 5.99 \pm 0.02$ Å, $c = 3.53 \pm 0.02$ Å from Kazakhstan [4, 6]; $a = 5.97 \pm 0.03$ Å, $c = 3.55 \pm 0.03$ Å from Eastern Siberia [7, 10] as well as from Eastern Tuva [3, 9]. The most probable space group is $P\bar{3}\text{-}C_{3i}^{i}$ (No. 147) with pronounced $P6_3\text{-}C_6^6$ (No. 173) symmetry. Interatomic distances in the (RE, Ca)-F polyhedra with coordination number 9 are (RE, Ca)-F = 2.30 Å (horizontal distance) and = 2.39 Å (inclined distance) [11], see also [12].

10.1.4.4 Physical Properties

Optical Properties

Luster is vitreous [2, 3, 9], vitreous or dull [5, 6, 7], also dull-glassy [4, 13]. In thin splinters gagarinite is transparent [2, 6], transparent or semitransparent [3], or semitransparent to opaque [4]. The color is nonhomogeneous, changing even in a single crystal from colorless, yellow to rusty-orange; this is due to fine-dispersed, cloudy inclusions of Fe-oxides [5]. White-rose or rose to brick-red colors are caused by inclusions of hematite [7]. The chemically different gagarinites from Tarbagatai Mountains, see p. 9, can be distinguished by their color: while the Y-gagarinite is colorless, the Na-Ce-gagarinite shows a rose color [14]. The following colorings are given: colorless, very rarely yellowish white [2], pale rose, rarely colorless [1], cream, less frequently pinkish yellow [4], white, if pure, to weakly rose or yellowish [6], white to slightly yellowish [3, 9], and yellow-rose [13]. The streak is white [2, 4, 6]. The color in transmitted light is transparent colorless [3, 4, 7, 9] or slightly yellowish, if altered [6], and yellowish or cloudily brown [5].

Gagarinite is uniaxial positive [2, 7, 10, 13], sometimes biaxial with $2V < 20°$ [4, 5, 6], $< 24°$ [3, 14]. Elongation is positive [4] with straight extinction [5], Refractive indices of analyzed samples from different localities are given in the following table:

Locality	n_e	n_o	Ref.
Kazakhstan	1.492	1.472	
(4 samples)	1.494	1.474	
	1.496	1.472	
	1.501	1.474	[6]
Tuva	1.493	1.472	[3, 9]
	1.490	1.475	[2]

Locality	n_e	n_o	Ref.
Eastern Siberia	1.493	1.472	[7, 10]
	1.492±0.001	1.472±0.001	[5]
Far East	1.498	1.472	[13]

The reflecting power of gagarinite is determined as R = 6% [3, 9].

Gagarinite is not luminescent on ultraviolet irradiation [4, 6].

Other Physical Properties

Hardness. H_M = 4 [13], 4 to 4.5 [6], 4 to 5 [2], or 4.7 [7, 10]; microhardness = 186 kg/mm² corresponding with H_M = 4 [3, 9], = 370 kg/mm² and H_M = 4.5 [4]. Gagarinite is very brittle [3, 13], or brittle [2, 4, 7]; its fracture is uneven [7].

The **density** of gagarinite varies from 4.18 for a sample from Tuva [2] up to 4.53 g/cm³ for samples from Kazakhstan and Transbaikalia [8], cf. [15]. Further data in g/cm³ of gagarinites from different localities in increasing order are: Eastern Tuva 4.2 [3, 9], Kazakhstan 4.21 [4], 4.2 and 4.3 [16], Eastern Siberia 4.29 [7, 10], 4.34 to 4.51 [5], and Far East 4.38 [13]. Four samples from Kazakhstan with 45.4, 54.8, 56.1, and 58.6% $(RE,Y)_2O_3$ have D = 4.22, 4.52, 4.27, and 4.21 g/cm³, respectively [6]. From the chemically various gagarinites from Tarbagatai Mountains (see p. 9), Ce-gagarinite has D = 4.38, Na-RE-gagarinite 4.52, and Y-gagarinite 4.32 and 4.33 g/cm³ [14].

Magnetism of gagarinite is weak [3, 7, 10], but may also be strong [9]. It is near to that of riebeckite, aegirine, or astrophyllite [4, 6] or similar to that of fergusonite [2].

Radioactivity is absent or very weak [4, 6].

References for 10.1.4.3 and 10.1.4.4:

[1] G. Raade, J. Haug (Mineral. Rec. **11** No. 2 [1980] 83/91, 84/5). – [2] N. A. Akelin, M. E. Kazakova (Dokl. Akad. Nauk SSSR **149** [1963] 672/4; Dokl. Earth Sci. Sect. **149** [1963] 111/3). – [3] V. S. Kudrin, M. A. Kudrina, T. I. Stolyarova (Geol. Mestorozhd. Redk. Elem. Vses. Nauchn. Issled. Inst. Mineral. Syr'ya No. 30 [1966] 131/9, 132, 134). – [4] A. V. Stepanov, E. A. Severov (Dokl. Akad. Nauk SSSR **141** [1961] 954/7; Dokl. Earth Sci. Sect. **139/141** [1961] 1290/3). – [5] N. E. Zalashkova, D. I. Agapova, N. B. Yusupov (Mineral. Issled. **2** [1972] 69/78, 71, 73, 76).

[6] A. V. Stepanov (Tr. Kaz. Issled. Inst. Mineral'n. Syr'ya No. 5 [1961] 147/61, 148/52). – [7] V. V. Arkhangel'skaya, T. I. Stolyarova (Tr. Mineral. Muzeya Akad. Nauk SSSR No. 21 [1972] 174/8, 174/6). – [8] N. E. Zalashkova, D. A. Mineev, N. B. Yusupov (in: D. A. Mineev, Osobennosti Raspredeleniya Redkikh Elementov v Izverzhennykh Gornykh Porodakh, Inst. Mineral. Akad. Nauk, Moscow 1970, pp. 16/32, 25, 27). – [9] V. S. Kudrin, M. A. Kudrina (Tr. Mineral. Muzeya Akad. Nauk SSSR No. 18 [1968] 80/97, 93). – [10] V. V. Arkhangel'skaya (Sb. Kratk. Soobshch. Mineral. Geokhim. No. 1 [1968/70] 108/12 from Ref. Zh. Geol. **1970** No. 10 V 216).

[11] A. A. Voronkov, N. G. Shumyatskaya, Yu. A. Pyatenko (Zh. Strukt. Khim. **3** [1962] 691/6; J. Struct. Chem. [USSR] **3** [1962] 665/9, 665, 669). – [12] Yu. A. Pyatenko, A. A. Voronkov (Zh. Strukt. Khim. **3** [1962] 720/1; J. Struct. Chem. [USSR] **3** [1962] 696/7). – [13] P. G. Nedashkovskii, E. I. Lagovskaya, V. S. Demchenko (in: I. N. Govorov, P. G. Nedashkovskii, Geokhimicheskie Tsikly Dal'nego Vostoka, Dal'nevost. Geol. Inst., Vladivostok 1969, pp. 29/36, 33). – [14] D. A. Mineev (Geokhimiya Apogranitov i Redkozemel'nykh Metasomatitov Severo-Zapadnogo Tarbagataya, Nauka, Moscow 1968, pp. 1/183, 93/4). – [15] Yu. L. Kapustin (in: L. N. Ovchinnikov,

N. A. Solodov, Mestorozhdeniya Litofil'nykh Redkikh Metallov, Nedra, Moscow 1980, pp. 210/25, 211).

[16] D. A. Mineev (Geokhimiya **1963** 1082/100; Geochemistry [USSR] **1963** 1129/49, 1145).

10.1.4.5 Chemical Behavior and Differential Thermal Analysis

Gagarinite is soluble in mineral acids such as hydrochloric, nitric, and concentrated sulfuric acid [1], see also [2, 3]. It is insoluble in cold hydrochloric and nitric acid, but completely soluble in hot hydrochloric acid [4].

The DTA curve of gagarinite from Kazakhstan shows a strong endothermic peak at 720 to 750°C, due to losses of H_2O and F, and a weak exothermic effect at about 900°C [1]. Four samples from the same region are characterized by curves with endothermic effects in the interval of 630 to 850°C with a minimum at 720 to 750°C, that sometimes is split; admixtures of bastnaesite cause additional effects [3]. Gagarinite from Eastern Siberia has a DTA curve with an endothermic peak at 750°C [5].

References for 10.1.4.5:

[1] A. V. Stepanov, E. A. Severov (Dokl. Akad. Nauk SSSR **141** [1961] 954/7; Dokl. Earth Sci. Sect. **139/141** [1961] 1290/3). – [2] P. G. Nedashkovskii, E. A. Lagovskaya, V. S. Demchenko (in: I. N. Govorov, P. G. Nedashkovskii, Geokhimicheskie Tsikly Dal'nego Vostoka, Dal'nevost. Geol. Inst., Vladivostok 1969, pp. 29/36, 33). – [3] A. V. Stepanov (Tr. Kaz. Issled. Inst. Mineral'n. Syr'ya No. 5 [1961] 147/61, 149, 151/2). – [4] N. E. Zalashkova, D. I. Agapova, N. B. Yusupov (Mineral. Issled. **2** [1972] 69/78, 76). – [5] V. V. Arkhangel'skaya, T. I. Stolyarova (Tr. Mineral. Muzeya Akad. Nauk SSSR No. 21 [1972] 174/8, 176).

10.1.5 Tysonite (Ce, La)F_3

Tysonite is also known as fluocerite. The minerals fluocerite and tysonite have been shown to be identical [1]. Though fluocerite has priority, see below, the name is so misleading that it seems preferable to use the name tysonite [2], see also [3, 4]. This will be done in the following mineral description though many authors up to date prefer the name fluocerite, see e.g. [5] and nearly all Russian authors, cited in this chapter.

References:

[1] P. Geijer (Geol. Foeren Stockholm Foerh. **43** [1921] 19/23, 23). – [2] E. N. Goddard, J. J. Glass (Am. Mineralogist **25** [1940] 381/404, 400, footnote). – [3] T. L. Sverdrup, P. C. Saebø, K. Ø. Bryn (Norsk Geol. Tidsskr. **45** [1965] 177/88, 179). – [4] P. Ramdohr, H. Strunz (Klockmanns Lehrbuch der Mineralogie, 16. Aufl., Enke, Stuttgart 1978, p. 489). – [5] E. B. Eckel (U. S. Geol. Surv. Bull. No. 1114 [1961] 1/399, 146).

10.1.5.1 Occurrence and Paragenesis. Alteration

Originally found in 1818 in granite **pegmatites** of Finbo and Brodbo near Falun, Dalarna, Sweden, and named fluocerite in 1845. In 1886, fluocerite, described as oxyfluoride, has been found in Österby, Dalarna, Sweden [1]. In the USA the first description of the mineral (named tysonite) was in 1880 of a sample with bastnaesite from the Pikes Peak Area at the base of Stove Mountain, west of Cheyenne Mountain, El Paso County, Colorado. Sometimes later it was also found in the Crystal Park, El Paso County, see [1, 2]. As a second locality in USA the South Platte-Lake George area, Douglas, Teller, and Park Counties, Colorado, has been named, where tysonite occurs in pegmatites with the RE minerals xenotime, monazite, yttrofluorite, gadolinite, allanite, and yttrotantalite [3]; particularly well-known is the Black Cloud pegmatite, Teller County, where tysonite occurs in fluorite lenses, associated with xenotime, monazite, gadolinite, and thorite [4]. Cf. "Rare Earth Elements" A 8, 1984, Chapter 11.7.2.3.3.2. In pegmatites tysonite has also been found in Norway together with bastnaesite in the granite pegmatite Jennyhaugen at Drag and in Hundholmen (both Tysfjord), Nordland, northern Norway [5, 6]; and in the lepidolite- and cassiterite-bearing pegmatite from Høydalen, Tørdal, Telemark, southern Norway [7]; here tysonite, together with cerianite, may be an alteration product of monazite [8], see also [6]. In certain parts of the Ukrainian Shield, tysonite has been found in chambered pegmatites of a granite pluton where it is associated with monazite and yttrosynchisite. Within the zone of blockquartz and feldspar in pegmatitic veins of the same pluton, tysonite occurs as inclusions in yttrofluorite, together with bastnaesite, yttrobastnaesite, and yttrosynchisite [9]. Tysonite also occurs in the muscovite type of granitic pegmatites in Inner Mongolia [10] and in pegmatites at Mpuye Hill, near Mubende, Uganda [11]. In four rare metal-containing granite massifs of amazonite facies in North Tyan'-Shan, tysonite is enriched in the apical parts of the massifs, especially in amazonite pegmatites of this zone; it sometimes forms pseudomorphs (partly or totally) after monazite [12], cf. "Seltenerdelemente" A 5, 1981, p. 442. Amazonite pegmatites of Ploskaya Mountain, Kola Peninsula, contain yttrofluorites, see p. 1, with mechanical admixtures of tysonite [13].

In western Tuva tysonite occurs in **aplites** of the Alashkii granite massif, but not in pegmatites, diabases, quartz veins and the granites themselves [14].

In the Permian granite Karakyz, upper Pskem river, Uzbekistan, tysonite was found as an accessory mineral in **quartz-greisen** veins and greisenized granite [15]. Tysonite of the tin deposit Berezov, Malyi Khingan, Far East (cf. "Zinn" A, 1971, pp. 396/7), is spatially related to Cretacious greisenized granite porphyries and liparites; it occurs together with cassiterite, fluorite, topaz, and sulfides as a **hydrothermal** product of ore formation [16]. As a hydrothermal mineral, tysonite also occurs at several places in the tin fields of Potgietersrus, Republic of South Africa, these are the Bobbejaankop granite, in the Zaaiplaats mine and on Groenvley 610, where tysonite is usually associated with bastnaesite and replaces parisite [17], cf. "Rare Earth Elements" A 8, 1984, Chapter 11.7.4.4; and on the farm Mutue Fides, 170 km NE Pretoria [18]. For a description and literature of these tin deposits, see "Zinn" A, 1971, pp. 381/2. In the Karonge rare-earth deposits, Republic of Burundi, Africa, whose mineralization is related to the hydrothermal stage of carbonatitic differentiation, tysonite has been found in very small amounts as fine coatings on the crystallized second generation of quartz [19]. Further hydrothermal deposits of tysonite are unnamed deposits in Central Asia, where it occurs together with monazite, xenotime, yttrofluorite, yttroparisite, and often replaces monazite [20]. Particularly in Central Kazakhstan tysonite has been found together with riebeckite, ilmenite, hematite, and rarely zircon, thorite, monazite, and fluorite in hydrothermal formations of the Kentsk granite massif [21].

In the USSR tysonite often occurs, together with zircon, Ce-pyrochlore and gagarinite, in **metasomatic deposits** in connection with deep-seated alkali magmatism and metasomatism, that is in biotite-riebeckite granitoids and their contact zones [22]. Thus in Eastern Siberia tysonite is associated with RE-fluorite and gagarinite in metasomatic granitoids of the subalkalic plumasite and alkalic agpaite type [23]. In biotite-riebeckite rocks and in late amphibole-fluorite-plagioclase veins it occurs together with gadolinite, thorite, titanite, and yttrofluorite [24], cf. [25]; in rocks formed during initial phases of Na metasomatism it replaces monazite [23]. In the Ukrainian Crystalline Shield tysonite occurs as an accessory mineral in metasomatic apogranites in the northwestern region; it is confined to veinlets of Al fluorides, such as websterite, prosopite, rarely thomsenolite, cryolite, and ralstonite, and to thin veinlets of quartz [26], see also [9].

Alteration of tysonite to bastnaesite is common and has been observed in the mineral deposits of Potgietersrus, Republic of South Africa [17], where tysonite from Mutue Fides is replaced by bastnaesite and an opaque earthy product along cleavage cracks or twinning planes [18]. Tysonite grains from occurrences in Eastern Siberia are surrounded and penetrated by bastnaesite and additionally penetrated in net-like patterns on fissures and cracks by Fe hydroxides and submicroscopic lamellas of hematite [24]. Tysonite from North Tyan'-Shan often is altered to fluorite and bastnaesite and its cleavage faces are penetrated by hematite [12]; see also "Seltenerdelemente" A 5, 1981, p. 442. Altered tysonite from the Kentsk granite massif, Kazakhstan, contains fluorite with inclusions of Fe oxides [21] and tysonites occurring in metasomatites of the Ukrainian Shield are replaced by RE oxides of cerianite type [9].

As a mineral with high difference in electronegativity of its cations ($>$300 kcal/g-atom) tysonite shows no metamictization [27].

References for 10.1.5.1:

[1] C. Hintze (Handbuch der Mineralogie, Vol. 1, Pt. 2, Veit, Leipzig 1915, pp. 1211/2674, 2568). – [2] E. B. Eckel (U. S. Geol. Surv. Bull. No. 1114 [1961] 1/399, 66/7, 146). – [3] E. W. Heinrich (Bull. Geol. Soc. Am. **69** [1958] 1579/80). – [4] E. W. Heinrich, E. B. Gross (Am. Mineralogist **45** [1960] 455/9, 456). – [5] H. Neumann (Norsk Geol. Tidsskr. **39** [1959] 231/6, 233).

[6] T. L. Sverdrup, P. C. Saebø, K. Ø. Bryn (Norsk Geol. Tidsskr. **45** [1965] 177/88, 177, 184). – [7] I. Oftedal (Norsk Geol. Tidsskr. **22** [1942] 218/9). – [8] S. Bergstøl, B. B. Jensen, H. Neumann (Lithos **10** [1977] 81/7, 81). – [9] E. P. Gurov, E. P. Gurova, L. G. Loginova, Yu. A. Labitzkaya (Zap. Vses. Mineral. Obshchestva **104** [1975] 455/8, 456, 458). – [10] C.-C. Kuo (Sci. Record [Peking] [2] **2** [1958] 400/13, 400, 405, 408, 410/1).

[11] R. O. Roberts (Ann. Rept. Geol. Surv. Uganda 1945/46 **1948** 31/2, 37 from Mineral. Abstr. **11** [1951] 404). – [12] L. G. Fel'dman, B. K. Surkov, T. I. Stolyarova (Tr. Mineral. Muzeya Akad. Nauk SSSR No. 22 [1973] 143/58, 145/6, 149/50). – [13] A. M. Vetoshkina, V. V. Gordienko, N. A. Elina, L. I. Polezhaeva (Mineral. Zh. **2** No. 4 [1980] 51/8, 55, 57). – [14] V. V. Lyakhovich, V. I. Noneshnikova (Tr. Inst. Mineral. Geokhim. Kristallokhim. Redk. Elem. No. 7 [1961] 182/211, 186). – [15] V. B. Shuvalov (Uzb. Geol. Zh. **19** No. 4 [1975] 70/3).

[16] V. I. Popova, L. F. Bazhenova (in: P. M. Tatarinov et al., Mineraly i Paragenezisy Mineralov Gornykh Porod, Nauka, Leningrad 1976, pp. 135/8). – [17] C. A. Strauss (Union S. Africa Dept. Mines Geol. Surv. Div. Geol. Surv. Mem. No. 46 [1954] 1/241, 61, 84, 98, 116/7, 154, 157/8). – [18] J. G. D. Steyn (Rep. S. Africa Dept. Mines Geol. Surv. Bull. No. 35 [1961] 1/31, 6, 19). – [19] L. Van Wambeke (Mineralium Deposita **12** [1977] 373/80, 374, 378). – [20] V. N. Zuev, A. V. Kosterin (Tr. Inst. Mineral. Geokhim. Kristallokhim. Redk. Elem. No. 3 [1959] 151/3).

[21] M. B. Chistyakova, M. E. Kazakova (Tr. Mineral. Muzeya Akad. Nauk SSSR No. 19 [1969] 236/8). – [22] Yu. L. Kapustin (in: L. N. Ovchinnikov, N. A. Solodov, Mestorozhdeniya Litofil'nykh Redkikh Metallov, Nedra, Moscow 1980, pp. 210/25, 210/1). – [23] V. V. Arkhangel'skaya (Dokl. Akad. Nauk SSSR **195** [1970] 1411/4; Dokl. Earth Sci. Sect. **195** [1970] 142/5). – [24] N. E. Zalashkova, D. I. Agapova, N. B. Yusupov (Mineral. Issled. **2** [1972] 69/78, 71). – [25] N. E. Zalashkova, D. A. Mineev, N. B. Yusupov (in: D. A. Mineev, Osobennosti Raspredeleniya Redkikh Elementov v Izverzhennykh Gornykh Porodakh, Inst. Mineral. Akad. Nauk, Moscow 1970, pp. 16/32, 24/5).

[26] E. P. Gurov, E. P. Gurova (Mineral. Sb. [Lvov] **28** No. 4 [1974] 41/3). – [27] A. S. Povarennykh (Zap. Vses. Mineral. Obshchestva **85** [1956] 593/7).

10.1.5.2 Chemistry

For a compilaton of analyses and the content of $(RE,Y)_2O_3$, mean and variance of 10 analyses, see "Seltenerdelemente" A 4, 1979, pp. 32/3, 5. The formula of tysonite is controversial. While Steyn [1] differentiates between a fluoride type (Colorado) and an oxyfluoride type (Österby and Uganda), Sverdrup et al. [2] think the oxyfluoride type to be caused by the impurities in tysonite that are not deduced before calculating the formula. On the other hand the calculation of 5 tysonite analyses from the literature by Fel'dman et al. after deduction of Ca, Fe, Mg, Al, Ti (as caused by impurities) results for all these samples from different localities in the formula $(RE, \pm Th)(O_{0.2}F_{2.6})$, that implies as an oxyfluoride with an anion defect of 0.4 in the unit cell (Z = 2) [3]. A nearly ideal confirmation with the theoretical fluoride formula has been found for tysonite of Eastern Siberia with $RE_{0.99}F_{3.01}$ [4, 5]; and also tysonites from the Far East with a crystal chemical formula $(Ce_{0.44}La_{0.21}Nd_{0.20}Pr_{0.05}Y_{0.04}Sm_{0.03}Gd_{0.02}Dy_{0.005}Th_{0.002})F_3$ [6] and from pegmatites in the Ukrainian Shield with $(La_{0.33}Ce_{0.48}Pr_{0.05}Nd_{0.12}Gd_{0.01}Y_{0.01})F_{3.03}$ [7] are calculated as fluorides with 3 F. Oxyfluoride formulas are given for tysonite from Uganda as $(Ce,La)_2O_{0.4}F_{5.2}$ [8], tysonite from the Kentsk granite, Central Kazakhstan, as $(La_{0.50}Ce_{0.39}Pr_{0.04}Nd_{0.04}\ Ca_{0.02}Th_{0.01})F_{2.55}O_{0.22}$ (after deduction of Fe_2O_3) [9], and tysonite from granites in North Tyan'-Shan as $(RE_{0.98}Th_{0.02})(F_{2.59}O_{0.215})$ after deduction of Fe_2O_3 and Ca as impurities [3].

As it is to be seen from the chemical formulas, tysonite may contain Th replacing Ce^{3+} compensated by the replacement of F^- by O^{2-} [3, 10]. Two samples of tysonite from Uzbekistan contain 3.54 and 2.92% Th [11], tysonite of North Tyan'-Shan 2.33% Th [3], and tysonite from Central Kazakhstan 1.37% Th [9]. The two samples from Uzbekistan in addition have 0.023 and 0.024% Sc [11]. As impurities are often given Fe_2O_3 [3, 5], Si, Fe [7], Si, Al, P, Mn [9], Ba, Sr [12], and Ca [13, 14]. See also above.

The following table of the **distribution of RE** in tysonite shows the predominance of Ce-earth:

Locality		$\Sigma RE \pm Y = 100$		Ref.
	ΣCe 1)	ΣY_1	ΣY_2	
Eastern Siberia				
sample 1	97.8	2.2 2)	0.0	[4, 5]
sample 2	99.5	0.5 3)	0.0	[4]
sample 1 *)	88.0	11.5	0.5 6)	[15]
sample 2 **)	87.4	12.0 4)	0.6 6)	[15]
North Tyan'-Shan	79.4	16.6 5)	4.0 7)	[3]
Ukrainian Shield				
sample from metasomatites	97.4	2.4 5)	0.2 8)	[10]
Ploskaya Mountain, Kola	82.1	16.2 4)	1.7 9)	[14]
Black Cloud pegmatite, Colorado	98.3	1.8 5)	0.0	[13]

*) Tysonite developed from monazite. – **) Tysonite from intergrowth with yttrofluorite. – 1) La, Ce, Pr, Nd. – 2) Sm, Eu, Gd, Dy. – 3) Sm and Dy. – 4) Sm through Dy + Y. – 5) Sm, Gd, Dy + Y. – 6) Er, Tm, Yb, Lu. – 7) Er, Yb. – 8) Ho through Yb. – 9) Ho through Lu.

Two samples from Uzbekistan have $RE_{Ce}/Y = 226.3$ and 211.8 with $RE_{Ce} = La + Ce + Pr + Nd + Sm + Gd$ [11]. Two samples from South Africa (Transvaal), the second of which is calculated after deduction of intergrown fluorite, have $Ce_2O_3 = 42.7$ and 42.9, La_2O_3 group = 55.9 and 41.3, and Y_2O_3 group = 1.3 and 15.8 for $\Sigma RE + Y = 100$ [1]. The distribution of RE in tysonites from pegmatites in the Ukrainian Shield, from Far East, and Central

Kazakhstan is given by their crystal chemical formulas, see p. 16. The distribution of RE in tysonites from pegmatites and metasomatites in the Ukrainian Shield is similar, with more La and less Nd and Pr in the tysonites of the first rocks [7]. With the exception of tysonites from Central Kazakhstan [9] and from South Africa [1] with La > Ce, all analyzed tysonites show a distinct predominance of Ce itself. As the fluorides of light and heavy RE are not isostructural (LaF_3 type is hexagonal, see p. 18, YF_3 type is orthorhombic), the distribution of RE in the hydrothermal tysonites from Tyan'-Shan and from South Africa (Transvaal) are exceptions. Both samples have a high content of heavy RE (Gd through Lu) and Y (see table, p. 16 for Tyan'-Shan and second sample from Transvaal, p. 16); this is not characteristic of tysonites [3]. For LaF_3 and YF_3 structures cf. "Seltenerdelemente" C 3, 1976, pp. 44, 50, 56, 148, 156.

References for 10.1.5.2:

[1] J. G. D. Steyn (Rep. S. Africa Dept. Mines Geol. Surv. Bull. No. 35 [1961] 1/31, 2, 13/4). – [2] T. L. Sverdrup, P. C. Saebø, K. Ø. Bryn (Norsk Geol. Tidsskr. **45** [1965] 177/88, 184/7). – [3] L. G. Fel'dman, B. K. Surkov, T. I. Stolyarova (Tr. Mineral. Muzeya Akad. Nauk SSSR No. 22 [1973] 143/58, 148/9 151/2). – [4] N. E. Zalashkova, D. A. Mineev, N. B. Yusupov (in: D. A. Mineev, Osobennosti Raspredeleniya Redkikh Elementov v Izverzhennykh Gornykh Porodakh, Inst. Mineral. Akad. Nauk, Moscow 1970, pp. 16/32, 24/5). – [5] N. E. Zalashkova, D. I. Agapova, N. B. Yusupov (Mineral. Issled. **2** [1972] 69/78, 73, 75).

[6] V. I. Popova, L. F. Bazhenova (in: P. M. Tatarinov et al., Mineraly i Paragenezisy Mineralov Gornykh Porod, Nauka, Leningrad 1976, pp. 135/8, 138). – [7] E. P. Gurov, E. P. Gurova, L. G. Loginova, Yu. A. Labitzkaya (Zap. Vses. Mineral. Obshchestva **104** [1975] 455/8). – [8] R. O. Roberts (Ann. Rept. Geol. Surv. Uganda 1945/46 **1948** 31/2, 37 from Mineral. Abstr. **11** [1951] 404). – [9] M. B. Chistyakova, M. E. Kazakova (Tr. Mineral. Muzeya Akad. Nauk SSSR No. 19 [1969] 236/8). – [10] E. P. Gurov, E. P. Gurova (Mineral. Sb. [Lvov] **28** No. 4 [1974] 41/3).

[11] V. B. Shuvalov (Uzb. Geol. Zh. **19** No. 4 [1975] 70/3). – [12] V. V. Lyakhovich, V. I. Noneshnikova (Tr. Inst. Mineral. Geokhim. Kristallokhim. Redk. Elem. No. 7 [1961] 182/211, 189). – [13] E. W. Heinrich, E. B. Gross (Am. Mineralogist **45** [1960] 455/9, 458). – [14] A. M. Vetoshkina, V. V. Gordienko, N. A. Elina, L. I. Polezhaeva (Mineral. Zh. **2** No. 4 [1980] 51/8, 55). – [15] V. V. Arkhangel'skaya (Dokl. Akad. Nauk SSSR **195** [1970] 1411/4; Dokl. Earth Sci. Sect. **195** [1970] 142/5).

10.1.5.3 Crystal Form and Structure

Tysonite is hexagonal [1, 2]. Crystal forms are $\{0001\}$, $\{10\bar{1}0\}$, $\{11\bar{2}0\}$, $\{30\bar{3}2\}$, $\{11\bar{2}1\}$, $\{22\bar{4}1\}$ [3], see drawings of tysonite crystals in [4]. The habit is short prismatic with weakly developed faces of the dipyramid, prismatic with well developed dipyramids, and dipyramidal with weakly developed prism, but also irregular grains are formed [4]. There occur lamellar, rarely tabular crystals or wedge-shaped ones by the development of prism and dipyramidic faces [5]; and prismatic crystals with smooth faces [6] and short prismatic ones with hexagonal habit [7], see also [8, 9].

In thin sections the cleavage is $\{0001\}$ and $\{10\bar{1}0\}$ [4, 10], see for perfect cleavage $\{0001\}$ [11, 12, 13], for basal and prismatic cleavage [14]. The cleavage may be imperfect $\{0001\}$ [15], imperfect after the dipyramids [5]. For a cleavage in two perpendicular directions see [16], for no cleavage [8, 9].

Under the microscope polysynthetic twinning after $\{10\bar{1}2\}$ can be observed [4, 10], cf. also [12, 16]. Twins with growth plane that forms an angle of 26° to 27° with [0001], probably imply twins after a dipyramid [4].

Parallel intergrowths of tysonite and bastnaesite [10, 17] and of tysonite and gagarinite occur [6]. Submicroscopic intergrowths of yttrofluorite and tysonite with dactyloscopic texture is comparable to the dissociation texture of a solid solution [18].

As the trifluorides of La and Ce, tysonite has a hexagonal structure with space group $P6_3/mcm-D_{6h}^3$ (No. 193) and Z = 6 [19, 20], cf. [18] and "Seltenerdelemente" C 3, 1976, p. 50. See, however, $P6_3/mmc-D_{6h}^4$ (No. 194) given by Strunz [22, 23]. Cell parameters of tysonites from different localities are given as follows:

Localities and parent rocks	a in Å	c in Å	Ref.
Pikes Peak area, Colorado	4.120	7.295	[1]
Österby, Sweden	~4.1	~7.2	[20]
Jennyhaugen, Drag, Norway	~4.115	~7.290	
Tørdal, Telemarken, Norway	~4.12	~7.3	[21]
Mutue Fides, South Africa	4.122	7.304	[10]
Kentsk granite, Kazakhstan	4.135	7.295	[5]
North Tyan'-Shan, two massifs	4.10±0.01	7.310±0.015	
	4.10±0.01	7.30±0.02	[4]
Ukrainian Shield, pegmatites			
inclusion in fluorite	4.14	7.28	
inclusion in yttrofluorite	4.11	7.23	
metasomatites	4.13	7.28	[8]

Drawings of the unit cell of tysonite see [1, 17] and of hexagonal LaF_3 type "Seltenerdelemente" C 3, 1976, p. 50.

References for 10.1.5.3:

[1] I. Oftedal (Z. Physik. Chem. B **5** [1929] 272/90, 276, 278, 284). – [2] E. S. Larsen, H. Berman (U.S. Geol. Surv. Bull. No. 848 [1934] 1/266, 71, 82). – [3] C. Palache, H. Berman, C. Frondel (Dana's System of Mineralogy, 7th Ed., Vol. 2, Wiley, New York 1951, pp. 1/1124, 48). – [4] L. G. Fel'dman, B. K. Surkov, T. I. Stolyarova (Tr. Mineral. Muzeya Akad. Nauk SSSR No. 22 [1973] 143/58, 144/7). – [5] M. B. Chistyakova, M. E. Kazakova (Tr. Mineral. Muzeya Akad. Nauk SSSR No. 19 [1969] 236/8).

[6] N. E. Zalashkova, D. I. Agapova, N. B. Yusupov (Mineral. Issled. **2** [1972] 69/78, 71). – [7] V. B. Shuvalov (Uzb. Geol. Zh. **19** No. 4 [1975] 70/3). – [8] E. P. Gurov, E. P. Gurova, L. G. Loginova, Yu. A. Labitzkaya (Zap. Vses. Mineral. Obshchestva **104** [1975] 455/8, 455/6). – [9] E. P. Gurov, E. P. Gurova (Mineral. Sb. [Lvov] **28** No. 4 [1974] 41/3). – [10] J. G. D. Steyn (Rep. S. Africa Dept. Mines Geol. Surv. Bull. No. 35 [1961] 1/31, 6, 8/10, 12, 20/1).

[11] R. O. Roberts (Ann. Rept. Geol. Surv. Uganda 1945/46 **1948** 31/2, 37 from Mineral. Abstr. **11** [1951] 404). – [12] E. W. Heinrich, E. B. Gross (Am. Mineralogist **45** [1960] 455/9, 457). – [13] V. V. Lyakhovich, V. I. Noneshnikova (Tr. Inst. Mineral. Geokhim. Kristallokhim. Redk. Elem. No. 7 [1961] 182/211, 189). – [14] P. Geijer (Geol. Foeren. Stockholm Foerh. **43** [1921] 19/23, 21). – [15] V. I. Popova, L. F. Bazhenova (in: P. M. Tatarinov et al., Mineraly i Paragenezisy Mineralov Gornykh Porod, Nauka, Leningrad 1976, pp. 135/8, 137).

[16] V. N. Zuev, A. V. Kosterin (Tr. Inst. Mineral. Geokhim. Kristallokhim. Redk. Elem. No. 3 [1959] 151/3). – [17] I. Oftedal (Norsk Geol. Tidsskr. **12** [1931] 459/67). – [18] V. V. Arkhangel'skaya (Dokl. Akad. Nauk SSSR **195** [1970] 1411/4; Dokl. Earth Sci. Sect. **195** [1970] 142/5). – [19]

I. Oftedal (Z. Physik. Chem. B **13** [1931] 190/200, 198/9). – [20] K. Schlyter (Arkiv Kemi **5** [1953] 73/82, 75).

[21] T. L. Sverdrup, P. C. Saebø, K. Ø. Bryn (Norsk Geol. Tidsskr. **45** [1965] 177/88, 185). – [22] H. Strunz (Rend. Soc. Mineral. Ital. **14** [1958] 326/8). – [23] H. Strunz, C. Tennyson (Mineralogische Tabellen, 6th Ed., Geest and Portig, Leipzig 1977, pp. 1/621, 157).

10.1.5.4 Physical Properties

Optical Properties

Pure tysonite has vitreous luster and is transparent [1, 2] or translucent [3]. Altered tysonite loses luster and transparency [1, 4] and becomes hardly greasy [5] or fatty [6]. But unaltered tysonite may be waxy to greasy too [4]. The color of fresh tysonite is greenish yellow, altered tysonite creamy greenish gray or dark cinnamon [1], unaltered tysonite colorless [3] or yellow in different tints, changing with alteration [4], also yellowish brown, if fresh, changing to reddish brown due to alteration [7]. Color has also been reported as weakly rose or clear yellow to colorless [6], yellow to light yellow [8], pale rose to colorless [2], pinkish to pinkish yellow [5], rose to rose-white [9], pinkish buff [10], waxy yellow [11], light yellowish brown [12], or medium brown [13]. The streak of fresh tysonite is white, that of the altered mineral is reddish [4]. The streak may also be light yellow to colorless [8], the powder is rose-white [9]. Under the microscope tysonite in thin sections appears transparent and colorless [1, 4, 6, 8, 11].

Optical character is uniaxial negative as can be seen from the following table with the refractive indices of tysonites from the various localities:

Parent rocks and locality	n_o	n_e	Ref.
Pegmatites			
Brodbo, Sweden	1.618	1.611	[12]
Ukrainian Shield			
as inclusion in fluorite	1.616	1.609	
as inclusion in yttrofluorite	1.618	1.612	[3]
North Tyan'-Shan	1.613±0.002	1.608±0.002	[4]
Pikes Peak, Colorado	1.613	1.607	[12]
Black Cloud, Colorado	1.615	1.607	[13]
	1.612	1.608	[14]
Aplite			
Alashskii massif, Western Tuva	1.618	1.611	[11]
Greisen			
Karakyz, Uzbekistan	1.616±0.002	1.609±0.002	
(2 samples)	1.617±0.002	1.609±0.002	[9]
Berezov, Malyi Khingan, Far East	1.615	1.608 to 1.609	[5]
Hydrothermal formations			
Kentsk granite, Central Kazakhstan	1.609	1.603	[1]
Sn deposit, Mutue Fides, S. Africa	1.614±0.002	1.608±0.002	[7]
Metasomatic rocks			
Ukrainian Shield	1.618	1.609	[2, 3]
Eastern Siberia	1.611±0.001	1.607±0.001	[6, 15]

Refractive indices of tysonites from Jennyhaugen, Drag, and of Tørdal, Telemark, both Norway, are near 1.610 with n_o somewhat higher than n_e [16]. Tysonite, found as pseudomorphs after monazite in a hydrothermal deposit of Central Asia, has $n_o = 1.611$ to 1.612 and $n_e = 1.605$ to 1.607. This tysonite sometimes is optically biaxial with $2V = 40°$ to 80°; dispersion of the optical axis is small, often $r < v$, sometimes $r > v$ [8].

Absorption spectra of tysonite may be seen in [17]. Tysonites from rare-metal granites and pegmatites in North Tyan'-Shan show blue luminescence in ultraviolet light, no thermoluminescence is observed [4]. Tysonites from pegmatites in the Ukrainian Shield show no photoluminescence while those from metasomatites in the same region are weakly photoluminescent [3], cf. [2].

Other Physical Properties

Hardness, $H_M = 5.0$ to 5.5 of fresh tysonite, decreases with alteration to 4.5 and 4.0 [4]. Further data given are $H_M = 5.5$ [8], ~5 [9], 4 to 5 [10], and ~4 [5]. Microhardness of fresh tysonite is 390 kg/mm² [1]. The fracture of fresh samples is conchoidal [1, 2, 3, 5]. The mineral is brittle [18, 19].

Density decreases with alteration, values D in g/cm³ given in decreasing order are: ~6.5, tysonite from aplite in West-Tuva [11], 6.01 to 6.14, from Pikes Peak area, Colorado [12], 5.943 ± 0.005 or, corrected for alteration products, 5.995, tysonite from Mutue Fides, South Africa [7], 5.93, from Kentsk granite, Central Kazakhstan [1], 5.83, partly altered tysonite (corrected to 6.10) from granites in North Tyan'-Shan [4], 5.8, for tysonites from Uganda [10], Central Asia [8], and Black Cloud pegmatite, Colorado [13], 5.73, tysonite from Brodbo, Sweden [12], 5.64, from East Siberia [6, 12], see also [20]. D(calc.) = 6.13 g/cm³ [21].

Tysonite is weakly electromagnetic [9]; for data on paramagnetic rotation in tysonite, see [22].

References for 10.1.5.4:

[1] M. B. Chistyakova, M. E. Kazakova (Tr. Mineral. Muzeya Akad. Nauk SSSR No. 19 [1969] 236/8). – [2] E. P. Gurov, E. P. Gurova (Mineral. Sb. [Lvov] **28** No. 4 [1974] 41/3). – [3] E. P. Gurov, E. P. Gurova, L. G. Loginova, Yu. A. Labitzkaya (Zap. Vses. Mineral. Obshchestva **104** [1975] 455/8). – [4] L. G. Fel'dman, B. K. Surkov, T. I. Stolyarova (Tr. Mineral. Muzeya Akad. Nauk SSSR No. 22 [1973] 143/58, 146/8). – [5] V. I. Popova, L. F. Bazhenova (in: P. M. Tatarinov et al., Mineraly i Paragenezisy Mineralov Gornykh Porod, Nauka, Leningrad 1976, pp. 135/8, 137).

[6] N. E. Zalashkova, D. I. Agapova, N. B. Yusupov (Mineral. Issled. **2** [1972] 69/78, 71). – [7] J. G. D. Steyn (Rep. S. Africa Dept. Mines Geol. Surv. Bull. No. 35 [1961] 1/31, 6/7). – [8] V. N. Zuev, A. V. Kosterin (Tr. Inst. Mineral. Geokhim. Kristallokhim. Redk. Elem. No. 3 [1959] 151/3). – [9] V. B. Shuvalov (Uzb. Geol. Zh. **19** No. 4 [1975] 70/3). – [10] R. O. Roberts (Ann. Rept. Geol. Surv. Uganda 1945/46 **1948** 31/2, 37 from Mineral. Abstr. **11** [1951] 404).

[11] V. V. Lyakhovich, V. I. Noneshnikova (Tr. Inst. Mineral. Geokhim. Kristallokhim. Redk. Elem. No. 7 [1961] 182/211, 189). – [12] P. Geijer (Geol. Foeren. Stockholm Foerh. **43** [1921] 19/23, 20/1). – [13] E. W. Heinrich, E. B. Gross (Am. Mineralogist **45** [1960] 455/9, 456). – [14] C. W. Wolfe (from [18]). – [15] N. E. Zalashkova, D. A. Mineev, N. B. Yusupov (in: D. A. Mineev, Osobennosti Raspredeleniya Redkikh Elementov v Izverzhennykh Gornykh Porodakh, Inst. Mineral. Akad. Nauk, Moscow 1970, pp. 16/32, 25).

[16] T. L. Sverdrup, P. C. Saebø, K. Ø. Bryn (Norsk Geol. Tidsskr. **45** [1965] 177/88, 185). – [17] A. N. Platonov, O. A. Krasil'shchikova, A. N. Tarashchan (Konst. Svoistva Mineral. No. 9 [1975] 58/64, 60/1). – [18] C. Palache, H. Berman, C. Frondel (Dana's System of Mineralogy, 7th

Ed., Vol. 2, Wiley, New York 1951, pp. 1/1124, 49). – [19] E. I. Semenov (in: K. A. Vlasov, Geochemistry and Mineralogy of Rare Elements and Genetic Types of Their Deposits, Vol. 2: Mineralogy of Rare Elements, Israel Program Sci. Trans., Jerusalem 1966, pp. 249/328, 254 [Russian Original, Moscow 1964]). – [20] Yu. L. Kapustin (in: L. N. Ovchinnikov, N. A. Solodov, Mestorozhdeniya Litofil'nykh Redkikh Metallov, Nedra, Moscow 1980, pp. 210/25, 211).

[21] I. Oftedal (Z. Physik. Chem. B **5** [1929] 272/90, 279). – [22] H. A. Kramers, J. Becquerel (Proc. Acad. Sci. Amsterdam **32** [1929] 1190/8, 1190/3).

10.1.5.5 Chemical Behavior and Differential Thermal Analysis

Tests with the solubility of tysonite in mineral acids are different: it is soluble in sulfuric acid especially with additional H_2O_2 [1], however, not soluble in sulfuric acid [2]. It is soluble in sulfuric and nitric acid, not in hydrochloric acid [3], and not attacked in the cold by sulfuric, nitric, and hydrochloric acid, but soluble after long heating in concentrated sulfuric acid [4].

The differential thermal curve of tysonite from North Tyan'-Shan shows a distinct exothermic effect of low intensity at 250 to 400°C with maximum at 359°C that is due to the oxidation of Ce^{3+} to Ce^{4+} and the transition of CeF_3 to CeO_2 [4]. Calcination of tysonites from metasomatites of the Ukrainian Shield at temperature of 1000°C for 15 min leads to a fundamental alteration, the interplanar distances of the heating products confirm the formation of cerianite and the presence of much thorianite in the mineral [5].

References for 10.1.5.5:

[1] V. I. Popova, L. F. Bazhenova (in: P. M. Tatarinov et al., Mineraly i Paragenezisy Mineralov Gornykh Porod, Nauka, Leningrad 1976, pp. 135/8, 137). – [2] V. V. Lyakhovich, V. I. Noneshnikova (Tr. Inst. Mineral. Geokhim. Kristallokhim. Redk. Elem. No. 7 [1961] 182/211, 189). – [3] V. N. Zuev, A. V. Kosterin (Tr. Inst. Mineral. Geokhim. Kristallokhim. Redk. Elem. No. 3 [1959] 151/3). – [4] L. G. Fel'dman, B. K. Surkov, T. I. Stolyarova (Tr. Mineral. Muzeya Akad. Nauk SSSR No. 22 [1973] 143/58, 146, 153). – [5] E. P. Gurov, E. P. Gurova (Mineral. Sb. [Lvov] **28** No. 4 [1974] 41/3).

10.1.6 Chukhrovite $(Ca,Y,Ce)_{>3}[(AlF_6)_2|SO_4]\cdot 10H_2O$

Chukhrovite has been found in the Mo-W deposit Karaoba, Central Kazakhstan, together with halloysite, gearksutite, fluorite, creedite, anglesite, limonite, and hydrohematite in the oxidation zone and in cavities of fresh ores between quartz or "gilbertite" [1, p. 15/6]. It here is thought to have been formed by the action on RE-fluorite of sulfuric acid derived from oxidation of pyrite with a shifting of RE distribution [2], cf. [1, p. 23]. It furthermore occurs in one of the hydrothermal fluorite-rare metal deposits of Siberia together with fluorite, gearksutite, rarely ralstonite, in cavities of tourmaline-fluorite rocks [3], where it is formed under supergene conditions by alteration of sellaite [4]. Recently chukhrovite has been found in druses of hydrothermal barite and fluorite veins in the Clara mine near Oberwolfach, middle Black Forest, Western Germany, as a hydrothermal primary mineral on fluorite, associated with pyrite, native sulfur, and jarosite [5]. The so-called "second cubic mineral" from the cryolite deposit of Ivigtut, Greenland, described by Bøggild [6], is probably identical with chukhrovite [1, pp. 24/5].

The general formula, given above, is from Strunz [7] on account of two analyses by Moleva in [1, p. 19]. These analyses lead to the special formulas $Ca_{3.03}Al_{1.62}RE_{0.98}(SO_4)_{1.00}(F_{11.46}OH_{0.30})\cdot 9.60H_2O$ and $Ca_{2.97}Al_{1.64}RE_{1.02}(SO_4)_{1.00}(F_{11.37}OH_{0.55})\cdot 9.72H_2O$ [1, pp. 18/20]. For octahedric complexes $[AlF_6]^{3-}$ see also synthetic chukhrovite [8] and structure, p. 22.

The distribution of RE and Y in the two Russian samples (1 Karaoba, 2 Siberia) is determined as follows:

	$(RE,Y)_2O_3 = 100$															Ref.
	Y	La	Ce	Pr	Nd	Sm	Eu	Gd	Tb	Dy	Ho	Er	Tm	Yb	Lu	
1	41.0	5.0	15.0	3.9	12.0	7.2	0.2	6.5	0.9	4.1	0.8	1.7	0.3	1.2	0.2	[1, p. 20]
2	20.7	8.9	27.1	6.3	14.8	7.0	—	6.6	1.0	4.0	—	2.2	—	1.4	—	[4]

These data imply that chukhrovite from Karaoba has Y > Ce, chukhrovite from Siberia Y < Ce [4]. Chukhrovite from Black Forest has no Y [9], besides Ce and Nd it contains La, Pr, Sm, Gd, Eu; as Ce is predominant, the mineral should be named chukhrovite - (Ce), while the mineral from Karaoba could be named chukhrovite - (Y) [5]. RE distribution in at.% of chukhrovite and its starting material, RE-fluorite, from Karaoba see [2].

Chukhrovite is isometric. The crystals have {100} and {111}, and cubic or cubooctahedral habit; more rarely are aggregates of isometric grains [1, p. 16], [10]; octahedral habit also occurs [4]. Crystal faces are partly bent by step formation [1, p. 16]. Cleavage is indistinct {111} [1, p. 17], or distinct {111} and {100} [4].

Lattice constant a = 16.80 ± 0.05 Å, Z = 8 (Karaoba) [1, p. 20], a = 16.73 ± 0.05 Å (Siberia) [4], a = 16.74(4) Å (Black Forest) [5]. Synthetic chukhrovite has a = 16.8 Å [8]. Space group is $Fd3\text{-}T_h^4$ (No. 203). The structure is built up by two Al octahedra and two (Ca, RE) octahedra, which form four-membered pseudocubic rings linked with each other. The largest cavities, between the (Ca, RE) octahedra, are filled with SO_4 tetrahedra which are connected with the H_2O apices of neighboring polyhedra by a hydrogen bond. The H_2O molecules form six-membered rings connected up into the three-dimensional network [10].

Luster is vitreous or weakly pearly [1, pp. 16/7], [4]. Chukhrovite is colorless or white porcelaneous, sometimes with lilac tint, the mineral is transparent or translucent [1, p. 16], cf. [4, 10]. The color is white, the mineral opaque [5]. Chukhrovite is anomalous anisotropic, and has zonal structure with alternating transparent and translucent zones (because of liquid inclusions) [1, p. 17], [4]. Refractive index n = 1.440 to 1.442 (Karaoba) [1, p. 18], n = 1.438 (Siberia) [4], n = 1.443 ± 0.002 (Black Forest) [5]. Weak luminescence in lilac tints with bluish shade have been observed [1, p. 17], [4].

Hardness $H_M \sim 3$. Fracture is uneven. The mineral is brittle [1, p. 17]. Density D = 2.274 to 2.389, mean from 7 determinations = 2.353 g/cm³ (Karaoba) [1, p. 17], D = 2.353 g/cm³ (Siberia) [4].

Chukhrovite is easily dissolved in dilute hydrochloric and sulfuric acid [1, p. 17]. The DTA curve of chukhrovite from Karaoba shows two endothermic effects at 280 and 780°C and an exothermic effect at 480°C. The first endothermic effect is due to the loss of crystal chemical water and perhaps some F as HF, the second is characteristic for many Al-fluorides, but its cause is not known; the exothermic effect at 480°C probably shows the formation of a new phase [1, pp. 17/8]. The DTA curve of chukhrovite from Siberia is, with endothermic effects at 270 and 770°C and an exothermic effect at 460°C, very similar to that of the mineral from Karaoba [4].

References for 10.1.6:

[1] L. P. Ermilova, V. A. Moleva, R. F. Klevtsova (Zap. Vses. Mineral. Obshchestva **89** [1960] 15/25). – [2] M. Fleischer (Indian Mineralogist **10** [1969/71] 36/9, 37/8, Table 1). – [3] M. I. Novikova, G. A. Sidorenko, N. N. Kuznetsova (Zap. Vses. Mineral. Obshchestva **95** [1966] 39/44,

39). – [4] M. I. Novikova (Zap. Vses. Mineral. Obshchestva **102** [1973] 200/2). – [5] K. Walenta (Chem. Erde **38** [1979] 331/9, 331, 333/4, 336).

[6] O. B. Bøggild (Medd. Groenland **149** No. 3 [1953] 1/442, 97). – [7] H. Strunz, C. Tennyson (Mineralogische Tabellen, 6th Ed., Geest and Portig, Leipzig 1977, pp. 1/621, 161). – [8] N. G. Sokolova, S. L. Konovalova (Zap. Vses. Mineral. Obshchestva **110** [1981] 160/2). – [9] K. Walenta (private communication from M. Fleischer, Am. Mineralogist **66** [1981] 879). – [10] G. B. Bokii, L. I. Gorogotskaya (Dokl. Akad. Nauk SSSR **163** [1965] 183/5; Dokl. Earth Sci. Sect. **163** [1965] 92/4).

10.2 Oxidic Compounds

10.2.1 Simple Oxides

10.2.1.1 Cerianite $(Ce,Th)O_2$

Occurrence, Paragenesis, and Chemistry

The first description of cerianite was of a sample from Lackner Township, Sudbury district, Ontario, where it occurs in carbonate rocks cutting a nepheline syenite [1]; for a further occurrence in Ontario, see "Rare Earth Elements" A 8, 1984, Chapter 11.7.2.2.1. As a rare accessory mineral it was found in nepheline syenites from Khodzhaachkan, Kirghizia [2], and could be determined in the slightly biotitic fine-grained granite dike of Kismórágy, Hungary, sometimes as an inclusion in magnetite grains in feldspar [14]. It also has been described from microcline pegmatite from Nesoya, Lutzow-Holmbukta, East Antarctica. Cerianite here is completely metamict [15]. For cerianite in the pyrochlore-zircon ore type of a postmagmatic, metasomatic mineralization in the deposit Borsuksai, western Kazakhstan, see "Rare Earth Elements" A 8, 1984, Chapter 11.7.6.2.1. In cleavelandite pegmatite dikes from Iveland, Norway, cerianite is a pseudomorph after monazite and has itself been replaced by tysonite (fluocerite), bastnaesite, and törnebohmite [3]. In the hydrothermal rare-earth deposit Karonge, Republic of Burundi, Africa, cerianite locally replaces bastnaesite [4], and in the zone of oxidation of a Siberian deposit with RE fluorocarbonates it forms pseudomorphs after bastnaesite [5]. Cerianite occurs as an alteration product of bastnaesite also in lavas in the Cerro Manomó carbonatite complex, eastern Bolivia [6]; cf. p. 97. As a weathering product of bastnaesite, cerianite has been identified in pegmatites of Madagascar [7], and as a secondary mineral after bastnaesite and/or allanite in Morro do Ferro on the Poços de Caldas plateau, Minas Gerais, Brazil, in soft weathered materials together with fine-grained hydromica, kaolinite, and limonite [8]; cf. "Rare Earth Elements" A 8, 1984, Chapter 11.7.3.1. Cerianite as an alteration product of allanite is the most wide-spread secondary RE mineral in the allanite pegmatite, Amherst County, Virginia, where it occurs as coatings on secondary monazite layers on allanite and directly on allanite, or intimately mixed with bastnaesite and halloysite in allanite. Similar cerianite with altered allanite and altered perrierite occurs also at other localities in Amherst and Bedford County, Virginia [9]. In pegmatite dikes at Høydalen, in Tørdal, Telemark, southern Norway, monazite is altered to a mixture of tysonite (fluocerite) and cerianite [10].

The cerianites from Ontario and Brazil are nearly pure CeO_2, while the cerianite from Iveland, Norway, has a content of 50 to 56% CeO_2 only, and 30 to 36% of other RE oxides [3]. RE composition of cerianite from Madagascar is $La_6Ce_{91}Pr_{0.9}Nd_{2.1}$ [7]. For the distribution pattern of La to Gd in cerianite from Iveland, Norway, see [11]. Besides the main elements Ce and Th, cerianite from Hungary contains the elements La, Nd, Ca, Pb, and U [14]. Two samples from Iveland contain 1.6 and 1.8% ThO_2 [3]; cerianite from Sudbury, Ontario, even 5.1±0.5% ThO_2 [1]. Cerianite is thought to be a solid solution series $(Ce^{IV},Th,U)O_2$–RE_2O_3 [7].

Physical Properties

There occur octahedral (Ontario) [1], thick tabular (Iveland) [3], or cubic crystals with {100} and {111}, and sometimes penetration twins on {111} (Kirghizia) [2]; also thin coatings (Siberia) [5].

Synthetic CeO_2 has a face-centered cubic fluorite structure, space group Fm3m-O_h^5 (No. 225), Z=4, a=5.411 Å; see "Seltenerdelemente" C 1, 1974, p. 234. The following lattice constants were determined for the minerals investigated:

Locality	a in Å	Ref.
Sudbury district, Ontario[1)]	5.42	[1]
Iveland, Norway		
Kåbuland	5.470±0.005	[3]
Birkeland	5.482±0.005	[3]
Siberia	5.42	[5]
Poços de Caldas, Brazil	5.411±0.004	[8]
Karonge, Burundi, Africa	5.410±0.007	[4]
East Antarctica	5.451[2)]	[15]

[1)] Ce:Th=16:1. – [2)] After a heat treatment, 1000°C for seven hours.

Cerianite is translucent [1] or semitransparent [5]. Its color is dark greenish amber (Ontario) [1] and (Kirghizia) [2]; reddish brown to orange-yellow (Norway) [3]; rose-red and yellow (Siberia) [5]. The mineral is isotropic, index of refraction $n>2.0$ [8]. Luster is nacreous. Cerianite is weakly anisotropic, n=1.775 (Siberia) [5]. It shows no fluorescence in ultraviolet radiation, in contrast with synthetic CeO_2 [8]. For IR spectra of synthetic CeO_2 see [12] and "Seltenerdelemente" C 1, 1974, p. 254.

Cerianite is very soft (Siberia) [5]. It has density D=6.033 g/cm³ (Kåbuland) and 5.89 g/cm³ (Birkeland) [3]. D(calc.)=7.19 g/cm³ [13].

References for 10.2.1.1:

[1] A. R. Graham (Am. Mineralogist **40** [1955] 560/4, 560/2). – [2] E. I. Semenov (Mineralogiya Redkikh Zemel', Nauka, Moscow 1963, pp. 1/412, 46). – [3] H. Neumann, S. Bergstøl (Norsk Geol. Tidsskr. **43** [1963] 247/55, 247/9, 252/4). – [4] L. Van Wambeke (Mineralium Deposita **12** [1977] 373/80, 374). – [5] A. P. Khomyakov (in: E. I. Semenov, R. P. Tikhonenkova, Mineralogiya i Geneticheskie Osobennosti Shchelochnykh Massivov, Nauka, Moscow 1964, pp. 56/82, 66).

[6] C. J. N. Fletcher, J. D. Appleton, B. C. Webb, I. R. Basham (Trans. Inst. Mining Met. [London] B **90** [1981] 37/50, 41). – [7] E. I. Semenov, R. L. Barinskii (Geokhimiya **1958** 314/33; Geochemistry [USSR] **1958** 398/419, 412/3). – [8] C. Frondel, U. B. Marvin (Am. Mineralogist **44** [1959] 882/4). – [9] R. S. Mitchell, G. E. Redline (Rocks Minerals **55** [1980] 245/9, 248/9). – [10] S. Bergstøl, B. B. Jensen, H. Neumann (Lithos **10** [1977] 81/7, 81).

[11] B. B. Jensen (Norsk Geol. Tidsskr. **47** [1967] 1/8, 4/5). – [12] A. S. Povarennykh (Konst. Svoistva Mineral. No. 13 [1979] 53/78, 58/9). – [13] E. I. Semenov (in: K. A. Vlasov, Geochemistry and Mineralogy of Rare Elements and Genetic Types of Their Deposits, Vol. 2: Mineralogy of Rare Elements, Israel Program Sci. Trans., Jerusalem 1966, pp. 249/328, 259 [Russian Original, Moscow 1964]). – [14] G. Pantó (Acta Geol. Acad. Sci. Hung. **19** [1975] 59/93, 69). – [15] Y. Matsumoto, A. Sakamoto (Mem. Natl. Inst. Polar Res. Spec. Issue [Japan] No. 21 [1982] 103/11 from C.A. **97** [1982] No. 75830).

10.2.1.2 RE-bearing Uraninite

History and Occurrence

"Cleveite" and "nivenite", originally thought to be distinct from uraninite (see mineral descriptions and occurrences [1, 2]) have been found by X-ray studies to be varieties of uraninites with relatively high RE content. The names therefore should be dropped [3].

Such uraninites with high RE content are found primarily in granite pegmatites from Garta near Arendal, Norway ("cleveite"), and from Llano County, Texas ("nivenite") [1]. Further occurrences in Norway are granite pegmatites from Auseimyren near Tvedestrand [4], dikes at two localities east of Isesjö, commune of Skjeberg, and at Lundestad, north of the Iddefjord, Østfold [5], and other occurrences at the southern coast of Norway [6]. Uraninites with high RE content occur also in pegmatites of the Chupa area, Northern Karelia [7, p. 21], in the pegmatite of Isaka near the north end of the Abukuma granite region, Honshu, Japan [8], in the Clear Creak pegmatite, Burnet County, Texas [9], and in pegmatites of the Halo mine, Bancroft area, Haliburton County, Ontario [10]; cf. "Uran" A 1, 1979, pp. 157/60. Y- and Nd-bearing uraninite is an accessory mineral in a biotite granodiorite in Mariposa County, California [11]; uraninite with up to 10% RE_2O_3 occurs as a magmatic mineral in granitic rocks of the Liberty Hill pluton, South Carolina [20]. Uraninites with high contents of RE occur in the Canadian uranium deposits of the Blind River – Elliot Lake – Quirke Lake region, South Ontario [10]; especially for Elliot Lake see also [12] and for Blind River deposits [13]. Furthermore, the uraninite deposits of the Singhbhum Shear Zone or copper belt, Bihar, India, contain RE-bearing uraninites [14] which occur as anhedral to subhedral disseminated grains. Most of these grains are in the interstices of silicates and other minerals that constitute the gangue [22]. For descriptions of the uranium deposits of Ontario and India see "Uran" A 1, 1979, pp. 146/57 and 261/2, respectively. Further localities with RE-rich uraninites are given in "Seltenerdelemente" A 4, 1979, pp. 122/3 and "Rare Earth Elements" A 8, 1984, Chapters 11.2.2.2 and 11.2.2.3.

Chemistry

Literature relating to analyses of RE-bearing uraninites (some of them named "cleveite" or "nivenite") and RE contents in uraninites is summarized in "Seltenerdelemente" A 4, 1979, pp. 9, 33, 122/3.

The high contents of RE ± Y in uraninites are due to replacement of U by RE and Y; Ce^{4+} and Pr^{4+} may replace U^{4+} without valence compensation, while the other RE cause a change of the oxidation state of U by replacement $3U^{4+} \rightarrow 2RE^{3+} + U^{6+}$ [7, p. 31].

Uraninites often have $\Sigma Y_2O_3 > \Sigma Ce_2O_3$; examples are uraninites from Isaka, Japan, with 14.60% Y-earth and 0.41% Ce-earth [8], 44 uraninites from the Lake Elliot area, Ontario, with 0.81 to 4.42, mean 2.09% Y_2O_3, and 0.0 to 1.10, mean 0.46% Ce_2O_3 [12], and 32 out of 33 samples of uraninite from the Indus River alluvium, near Hazro, Pakistan, with 0.05 to 2.52, mean 0.67% Y_2O_3, and 23 of these 33 samples with 0.01 to 0.42, mean 0.11% Ce_2O_3 [21]. In uraninites from 4 localities in the Singhbhum Shear Zone, Bihar, India, the heavy RE, particularly Gd, Dy, and Er predominate; uraninite from Bhatin contains more Eu than that from Narwapahar, Eu/Gd being 0.305 and 0.146, respectively [14]. By contrast, in Chinese uraninites and pitchblendes mainly Ce-group elements are predominant [15, p. 571]. Depletion of Eu in pegmatitic uraninite from Chupa, Northern Karelia, and of normally abundant Eu in pitchblende from Great Bear Lake, Canada, from Jáchymov (Joachimsthal), ČSSR, from Katanga, Zaire, and from Morogoro, Tanzania, is shown by RE distribution patterns of these minerals [16].

It seems that the content of RE in uraninites is mainly governed by temperature, the highest content of RE_2O_3 being observed in metamorphic uraninites, the lowest in hydrothermal pitchblendes from China [15, p. 568]. Uraninites from metamorphic rocks in the Elliot Lake area and in pegmatites of the Bancroft area, both Ontario, have higher RE_2O_3 contents than uraninites from sandstones or vein-type deposits [10]. For $(RE+Y)_2O_3$ contents and distribution of the single RE in pegmatitic and hydrothermal uraninites see also "Seltenerdelemente" A 4, 1979, pp. 122/3. Distribution of RE in uraninites from Ontario is given as follows [10]:

Locality	La_2O_3	CeO_2	Pr_2O_3	Nd_2O_3	Sm_2O_3	Gd_2O_3	Dy_2O_3	Er_2O_3	Y_2O_3
Quirke mine	3.0	14.5	3.0	14.1	6.0	7.2	9.1	5.0	38.1
	2.5	12.4	2.5	14.1	6.6	7.8	10.1	5.2	37.8
Denison mine	1.2	10.6	3.1	12.1	6.4	7.8	12.4	6.1	40.7
Halo mine, Bancroft	1.1	10.1	1.8	13.4	5.2	5.8	9.1	6.9	50.4
area	1.0	10.2	2.3	13.6	5.1	5.9	8.6	5.8	46.6

For variations of content and distribution of RE in zones of single uraninite crystals from core to rim, see [4] and "Seltenerdelemente" A 4, 1979, p. 122.

For isotopic composition of Xe and Kr in cleveite from eastern USSR, see [17].

Crystal Forms and Structure

The dependence of crystal forms on RE content has been studied for uraninite crystals from RE pegmatites, Chupa area, Northern Karelia; see "Seltenerdelemente" A 4, 1979, p. 122.

"Cleveite" crystals from Norway (Auselmyren near Tvedestrand, with 2.75% RE_2O_3 and Arendal, with 1.34% RE_2O_3 [18]) have lattice constants a = 5.48 Å as also has "cleveite" from Svinör, west of Lindesnes, southern Norway [6]. Uraninite from a biotite granodiorite, Mariposa County, California, with formula $U_{1.73}Th_{0.14}Pb_{0.04}Y_{0.09}Nd_{0.02}O_4$, and containing small amounts of La, Pr, Ce, Sm, and Eu, has $a = 5.45 \pm 0.01$ Å [11].

Besides the RE also other substituting ions, such as Th and others, and the state of oxidation of U influence the cell edge variations [22]. Various RE differ in their influence on cell edges of uraninite. Thus in three uraninites from Singhbhum uranium deposits, where trivalent Eu, Gd, Dy, and Y, whose ionic radii in eightfold coordination exceed that of U^{4+}, predominate, the cell constants are enlarged to 5.438, 5.440, and 5.455 Å [14]. Three pitchblendes from China, containing 1.01 to 1.03, mean 1.02% RE_2O_3, have unit parameters a = 5.396 to 5.445, mean 5.416 Å, while seven Chinese uraninites with 1.36 to 3.62, mean 2.25% RE_2O_3, have unit parameters a = 5.437 to 5.466, mean 5.455 Å. Thus the cell edge increases with increasing content of RE. As the main RE in pitchblende and uraninites are Ce-group elements with ionic radii larger than that of U^{4+}, the RE_2O_3 content generally is positively correlated to the cell size [15, pp. 566/7, 571, 574]. Contrary to these results, 14 uraninites from Northern Karelia with RE_2O_3 = 1.83 to 9.78, mean 4.21%, have unit parameters a = 5.413 to 5.484, mean 5.47 Å; when these 14 specimens are classified into 4 groups with increasing RE contents of 2.0, 4.3, 6.2, and 11.3 at. %, the cell parameters (corrected for the mean contents of Pb and Th) decrease from $5.477 \rightarrow 5.476 \rightarrow 5.472 \rightarrow 5.471$ Å. The authors interpret this result in terms of the predominance of the lanthanides of the Y-group, Eu^{3+} to Lu^{3+} (replacing U^{4+}) with ionic radii smaller than that of U^{4+} [7, pp. 24/5, 34/5], [19][1)].

[1)] This is consistent only with the values of ionic radii from Ahrens, not with those of Whittaker and Muntus for eightfold coordination, see "Seltenerdelemente" A 4, 1979, p. 2.

Physical Properties

Radiant total reflectivity in % and (RE_2O_3 content in %): Two Chinese pitchblendes 10.5 (1.01), 12.75 (1.03); four uraninites 15.6 (1.38), 15.6 (3.17), 17 (3.62), 18.56 (2.03) [15, pp. 566/7].

For "cleveite" from Auselmyren, Norway, D = 8.24 g/cm^3 [4].

References for 10.2.1.2:

[1] C. Hintze (Handbuch der Mineralogie, Vol. 1, Pt. 3, De Gruyter, Berlin 1930, pp. 1/4565, 4175/7). – [2] M. V. Soboleva, I. A. Podovkina (AEC-TR-4487 [1961] 1/455, 54/7 [Russian Original 1957]; N.S.A. **15** [1961] No. 23739). – [3] C. Frondel (U.S. Geol. Surv. Bull. No. 1064 [1958] 1/400, 12/3). – [4] R. Bakken, E. Gleditsch (Am. J. Sci. [5] **36** [1938] 95/106, 95, 98, 100). – [5] O. A. Broch (Norg. Geol. Unders. No. 141 [1934] 1/119, 90, 91, 112).

[6] V. M. Goldschmidt, L. Thomassen (Videnskapsselskap. Skrifter I Mat. Naturv. Klasse **1923** No. 2, pp. 1/48, 11, 31/3). – [7] V. A. Leonova (Zap. Vses. Mineral. Obshchestva **88** [1959] 21/38). – [8] T. Iimori (Am. J. Sci. **239** [1941] 819/21). – [9] W. W. Crook (Mineral. Rec. **8** [1977] 88/90). – [10] S. M. Roscoe (Geol. Surv. Can. Paper No. 68-40 [1969] 1/205, 134/6).

[11] K. G. Snetsinger, G. Polkowski (Am. Mineralogist **62** [1977] 587/8). – [12] N. J. Theis (Bull. Geol. Surv. Can. No. 304 [1979] 1/50, 15, 31, 44/6). – [13] A. H. Lang (Can. Mining J. **77** No. 6 [1956] 71/6, 73/4). – [14] N. K. Rao, G. V. U. Rao (J. Geol. Soc. India **21** [1980] 387/97, 387/8, 393/5). – [15] G. Xu, A. Wang, Q. Gu, J. Zhang, Z. Zhang, Y. Huang (Bull. Mineral. **104** [1981] 565/74).

[16] M. N. Rao (Earth Planet. Sci. Letters **2** [1967] 394/6). – [17] V. B. Sharif-Zade, Yu. A. Shukolyuakov, E. K. Gerling (Geokhimiya **1972** 314/22; Geochem. Intern. **9** [1972] 199/207, 202). – [18] E. Føyn (Skrifter Norske Videnskaps Akad. Oslo I Mat. Naturv. Klasse **1939** No. 4, pp. 1/60, 17). – [19] V. A. Leonova (Dokl. Akad. Nauk SSSR **126** [1959] 1342/5; Dokl. Earth Sci. Sect. **124/129** [1959] 415/7). – [20] J. A. Speer, T. N. Solberg, S. W. Becker (Econ. Geol. **76** [1981] 2162/75, 2165, 2167).

[21] P. R. Simpson, J. W. F. Bowles (U.S. Geol. Surv. Profess. Papers No. 1161-A-BB [1981] S1/S8, S3) – [22] S. C. Sarkar (Mineralium Deposita **17** [1982] 257/78, 264/5, 267, 272).

10.2.1.3 Mozambikite $\sim(Th, RE, U)(O, OH)_2$

Mozambikite has been found in the granite pegmatite of Muiane, Alto-Ligonha, Mozambique. It contains lamellar inclusions of lepidolite. Its faces are corroded by albite. The formula given above is adopted from [2], it corresponds to the analysis with RE_2O_3 = 8.60% given by [1]. X-ray spectrographic analysis showed Th as predominant element, Si, U, Zr, and Y as important elements, and Pb, Er, Gd, (Sm?), Mn in traces [1].

Mozambikite forms yellow-brown octahedra. Its structure with a = 5.704 Å is not changed by heating up to 1000°C. Mozambikite is isotropic, n = 1.811 after heating, ranging from 1.690 to 1.703 before heating. Density D = 5.24 g/cm^3. The mineral loses H_2O at 115°C [1].

The data on this mineral are still insufficient. The mineral, in some of its properties, agrees with thorite or thorogummite [3].

References for 10.2.1.3:

[1] J. M. Cotelo Neiva, J. M. Correia Neves (21st Intern. Geol. Congr. Rept. Session Norden, Copenhagen 1960, Pt. 17, pp. 53/62, 58). – [2] H. Strunz, C. Tennyson (Mineralogische Tabellen, 6th Ed., Geest and Portig, Leipzig 1977, pp. 1/621, 212). – [3] M. Fleischer (Am. Mineralogist **45** [1960] 1316/7).

10.2.1.4 "Thucholite"

"Thucholite" is an asphalt-like material which sometimes contains considerable concentrations of RE and Y; $(RE,Y)_2O_3$ contents with literature references of analyses are given in "Seltenerdelemente" A 4, 1979, pp. 8, 31. For mineral description of "thucholite" see "Seltenerdelemente" A 1, 1938, p. 109 and [1, 2]. As it is not a distinct mineral, but a mixture of hydrocarbons and uraninite with some sulfides [3, 4], its properties vary widely and will not be further discussed here. Only some examples of typical occurrences of RE-containing "thucholite" are given in the following.

As detailed in "Seltenerdelemente" A 1, 1938, p. 109 and in [1, 5], "thucholite" was reported as occurring in pegmatites and in biotite gneiss from Canada. In addition, it occurs in pegmatite from the mica mine Pieds des Monts, Charlevoix County, Quebec [6], and in granite pegmatites from Amaki village, Fukuoka prefecture, Japan [7]. It furthermore is found in many Canadian U and Th deposits of vein type as those of the Beaverlodge area, Saskatchewan (see "Uran" A 1, 1979, pp. 160/4 and [8]) and of conglomerate type, such as the Quirke Lake deposits, the Agnew Lake mine [9], and the Blind River deposits [8] (see "Uran" A 1, 1979, pp. 149, 155). In the conglomerate-type deposits of Witwatersrand, Republic of South Africa, "thucholite" also is one of the main and constant ore minerals [10, 11], "Uran" A 1, 1979, p. 205. Examples of further occurrences are an andalusite-sericite rock of Boliden, Sweden [12], non-granitized parts of the Precambrian quartzite in the Västervik area, Län Kalmar, Sweden [13], and old copper workings at Moonta, South Australia [14]. In a chalcopyrite zone of the old Pb-Zn mine at Laxey, Isle of Man, Great Britain, a highly radioactive uraninite-hydrocarbon aggregate of thucholite type has been found [3] and in Walt Canyon near Rocky Arroyo, 64 km NW of Carlsbad, Eddy County, New Mexico, thucholite-like material occurs in dolomite [15].

References for 10.2.1.4:

[1] C. W. Hintze (Handbuch der Mineralogie, Suppl. Vol. 1, de Gruyter, Berlin 1936/38, pp. 1/760, 688/90). – [2] M. V. Soboleva, I. A. Pudovkina (AEC-TR-4487 [1961] 1/455 [Russian Original, Moscow 1957]; N.S.A. **15** [1961] No. 23739). – [3] C. F. Davidson, S. H. U. Bowie (Bull. Geol. Surv. Gt. Britain No. 3 [1951] 1/18, 3, 4/5, 11/2). – [4] P. Ramdohr, H. Strunz (Klockmanns Lehrbuch der Mineralogie, 16th Ed., Enke, Stuttgart 1978, pp. 1/876, 548). – [5] H. V. Ellsworth (Can. Dept. Mines Geol. Surv. Econ. Geol. Ser. No. 11 [1932] 1/272, 171/2, 175, 178/81).

[6] H. S. Spence (Am. Mineralogist **25** [1940] 711/8, 717). – [7] K. Kimura, T. Iimori (Nippon Kwagaku Kwaishi **58** [1937] 1135/43; C.A. **1938** 1213 from [3]). – [8] A. H. Lang, J. W. Griffith, H. R. Steacy (Geol. Surv. Can. Econ. Geol. Ser. No. 16 [1962] 1/324, 46, 50, 136, 152, 165, 170, 173). – [9] S. M. Roscoe (Geol. Surv. Can. Paper No. 68-40 [1969] 1/205, 125, 139). – [10] C. E. Feather, G. M. Koen (Minerals Sci. Eng. **7** [1975] 189/224, 200).

[11] M. Schidlowski (Contrib. Mineral. Petrol. [Berlin] **12** [1966] 365/80). – [12] E. Grip, O. H. Ödman (Sveriges Geol. Undersoekn. Arsbok **38** No. 6 [1944] 1/14, 3/7). – [13] W. Uytenbogaardt (21st Intern. Geol. Congr. Rept. Session Norden, Copenhagen 1960, Pt. 15, pp. 114/22, 118/9). – [14] A. W. G. Whittle (S. Australia Geol. Surv. Bull. No. 30 [1954] 126/51, 145). – [15] S. A. Northrop (Minerals of New Mexico, 2nd Ed., New Mexico Univ. Press, Albuquerque 1959, pp. 1/665, 509/10).

10.2.1.5 Hibonite $(Ca,RE)(Al,Fe^{3+},Ti,Si,Mg,Fe^{2+})_{12}O_{19}$

Occurrence and Paragenesis

Hibonite first was described by Curien et al. from an alluvial deposit and from a metamorphic limestone rich in plagioclase associated with corundum, spinel, and thorianite at Esiva region, Fort Dauphin, Madagascar [1]. Somewhat later it was found in plagioclasites (anorthosites) intercalated between pyroxenites of the Ambindandrakambra region and of Andakato, Madagascar [2]. In the USSR hibonite occurs in Gornaya Shoriya in the basin of the river Mras-Su, Kemerovsk Oblast', associated with vesuviane and minor amounts of hercynite, corundum, andalusite, disthene, diopside, rutile, titanite, and magnetite [3, 4]. Kuzmin [5] described the mineral from this occurrence as högbohmite, see [6]. Hibonite occurs as small inclusions in corundum in Precambrian granulite facies gneisses of the Anabar Shield, Yakutian ASSR, together with garnet, pyroxene, and phlogopite. Paleogene placers in Kazakhstan, Neogene Ti-Zr placers of the East European Platform, and recent sands on the Black Sea shore contain hibonite, frequently intergrown with corundum [7]. Hibonite, included in garnet, has been described from a calc-silicate rock in the Furua granulite complex, southern Tanzania. It is unstable in a clinozoisite-zoisite environment, where it reacts to form spinel, RE-bearing clinozoisite, and sphene [8], cf. [14]. Hibonites are also found in carbonaceous chondrites, but RE have not been determined in these meteorite minerals, see e.g. [9, 10], or were searched for but not detected [15]. Only in 1982 there were found RE in hibonite of an unusual, hibonite-rich inclusion (HAL) of the Allende meteorite [16].

In the following sections chemistry and properties are only given for RE-containing hibonites.

Chemistry

The formula, given above, has been deduced from an analysis of the mineral of Evisa, Madagascar, by [1]. Microprobe analyses of 14 grains of hibonite from the Furua granulite complex, Tanzania, result in the average formula $(Ca_{0.80}RE_{0.18})[Al_{10.38}Fe^{3+}_{0.25}(Ti,Si)_{0.60}(Fe^{2+},Mg)_{0.79}]O_{19}$ or, generalized, $(Ca_{1-x}RE_x)[(Al,Fe^{3+})_{12-2a+x}(Ti,Si)_{a-x}(Fe^{2+},Mg)_a]O_{19}$ with $a=0.8$ and $x=0.2$. Individual mineral analyses show a cation substitution $Ca+Fe^{3+} \rightleftharpoons RE+Fe^{2+}$ and $Ti+Fe^{2+} \rightleftharpoons 2Al$, that may be coupled to $Ca+Ti+Fe^{3+} \rightleftharpoons RE+2Al$ [8], cf. [14]. Hibonites from gneisses of the Anabar Shield, Yakutian ASSR, contain 0.38, 1.27, 1.79, and 1.88% $(La,Ce)_2O_3$; their formulas are given in the following way [7]:

1. $(Ca_{1.00}(La,Ce)_{0.01})(Al_{11.28}Ti_{0.56}Mg_{0.19})O_{19}$
2. $(Ca_{0.96}La_{0.01}Ce_{0.04})(Al_{11.14}Ti_{0.62}Mg_{0.31})O_{19}$
3. $(Ca_{0.96}La_{0.01}Ce_{0.05})(Al_{11.35}Ti_{0.42}Mg_{0.24})O_{19}$
4. $(Ca_{0.93}La_{0.02}Ce_{0.05})(Al_{10.93}Ti_{0.82}Mg_{0.30})O_{19}$

The following RE contents in ppm have been determined in two samples of hibonite, (1) core + clear margin and (2) clear margin, of the inclusion HAL from Allende meteorite (the data determined via RNAA irradiation have been selected) [16]:

	(1)	(2)
La	11.59	12.74
Ce	<0.11	<0.14
Nd	14.95	16.45
Sm	2.94	3.59
Eu	1.10	0.96
Gd	<5.4	<7.8

	(1)	(2)
Tb	0.27	0.33
Dy	1.34	1.46
Tm	0.04	0.04
Yb	0.26	0.24
Lu	0.05	0.05

According to [11], the distribution of RE and Y in Madagascar hibonite is in %: La 24, Ce 41, Pr 2.5, Nd 15, Sm 0.5, Eu 1.2, Gd 0.5, Dy 1.2, Er 1.2, Yb 3.1, Y (10). In contrast, hibonites from the Anabar Shield contain only La and Ce, see formulas above. Hibonite from Tanzania contains (mean of 14 microprobe analyses) 2.1% Ce_2O_3, 1.8% La_2O_3, and 0.2% Nd_2O_3 [8, 14]. The percentual composition of the hibonite samples from the Allende inclusion HAL (according to the data given p. 29) approximates to the following values:

La	Ce	Nd	Sm	Eu	Gd	Tb	Dy	Tm	Yb	Lu
30.5	0.3	39.3	7.7	2.9	14.2	0.7	3.5	0.1	0.7	0.1
29.1	0.3	37.6	8.2	2.2	17.8	0.8	3.3	0.1	0.5	0.1

A strong negative Ce anomaly in the chondrite normalized pattern of this hibonite is noteworthy [16].

Hibonite from Madagascar contains 1000 ppm Th and 100 ppm U [1]. Hibonites from gneisses of the Anabar Shield have no Fe [7].

Crystal Form and Structure

Hibonite occurs as hexagonal prisms flattened parallel (0001) or as acute pyramids [1], as tabular or lamellar crystals [3, 7] or as laths and stringers [8]. Hexagonal with $\{11\bar{2}0\}$, $\{10\bar{1}0\}$, $\{10\bar{1}3\}$, $\{10\bar{1}2\}$, $\{30\bar{3}5\}$, $\{10\bar{1}1\}$, and some other forms [4] from [6, 12] and [5] (the last reference with corrections by [6]). Cleavage is perfect {0001} [1, 8]. Unit cell parameters in Å of analyzed samples are: from Madagascar, a = 5.61, c = 22.16, space group $P6_3/mmc\text{-}D_{6h}^4$ (No. 194) [1] and a = 5.597 ± 0.002, c = 22.50 ± 0.01 [9]; from Gornaya Shoriya, a = 5.59, c = 22.2 [13]; from Tanzania, a = 5.61, c = 22.18 [8], and from Anabar Shield, a = 5.60, c = 22.22 [7]; c/a = 3.95 [4]; Z = 2 [1].

Absorption in the red region of the spectrum indicates the presence of Ti^{3+} that diadochically replaces Al in octahedral coordination [7].

Optical Properties

Hibonite from Madagascar has vitreous luster, it is uniaxial negative, its color is brown to black or reddish brown. It is translucent in fine splits and has pleochroism O = grayish brown, E = grayish [1]. Hibonite from Gornaya Shoriya is yellow and cinnamon to dark brown, in thin sections pale yellow with weakly green tint; its pleochroism is distinct O = pale yellow, E = yellow [5]; or its color is yellow-brown or dark brown, nearly black and its pleochroism O = brownish yellow, E = yellow [3]. Hibonite from the Anabar Shield is pale blue to dark blue. It has a distinct pleochroism in these colors with absorption E > O [7]. Hibonite from Tanzania shows yellowish brown color in thin sections which decreases in intensity with rising RE contents; it has no pleochroism, the elongation is positive [8]. Refractive indices of these hibonites are:

Locality	n_o	n_e	Ref.
Madagascar	1.807 ± 0.002	1.79 ± 0.01	[1]
Gornaya Shoriya	1.851 ± 0.002	1.802 ± 0.003	[5]
	1.846	1.804	[13]
Anabar Shield	1.770 ± 0.005	1.785 ± 0.005	[7]

No cathodoluminescence was detected [7, 8].

Other Physical Properties

Hardness H_M = 7.5 to 8 [1], 8.5 [5]. The mineral is brittle [7], its fracture is subconchoidal. Density D = 3.84 ± 0.01 g/cm³ (Madagascar) [1]; 3.80 to 3.82 g/cm³ [5], 3.88 g/cm³ (Gornaya Shoriya) [13]. Hibonite reveals very weak radioactivity [1].

Chemical Behavior

Hibonite dissolves slowly in a mixture of sulfuric and phosphoric acids [1].

References for 10.2.1.5:

[1] H. Curien, C. Guillemin, J. Orcel, M. Sternberg (Compt. Rend. **242** [1956] 2845/7). – [2] L. Delbos (Compt. Rend. **244** [1957] 214/5). – [3] A. I. Letuvninkas (Geol. Geofiz. **1971** No. 1, pp. 58/64, 62). – [4] F. V. Chukhrov, E. M. Bonshtedt-Kupletskaya (Mineraly Spravochnik, Vol. 2, Pt. 3, Nauka, Moscow 1967, pp. 1/676, 195/7). – [5] A. M. Kuz'min (Geol. Geofiz. **1960** No. 4, pp. 63/75, 62/7, 70/1).

[6] T. A. Yakovlevskaya (Zap. Vses. Mineral. Obshchestva **90** [1961] 458/61, 460). – [7] E. I. Eremenko, S. M. Sandomirskaya, G. K. Eremenko, Yu. A. Polkanov, G. S. Rumyantsev (Mineral. Zh. **3** [1981] 75/8). – [8] P. Maaskant, J. J. M. M. M. Coolen, E. A. J. Burke (Mineral. Mag. [London] **43** [1980] 995/1003, 995/7, 999/1001). – [9] K. Keil, L. H. Fuchs (Earth Planet. Sci. Letters **12** [1971] 184/90, 186/7). – [10] J. D. Macdougall (Earth Planet. Sci. Letters **42** [1979] 1/6, 3).

[11] E. I. Semenov (Mineralogiya Redkikh Zemel', Nauka, Moscow 1963, pp. 1/412, 86/7). – [12] E. I. Nefedov (from [4]). – [13] E. I. Nefedov (from [6]). – [14] J. J. M. M. M. Coolen (GUA Papers Geol. No. 13 [1980] 1/258, 105). – [15] J. M. Allen, L. Grossman, A. M. Davis, I. D. Hutcheon (Proc. 9th Lunar Planet. Sci. Conf., Houston, Tex., 1978, Vol. 1, pp. 1209/33, 1223/4).

[16] A. M. Davis, T. Tanaka, L. Grossman, T. Lee, G. J. Wasserburg (Geochim. Cosmochim. Acta **46** [1982] 1627/51, 1628/9, 1638/9).

10.2.2 Complex Oxides with Ti (and sometimes Nb)

10.2.2.1 Davidite $(RE,U)_x(U,Fe^{2+})(Ti,Fe^{3+})_6O_{13-x}$

Synonyms: Ufertite [1], see also [2], and ferutite [3], see also [4], are davidites; these names therefore should be dropped [5]. The names mavudzite for the mineral from Mozambique [6] and miomirite for a Pb-bearing davidite from Nežilovo, Macedonia [7], are also unnecessary [8].

The name davidite has been used for the first described prototype mineral in the Olary district, South Australia, see p. 32. But later here and in other parts of South Australia it often has been used to a mixture of complex iron-uranium titanates (daviditic mixtures) [9].

In the following sections only the pure mineral davidite will be described.

References:

[1] N. I. Balashov (Byul. Nauchn. Tekhn. Inform. Mineral. Geol. Okhrany Nedr SSSR **1961** No. 2, pp. 10/2 from C.A. **57** [1962] 5607; Ref. Zh. Geol. **1962** No. 1 V 133). – [2] N. A. Kulik (Zap. Vses. Mineral. Obshchestva **94** [1965] 114/9). – [3] Yu. V. Kazitzyn (Zap. Vses. Mineral. Obshchestva **83** [1954] 424/7, 425). – [4] B. Contag, H. Strunz (Naturwissenschaften **48** [1961] 597). – [5] A. G. Zhabin, A. A. Voronkov, E. B. Khalezova, M. E. Kazakova (Tr. Inst. Mineral. Geokhim. Kristallokhim. Redk. Elem. No. 15 [1963] 110/20, 119/20).

[6] A. V. P. Coelho (Rev. Junta Missoes Geogr. Invest. Ultramar **2** [1954] 209/19 from Am. Mineralogist **41** [1956] 164). – [7] V. Vujanović (Radovi Inst. Geol. Rud. Istraz. Ispit. Nukl. Drugih Mineral. Sirovina **5** [1969] 147/55 from Am. Mineralogist **58** [1973] 560; Zap. Vses. Mineral. Obshchestva **100** [1971] 619). – [8] M. Fleischer (Am. Mineralogist **41** [1956] 164, **58** [1973] 560*)). – [9] B. Campana, D. King (S. Australia Geol. Surv. Bull. No. 34 [1958] 1/133, 51).

---

*) Here erroneously miromirite.

10.2.2.1.1 Occurrence and Paragenesis

As a hypogene **metasomatic mineral** davidite is widely distributed in U deposits of Australia. The first description of davidite by Mason 1906, was of the mineral from Radium Hill, Olary district, South Australia, where it occurs in high-temperature hydrothermal lodes containing quartz, biotite, hematite, ilmenite, magnetite, rutile, and iron and copper sulfides [1], cf. [2, 3]. Further occurrences in South Australia with davidite as replacement mineral are in the districts Mount Victoria and Houghton; davidite of high-hydrothermal origin was found at Spring Hill, Outalpa Springs, and in the Crocker's Well area [4]. As a metasomatic U mineral davidite was found in a Russian deposit of the U-Ti formation of the Precambrian together with ilmenite, chromite, RE titanite, U oxides, pyrite, chalcopyrite, marcasite, and secondary U minerals [5]. As a typomorphic high-temperature mineral davidite occurs in association with metasomatic albite, ilmenite, rutile, uraninite, and magnetite [23]. For these and further occurrences of davidite in South Australia, see "Rare Earth Elements" A 8, 1984, Chapter 11.7.8. For davidite in the "vein type" deposits of the Singhbhum copper belt, Bihar, India (cf. "Uranium" A 1, 1979, pp. 261/3), see [27].

In the Vishnevye Gory, Ural Mountains, strongly albitized microcline veins that cut a fenitized aureole around an intrusive complex contain davidite, titanite, and rutile [6]. Metasomatic veins in connection with shear planes and cutting a complex of norites and anorthosites at the Mavuzi River, a tributary of the Zambezi River, Tete district, Mozambique, contain davidite in four distinctive types of gangue: scapolite, calcite, pyroxene, and a "felsic" rock [7]. Davidite in scapolite-albite veins, cutting biotite gneisses and schists, together with rutile, hematite, magnetite, and sulfides has been described as "ufertite" of a Russian deposit [7, 9]. Davidites as segregations in albitites of a Russian locality are sometimes associated with barite, hematite, pyrite, chalcopyrite, and chlorite [10].

Davidite in a **contact zone** between a metaspessartite dike and quartz monzonite occurs on the west side of Quijotoa Mountains, Pima County, Arizona [11]. Davidite as an impregnation in a brecciated zone between a pegmatite and the amphibolite bedrock was found north of Bringebærkastet on the Island of Langøy, east of Kragerø, Norway [12].

A **regional metamorphic** deposit in Australia is the Isa-Cloncurry district, Queensland, cf. "Uranium" A 1, 1979, p. 242. This deposit has 3 types of davidite as one of the principal U minerals beside uraniferous allanite and brannerite in schists and granulites of the Argylla Series [13]. In the Harriman State Park, NW of New York, davidite has been found in Precambrian gneisses [26]. The first reported occurrence of davidite in Italy is a biotite-rich anatexite of Pianche di Vinadio, Piedmont, Maritime Alps, where it occurs together with quartz, albite, anatase, rutile, and brookite [24].

Deposits of davidite in **pegmatites** are in Tuftan, Iveland, southern Norway, where it occurs with garnet, allanite, zircon, beryl, gadolinite, thortveitite, euxenite, blomstrandine, magnetite, ilmenite, rutile, xenotime, pyrite, and molybdenite [12]. Disseminated and as accumulations, davidite was found associated with magnetite, ilmenite, titanite, rutile, anatase, and fluorite in pegmatites of the Bektauata Mountains, Kazakhstan [14]. In South Australia pegmatites with davidite are known from Billeroo and Wiperaminga Hill, Barrier Ranges, and at Mount Pleasant [4], cf. [28].

In Sweden davidite together with thorite, surrounded by quartz and muscovite, occurs in an **aplite vein** on the island of Björkö, north of Västervik [15]. Veinlets of davidite cut alkali syenites in Kivu, Zaire (Congo-Kinshasa) [16]. Davidite, scattered or concentrated in small nodules, in a lens of sodic aplite that is included in a highly tourmalized sodic gneiss, occurs in Villanueva, Badajoz, Spain [17]. **Pegmatoid quartz** from the Lohninger quarry, Rauris valley, Salzburg, Austria, contains davidite [18]. Davidite from Nežilovo, Macedonia, occurs in quartz veins together with sulfides of Zn and Pb [19].

As **placer mineral** davidite occurs together with other radioactive minerals in the placer deposits of the Red River Valley, Idaho County, Idaho [20], cf. "Rare Earth Elements" A 8, 1984, Chapter 11.7.2.3.3.1. In Crni Kamen, Macedonia, davidite was found in humic soil, probably derived from a pegmatitic vein [21], cf. [22].

A plumboan davidite, designated as a new variety, was discovered in uranium ores in Saisi, China [25].

Inclusions. Davidites from nearly all of the occurrences, named above, contain inclusions of the three minerals ilmenite, hematite, and rutile [2], or only hematite with ilmenite [13], ilmenite with rutile [9, 18], or rutile alone [4]. Besides ilmenite or magnetite, hematite, and rutile, there are also silicates and, to a smaller degree, pyrite, chalcopyrite, and occasionally molybdenite as inclusions in South Australian davidites [3].

Alteration. The alteration of davidite from scapolite-albite veins, USSR, begins with the formation of small intergrowths of ilmenite and hematite. The liberated RE are fixed in allanite, and uranium also is partly fixed in newly formed minerals or it migrates. The formation of rutile and leucoxene as alteration products of ilmenite, and the leaching and redeposition of hematite occur later [9]. Rutile, titanite, and leucoxene are alteration products of davidite from Quijotoa Mountains, Pima County, Arizona [11]. Metasomatic davidite from a U-Ti formation, USSR, often shows rims of leucoxene stained by Fe-hydrates and including skeletal crystals of rutile [5]; and davidite in metasomatic albites, USSR, is sometimes altered to leucoxene [10]. Of two South Australian davidites one, from Radium Hill, has rims of carnotite incrustations; the other, from Houghton, is altered to an earthy mass of leucoxene, limonite, uranium ochres, and carnotite [2]. For a weathering of davidite from the ore deposits of Radium Hill and Mount Victoria to carnotite and to a lesser extent to autunite and torbernite, see [28]. Davidite from Isa-Cloncurry district, Queensland, tends to become altered to carnotite and gummite [13].

References for 10.2.2.1.1:

[1] C. Frondel (U.S. Geol. Surv. Bull. No. 1064 [1958] 1/400, 337, 340). – [2] A. W. G. Whittle (S. Australia Geol. Surv. Bull. No. 30 [1954] 126/51, 137/41). – [3] A. W. G. Whittle (J. Geol. Soc. Australia **2** [1955] 21/45, 31/2, 42). – [4] A. W. G. Whittle (Econ. Geol. **54** [1959] 64/81, 67/72, 80). – [5] E. V. Petrosyan (Sov. Geol. No. 11 [1961] 64/72, 69/70).

[6] A. G. Zhabin, A. A. Voronkov, E. B. Khalezova, M. E. Kazakova (Tr. Inst. Mineral. Geokhim. Kristallokhim. Redk. Elem. No. 15 [1963] 110/20, 113). – [7] C. F. Davidson, J. A. E. Bennett (Mineral. Mag. **29** [1950/52] 291/303, 294, 296/302). – [8] N. I. Balashov (Byul. Nauchn. Tekhn. Inform. Min. Geol. Okhrany Nedr SSSR **1961** No. 2, pp. 10/2 from C.A. **57** [1962] 5607; Ref. Zh. Geol. **1962** No. 1 V 133). – [9] N. A. Kulik (Zap. Vses. Mineral. Obshchestva **94** [1965] 114/9, 114/6, 118). – [10] L. I. Davydova, G. N. Shaposhnikov (Zap. Vses. Mineral. Obshchestva **95** [1966] 474/6).

[11] A. Pabst, R. W. Thomssen (Bull. Geol. Soc. Am. **70** [1959] 1739). – [12] H. Neumann, T. L. Sverdrup (Norsk Geol. Tidsskr. **40** [1960] 277/88, 277/8, 287). – [13] L. J. Lawrence, G. T. See, F. McBride, H. Hofer (Econ. Geol. **52** [1957] 140/8, 140, 142). – [14] B. F. Ablanov (in: P. G. Sagunov et al., Geologiya Geokhimiya i Mineralogiya Mestorozhdenii Redkikh Elementov Kazakhstana, Alma-Ata 1966, pp. 152/7 from C.A. **67** [1967] No. 119032). – [15] E. Welin, W. Uytenbogaardt (Arkiv Mineral. Geol. **3** [1965] 277/92, 279/80).

[16] L. Van Wambeke (Mineralium Deposita **3** [1968] 178/81, 178). – [17] A. Arribas (Estud. Geol. [Madrid] **19** No. 1/4 [1963/64] 33/52, 33). – [18] H. Meixner (Karinthin No. 81 [1979] 144/7). – [19] V. Vujanović (Radovi Inst. Geol. Rud. Istraz. Ispit. Nukl. Drugih Mineral. Sirovina **5** [1969] 147/55 from Am. Mineralogist **58** [1973] 560; Zap. Vses. Mineral. Obshchestva **100** [1971] 619). – [20] F. C. Armstrong, P. L. Weis (U.S. Geol. Surv. Bull. No. 1046-C [1957] 25/36, 28, 32).

[21] A. Damjanović, M. Vukasović (Radovi Inst. Geol. Rud. Istraz. Ispit. Nukl. Drugih Mineral. Sirovina **1** [1961] 11/5, 11, 15 [English Summary]). – [22] L. Barić (Tschermaks Mineral. Petrogr. Mitt. [3] **10** [1965] 368/78, 377). – [23] A. A. Chernikov (Sov. Geol. No. 4 [1981] 85/98, 89). – [24] G. C. Piccoli (Riv. Mineral. Ital. **1981** No. 1, pp. 19/22, 19). – [25] T.-M. Wang (K'o Hsueh T'ung Pao **26** [1981] 618/21 from C.A. **96** [1982] No. 38453).

[26] J. Gillespie (Rocks Minerals **56** [1981] 70/2). – [27] S. C. Sarkar (Mineralium Deposita **17** [1982] 257/78, 265, 269, 270). – [28] B. Campana, D. King (S. Australia Geol. Surv. Bull. No. 34 [1958] 1/133, 51/3).

10.2.2.1.2 Chemistry

A compilation of davidite analyses, and the mean and variance of Y_2O_3 and $RE_2O_3+Y_2O_3$ contents of 54 samples is given in "Seltenerdelemente" A 4, pp. 17/8 and 9. The strong variations in analytical results probably are caused by the numerous, often submicroscopic, inclusions in davidite [1].

As calculations by Pabst [2] indicate the unit cell content of davidite to be $X_6Z_{15}O_{36}$, analyses of davidite from various localities have been normalized to 36 oxygen atoms. These localities include davidite from Vishnevye Gory, which lacks U and Th and has $X=Fe^{2+}$, RE, Ca, Mg, Mn, Na, K, Pb, Al, Si, Nb, Zr with $Z=Ti$, Fe^{3+} [3]; davidite from Mavuzi, Tete district, Mozambique, with $X_{3.19}Z_{17.18}(O,OH)_{36.00}$, where X = U, RE, Fe^{2+}, Pb, Th, Ca, Mg, Mn, Na, K, and Z = Ti, Fe^{3+}, Cr, V, Al; if all U is treated as U^{4+} and RE as Ce_2O_3, V as V^{3+} and Cr as Cr^{3+}; this formula can be adjusted to $X_{5.38}Z_{15.00}(O,OH)_{36.00}$ if rather less than half of the determined ferric iron is assigned to the X rather than to Z cations [4]. Davidite from the Rauris valley, Salzburg, Austria, with $(Fe^{2+}_{4.52}$ "Ce-La"$_{0.59}U_{0.89})(Ti_{10.61}Fe^{3+}_{4.39})(O,OH)_{36}$ has also a formula with 36 oxygen atoms [5]. For davidite (ferutite) from Mozambique a formula with 24 oxygen is calculated as $(U^{4+},Ce^{4+},La)_x(Fe^{2+},Mg^{2+})_{4-x}Ti^{4+}_{10-2x}(Fe^{3+},Cr^{3+},V^{3+})_{2x}O_{24}$ with $x\approx1$ or ideal formula $UFe^{2+}_3Fe^{3+}_2Ti_8O_{24}$ [6]. The chemical formula of davidite from the Bektauata Mountains, Kazakhstan, calculated with 33 oxygen, is given as $RE_{1.05}Ta_{0.20}Nb_{0.03}Fe^{3+}_{6.28}Fe^{2+}_{1.31}Ti_{9.80}Mn_{0.45}O_{33}$ [7]. Normalized to 38 oxygen, davidite from Quijotoa Mountains, Arizona, has $RE_{0.91}U_{0.33}Ca_{0.20}Y_{0.25}(Sr,Th,Pb)_{0.09}Ti_{12.53}Fe_{5.97}Mg_{0.24}Cr_{0.21}Al_{0.14}(Mn,V,Zr,Nb,Sc)_{0.23}O_{38}$ [8]. See also formula p. 31, where $x\approx1/3$ of the cell content found by structure determination (see p. 37) [9]. An analysis of davidite from Crni Kamen, Macedonia, is calculated with 7 (O,OH) as $(U^{6+}_{0.133}Fe^{2+}_{0.675}Zr_{0.007}Mn_{0.014}Na_{0.011}Ca_{0.004}Th_{0.004}RE_{0.032}Pb_{0.115})(Si_{0.022}Ti_{2.500}Cr_{0.007}V_{0.014}Fe^{3+}_{0.424})(O,OH)_7$ [21]. The plumboan davidite from China has the formula $U_{0.85}Pb_{0.77}RE_{0.07}Zr_{0.29}Sr_{0.11}Ba_{0.02}Mn_{0.09}V_{0.47}Fe^{2+}_{4.06}Fe^{3+}_{2.04}Ti_{13.13}O_{38}$ [23].

U and RE replace each other in davidite; the sum of both is nearly the same in all analyses so that samples low in U are rich in RE and vice versa [1, 10, 11]; besides U and RE, Ti, V, Cr, Al are probably interchangeable [12].

Si, Al, and Ca in South Australian davidites [10] and Si, Al, Ca, K, Na, Sr in davidite from Badajos, Spain [20], are thought to be impurities. Trace elements (<1%) in davidite from Kivu, Zaire, are Sr (0.9%), Zr (0.7%), Sn (0.1%), Cr and V [13]. Davidite (miomirite) from Macedonia contains 1% Y and (in ppm) Mg 350, Sn 159, Zr 1500, Cr 316, Sc 140, La 159, and Ba 32 [14]; davidite from Spain contains B, Ba, Bi, Mn, Mo, Ni, Sn [20], and davidite from Pianche di Vinadio, Italy, Sr, Zn, Y [22]. Davidite from Radium Hill in addition contains 0.012, perhaps even 0.02% Sc_2O_3 [15]. Trace elements found in davidite from Radium Hill and in two different types of davidite from Isa-Cloncurry district, Queensland, are shown in the following table (+ present, − absent) [16]:

Locality	Trace elements																			
	Sn	Pb	Zn	Ge	As	Mo	Ta	Be	Cd	In	Au	Ag	Cu	Sb	Zr	Ru	Rh	La	W	Cr
Radium Hill	+	+	+	−	+	+	−	+	−	+	−	+	+	+	+	−	−	+	+	+
Isa-Cloncurry district																				
Type 1	+	+	+	−	−	−	−	+	−	+	−	+	+	+	+	+	+	+	+	+
Type 3	+	+	+	−	−	−	−	+	−	+	−	+	+	+	+	+	+	+	+	+

Distribution of RE. Davidite has, as was first shown by Dixon and Wylie for davidite from Radium Hill, a special distribution of RE with enrichment towards both ends of the element series [17]. This was confirmed by further analyses of davidite from different localities, showing distinct maxima of La and Yb [3]; see the following table, listing percent of RE oxides. For further data on percent of RE oxides in total $RE_2O_3+ThO_2$, see [19].

Locality	$\Sigma RE_2O_3 \pm Y_2O_3 = 100$								Ref.
	Y	La	Ce	Pr	Nd	Sm	Eu	Gd	
Radium Hill, South Australia	—	48	29	1.3	1.7	2.2	0.9	1.2	[17]
Radium Hill, South Australia	9.8	46.3	34	1.0	1.7	0.2	0.03	0.27	[12]
South Australia	18.3	35.4	26.8	1.0	3.3	0.3	0.1	0.6	[12]
Tete District, Mozambique	5.0	49	39	0.7	2.3	0.2	0.03	0.23	[12]
Vishnevye Gory, Urals	11	38	32	1.5	3.6	1.0	0.1	2.6	[3]
Island of Björkö, Sweden	23.4	32.7	18.7	—	2.8	0.9	0.0	0.9	[18]
Tuftan, Iveland, Norway	23	22	32	2.4	2.8	<0.2	0.4	<0.1	[11]
Arizona, USA	40	25.5	23.5	0.6	2.5	0.3	—	0.5	[3]

Locality	$\Sigma RE_2O_3 \pm Y_2O_3 = 100$							Re-marks	Ref.
	Tb	Dy	Ho	Er	Tm	Yb	Lu		
Radium Hill, South Australia	0.6	0.9	0.9	6.5	0.6	6.1	0.6	—	[17]
Radium Hill, South Australia	0.21	1.7	0.2	2.1	0.2	2.0	0.4	1)	[12]
South Australia	0.5	4.0	1.3	3.2	0.3	4.3	0.8	2)	[12]
Tete District, Mozambique	0.12	0.7	0.2	0.6	0.1	1.2	0.3	—	[12]
Vishnevye Gory, Urals	0.4	3.0	0.7	2.4	0.4	2.3	0.4	—	[3]
Island of Björkö, Sweden	—	1.9	—	1.9	0.9	15.9	—	3)	[18]
Tuftan, Iveland, Norway	0.2	1.6	0.55	2.5	1.2	9.5	1.7	—	[11]
Arizona, USA	0.1	1.7	0.4	2.1	0.3	3.7	0.4	—	[3]

1) Mean of 3 analyses; $Y_2O_3=9.0$, 9.0, and 11.4%. – 2) Mean of 5 analyses from: 1 Kalabity Hill, NNW of Radium Hill, 1 Radium Hill, 213 m level, 2 Mount Pleasant, 1 Hundred of Jutland. Y_2O_3 in these samples 15.0 to 21.0%. – 3) Calculated from the chemical analysis of the oxides.

The table shows that, with the exception of the davidite from Norway, all davidites have La > Ce, see also [19]. Davidite from Kivu, Zaire, has La > Ce ≫ Nd > Pr > Sm and Y ≫ Yb > Er ≈ Gd > Dy > Lu [13]. In davidite from Spain, La and Ce predominate, Gd, Yb, Eu are less abundant, and Er and Y are present [20].

References for 10.2.2.1.2

[1] A. W. G. Whittle (J. Geol. Soc. Australia **2** [1955] 21/45, 31/2, 38/9). – [2] A. Pabst (Am. Mineralogist **46** [1961] 700/18, 707). – [3] A. G. Zhabin, A. A. Voronkov, E. B. Khalezova, M. E. Kazakova (Tr. Inst. Mineral. Geokhim. Kristallokhim. Redk. Elem. No. 15 [1963] 110/20, 118/9). – [4] J. A. T. Smellie, N. Cogger, J. Herrington (Chem. Geol. **22** [1978] 1/10, 6). – [5] H. Meixner (Karinthin No. 81 [1979] 144/7).

[6] B. Contag, H. Strunz (Naturwissenschaften **48** [1961] 597). – [7] B. F. Ablanov (in: P. G. Sagunov et al., Geologiya Geokhimiya i Mineralogiya Mestorozhdenii Redkikh Elementov Kazakhstana, Alma-Ata 1966, pp. 152/7 from C.A. **67** [1967] No. 119032. – [8] B. M. Gatehouse, I. E. Gray, P. R. Kelly (Am. Mineralogist **64** [1979] 1010/7, 1011). – [9] B. Contag (in: H. Strunz, Mineralogische Tabellen, 6th Ed., Geest and Portig, Leipzig 1977, pp. 1/621, 187/8). – [10] A. W. G. Whittle (Econ. Geol. **54** [1959] 64/81, 76/7).

[11] H. Neumann, T. L. Sverdrup (Norsk Geol. Tidsskr. **40** [1960] 277/88, 280/1, 283). – [12] J. D. Hayton (Econ. Geol. **55** [1960] 1030/8, 1034, 1037). – [13] L. Van Wambeke (Mineralium Deposita **3** [1968] 178/81, 180). – [14] V. Vujanović (Radovi Inst. Geol. Rud. Istraz. Ispit. Nukl. Drugih Mineral. Sirovina **5** [1969] 147/55 from Am. Mineralogist **58** [1973] 560). – [15] R. C. Vickery (Australian J. Sci. **16** [1953] 112).

[16] L. J. Lawrence, G. T. See, F. McBride, H. Hofer (Econ. Geol. **52** [1957] 140/8, 145). – [17] P. Dixon, A. W. Wylie (Nature **167** [1951] 526). – [18] E. Welin, W. Uytenbogaardt (Arkiv Mineral. Geol. **3** [1965] 277/92, 288). – [19] J. R. Butler, R. Hall (Econ. Geol. **55** [1960] 1541/50, 1545). – [20] A. Arribas (Estud. Geol. [Madrid] **19** No. 1/4 [1963/64] 33/51, 40, 42).

[21] A. Damjanović, M. Vukasović (Radovi Inst. Geol. Rud. Istraz. Ispit. Nukl. Drugih Mineral. Sirovina **1** [1961] 11/5, 12). – [22] G. C. Piccoli (Riv. Mineral. Ital. **1981** No. 1, pp. 19/22, 22). – [23] T.-M. Wang (K'o Hsueh T'ung Pao **26** [1981] 618/21 from C.A. **96** [1982] No. 38453).

10.2.2.1.3 Crystal Form and Structure

Trigonal. Axial ratio (after the structural data of Pabst [1]) c/a = 2.013.

Davidite mostly occurs in masses [2, 3, 4] or as imperfect crystals [5]. Lamellar crystals have been found in Italy [24], and well-formed crystals in Mozambique [6] and Spain [7], see **Fig. 2**.

Some forms are due to pseudomorphous replacement [2]. Habit of the crystals of Mozambique is tabular to pyramidal. Occasional rotation twinning about the axis [$11\bar{2}0$] [6]. Davidites from a Russian regional-metamorphic U deposit are tabular and short prismatic [8] and crystals from Spain are tabular, prismatic, or pyramidal with interpenetration twins after ($11\bar{2}0$) [7]. Further twinning planes see [1]. No cleavage [4, 9, 10]. The crystals sometimes show zoned or rythmic variation of radioactivity [4].

With the exception of davidite from Vishnevye Gory, Ural Mountains [11], from Kivu, Zaire [12], and from Italy [24], davidites are metamict. Diffraction patterns are observable only after heating. Cell units of davidites of various localities are given in the following table:

Locality	a (Å)	c (Å)	Ref.
Arizona	10.36	20.85	[13]
Arizona and Norway	10.37 ± 0.02	20.87 ± 0.02	[1]
Radium Hill, Australia	10.36	20.84	[11]
Mozambique	10.35	20.91	[11]

Locality	a (Å)	c (Å)	Ref.
USSR	10.35	20.74	[11]
USSR (ufertite)	10.36	20.82	[11]
Vishnevye Gory, Urals	10.405 ± 0.01	20.88 ± 0.02	[11]
Kivu, Zaire	10.41 ± 0.03	21.00 ± 0.06$^{+)}$	[12]
Mozambique (ferutite)	37.25	20.70	[14]
Kivu, Zaire	37.56 ± 0.05	21.00 ± 0.06$^{++)}$	[12]

$^{+)}$ Indexed after [1]. – $^{++)}$ Indexed after [15].

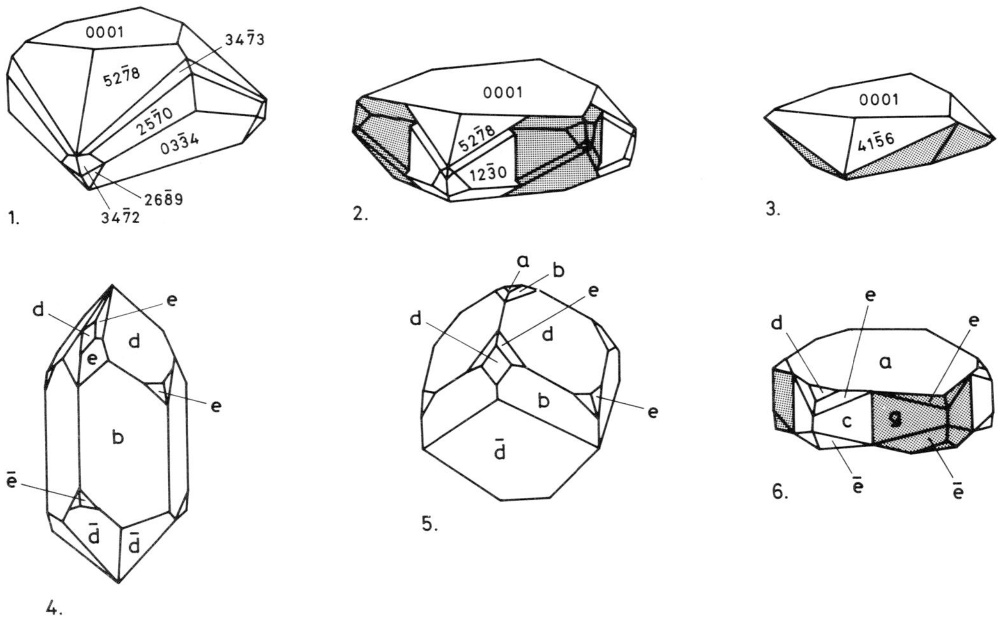

Fig. 2. Crystals of davidite.

1. Single crystal
2. Interpenetrant twin
3. Rotation twin about [11$\bar{2}$0]
4. Single crystal
5. Single crystal
6. Interpenetrant twin

One individual of each twin is stippled.

1., 2., and 3. Mozambique from [6]. – 4., 5., and 6. Spain; indices and angles are not given. a = basis pinacoids, b = hexagonal prisms, c = ditrigonal prism, d = trigonal pyramids, and e = ditrigonal pyramids [7].

Rhombohedral indexing of davidites from Arizona and Norway give a_{rh} = 9.178 Å. α_{rh} = 68°48′ [1], the values of davidite from the Quijotoa Mountains, Pima County, Arizona, are a_{rh} = 9.190(3) Å, α_{rh} = 68.73(1)° [16]. Space group R$\bar{3}$-C_{3i}^2 (No. 148) or R3-C_3^4 (No. 146) [1, 11, 13], R$\bar{3}$-C_{3i}^2 (No. 148) [16]. Z_{rh} = 3 [1], Z_{rh} = 1 [16]. – For further X-ray diffraction studies of previously heated davidite, see [6, 17 to 21] and [4]. The last reference also deals with natural davidite. The relationship of davidite to senaite has been shown by [22]. Davidite is isostructural with crichtonite group-minerals. Its structure is based on a close-packed anion framework with RE occupying an anion site. The large cations in M(0) are large RE + Ca + (Sr, Th, Pb); the sites of small cations are variously tetrahedral M(2) with Fe + Mg, octahedral M(1) with U + Y (incl. Sc, Zr) + small RE, M(3), M(4), or M(5) with Ti, Fe, and Cr. This means that U and the dominant RE (La^{3+}, Ce^{3+}) occupy different sites. Small RE such as Er and Yb, along with Y and Ce^{4+}, substitute for U in the octahedral site [16]. On the distribution of the RE to different places in the davidite lattice see also Butler, who suggests that La^{3+} and Ce^{3+} (with lesser amounts of Pr^{3+} and Nd^{3+}) occupy a structural position different from that occupied by U^{4+}, Y^{3+}, Dy^{3+}, Er^{3+}, and Lu^{3+} [23], and Welin and Uytenbogaardt, who associate Sc with Y and Yb, in places different from those occupied by La and Ce [19]. This distribution of the RE on different sites explains the unusual distribution patterns of davidite (see p. 35) [16].

References for 10.2.2.1.3:

[1] A. Pabst (Am. Mineralogist **46** [1961] 700/18, 702, 707/9). – [2] A. W. G. Whittle (J. Geol. Soc. Australia **2** [1955] 21/45, 32, 38). – [3] A. W. G. Whittle (S. Australia Geol. Surv. Bull. No. 30 [1954] 126/51, 137). – [4] A. W. G. Whittle (Econ. Geol. **54** [1959] 64/81, 72, 74/5). – [5] F. V. Chukhrov, E. M. Bohnshtedt-Kupletskaya (Mineraly Spravochnik, Vol. 2, Pt. 3, Nauka, Moscow 1967, pp. 1/676, 291).

[6] F. A. Bannister, J. E. T. Horne (Mineral. Mag. **29** [1950/52] 101/12, 102/4, 108/9). – [7] A. Arribas (Estud. Geol. [Madrid] **19** No. 1/4 [1963/64] 33/51, 37/9). – [8] N. A. Kulik (Zap. Vses. Mineral. Obshchestva **94** [1965] 114/9, 115). – [9] H. Neumann, T. L. Sverdrup (Norsk Geol. Tidsskr. **40** [1960] 277/88, 280). – [10] E. V. Petrosyan (Sov. Geol. No. 11 [1961] 64/72, 70).

[11] A. G. Zhabin, A. A. Voronkov, E. B. Khalezova, M. E. Kazakova (Tr. Inst. Mineral. Geokhim. Kristallokhim. Redk. Elem. No. 15 [1963] 110/20, 113/4, 117). – [12] L. Van Wambeke (Mineralium Deposita **3** [1968] 178/81, 179/80). – [13] A. Pabst, R. W. Thomssen (Bull. Geol. Soc. Am. **70** [1959] 1739). – [14] B. Contag, H. Strunz (Naturwissenschaften **48** [1961] 597). – [15] B. Contag (Diss. T.U. Berlin 1962 from [12]).

[16] B. M. Gatehouse, I. E. Gray, P. R. Kelly (Am. Mineralogist **64** [1979] 1010/7, 1011, 1013/5). – [17] P. F. Kerr, H. D. Holland (Am. Mineralogist **36** [1951] 563/72, 570). – [18] N. I. Balashov (Byul. Nauchn. Tekhn. Inform. Mineral. Geol. Okhrany Nedr SSSR **1961** No. 2, pp. 10/2 from C.A. **57** [1962] 5607; Ref. Zh. Geol. **1962** No. 1 V 133). – [19] E. Welin, W. Uytenbogaardt (Arkiv Mineral. Geol. **3** [1965] 277/92, 287, 289). – [20] L. I. Davydova, G. N. Shaposhnikov (Zap. Vses. Mineral. Obshchestva **95** [1966] 474/6).

[21] A. Damjanović, M. Vukasović (Radovi Inst. Geol. Rud. Istraz. Ispit. Nukl. Drugih Mineral. Sirovina **1** [1961] 11/5, 12/4). – [22] R. C. Rouse, D. R. Peacor (Am. Mineralogist **53** [1968] 869/79, 873/5). – [23] J. R. Butler (Econ. Geol. **56** [1961] 442/4). – [24] G. C. Piccoli (Riv. Mineral. Ital. **1961** No. 1, pp. 19/22, 20).

10.2.2.1.4 Optical and Other Physical Properties

Optical Properties

The luster ranges from highly lustrous for fresh davidites [1] to strongly resinous [2] or pitchy for metamict davidites [3, 4]. It is vitreous for pure davidites, becoming metallic with the increase of inclusions [5]. Intermediate stages are semi- or submetallic [6, 7, 8], brilliant [9] or shining vitreous [10], polymetallic, on fresh fracture faces near resinous [11]. Color is mostly black [1, 2, 4, 7, 10, 11, 12], but also black with brown tint [3] or black to steel gray [9], or gray-black [8], and reddish brown to cinnamon [13]. Streak or powder black [2, 3, 12], black or very dark gray [10], and gray-black [7, 8]. In transmitted light opaque [5] with the exception of very thin splinters that are translucent and clove-brown [1, 8, 10] or maroon-brown [6]. Refractive index of davidite from South Australia is $n>2.2$ of pegmatitic davidite, $n\approx2.3$ of metasomatic davidite [1]; $n=2.11$ in "ufertite", USSR [3].

Davidite is isotropic in reflected light [5, 7, 8, 10] with complete extinction [1]. Weak anisotropy can be seen in polarized light [12, 13]. Color in reflected light is gray, grayish and grayish white [9], gray [6, 10, 14], light gray [1, 5], gray with brownish tint [3], or weakly yellow [2]. Inner reflections are dark brown, weakly discernible [15], yellow, brownish yellow and brown [13], or reddish brown [25]. The reflecting power varies, even in a single grain [15]. It varies from 19 to 21% in davidites from USSR [16], is 17% for davidites from South Australia [5, 14] and, in white light and air, 19.5% for davidite from Mozambique [8], 16% for davidite from Bjorkö, Sweden [17], 17.9% for davidite from Crni Kamen, Macedonia [26], and mean 17.8% for davidites from 3 different localities [25]. The variation for different colors is given as follows:

Locality	Reflecting power in air in %				Ref.
	white	blue	green	red	
—	17.3	23.7	21.5	16.5	[18]
Spain	—	—	16.5	15	[6]
South Australia (pegmatitic)	18.2	—	—	19.9*)	[1]

*) 6.5% in oil [1]. The reflecting power of davidite from Kivu, Zaire also decreases on oil immersion [12], cf. [15].

Reflectivity of the plumboan davidite from China in % (at different wavelengths in nm): 20.91 (470), 20.55 (550), 19.96 (590), and 19.02 (663) [27].

Other Physical Properties

Hardness $H_M=5$ to 6, pegmatitic davidite, South Australia [1], 5.6, variable, metasomatic davidite, South Australia [5]. $H_M\approx6$ for davidite from U-Ti formation, USSR [2], from albitites, USSR [11], and from a quartz vein, Austria [7]. $H_M=6$ for davidite from metasomatic veins, Mozambique [8], from pegmatite, Iveland, Norway [10], and derived from a pegmatitic vein, Crni Kamen, Macedonia [26]. $H_M=6.5$ to 7.5, davidite from metasomatic veins, USSR [3]. The plumboan davidite from China has $H_M=6$ [27]. The microhardness of davidite shows differences with the degree of alteration [6] in different zones of a single grain [19]. Data given are in HV kg/mm^2 with 100 g load: 693 to 762 [20], 690 to 1100 [19], 707 to 803, mean 745, for davidites from 3 localities [25], 711 [13] and 814 to 885, mean 830 [29]. With load 100 to 200 g HV = 774 to 841 kg/mm^2 [16].

Fracture conchoidal, when pure, subconchoidal or uneven and hackly, when rich in inclusions [5]. For conchoidal fracture see [1, 2, 3, 7, 11], for subconchoidal [8, 10], and for three types of davidite with conchoidal, irregular and irregular to hackly fracture [9]. Very brittle [16].

Density D(meas.) in g/cm^3 of analyzed samples from various localities is listed in the following table:

Locality	D(g/cm^3)	Ref.
Badajos, Spain	4.35	[6]
Iveland, Norway	4.29	[10]
Björkö, Sweden	4.4	[17]
Crni Kamen, Macedonia	4.85	[26]

Locality	D(g/cm³)	Ref.
Vishnevye Gory, Urals	4.60	[22]
USSR	4.33	[22]
USSR	4.38	[3]
USSR	4.37	[2]
USSR	4.62	[11]
China*)	4.39	[27]
Mozambique	4.46	[8]
	4.41 and 4.42	[24]
Kivu, Zaire	4.62 ± 0.03	[12]
Pima County, Arizona	4.42	[4, 23]
South Australia	4.4 to 4.5	[1, 5]
Radium Hill, South Australia	4.42, 4.43, and 4.50	[24]
Isa-Cloncurry district, Queensland	4.32, 4.26, and 4.89	[9]

*) Plumboan davidite.

Further data from samples not analyzed, D in g/cm^3: 4.37, Iveland, Norway [23], 4.98 ± 0.02, Rauris, Austria [7], 4.49, Crockers Well and 4.38, Billeroo (both South Australia) [23]. D(calc.) = 4.72 g/cm^3 [21].

Magnetism. Weakly magnetic [1]. Magnetic susceptibility (at room temperature) = 100×10^{-6} cm^3/g [2, 28].

References for 10.2.2.1.4:

[1] A. W. G. Whittle (Econ. Geol. **54** [1959] 64/81, 72/3). – [2] E. V. Petrosyan (Sov. Geol. No. 11 [1961] 64/72, 69/70). – [3] N. I. Balashov (Byul. Nauchn. Tekhn. Inform. Mineral. Geol. Okhrany Nedr SSSR **1961** No. 2, pp. 10/2 from C.A. **57** [1962] 5607; Ref. Zh. Geol. **1962** No. 1 V 133). – [4] A. Pabst, R. W. Thomssen (Bull. Geol. Soc. Am. **70** [1959] 1739). – [5] A. W. G. Whittle (S. Australia Geol. Surv. Bull. No. 30 [1954] 126/51, 137).

[6] A. Arribas (Estud. Geol. [Madrid] **19** No. 1/4 [1963/64] 33/51, 37/8). – [7] H. Meixner (Karinthin No. 81 [1979] 144/7). – [8] F. A. Bannister, J. E. T. Horne (Mineral. Mag. **29** [1950/52] 101/12, 103). – [9] L. J. Lawrence, G. T. See, F. McBride, H. Hofer (Econ. Geol. **52** [1957] 140/8, 140/2, 144). – [10] H. Neumann, T. L. Sverdrup (Norsk Geol. Tidsskr. **40** [1960] 277/88, 280/1).

[11] L. I. Davydova, G. N. Shaposhnikov (Zap. Vses. Mineral. Obshchestva **95** [1966] 474/6). – [12] L. Van Wambeke (Mineralium Deposita **3** [1968] 178/81, 178/9). – [13] V. Vujanović (Radovi Inst. Geol. Rud. Istraz. Ispit. Nukl. Drugih Mineral. Sirovina **5** [1969] 147/55 from Zap. Vses. Mineral. Obshchestva **100** [1971] 619). – [14] A. W. G. Whittle (J. Geol. Soc. S. Australia **2** [1955] 21/45, 32). – [15] P. Ramdohr (Die Erzmineralien und ihre Verwachsungen, 3rd Ed., Akademie-Verlag, Berlin 1960, pp. 1/1089, 917).

[16] F. V. Chukhrov, E. M. Bohnshtedt-Kupletskaya (Mineraly Spravochnik, Vol. 2, Pt. 3, Nauka, Moscow 1967, pp. 1/676, 292/3). – [17] E. Welin, W. Uytenbogaardt (Arkiv Mineral. Geol. **3** [1965] 277/92, 286). – [18] I. M. Gray, A. P. Millman (Econ. Geol. **57** [1962] 325/49, 334). – [19] N. A. Kulik (Zap. Vses. Mineral. Obshchestva **94** [1965] 114/9, 115). – [20] B. B. Young, A. P. Millman (Bull. Inst. Mining Met. [London] No. 689 [1964] 437/66, 449).

[21] B. M. Gatehouse, I. E. Gray, P. R. Kelly (Am. Mineralogist **64** [1979] 1010/7, 1011). – [22] A. G. Zhabin, A. A. Voronkov, E. B. Khalezova, M. E. Kazakova (Tr. Inst. Mineral. Geokhim.

Kristallokhim. Redk. Elem. No. 15 [1963] 110/20, 117). – [23] A. Pabst (Am. Mineralogist **46** [1961] 700/18, 705/6). – [24] J. D. Hayton (Econ. Geol. **55** [1960] 1030/8, 1033). – [25] S. H. U. Bowie, K. Taylor (Proc. 2nd Intern. Conf. Peaceful Uses At. Energy, Geneva 1958, Vol. 3, pp. 527/40, 528, 532, 537).

[26] A. Damjanović, M. Vukasović (Radovi Inst. Geol. Rud. Istraz. Ispit. Nukl. Drugih Mineral. Sirovina **1** [1961] 11/5, 11). – [27] T.-M. Wang (K'o Hsueh T'ung Pao **26** [1981] 618/21 from C.A. **96** [1982] No. 38453). – [28] E. I. Semenov, A. P. Khomyakov (Diagnost. Diagnost. Svoistva Mineral. **1981** 88/93, 90/1). – [29] A. S. Povarennykh, A. D. Lebedeva (Konst. Svoistva Mineral. No. 6 [1972] 117/21, 118).

10.2.2.1.5 Chemical Behavior and Differential Thermal Analysis

Not noticeably soluble in hot hydrochloric, nitric, or sulfuric acids or in a combination of the first two; only soluble in hot concentrated HCl [1]. Dissolves in boiling sulfuric acid [2]. Inert to common etch reagents [3]. Etching with HF vapor 30 to 50 s shows zonality and inclusions of rutile and ilmenite [4].

The heating curve of davidite from Bektauata Mountains, Kazakhstan, exhibits three endothermic effects. The first, at 200°C, reflects loss of water from hygroscopic sites; others, at 500 and 600°C, are caused by loss of crystal water. Two exothermic peaks, at 400 and 800°C, are due to oxidation of Fe^{2+} to Fe^{3+} and Mn^{2+} to Mn^{3+} and to a phase transformation, respectively [5]. An endothermic effect at 200°C and an exothermic effect at 530 to 670°C are found in the heating curves of davidite (ufertite) from a Russian deposit [2]. Davidites from Mozambique and from Olary, South Australia, exhibit an exothermic peak at ~680°C and a lower one at 800°C [1]. The curves of davidites from Crocker's Well, South Australia, show peaks at 660°C and (lower) at 820°C, while the curve for Arizona davidite shows no peaks [6]. The curve of davidite from Bajados, Spain, corresponds to that of davidite from Crocker's Well [7].

References for 10.2.2.1.5:

[1] P. F. Kerr, H. D. Holland (Am. Mineralogist **36** [1951] 563/72, 566/7). – [2] N. I. Balashov (Byul. Nauchn. Tekhn. Inform. Mineral. Geol. Okhrany Nedr SSSR **1961** No. 2, pp. 10/2 from C.A. **57** [1962] 5607; Ref. Zh. Geol. **1962** No. 1 V 133). – [3] A. W. G. Whittle (S. Australia Geol. Surv. Bull. No. 30 [1954] 126/51, 137). – [4] N. A. Kulik (Zap. Vses. Mineral. Obshchestva **94** [1965] 114/9, 115/6). – [5] B. F. Ablanov (in: P. G. Sagunov et al., Geologiya Geokhimiya i Mineralogiya Mestorozhdenii Redkikh Elementov Kazakhstana, Alma-Ata 1966, pp. 152/7 from C.A. **67** [1967] No. 119032).

[6] A. Pabst (Am. Mineralogist **46** [1961] 700/18, 713). – [7] A. Arribas (Estud. Geol. [Madrid] **19** No. 1/4 [1963/64] 33/51, 38).

10.2.2.2 Loveringite $(Ca,RE,\ldots)(Ti,Fe,Cr,Mg,\ldots)_{21}O_{38}$

Loveringite was found in U-bearing phases of the Jimberlana intrusion near Norseman, Western Australia, in bronzite- and plagioclase-augite-hypersthene cumulate layers. It occurs as isolated crystals, mostly anhedral, occasionally as needles [1, 2]. Formula deduced from microprobe analyses (see "Seltenerdelemente" A 4, 1979, pp. 9, 24) is $[Ca_{0.72}RE_{0.33}(Y,Th,U,Pb)_{0.05}][Ti_{12.48}Fe_{3.38}Cr_{2.24}Mg_{0.92}Zr_{0.58}Al_{0.39}V_{0.21}Mn_{0.04}]O_{38}$ [2]. Minor elements are B, F, P, Sr, Nb, Ba, and Hf [1].

Rhombohedral cell parameters $a_{rh}=9.117(4)$ Å, $\alpha=69.07(1)°$, Z=1. Space group R3-C_{3i}^2 (No. 148) [2, 3]. The structure is based on a close-packed anion framework with a nine-layer stacking sequence and with the large cations Ca and RE occupying anion sites [2]. Cation substitution and coordination [1], cf. [2]:

Site	No.	Metal-oxygen distance	Ions
M0	1	2.775 Å	Ca, U, Th, RE (large)
M1$^{x)}$	1	2.167 Å	Zr, RE (small), Y, Fe^{2+}, Mn
M2$^{+)}$	2	1.983 Å	Mg, Fe
M3$^{x)}$	6	1.975 Å	Fe, Cr, V
M4$^{x)}$	6	1.972 Å	Ti
M5$^{x)}$	6	1.971 Å	Ti

$^{x)}$ Sixfold coordination. – $^{+)}$ Fourfold coordination.

The assignment of RE to two different sites is based on the RE distribution pattern which shows depletion from La to Eu, a sharp reversal between Eu and Ho and a second depletion between Ho and Lu [1, 2, 3]. It is possible that Ce^{4+} (and Hf^{4+}) also occupy M1 sites [2]. Loveringite is isostructural with senaite, crichtonite [2], and davidite [3].

Color black with submetallic luster. Opaque and grayish white in reflected light. Reflectivity in white light 17% in air and 5.9% in oil. Streak iron-gray. Microhardness VHN (load 10 g) 421 to 464 kg/mm² [3].

References for 10.2.2.2:

[1] I. H. Campbell, P. R. Kelly (Mineral. Mag. **42** [1978] 187/93, 188/9, 191/2). – [2] B. M. Gatehouse, I. E. Grey, I. H. Campbell, P. Kelly (Am. Mineralogist **63** [1978] 28/36, 29/31, 34). – [3] P. R. Kelly, I. H. Campbell, I. E. Grey, B. M. Gatehouse (Can. Mineralogist **17** [1979] 635/8, 636/7).

10.2.2.3 Unnamed Mineral $(Pb,Sr,Y)_{2.13}(Ti,Fe,Re,Mn)_{20.94}O_{38}$

The mineral occurs in quartz crystals developed on a granitic rock with altered Fe-Mg minerals. Locality unknown, possibly the Valais, Switzerland, or the Aosta Valley, northern Italy. The formula, calculated from the mean of 4 series of microprobe analyses with $Y_2O_3=2.49\%$ and $ReO_2=2.63\%$, is given as $(Pb_{0.87}Sr_{0.83}Y_{0.43})(Ti_{14.43}Fe_{6.05}Re_{0.23}Mn_{0.23})O_{38}$.

Hexagonal euhedral grains, flattened along the c axis. Trigonal with $a_{hex}=10.44$ Å, $c_{hex}=20.82$ Å. Space group R$\bar{3}$-C_{3i}^2 (No. 148); Z=3.

Luster is metallic. The mineral is opaque, black. In reflected light its color is white to light cream-white; under crossed nicols the mineral is isotropic. Reflectivity is 17.5 to 19.7%. Microhardness VHN (load 30 g)=1100 kg/mm² and VHN (load 100 g)=1185 kg/mm². Density D(meas.)=5 g/cm³, D(calc.)=5.02 g/cm³.

Reference for 10.2.2.3:

H. Sarp, J. Bertrand, J. Deferne, B. W. Liebich (Neues Jahrb. Mineral. Monatsh. **1981** 433/42).

10.2.2.4 Knopite (Ca, RE)(Ti, Fe^{3+}, Nb)(O, OH)$_3$

Knopite is a variety of perovskite with $(RE,Y)_2O_3 \geqq 5\%$ and $(Nb \pm Ta)_2O_5 < 1.9\%$, see in the following sections. There exists a gradual transition between perovskite and knopite; it is proposed to classify perovskites with 94 to 85 mol% $CaTiO_3$ [1] or with 4 to 5% Ce-earth [2] as knopite.

Knopite with such a high RE content occurs in contact metamorphosed limestones on the island of Alnö, and the nearby mainland of Sweden [3], see also [4]; in apatite-nepheline ores of the Khibina massif, and in an ore-bearing pegmatite in pyroxenite of the Africanda massif, both massifs in Kola Peninsula, USSR [5]; and in melilite-bearing ejected blocks of the black pozzolan formation at Corcolle, Alban Hills, Italy [6]. For knopite from knopite-titanomagnetite-containing rocks, see "Rare Earth Elements" A 8, 1984, Chapter 11.2.2.1. Knopite from Khibina usually has rims of titanite, titanomagnetite, and other minerals [5].

For a compilation of knopite (and cerian perovskite) analyses and content of $(RE,Y)_2O_3$ in these minerals, see "Seltenerdelemente" A 4, 1979, pp. 23, 9. Further data on the content of Nb, Ta, Ti and of Na, Ca, and RE_2O_3, are given in "Tantal" A 1, 1969, pp. 23, 127. The analysis of knopite from the Khibina massif, Kola, leads to the formula

$$(Na_{0.086}Ca_{0.795}Sr_{0.021}RE_{0.07})(Ti_{0.99}Fe^{3+}_{0.001}Nb_{0.019})(O_{2.96}(OH)_{0.04}) \text{ [5]},$$

and the analysis of knopite from Corcolle, Alban Hills, to

$$(Na_{0.043}K_{0.026}Ca_{7.278}Sr_{0.026}La_{0.052}(Ln+Y)_{0.288})(Si_{0.369}Al_{0.214}Fe^{3+}_{0.574}Nb_{0.027}Zr_{0.057}P_{0.005}Ti_{6.891})((OH)_{0.519}O_{23.481}).$$

The (recalculated) distribution of La, Ln, and Y for this specimen in %, yields La 15.2, Ce 60, Pr 5, Nd 5, Sm 5, Eu 0.2, Gd 1, Tb 5, Dy 2, Ho 0.2, Yb 0.02, and Y 1.6 [6]. The RE and Y distribution, in %, of two knopites from Kola is given as:

	Ce_2O_3	La_2O_3	Nd_2O_3	Pr_2O_3	Y_2O_3
Khibina	52.0	24.1	14.6	7.0	1.3
Africanda	56.0	22.6	13.2	7.0	0.9

Further lanthanides found are Sm and Eu [5].

The forms of knopite are octahedral and cubooctahedral crystals [1], knopite from Khibina forms xenomorphic and isometric aggregates; crystals with twin intergrowths after {111} are common [5]; knopite from Alban Hills occurs in octahedral crystals [6]. Cubic cell $a = 15.26 \pm 0.02$ Å (Alban Hills) [6], $a = 7.65 \pm 0.01$ Å (Khibina) and $a = 7.66 \pm 0.01$ Å (Africanda) [5].

Color black and lead gray [1], black with metallic luster, in thin splinters brown; refractive index $n \geqq 2.38$ [6]. Color black, opaque, only fragments dark brown, streak black to pale brown; refractive index $n = 2.36 \pm 0.01$ [5].

Hardness $H_M = 5.5$ to 6; the fracture is shelly; density D = 4.16 to 4.24 g/cm^3 [5].

Knopite dissolves on boiling in hydrochloric, nitric, and concentrated sulfuric acid [5].

References for 10.2.2.4:

[1] F. V. Chukhrov, E. M. Bohnshtedt-Kupletskaya (Mineraly Spravochnik, Vol. 2, Pt. 3, Nauka, Moscow 1967, pp. 1/676, 422). – [2] P. Ramdohr, H. Strunz (Klockmanns Lehrbuch der Mineralogie, 16th Ed., Enke, Stuttgart 1978, pp. 1/876, 519). – [3] C. Palache, H. Berman, C. Frondel (Dana's System of Mineralogy, Vol. 1, 7th Ed., Wiley, New York – London 1944, pp. 1/834, 733). – [4] C. Hintze (Handbuch der Mineralogie, Vol. 2, De Gruyter, Leipzig 1897, pp. 1/1842, 1649/50). – [5] L. L. Shilin, M. T. Yanchenko (Dokl. Akad. Nauk SSSR **144** [1962] 639/42; Dokl. Earth Sci. Sect. **144** [1962] 139/42).

[6] B. Turi (Atti Accad. Nazl. Lincei Rend. Classe Sci. Fis. Mat. Nat. [8] **44** [1968] 801/11, 801, 807/9).

10.2.2.5 **Yttrocrasite** (Y,Th) Ti_2 $(O,OH)_6$

Occurrences and Paragenesis

The first description of yttrocrasite, by Hidden and Warren [1], was of a gadolinite-bearing pegmatite from Burnet County, Texas. The mineral also occurs in the Clear Creek pegmatite from the same area [2, 3], associated with gadolinite, behoite, rowlandite, tengerite, yttrialite, allanite, samarskite, fergusonite, bastnaesite, cyrtolite, uraninite, thorite, "gummite", thorogummite, and texasite [4]. Whittle describes a small occurrence of yttrocrasite from Crocker's Well, South Australia [5]; and Ramdohr mentions Graveggia, Piemont, northern Italy, as a locality of yttrocrasite [6]. The occurrence of yttrocrasite in the pegmatite of the Southern Pacific silica quarry at Nuevo, Riverside County, California, is doubtful [7]. The samples from Mitwaba, Zaire, and Ruanda, both central Africa, initially identified as yttrocrasite [8, 9, 10] were later shown to be priorite (= blomstrandine) and fergusonite, respectively [11].

Yttrocrasite often has a brown alteration coating [1, 2], perhaps a variety of bastnaesite [3].

Chemistry

Observed $(RE+Y)_2O_3$ contents of yttrocrasite are tabulated in "Seltenerdelemente" A 4, 1979, p. 7; references for the analyses are cited on p. 35 of the same volume. From these analyses the general formula $(Y,Th)Ti_2(O,OH)_6$ is derived [2]; yttrocrasite from Clear Creek pegmatite was found to have the special formula

$$(Y_{0.601}La_{0.003}Ce_{0.013}Pr_{0.002}Nd_{0.0007}Sm_{0.017}Gd_{0.0039}Tb_{0.001}Dy_{0.018}Ho_{0.002}Er_{0.029}Tm_{0.002}Yb_{0.005}$$
$$Lu_{0.003}Th_{0.118}U_{0.029}Mn_{0.007}Ca_{0.104}Pb_{0.006})(Ti_{1.838}Fe_{0.092}W_{0.062}Ta_{0.002})(O_{5.058}OH_{0.942})\ [4].$$

Peacor et al. from the same analysis, and also normalized to 1.838 Ti atoms, get the following atomic distribution: Y 0.523, La 0.003, Ce 0.011, Pr 0.002, Nd 0.006, Sm 0.015, Gd 0.034, Tb 0.001, Dy 0.041, Ho 0.002, Er 0.025, Tm 0.002, Yb 0.004, Lu 0.003, Th 0.102, U 0.025, Mn 0.006, Ca 0.090, Pb 0.005, Ti 1.838, Fe 0.060, W 0.023, Ta 0.001, Nb 0.0004 [13]. Both, metamict and non-metamict material from the Clear Creek pegmatite, have essentially the same composition. Yttrocrasite is thought to be the Ti-rich, Nb- and Ta-poor end-member of the RE multiple oxides of the euxenite group [4].

The analysis given by Crook yields a ratio Y to RE (La to Lu) of 6.1:1, $ThO_2=9.12$, $U_3O_8=2.36\%$ [4].

Crystal Forms and Structure

Orthorhombic dipyramidal. Prismatic crystals with {100}, {010}, and {001} [4], no observable cleavage [2, 4, 5]. Unit cell parameters a = 12.862(7), b = 4.810(2), c = 4.571(3) Å, space group Pbcn-D_{2h}^{14} (No. 60), V = 282.79 $Å^3$ [4], see here also discussion on structure as compared to euxenite and fersmite[1)].

[1)] See however [13].

Optical Properties

Luster brightly pitchy to resinous [1], resinous, becoming dull [5], resinous [2]. Opaque, transparent in thin fragments [4]. Color black [1, 2, 12], black to faintly brownish black [5], green-brown and streak pale brown to greenish gray [4] or greenish black [2]. Color in transmitted light amber to light yellow, nearly colorless in very thin slivers [1], amber colored [5]. Pleochroism X = colorless, amber to pale brown, Y = brown, Z = brown, green-brown to opaque, with absorption X > Y > Z. Biaxial negative, $n_x = 2.131(2)$, $n_y = 2.137(2)$, $n_z = 1.142(2)$, $2V_x = 60°$ to 70° [4]. But also in part metamict and only in part feebly double reflecting [1] with $n' \approx 2.13$ [12] and $n' = 2.12$ to 2.15 [5].

Other Physical Properties

Hardness $H_M = 5.5$ to 6 [1, 2, 4, 5], 6 [12]. Fracture is uneven to subconchoidal [1, 4, 5], slightly conchoidal [2], or conchoidal [12]. Density changes with state of alteration: fresh samples D = 4.8, altered samples 4.26 g/cm^3 [2], D = 5.315(5) on unaltered material, 4.624(5) to 5.149(5) g/cm^3 on metamict samples [4]. Earlier determined values are 4.8043 g/cm^3 [1] and 4.8 [5, 12].

Radioactivity is in agreement with analytical data for U and Th [1].

Chemical Behavior

Yttrocrasite is easily decomposed by hydrofluoric acid; if finely powdered it is soluble in boiling concentrated sulfuric acid [1].

References for 10.2.2.5:

[1] W. E. Hidden, C. H. Warren (Am. J. Sci. [4] **22** [1906] 515/9, 515/6). – [2] W. W. Crook III (Mineral. Rec. **7** [1976] 182/3). – [3] W. W. Crook III (Mineral. Rec. **8** [1977] 88/90). – [4] W. W. Crook III (Am. Mineralogist **62** [1977] 1009/11). – [5] A. W. G. Whittle (S. Australia Geol. Surv. Bull. No. 30 [1954] 126/51, 136).

[6] P. Ramdohr, H. Strunz (Klockmanns Lehrbuch der Mineralogie, 16th Ed., Enke, Stuttgart 1978, pp. 1/876, 542). – [7] J. Murdoch, R. W. Webb (Calif. Bur. Mines Bull. No. 173 [1956] 1/452, 351). – [8] L. Van Wambeke (Bull. Soc. Belge Geol. **64** [1955] 432/45, 437/9). – [9] D. Ledent, E. Picciotto, G. Poulaert (Bull. Soc. Belge Geol. **65** [1956] 233/50, 234). – [10] P. Eberhardt, J. Geiss, H. R. von Gunten, F. G. Houtermans, P. Signer (Bull. Soc. Belge Geol. **65** [1956] 251/6).

[11] L. Van Wambeke (Bull. Soc. Belge Geol. **66** [1957] 35/53, 35). – [12] E. S. Larsen, H. Berman (U.S. Geol. Surv. Bull. No. 848 [1934] 1/266, 63). – [13] D. R. Peacor, W. B. Simmons Jr., E. J. Essene, E. W. Heinrich (Am. Mineralogist **67** [1982] 156/69, 163/4).

10.2.2.6 **Brannerite** (U, Ca, Th, Y)(Ti, Fe)$_2$O$_6$

Synonyms. Cordobaite from Fuenteovejuna, Cordoba, Spain, is brannerite [1]; also the names absite for thorian brannerite [2] and for lodochnikite [3], sometimes containing RE (see [4]) are to be dropped [5]; for absite see also [6], for lodochnikite [7].

References:

[1] D'A. George (RMO-563 [1949] 1/198, 79); N.S.A. **5** [1951] No. 833). – [2] A. W. G. Whittle (S. Australia Geol. Surv. Bull. No. 30 [1954] 126/51, 142). – [3] Ya. D. Gotman (Zap. Vses. Mineral.

Obshchestva **87** [1958] 197/200). – [4] G. A. Il'inskii (Vestn. Leningr. Univ. **1961** No. 6, pp. 134/8 from Ref. Zh. Geol. **1961** No. 12 V 166). – [5] M. M. Povilaitis (Zap. Vses. Mineral. Obshchestva **92** [1963] 113/23, 122).

[6] M. Fleischer (Am. Mineralogist **41** [1956] 166). – [7] M. Fleischer (Am. Mineralogist **43** [1958] 1007).

10.2.2.6.1 Paragenesis, Inclusions, and Alteration

Occurrences and Paragenesis

A compilation of occurrences of brannerite by Holmes shows that this mineral occurs in granitic and dioritic rocks, in pegmatites, in quartz dikes of pegmatitic or metamorphic derivation, in sulfide-bearing conglomerates, and in placers [1]; that means it is usually found in magmatic and postmagmatic stages of intrusive processes and in sedimentary derivates of these rocks [2], see also [3]. Furthermore, hydrothermal veins with Mo, W, Bi, Au, and other hydrothermal formations contain brannerite, see in the following.

In **granites** and aplites brannerite occurs as an accessory mineral together with zircon and ilmenite [4]. Silicified parts of a fine-grained quartzsyenite of Topolčane, near Prilep, Yugoslavia, contain brannerite [5], cf. [6].

Brannerite in **pegmatites** has been described from Fuenteovejuna, Spain [7]; from Hornachuelos, Sierra Morena, Spain, together with uraninite, autunite, and radioactive ochers [8]; and from Rajasthan, India, where it occurs also in the country rocks of the pegmatites; that are schists of the Ajabgarh series [9]. Brannerite occurs in feldspar veins in schists, sandstones, and hornstones in Central Asia [10], and in **quartz veins,** cutting late Mesozoic quartz monzonite, south of Colville, Mono County, California [11], see also [12, p. 257]. Together with U, Th, and other RE minerals brannerite was found in quartz-albite veins, transecting granite of the Bokan Mountain area, southeastern Alaska [57]. A leucogranite that is a metasomatic alteration product of granitic gneisses in Wolowa Gora near Kowary, Sudetes, Poland, contains brannerite [13].

Brannerite of **hydrothermal origin** was found intimately associated with quartz and gold and sometimes galenite in veins near La Gardette, Isère, France [14], see also [58]. Associated with gold, brannerite occurs at the Little Nell prospect, Bucks Lake quadrangle, Plumas County, California, where it is confined to a narrow zone along a dike of partly sericitized albite phenocrysts and quartz in a sericitized groundmass with scattered minute pyrite cubes [15]. RE-containing brannerite ("lodochnikite") occurs in hydrothermal veins with barite, siderite, hematite, galenite, and calcite in alkali- and nepheline syenites (locality not given) [16]. In Mindala, Zambia, brannerite occurs in calcite veins with quartz, feldspar, anhydrite, uraninite, and also in albite-calcite veins with anatase and uraninite [17]. For brannerite in Alpine tension joints, see [60]. Brannerite of hydrothermal genesis was observed as cement of breccias with granitic and metamorphic rocks at a locality not named in the USSR [18].

In **Mo-W deposits**, associated with leucocratic granites, brannerite occurs in apical parts of the ore-bearing granites and in syngenetic quartz- and microcline-quartz veins [19, 20] together with beryl, wolframite, scheelite, molybdenite, ilmenorutile, sphalerite, aikinite, and tourmaline [4]. Examples of that kind are hydrothermal rare-metal deposits in Kazakhstan with brannerite as an accessory mineral associated with molybdenite, beryl, wolframite, scheelite, monazite, and bertrandite [21]. The California Vein of the Mount Antero region, Chaffee County, Colorado, is a quartz vein with molybdenite, beryl, muscovite, and minor amounts of brannerite [22], [12, pp. 256/7]. The Mo deposit of Château-Lambert, Vosges, France [23], contains brannerite together with molybdenite, chalcopyrite, and uraninite [12, pp. 50, 257], as well a molybdenite paragenesis at Bou Azzer, southern Morocco [24, 25], and the Climax molybdenite deposit, Colorado [26]. A very young brannerite occurs on the **loellingite-Bi-Au veins** of Knappenberg, Carinthia, Austria [27]. In meso- to epithermal **U-Pb-Zn-Co-Ni veins** in fractures and faults of the Helvetic thrust mass, Mürtschenalp, Canton Glarus, Switzerland, brannerite occurs together with anatase, rutile, pitchblende, and numerous sulfides [28]. On the **fluorite veins** of Wölsendorf, Bavaria, brannerite has been found as U mineral of the pitchblende paragenesis [29].

In Kansanshi, Zambia, brannerite, associated with **Cu-minerals,** molybdenite, calcite, quartz, and pyrite, occurs disseminated in feldspar-bearing quartzite [30]; and in South Australia in a small Cu sulfide ore body at Nichol's Nob in the northern Flinders Ranges associated with uraninites in small veins and in a calcite gangue [31, p. 143], [32, pp. 24, 33/4]. Brannerite exists as disseminated, hydrothermally formed grains in a sericitized, pyritic, rhyolithic porphyry copper prospect in the White Signal district, SW New Mexico [33]. Uranium nodules in the Cu deposit of Mitterberg, Styria, Austria, contain brannerite as the second U carrier besides uraninite [34].

Brannerite in **metamorphic rocks** was found in the form of lenticular or roughly spherical nodules in a granite gneiss north from San Bernardino Mountains, San Bernardino County, California, associated with rutile and sodic plagioclase [35], [12, pp. 49/50, 257], and in minor amounts in a small shear zone in Precambrian schists and granite on the east side of the Rio Grande, Caballo Mountains district, Sierra County, New Mexico [36]. At the Canadian Pacific Railway, on the north side of Christie Lake, eastern Ontario, brannerite occurs in a Precambrian tremolite rock [37]; and in Åmot, Modum district, Norway, it occurs together with ilmenite and rutile in small fissures in an albite-actinolite greenschist [38]. In the uranium deposit Jabiluka, Pine Creek geosyncline, Northern Territory, Australia, brannerite was observed together with coffinite in breccia clasts and in the metamorphic country rock [39].

In the USSR brannerite was found in zones of high-temperature **metasomatism** in paragenesis with coffinite (nenadkevite), Y-bearing titanite, uraninite, U-bearing malakon, and apatite, in albitites with nasturan, "uranschwärze", uranophane, and iriginite [4]. As impregnations in quartz-pyrite-sericite metasomatic ore bodies it occurs in ignimbrites and tuffs of Devonian age [40]. Related to albitized rocks brannerite (locally named "absite") was found in South Australia at Torrens Gorge near Houghton, at Crocker's Well, and at Myponga [32, pp. 24/5, 33/4, 42/3], see especially for "absite" also [31, p. 142]. For further occurrences of brannerite in South Australia, see "Rare Earth Elements" A 8, 1984, Chapter 11.7.8. Brannerite also occurs in metasomatic siderite ores at Olsa near Friesach, Carinthia, Austria [41].

In **sediments** brannerite is associated with clausthalite, galena, and uraninite in uraniferous sandstone of late Devonian age from Pen Haven Junction, Carbon County, Pennsylvania [42]; and it occurs in carbonate beds and linses 32 km north of Mount Isa, Queensland, Australia [12, pp. 50, 258]. The first find of brannerite was a gold **placer** near the head of Kelley Gulch in western Custer County, Idaho, where the mineral occurs in form of rounded grains and crystals [43]. At Florence, Montana, brannerite has been found in heavy sands, consisting largely of rutile, ilmenite, and garnet, and in Elk City, Idaho, in a placer deposit with ilmenite, rutile, garnet, magnetite, davidite, betafite, and euxenite [44]; cf. "Rare Earth Elements" A 8, 1984, Chapter 11.7.2.3.3.1. Radioactive placers in the Big Meadow area, Valley and Boise Counties, Idaho, contain brannerite besides allanite [59].

As very important U mineral brannerite occurs in uranium deposits of the conglomerate type, see "Uran" A 1, 1979, p. 17. Thus brannerite is one of the main minerals in uranium-gold deposits of Witwatersrand, Republic of South Africa [45], ("Uran" A 1, 1979, p. 205), where it occurs in paragenesis with uraninite, rutile, and sulfides [46, p. 10]. It forms during a long temperature range in the presence of TiO_2 and UO_2 [27] and is an alteration product of uraninite and rutile [47], and uraninite and ilmenite or titaniferous magnetite [48, 61]; see for "Pronto reaction" below. It often is included as veinlets or infiltrations in other minerals such as leucoxene, sericite, chlorite, quartz, pyrite [45, 48]. A further locality of brannerite in South Africa is the Dominion Reef [46, pp.21, 25]. In the Blind River area, Ontario, together with Elliot Lake one of the largest U-bearing provinces (see "Uran" A 1, 1979, pp. 146/54), brannerite first was found by Nuffield [49]. It is the main U mineral of several mines in the Blind River district and occurs together with rutile, anatase, and uraninite [50], see also [51]. It is formed by the "Pronto reaction" (named after the Pronto Mine in this area): $UO_2 + (2 \text{ to } 3)\ TiO_2 = UTi_{2\text{ to }3}O_{6\text{ to }8}$ with RE as diluting elements [52], [46, pp. 17/8]. Brannerite in the Elliot Lake area [53] occurs as secondary mineral in ovoid grains or aggregates together with uraninite, anatase, rutile, and monazite [54]. Brannerite in a conglomerate with rutile, uraninite, zircon, and pyrite in Serra de Jacobina, Brasilia, see [46, p. 26].

Brannerite without RE (as it is explicitly asserted) are described from pegmatites in the Leventina granite gneiss in two quarries near Lodrino in the Riviera Valley, Canton Tessin, Switzerland [55] and in a U mineralization in a Permian volcanic tuffite from Boarezzo, Valganna, Varese, Italy [56].

Inclusions and Alteration

Inclusions in brannerite are rutile in samples of San Bernardino, California [35]; plagioclase, often saussuritized, and dark minerals in brannerite from Åmot, Norway [38]; and microscopic anatase, partly as fine powder, in crystals from the Pronto Mine, Blind River district, Ontario [52]. Massive brannerite and fine needles of this mineral from Witwatersrand, Republic of South Africa, are penetrated by PbS particles, by pyrrhotine tablets, and clouds of anatase particles [47]. Minute cracks are filled with quartz as in brannerite from Kelley Gulch, Idaho [43], or with leucoxene as in brannerite from Nichol's Nob, Flinders Ranges, South Australia [31]. See also alteration of brannerite in the following.

Brannerite nearly always is altered to secondary minerals which form coatings on the crystals, occur in veinlets or as pseudomorphs. The color of these secondary products is yellowish gray to yellow; anatase is identified in them by X-rays [19]. Brownish yellow coatings are observed on brannerite from Kelley Gulch, Idaho [43], and from the Mount Antero region, Chaffee County, Colorado [22]. Encrustations are brown, yellow-brown, and olive-green on brannerite from Fuenteovejuna, Spain [7], yellowish green on brannerite of Wolowa Gora near Kowary, Sudetes, Poland [13], and yellow to red-brown on brannerite from Crocker's Well, South Australia [44]. Dense white to yellowish white, radioactive products occur on brannerite from Blind River area, Ontario [49]. Alteration to anatase show brannerites from Olsa near Friesach, Carinthia [41], from Plumas County, California [15], and from Pronto Mine, Blind River district, Ontario [27]. Brannerite from Canadian U deposits invariably is altered to anatase or rutile [50]. Besides anatase there are found Ti hydroxides as oxidation products of brannerite [18]. Brannerite from Bou Azzer, south Morocco, is surrounded by small rims of titanite [24]. "Gummite", a further alteration product of brannerite, is found on samples from Nichol's Nob, Flinders Ranges, South Australia [31], and curite and probably wölsendorfite are found on brannerite from Olsa near Friesach, Carinthia [41]; in addition uranium black and iriginite are also described as alteration products of brannerite [2]. Pseudomorphs of pechblende after brannerite are described from deposits in Kansanshi, Zambia, Africa [30]; for pseudomorphs of aeschynite after brannerite from Russian or of pyrite and chalcopyrite from Japanese deposits, see [4].

References for 10.2.2.6.1:

[1] S. W. Holmes (from O. Brotzen, Econ. Geol. **53** [1958] 489/92, 490). – [2] M. V. Soboleva, I. A. Pudovkina (AEC-TR-4487 [1961] 1/455, 398 [Russian Original, Moscow 1957]; N.S.A. **15**

[1961] No. 23739). – [3] A. A. Chernikov (Sov. Geol. No. 4 [1981] 85/98, 88). – [4] F. V. Chukhrov, E. M. Bohnshtedt-Kupletskaya (Mineraly Spravochnik, Vol. 2, Pt. 3, Nauka, Moscow 1967, pp. 1/676, 438/9). – [5] M. Vukasović, N. Mihailović-Vlajić (Radovi Inst. Geol. Rud. Istraz. Ispit. Nukl. Drugih Mineral. Sirovina **2** [1963] 67/8).

[6] L. Barić (Tschermaks Mineral. Petrogr. Mitt. [3] **10** [1965] 368/78, 377). – [7] D'A. George (RMO-563 [1949] 1/198, 78/9; N.S.A. **5** [1951] No. 833). – [8] M. Alia (Proc. 1st Intern. Conf. Proc. Peaceful Uses At. Energy, Geneva 1955 [1956], Vol. 6, pp. 196/7). – [9] G. V. Umamaheswararao, R. Krishnaswamy (J. Sci. Ind. Res. [India] A **15** [1956] 401/2). – [10] L. V. Chernysheva, I. I. Kupriyanova, G. A. Sidorenko (Sb. Kratk. Soobshch. Mineral. Geokhim. **1** [1968/70] 83/8 from Ref. Zh. Geol. **1970** No. 10 V 234).

[11] A. Pabst (Am. Mineralogist **39** [1954] 109/17, 109). – [12] E. W. Heinrich (Mineralogy and Geology of Radioactive Raw Materials, McGraw-Hill, New York 1958, pp. 1/654). – [13] J. Lis, M. Stepniewski, H. Sylwestrzak (Biul. Inst. Geol. [Warsaw] No. 193 [1965] 207/23, 214 [Polish], 227/8 [English Summary]). – [14] J. Geffroy (Bull. Soc. Franc. Mineral. Crist. **86** [1963] 129/32, 130). – [15] A. Pabst, M. C. Stinson (Bull. Geol. Soc. Am. **71** [1960] 2071/2).

[16] G. A. Il'inskii (Vestn. Leningr. Univ. **1961** No. 6, pp. 134/8 from Ref. Zh. Geol. **1961** No. 12 V 166). – [17] F. Mendel'son (from [4]). – [18] Ya. D. Gotman, V. M. Polyakova, A. K. Miguta (Tr. Mineral. Muzeya Akad. Nauk SSSR No. 20 [1971] 76/87, 76, 85). – [19] M. M. Povilaitis (Tr. Mineral. Muzeya Akad. Nauk SSSR No. 12 [1961] 64/79, 64, 77/9). – [20] M. M. Povilaitis, T. A. Yakolevskaya, D. N. Knyazeva, I. D. Belyaeva (Zap. Vses. Mineral. Obshchestva **97** [1968] 150/61, 150).

[21] V. L. Mel'nikova (Izv. Akad. Nauk Kaz.SSR Ser. Geol. **25** [1968] 62/5). – [22] J. W. Adams (U.S. Geol. Surv. Bull. No. 982-D [1953] 1/119, 104, 111). – [23] G. Branche, J. Chervet, C. Guillemin (Bull. Soc. Franc. Mineral. Crist. **74** [1951] 457/88, 483). – [24] G. Jouravsky (Compt. Rend. **234** [1952] 230/1). – [25] D. Ledent (Compt. Rend. **250** [1960] 1309/11).

[26] J. W. Vanderwilt, R. U. King (Trans. AIME Mining Eng. **7** [1955] 41/53, 48). – [27] P. Ramdohr (Z. Deut. Geol. Ges. **130** [1979] 439/58, 444/5). – [28] K. Bächtiger (Beitr. Geol. Schweiz Geotech. Ser. **38** [1963] 1/113, 34, 35/6). – [29] E. Seeliger, H. Strunz (Neues Jahrb. Mineral. Abhandl. **103** [1965] 163/78, 164). – [30] A. Grimmer (Geologie [Berlin] **12** [1963] 1169/81, 1174/5).

[31] A. W. G. Whittle (S. Australia Geol. Surv. Bull. No. 30 [1954] 126/51). – [32] A. W. G. Whittle (J. Geol. Soc. Australia **2** [1955] 21/45). – [33] C. S. Ferris, C. O. Ruud (Quart. Colo. School Mines **66** [1971] 1/35, 8, 19). – [34] W. Paar, V. Köppel (Neues Jahrb. Mineral. Abhandl. **131** [1978] 254/71, 257/9). – [35] D. F. Hewett, J. Stone, H. Levine (Am. Mineralogist **42** [1967] 30/8, 34, 37).

[36] F. S. Boyd (Field Conf. Guidebook New Mexico Geol. Soc. **6** [1955] 123 from S. A. Northrop, Minerals of New Mexico, 2nd Ed., Univ. New Mexico Press, Albuquerque 1959, pp. 1/655, 150). – [37] D. A. Moddle (Can. Mineralogist **6** [1957/61] 155/7). – [38] T. Van Autenboer, F. J. Skjerlie (Norg. Geol. Unders. No. 200 [1957] 5/7). – [39] R. A. Binns, J. McAndrew, S. S. Sun (Proc. Intern. Uranium Symp. Pine Creek Geosyncline, Vienna 1979 [1980], pp. 543/62, 543, 545, 548). – [40] A. V. Zavarizin (Geol. Rudn. Mestorozhd. **3** No. 6 [1961] 54/8, 55).

[41] A. El Goresy, H. Meixner (Neues Jahrb. Mineral. Abhandl. **103** [1965] 94/8, 96). – [42] R. B. Finkelman, H. Klemic (U.S. Geol. Surv. Res. J. **4** [1976] 715/6). – [43] F. L. Hess, R. C. Wells (J. Franklin Inst. **189** [1920] 225/37, 225). – [44] H. H. Adler, J. A. Puig (Am. Mineralogist **46** [1961] 1086/96, 1089/90). – [45] C. E. Feather, G. M. Koen (Minerals Sci. Engin. **7** [1975] 189/224, 192, 203).

[46] P. Ramdohr (Abhandl. Deut. Akad. Wiss. Berlin Kl. Chem. Geol. Biol. **1958** No. 3, pp. 1/35). – [47] M. Schidlowski (Neues Jahrb. Mineral. Abhandl. **105** [1966] 310/24, 312/3, 317/20; Geol. Jahrb. Beih. **85** [1970] 1/80, 49, 51/2). – [48] C. E. Feather, J. A. Snegg (Trans. Geol. Soc. S. Africa **81** [1978] 255/60, 256). – [49] E. W. Nuffield (Am. Mineralogist **39** [1954] 520/2). – [50] A. H. Lang, J. W. Griffith, H. R. Steacy (Geol. Surv. Can. Econ. Geol. Ser. No. 16 [1962] 1/324, 45, 61, 63, 69, 135, 137, 139, 144).

[51] A. H. Lang (Can. Mining J. **77** No. 6 [1956] 71/6, 73). – [52] P. Ramdohr (Neues Jahrb. Mineral. Monatsh. **1957** 217/22, 217, 220). – [53] N. J. Theis (Bull. Geol. Surv. Can. No. 304 [1979] 1/50, 31/2). – [54] S. M. Roscoe (Geol. Surv. Can. Paper No. 68-40 [1969] 1/205, 125, 136, 138). – [55] F. Bianconi, A. Simonetti (Schweiz. Mineral. Petrogr. Mitt. **47** [1967] 887/934, 894).

[56] G. Frenzel, J. Ottemann, W. Kurtze (Neues Jahrb. Mineral. Abhandl. **124** [1975] 75/102, 93/5). – [57] M. H. Staatz (Econ. Geol. **73** [1978] 512/23, 516/7). – [58] E. Asselborn (Lapis **6** No. 2 [1981] 9/14, 14). – [59] D. Truesdell, R. Wegrzyn, M. Dixon (GJBX-15-77 [1977] 1/53, 8, 14; INIS Atomindex **8** [1977] No. 337432). – [60] A. Kipfer (Schweizer Strahler **3** [1973] 133/69, 141/5, 162, 169).

[61] C. E. Feather (U.S. Geol. Surv. Profess. Papers No. 1161-A-BB [1981] Q1/Q23, Q12/Q15).

10.2.2.6.2 Chemistry

A compilation of brannerite analyses and the variance and mean of $RE_2O_3 + Y_2O_3$ contents of 26 samples is given "Seltenerdelemente" A 4, 1979, pp. 14/5 and 9. Electron microprobe analyses of 20 Elliot Lake brannerites show 0.36 to 2.92, mean 1.44% Y_2O_3, and 0.07 to 1.58, mean 0.66% Ce_2O_3 [1]. While crystalline brannerite contains ΣRE about 10%, brannerite from hydrothermal occurrences has $\Sigma RE < 0.n$ [2], see also p. 57.

Calculations of 20 analyses of brannerite from various localities as AB_2O_6, where Y and RE are elements of group A, are compiled by Povilaitis [3]. Newer analyses (always one) are calculated as follows: brannerite from Kazakhstan $(U^{6+}_{0.39}U^{4+}_{0.22}Ca_{0.54}Y_{0.02}Th_{0.03}Pb_{0.11})(Fe^{3+}_{0.01}Fe^{2+}_{0.04}Mg_{0.03}Ti_{1.8})O_6$ [4]; brannerite of hydrothermal genesis $(Ca_{0.29}U^{6+}_{0.21}U^{4+}_{0.57}RE_{0.008}Pb_{0.0008})(Ti,W,Nb,Zr,Fe^{3+})_{2.00}O_6$ [5]; brannerite from feldspar veins, Central Asia, $(U^{6+}_{0.53}U^{4+}_{0.11}Ca_{0.53})(Ti_{1.30}Si_{0.22}Fe^{3+}_{0.14}Ce^{3+}_{0.12}Al_{0.06}Mg_{0.05}Pb_{0.03}Ta_{0.01}Th_{0.01})O_{5.97}F_{0.03}$ [6].

With the exception of brannerite from Central Asia, see above, all these formulas have Y and RE in group A, substituting for U, as it is also recognized by Kaiman [7]; see also [8]. Trace elements in brannerites are Ba, Sr, Ni, Bi, Sn, Ta, Nb, Sc, P [9] and in addition Mn, V, Zr [3, 8] and He [10]. SiO_2 often is present in the form of included mineral fragments [11].

Most brannerite show predominance of Y and Y-earth, sometimes with special enrichment of Sm, Gd, Dy, or as in "lodochnikite" Gd, Dy, Er, and Yb [12] but in some brannerites Ce-earth are predominant [13], as it can be seen also from the compilation of Povilaitis [3] and in the following table.

Occurrence	$\Sigma RE = 100$								Ref.
	Y	La	Ce	Pr	Nd	Sm	Eu	Gd	
Greisen, Siberia	(44)	0.8	4.7	1.5	5.5	7.8	–1)	4.8	[14]3)
Greisen, Kazakhstan	(50)	2.4	8.3	—	1.2	9.5	–1)	1.2	[14]3)

Occurrence	ΣRE = 100								Ref.
	Y	La	Ce	Pr	Nd	Sm	Eu	Gd	
Crocker's Well, South Australia	(50)	2.0	2.0	0.5	4.9	3.0	–[1]	3.9	[14][3]
Feldspar-quartz dike, Siberia	36.2	3.6	7.7	1.1	9.8	5.6	–[1]	8.8	[14]
Hydrothermal deposit, USSR	18.2	7.0	20.0	3.9	18.4	8.5	+[2]	6.2	[5]

Occurrence	ΣRE = 100							Ref.
	Tb	Dy	Ho	Er	Tm	Yb	Lu	
Greisen, Siberia	0.9	7.5	1.4	5.2	1.8	12.0	1.9	[14][3]
Greisen, Kazakhstan	—	7.1	2.4	3.6	—	12.0	2.4	[14][3]
Crocker's Well, South Australia	0.5	3.9	1.0	3.0	—	6.9	0.5	[14][3]
Feldspar-quartz dike, Siberia	—	12.2	1.2	6.3	0.5	6.4	0.5	[14]
Hydrothermal deposit, USSR	0.8	5.6	0.8	4.2	0.6	5.6	+[2]	[5]

[1] Not determined. – [2] Present. – [3] Calculated from analyses of [12].

In brannerite from La Gardette, Isère, France, Y and Yb are predominant, Sm, Pr, Eu, Er, Lu occur in traces, and La, Ce are not detected [15]. Brannerite from Åmot, Modum district, Norway, contains only Y, while La and lanthanides could not be found [16]. – For isotope composition of Xe and Kr in brannerite from Middle Asia see [17].

References for 10.2.2.6.2:

[1] N. J. Theis (Bull. Geol. Surv. Can. No. 304 [1979] 1/50, 47/9). – [2] A. A. Chernikov (Sov. Geol. No. 4 [1981] 85/98, 88/9). – [3] M. M. Povilaitis (Zap. Vses. Mineral. Obshchestva **92** [1963] 113/23, 113/8). – [4] V. L. Mel'nikova (Izv. Akad. Nauk Kaz.SSR Ser. Geol. **25** [1968] 62/6, 65/6). – [5] Ya. D. Gotman, V. M. Polyakova, A. K. Miguta (Tr. Mineral. Muzeya Akad. Nauk SSSR No. 20 [1971] 76/87, 79/81).

[6] L. V. Chernysheva, I. I. Kupriyanova, G. A. Sidorenko (Sb. Kratk. Soobshch. Mineral. Geokhim. **1** [1968/70] 83/8 from Ref. Zh. Geol. **1970** No. 10 V 234). – [7] S. Kaiman (Can. Mineralogist **6** [1957/61] 389/90). – [8] F. V. Chukhrov, E. M. Bohnshtedt-Kupletskaya (Mineraly Spravochnik, Vol. 2, Pt. 3, Nauka, Moscow 1967, pp. 1/676, 436). – [9] E. W. Heinrich (Mineralogy and Geology of Radioactive Raw Materials, McGraw-Hill, New York 1958, pp. 1/654, 49). – [10] R. C. Wells (J. Franklin Inst. **189** [1920] 779/80).

[11] C. S. Ferris, C. O. Ruud (Quart. Colo. School Mines **66** [1971] 1/35, 15). – [12] E. I. Semenov (Mineralogiya Redkikh Zemel', Nauka, Moscow 1963, pp. 1/412, 85/6). – [13] M. M. Povilaitis (Tr. Mineral. Muzeya Akad. Nauk SSSR No. 12 [1961] 64/79, 70). – [14] M. M. Povilaitis, T. A. Yakolevskaya, D. N. Knyazeva, I. D. Belyaeva (Zap. Vses. Mineral. Obshchestva **97** [1968] 150/61, 156). – [15] J. Geffroy (Bull. Soc. Franc. Mineral. Crist. **86** [1963] 129/32, 131).

[16] T. Van Autenboer, F. J. Skjerlie (Norg. Geol. Unders. No. 200 [1957] 5/7). – [17] V. B. Sharif-Zade, Yu. A. Shukolyuakov, E. K. Gerling (Geokhimiya **1972** 314/22; Geochem. Intern. **9** [1972] 199/207, 202).

10.2.2.6.3 Crystal Forms and Crystal Structure

Monoclinic [1], with a:b:c = 2.580:1:1.889, β = 119°30′ [2], 1.186:1:1.182, β = 114°22′ ("Iodochnikite") [3].

Forms and angles in accordance with structure are given by Patchett, Nuffield [4]:

1		2		3	
hk0	Φ	hk0	Φ	hk0	Φ
010	0°00′	100	84°58′	100	90°00′
160	18°12′	510	66°46′	510	65°27′
130	31°42′	310	53°16′	310	52°43′
120, 590	45°35′	210, 950	39°23′	210	41°12′
110	61°34′	110	23°24′	110	23°39′
210	75°05′	130	9°53′	130	8°18′
310	79°37′	150	5.21′	150	5°01′
610	84°58′	010	0°00′	010	0°00′

1 The measured angles of Pabst [5]. – 2 Angles and indices of Pabst [5] transformed to the structural setting. – 3 Angles calculated from the structural constants, see p. 53.

Crystals from syngenetic quartz-feldspar veins in connection with Mo-W deposits, USSR, have the faces b(010), a(100), l(120), n(210), y(310), v(510), d(011), k($\bar{1}$12), h($\bar{2}$12), see **Fig. 3** [2]. Forms of "Iodochnikite" from [3] (axial ratio see above) in the orientation of Patchett, Nuffield [4] are given by Chukhrov, Bohnshtedt-Kupletskaya [6]. Orthorhombic or tetragonal forms may be the result of pseudomorphism [7], see e.g. brannerite of the Pronto Mine, Blind River district, Ontario, with forms that are due to pseudomorphs of brannerite after rutile and uraninite [8].

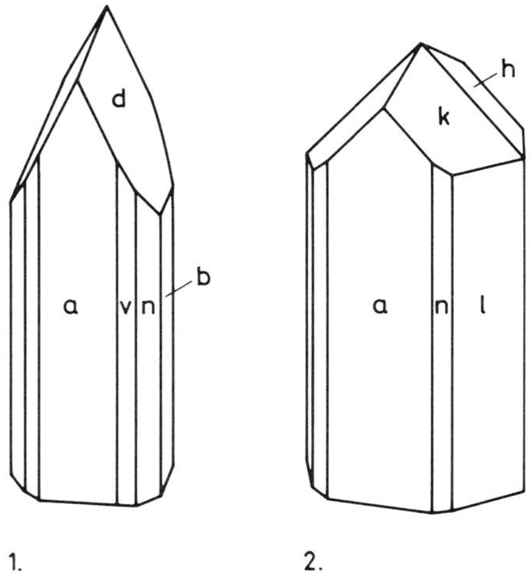

Fig. 3. Crystal forms of brannerite from quartz-feldspar veins, USSR, from [2].
1. Very common.
2. More seldom.

Habit is often prismatic [2], especially stout prismatic [1, 9] or ("Iodochnikite") short prismatic [3]. There also occur slender prisms [5] and needles (together with grains) [10, 11, 12], or needlelike and thinly columnar crystals [13]. Also pinacoidal or irregular elongated crystals [14], rectangular, elongated euhedral crystals [15], laminar crystals and platy grains [16], and often rounded crystals and grains [7]. Literature and new data show that brannerite from high-temperature deposits is deposited in form of prismatic crystals, brannerite from intermediate temperature deposits occurs as formless masses with transitions to finegrained, needlelike aggregates [17].

A deep lengthwise striation is observed on the prism faces of brannerite [2, 18]. No cleavage [7, 19, 20].

Brannerite is **metamict**, therefore X-ray powder pattern can be gained only after heating [6], see also [21, 22]. As probably single occurrence, brannerite from Knappenberg, Carinthia, Austria, is not isotropisized [23].

Cell constants, space group and Z are given as follows for natural brannerites and synthetic UTi_2O_6:

Cell parameter	Locality or source				
	Central Asia	Rajasthan, India	Wölsendorf, Bavaria	synthetic UTi_2O_6	synthetic UTi_2O_6
a in Å	9.81	10.689	9.89	9.87	5.27
b in Å	3.73	13.248	3.70	3.76	3.54
c in Å	6.91	10.258	6.87	6.95	6.31
Angles					
α	—	97°45′	—	—	77°30′
β	118°24′	104°53′	119°10′	119°30′	108°40′
γ	—	110°53′	—	—	100°10′
Space group$^{+)}$	—	$P1\text{-}C_1^1$ or $P\bar{1}\text{-}C_i^1$	—	$C2m\text{-}C_{2h}^3$	$P1\text{-}C_1^1$ or $P\bar{1}\text{-}C_i^1$
Z	—	8	2	2	1
References	[24]	[25]	[26]	[4]	[27]

$^{+)}$ $P1\text{-}C_1^1$ (No. 1), $P\bar{1}\text{-}C_i^1$ (No. 2), and $C2/m\text{-}C_{2h}^3$ (No. 12, standard setting B2/m).

References for 10.2.2.6.3:

[1] D'A. George (RMO-563 [1949] 1/198, 78; N.S.A. **5** [1951] No. 833). – [2] M. M. Povilaitis, T. A. Yakovlevskaya, D. N. Knyazeva, I. D. Belyaeva (Zap. Vses. Mineral. Obshchestva **97** [1968] 150/61, 150, 152/3). – [3] G. A. Il'inskii (Vestn. Leningr. Univ. **1961** No. 6, pp. 134/8 from Ref. Zh. Geol. **1961** No. 12 V 166). – [4] J. E. Patchett, E. W. Nuffield (Can. Mineralogist **6** [1957/61] 483/90, 485, 487). – [5] A. Pabst (Am. Mineralogist **39** [1954] 109/17, 110).

[6] F. V. Chukhrov, E. M. Bohnshtedt-Kupletskaya (Mineraly Spravochnik, Vol. 2, Pt. 3, Nauka, Moscow 1967, pp. 1/676, 433, 435). – [7] F. L. Hess, R. C. Wells (J. Franklin Inst. **189** [1920] 225/37, 225). – [8] P. Ramdohr (Neues Jahrb. Mineral. Monatsh. **1957** 217/22, 217). – [9] A. W. G. Whittle (S. Australia Geol. Surv. Bull. No. 30 [1954] 126/51, 143). – [10] G. Jouravsky (Compt. Rend. **234** [1952] 230/1).

[11] M. Schidlowski (Neues Jahrb. Mineral. Abhandl. **105** [1966] 310/24, 312). – [12] S. M. Roscoe (Geol. Surv. Can. Paper No. 68-40 [1969] 1/205, 136). – [13] P. Ramdohr (Abhandl. Deut. Akad. Wiss. Berlin Kl. Chem. Geol. Biol. **1958** No. 3, pp. 1/35, 25). – [14] J. Lis, M. Stepniewski,

H. Sylwestrzak (Biul. Inst. Geol. [Warsaw] No. 193 [1965] 207/23, 214 [Polish], 227/8 [English Summary]). – [15] J. W. Vanderwilt, R. U. King (Trans. AIME Mining Eng. **7** [1955] 41/53, 48).

[16] A. V. Zavarizin (Geol. Rudn. Mestorozhd. **3** No. 6 [1961] 54/8, 55). – [17] A. A. Chernikov (Sov. Geol. No. 4 [1981] 85/98, 88). – [18] M. M. Povilaitis (Tr. Mineral. Muzeya Akad. Nauk SSSR No. 12 [1961] 64/79, 65). – [19] M. M. Povilaitis (Zap. Vses. Mineral. Obshchestva **92** [1963] 113/23, 114). – [20] J. Geffroy (Bull. Soc. Franc. Mineral. Crist. **86** [1963] 129/32, 130).

[21] V. L. Mel'nikova (Izv. Akad. Nauk Kaz.SSR Ser. Geol. **25** [1968] 62/6, 65/6). – [22] Ya. D. Gotman, V. M. Polyakova, A. K. Miguta (Tr. Mineral. Muzeya Akad. Nauk SSSR No. 20 [1971] 76/87, 81/3). – [23] P. Ramdohr (Z. Deut. Geol. Ges. **130** [1979] 439/58, 445). – [24] L. V. Chernysheva, I. I. Kupriyanova, G. A. Sidorenko (Sb. Kratk. Soobshch. Mineral. Geokhim. **1** [1968/70] 83/8 from Ref. Zh. Geol. **1970** No. 10 V 234). – [25] M. D. Karkhanavala (J. Sci. Ind. Res. [India] B **18** [1959] 7/10).

[26] E. Seeliger, H. Strunz (Neues Jahrb. Mineral. Abhandl. **103** [1965] 163/78, 164). – [27] Ö. Öztunali (Neues Jahrb. Mineral. Monatsh. **1959** 187/8).

10.2.2.6.4 Optical Properties

The luster is vitreous or glassy on fresh surfaces and greasy to resinous on altered samples [1, 2]; it changes with the intensity of alteration from resinous to dull [3, 4, 5] and [2] ("absite"). For vitreous luster see [6, 7], resinous [8, 9], greasy or resinous [7], and fatty luster [10]. Submetallic luster occurs on freshly broken surfaces [11]. The mineral is translucent [8, 12, 13] in very thin sections [10], but also opaque [3, 7].

The color is jet black, when fresh, but usually altered to brown, yellow-brown and olive-green [1, 2], it is nearly metallic black to dark yellowish brown [12, 14], black through brown to light tan [5], jet black and occasionally brown [15], black and olive-green [16], the color may be jet black, when fresh, with lemon-yellow and olive-green shades, when altered [17]. The color is black, black with cinnamon shades or greenish brown [3]. See for black crystals of brannerite [6, 7, 8, 9, 11, 18], with variations of lustreous black [19, 20], resinous black [10], or pitch black [13]. Colors are olive-green to red-brown [7], also yellow-brown, olive-green, and honey colored [2] ("absite"). The streak or powder are dark greenish brown, when fresh, yellowish brown after alteration [1]; black, black with cinnamon shade, greenish brown [3], dark brown [9], dark greenish brown [14], brown-greenish [8], and chocolate brown [17].

The color in very thin splinters is brown-red to colorless in the thinnest splinters [21], yellowish green [14], or pale yellow to brown [7]. Fine needles are yellowish brown [8]. The color in transmitted light is dark brown, when fresh, alteration products are yellowish green [13] or show zones of different intensity [10]. Further colors in transmitted light are pale brown to pale green [12], brownish yellow [17], yellow-green; and (in the case of „absite") greenish yellow [2].

Brannerite is isotropic on account of its being metamict [22] in fresh and altered state [13], only sometimes it shows weak polarization [18]. The refractive index varies with the degree of alteration and decreases with increasing H_2O content [23], see also [7]. The following values in decreasing order are given for analyzed samples:

Locality	Refractive index	Ref.
Château Lambert, Vosges	2.33 ± 0.01	[8]
Central Asia	2.33	[24]
Kelley Gulch, Idaho	2.30 ± 0.02	[14]

Locality	Refractive index	Ref.
San Bernardino, California	2.2 to 2.3	[12]
USSR ("Iodochnikite")	2.11 > n < 2.18	[25]
USSR	2.06 > n < 2.11	[9]
Rajasthan, India	> 2.0	[17]
Nichol's Nob, South Australia	2.0 to 2.2	[16]
Torrens Gorge, South Australia	2.04 to 2.05	[16]
Crocker's Well, South Australia	1.94 to 1.96	[16]

Samples which have not been analyzed such as brannerite from Pronto Mine, Blind River district, Ontario [21], and from Wölsendorf, Bavaria, West Germany [26], have $n \approx 2.2$. Differences in refractive indices of brannerite from various localities before and after heating are given by Adler, Puig [7] (all data with the exception of *) ±0.03):

Degree of heating	Crockers's Well, South Australia	Fuenteovejuna, Spain	Bou Azzer, Morocco	Florence, Montana	Elk City, Idaho
Unheated	1.95	2.21	2.22	2.26	2.26
Heated to 350°C	2.06	2.28	2.23	2.30	2.30
Heated to 960°C	2.2 to 2.3*)	2.39	2.39	2.39	2.39

In reflected light the color of brannerite from India is light jet black if fresh, lemon-yellow or olive-green if altered [17]. Brannerites from 4 localities are light gray with reddish brown inner reflexes [27]; from White Signal, New Mexico, fairly bright, medium gray; in oil immersion red-brown to yellowish at thin edges [6]. The color in reflected light is gray for brannerite from a Mo-W deposit, USSR [10], from a hydrothermal rare-metal deposit in Kazakhstan (here with reddish brown inner reflexes) [18], and from La Gardette, Isère, France, with brown inner reflexes [20]. Further colors of inner reflexes are yellow-brown in brannerites from quartz-pyrite-sericite veinlets [28], and light brownish gray, bluish gray to bluish white, partly with yellow-brown tint in brannerites from Witwatersrand conglomerates, Republic of South Africa [29]. Reflectivity in air in white light 15 to 16% [3], 13.7 to 16.0%, mean 14.8% [27], 13 to 15% [9], 14.9% [18]. Brannerite from La Gardette, Isère, France [20], and from USSR [28] 15%. Brannerite from a Mo-W deposit, USSR, 15.4% in yellow light [10]. Reflective power of brannerite for different wavelengths: white 16.6 to 17.4%, blue 21.7 to 22.1%, green 19.6 to 20.4%, and red 16.6 to 17.4% [30].

Brannerite shows no luminescence [9], and „absite" no fluorescence [2].

References for 10.2.2.6.4:

[1] D'A. George (RMO-563 [1949] 1/198, 78; N.S.A. **5** [1951] No. 833). – [2] A. W. G. Whittle (S. Australia Geol. Surv. Bull. No. 30 [1954] 126/51, 142/3). – [3] M. M. Povilaitis (Zap. Vses. Mineral. Obshchestva **92** [1963] 113/23, 114). – [4] M. M. Povilaitis, T. A. Yakovlevskaya, D. N. Knyazeva, I. D. Belyaeva (Zap. Vses. Mineral. Obshchestva **97** [1968] 150/61, 151). – [5] A. H. Lang, J. W. Griffith, H. R. Steacy (Geol. Surv. Can. Econ. Geol. Ser. No. 16 [1962] 1/324, 45).

[6] C. S. Ferris, C. O. Ruud (Quart. Colo. School Mines **66** [1971] 1/35, 2/3). – [7] H. H. Adler, J. A. Puig (Am. Mineralogist **46** [1961] 1086/96, 1089/90). – [8] G. Branche, J. Chervet, G. Guillemin (Bull. Soc. Franc. Mineral. Crist. **74** [1951] 457/88, 483/4). – [9] Ya. D. Gotman, V. M. Polyakova, A. K. Miguta (Tr. Mineral. Muzeya Akad. Nauk SSSR No. 20 [1971] 76/87, 78). – [10] M. M. Povilaitis (Tr. Mineral. Muzeya Akad. Nauk SSSR No. 12 [1961] 64/79, 66).

[11] J. W. Adams (U.S. Geol. Surv. Bull. No. 982-D [1953] 1/119, 111). – [12] D. F. Hewett, J. Stone, H. Levine (Am. Mineralogist **42** [1967] 30/8, 35/6). – [13] J. Lis, M. Stepniewski, H. Sylwestrzak (Biul. Inst. Geol. [Warsaw] No. 193 [1965] 207/23, 214 [Polish], 227/8 [English Summary]). – [14] F. L. Hess, R. C. Wells (J. Franklin Inst. **189** [1920] 225/37, 225). – [15] D. A. Moddle (Can. Mineralogist **6** [1957/61] 155/7).

[16] A. W. G. Whittle (J. Geol. Soc. Australia **2** [1955] 21/45, 33). – [17] G. V. Umamaheswararao, R. Krishnaswamy (J. Sci. Ind. Res. [India] A **15** [1956] 401/2). – [18] V. L. Mel'nikova (Izv. Akad. Nauk Kaz.SSR Ser. Geol. **25** [1968] 62/6, 63). – [19] A. Pabst (Am. Mineralogist **39** [1954] 109/17, 110). – [20] J. Geffroy (Bull. Soc. Franc. Mineral. Crist. **86** [1963] 129/32, 130).

[21] P. Ramdohr (Neues Jahrb. Mineral. Monatsh. **1957** 217/22, 217). – [22] F. V. Chukhrov, E. M. Bohnshtedt-Kupletskaya (Mineraly Spravochnik, Vol. 2, Pt. 3, Nauka, Moscow 1967, pp. 1/676, 435). – [23] E. W. Heinrich (Mineralogy and Geology of Radioactive Raw Materials, McGraw-Hill, New York 1958, pp. 1/654, 51). – [24] L. V. Chernysheva, I. I. Kupriyanova, G. A. Sidorenko (Sb. Kratk. Soobshch. Mineral. Geokhim. **1** [1968] 83/8 from Ref. Zh. Geol. **1970** No.10 V 234). – [25] G. A. Il'inskii (Vestn. Leningr. Univ. **1961** No. 6, pp. 134/8 from Ref. Zh. Geol. **1961** No. 12 V 166).

[26] E. Seeliger, H. Strunz (Neues Jahrb. Mineral. Abhandl. **103** [1965] 163/78, 164). – [27] S. H. U. Bowie, K. Taylor (Proc. 2nd Intern. Conf. Peaceful Uses At. Energy, Geneva 1958, Vol. 3, pp. 527/40, 528, 537). – [28] A. V. Zavarizin (Geol. Rudn. Mestorozhd. **3** No. 6 [1961] 54/8, 55). – [29] M. Schidlowski (Neues Jahrb. Mineral. Abhandl. **105** [1966] 310/24, 314). – [30] I. M. Gray, A. P. Millman (Econ. Geol. **57** [1962] 325/49, 332).

10.2.2.6.5 Other Physical Properties

Hardness, H_M varies with the degree of alteration and therefore shows a wide range of values increasing from 4 to 6.5 in brannerites from different localities [1]. Some examples are as follows:

H_M	Locality	Ref.
4	Eastern Ontario	[2]
4 to 5	Torrens Gorge, South Australia	[3]
	Fuenteovejuna, Spain	[4]
	Åmot, Modum district, Norway	[27]
4.5	Kelley Gulch, Idaho	[5]
≈5	Rajasthan, India	[6]
	USSR	[7]
5	Château Lambert, Vosges	[8]
	San Bernardino, California	[9]
	Central Asia	[10]
5.1	USSR	[11]
5 to 6	Nichol's Nob, South Australia	[3]
	Crocker's Well, South Australia	[3]
5.5 to 6	La Gardette, Isère, France	[12]
5.6	Crocker's Well, South Australia ("absite")	[13]
6	Southwestern New Mexico	[14]

Microhardness, HV in kg/mm² of brannerite (load 100 g) is: 572 to 693 [15], 710 to 730, mean of samples from 4 localities 720 [16], 387 [11]. Load 70 to 100 g: 611 to 662 in samples with conchoidal fracture and resinous luster; 521 to 636 in samples with uneven fracture and dull luster [17].

Fracture of brannerite is mostly conchoidal, probably because of its metamict state [14], see e.g. [4, 6, 8, 9, 12, 18]. It is also subconchoidal [2], flat conchoidal or uneven [19], or subconchoidal to irregular [13] ("absite"). Lamellar crystals break after the prism [11]. The mineral is very brittle [2] and [13] ("absite"), or brittle [11].

Density changes with variation of chemistry, hydration, and inclusions [14]. For dependence of D on degree of alteration, especially hydration, see [4, 13]; increase of D from 5.49 to 5.70 g/cm³ with increase of U [17]. High-temperature brannerites with high contents of Th and RE have D = 5.1 to 5.9 g/cm³, while hydrothermal brannerites with very low contents of Th and RE have D = 4.2 to 4.9 [20], cf. pp. 50 and 53.

A compilation of analyses and of D values of 17 (RE $\pm$ Y)-containing brannerites results in a variance of D from 4.2 to 5.88, mean 5.19 g/cm³ [1]. Further determinations of D of analyzed brannerites yield the following values in g/cm³: brannerites from South Australia, Crocker's Well ("absite") 4.0 to 4.2; Torrens Gorge 4.2; and Nichol's Nob 4.3 [3]; hydrothermal brannerites from USSR 4.32 [11] and 4.7 [7]; brannerites from tremolite rock, eastern Ontario, 4.85 [2]; from feldspar veins, Central Asia, 4.95 [10]; and from a Mo-W deposit, USSR, 5.49 and 5.70 [17]. Further data in g/cm³ for samples not analyzed in increasing order are 2.26 $\pm$ 0.03, Florence, Montana; 4.13, Crocker's Well, South Australia [21]; 4.39, Wolowa Gora, near Kowary, Sudetes [18]; 4.9, Plumas County, California [22]; 5.01, La Gardette, Isère, France [12]; and 5.17, Fuenteovejuna, Spain [21]. Synthetic brannerite, UTi_2O_6, D(meas.) = 6.31 to 6.40, mean of three determinations 6.35 g/cm³, D(calc.) = 6.36 g/cm³ [23]; D(meas.) = 6.1, D(calc.) = 6.00 g/cm³ [24].

Magnetism was only seldom determined. Brannerite from a rare metal deposit in Kazakhstan is weakly magnetic [7], brannerite from a Mo-W deposit, USSR, electromagnetic [19].

Radioactivity is very strong [25, 26, 27].

References for 10.2.2.6.5:

[1] M. M. Povilaitis (Zap. Vses. Mineral. Obshchestva **92** [1963] 113/23, 114, 115/7). – [2] D. A. Moddle (Can. Mineralogist **6** [1957/61] 155/7). – [3] A. W. G. Whittle (J. Geol. Soc. Australia **2** [1955] 21/45, 33). – [4] D'A. George (RMO-563 [1949] 1/198, 78; N.S.A. **5** [1951] No. 833). – [5] F. L. Hess, R. C. Wells (J. Franklin Inst. **189** [1920] 225/37, 225).

[6] G. V. Umamaheswararao, R. Krishnaswamy (J. Sci. Ind. Res. [India] A **15** [1956] 401/2). – [7] V. L. Mel'nikova (Izv. Akad. Nauk Kaz.SSR Ser. Geol. **25** [1968] 62/6, 63). – [8] G. Branche, J. Chervet, C. Guillemin (Bull. Soc. Franc. Mineral. Crist. **74** [1951] 457/88, 483). – [9] D. F. Hewett, J. Stone, H. Levine (Am. Mineralogist **42** [1967] 30/8, 35). – [10] L. V. Chernysheva, I. I. Kupriyanova, G. A. Sidorenko (Sb. Kratk. Soobshch. Mineral. Geokhim. **1** [1968] 83/8 from Ref. Zh. Geol. **1970** No. 10 V 234).

[11] Ya. D. Gotman, V. M. Polyakova, A. K. Miguta (Tr. Mineral. Muzeya Akad. Nauk SSSR No. 20 [1971] 76/87, 78). – [12] J. Geffroy (Bull. Soc. Franc. Mineral. Crist. **86** [1963] 129/32, 130). – [13] A. W. G. Whittle (S. Australia Geol. Surv. Bull. No. 30 [1954] 126/51, 142/3). – [14] C. S. Ferris, C. O. Ruud (Quart. Colo. School Mines **66** [1971] 1/35, 2/3). – [15] B. B. Young, A. P. Millman (Bull. Inst. Min. Met. No. 689 [1964] 437/66, 448).

[16] S. H. U. Bowie, K. Taylor (Proc. 2nd Intern. Conf. Peaceful Uses At. Energy, Geneva 1958, Vol. 3, pp. 527/40, 532). – [17] M. M. Povilaitis, T. A. Yakovlevskaya, D. N. Knyazeva, I. D. Belyaeva (Zap. Vses. Mineral. Obshchestva **97** [1968] 150/61, 152, 157). – [18] J. Lis, M. Step-

niewski, H. Sylwestrzak (Biul. Inst. Geol. [Warsaw] No. 193 [1965] 207/23, 214 [Polish], 277/8 [English Summary]). – [19] M. M. Povilaitis (Tr. Mineral. Muzeya Akad. Nauk SSSR No. 12 [1961] 64/79, 66). – [20] A. A. Chernikov (Sov. Geol. No. 4 [1981] 85/98, 88).

[21] H. H. Adler, J. A. Puig (Am. Mineralogist **46** [1961] 1086/96, 1089/90). – [22] A. Pabst, M. C. Stinson (Bull. Geol. Soc. Am. **71** [1960] 2071/2). – [23] J. E. Patchett, E. W. Nuffield (Can. Mineralogist **6** [1957/61] 483/90, 486/7). – [24] Ö. Öztunali (Neues Jahrb. Mineral. Monatsh. **1959** 187/8). – [25] F. V. Chukhrov, E. M. Bohnshtedt-Kupletskaya (Mineraly Spravochnik, Vol. 2, Pt. 3, Nauka, Moscow 1967, pp. 1/676, 435).

[26] P. Ramdohr (Neues Jahrb. Mineral. Monatsh. **1957** 217/22, 217). – [27] T. Van Autenboer, F. J. Skjerlie (Norg. Geol. Unders. No. 200 [1957] 5/7).

10.2.2.6.6 Chemical Behavior

Brannerite dissolves in hot nitric and concentrated sulfuric acid, it is insoluble in a mixture of hydrochloric and nitric acid [1], see also [2]. It melts in $K_2S_2O_7$, and also with $KHSO_4$ and $NaHSO_4$ [3]. For etching behavior see [3, 4]. In nature brannerite is almost stable under varying pH conditions, it breaks down under the action of F [5].

Thermal curves of brannerite from 5 different localities (Fuenteovejuna, Spain; Crocker's Well, South Australia; Bou Azzer, Morocco; Florence, Montana, and Elk City, Idaho) show endothermic peaks between 100 and 300°C, due to release of water, and two exothermic peaks above 550°C, due to recrystallization. Dissimilarities of these peaks in position, shape, area, and amplitude characterize the samples from each locality [6]. Exothermal effects at 660 and 710°C for a brannerite from Central Asia are due to recrystallization of the metamict mineral (660°C) and to oxidation of U^{4+} (710°C) [7]. Heating curves of brannerite from a Mo-W deposit, USSR, show an endothermic peak at 180°C and one exothermic peak at 670°C [4, 8]; heating curves of brannerite deposited hydrothermally as cement of breccias have an endothermic effect at 200°C and an exothermic peak between 640 and 700°C [2].

A **thermogravimetric analysis** of one of the samples from the Mo-W deposit, USSR, shows that the weight changes in the temperature intervals 100 to 400°C and 700 to 1100°C [8].

Heating tests under the microscope show alteration of a metamict brannerite between 200 and 600°C; further heating causes the material to darken and to become completely opaque at 1000°C [9].

References for 10.2.2.6.6:

[1] M. V. Soboleva, I. A. Pudovkina (AEC-TR-4487 [1961] 1/455, 398 [Russian Original, Moscow 1957]; N.S.A. **15** [1961] No. 23739). – [2] Ya. D. Gotman, V. M. Polyakova, A. K. Miguta (Tr. Mineral. Muzeya Akad. Nauk SSSR No. 20 [1971] 76/87, 78, 83). – [3] F. F. Chukhrov, E. M. Bohnshtedt-Kupletskaya (Mineraly Spravochnik, Vol. 2, Pt. 3, Nauka, Moscow 1967, pp. 1/676, 437). – [4] M. M. Povilaitis (Tr. Mineral. Muzeya Akad. Nauk SSSR No. 12 [1961] 64/79, 66, 71). – [5] A. V. Zavarizin (Izv. Vysshikh Uchebn. Zavedenii Geol. Razvedka **23** No. 3 [1980] 54/62 from C.A. **93** [1980] No. 29333).

[6] H. H. Adler, J. A. Puig (Am. Mineralogist **46** [1961] 1086/96, 1090/6). – [7] L. V. Chernysheva, I. I. Kupriyanova, G. A. Sidorenko (Sb. Kratk. Soobshch. Mineral. Geokhim. **1** [1968/70] 83/8 from Ref. Zh. Geol. **1970** No. 10 V 234). – [8] M. M. Povilaitis, T. A. Yakovlevskaya, D. N. Knyazeva, I. D. Belyaeva (Zap. Vses. Mineral. Obshchestva **97** [1968] 150/61, 157/8, 159). – [9] J. Lis, M. Stepniewski, H. Sylwestrzak (Biul. Inst. Geol. [Warsaw] No. 193 [1965] 207/23, 216 [Polish], 227/8 [English Summary]).

10.2.2.7 Loparite $(Na,Ce,Ca)(Ti,Nb)O_3$

For a mineral description see "Tantal" A 1, 1969, pp. 132/4. For a compilation of analyses and $(RE,Y)_2O_3$ content, mean and variance of 57 analyses, see "Seltenerdelemente" A 4, 1979, pp. 24, 7.

10.2.2.8 Dysanalyte $(Ca,Na,Ce)(Ti,Nb,Fe)O_3$

For a mineral description see "Tantal" A 1, 1959, pp. 130/2. For a compilation of analyses and content of $(RE,Y)_2O_3$, mean and variances of 25 analyses, see "Seltenerdelemente" A 4, 1979, pp. 18, 9.

10.2.2.9 Latrappite $(Ca,Na,RE)_{0.98}(Nb,Ti,Fe,Mg,Mn)_{1.01}O_3$

For a mineral description see "Tantal" A 1, 1969, pp. 130/2. For the formula and one analysis, see E. H. Nickel, R. C. McAdam (Can. Mineralogist **7** [1962/63] 683/97, 689). For occurrences see also D. D. Gold (Papers Proc. General Meeting Intern. Mineral Ass. 4th, New Delhi 1964, pp. 109/25, 117).

10.2.2.10 Murataite $(Na,Y,Er)_4(Zn,Fe)_3(Ti,Nb)_6O_{18}(O,OH)_4$

Murataite was found in a pegmatite in the St. Peters Dome area, El Paso County, Colorado. The analysis of this murataite leads to the ideal formula given above [1]. A second find of murataite was in the oligoclase zone of rare-metal alkali syenite pegmatites of the Burpala massiv, northern Pribaikalia. The mineral here occurs as thin cubic crystals or rounded grains together with landauite and brookite. The analysis of this murataite yields the formula $(Na_{1.89}Y_{1.81}Yb_{0.04}Fe_{0.32})(Zn_{1.98}Ca_{1.19})Ti_6O_{19.19}$ or generalized $(Na,Y,Yb,Fe)_4(Zn,Ca)_3Ti_6O_{20}$ [2]. Both samples have a predominant heavy rare earth composition [1, 2]. The RE distribution of murataite from El Paso County is Gd 16.2, Dy 19.9, Ho 5.1, Er 24.4, Tm 3.1, Yb 25.5, and Lu 2.9% [1].

The mineral is cubic, and has a = 14.863(5) Å, Z = 8, possible space groups are Fm3m-O_h^5 (No. 225), F432-O^3 (No. 209) or Fm3-T_h^3 (No. 202), V = 3283 $Å^3$ [1]; murataite from the Burpala massif has a = 14.87 ± 0.02 Å [2].

The color is black, and the luster metallic. In thin section the mineral is translucent, brown, sometimes with very weak anisotropism. The refractive index n(estimated) = 2.13. In polished section it is gray with reflection pleochroism slight in air, but distinct in oil. Reflectance (546 nm wavelength) in air = 13.6 ± 0.14% [1]. In thin sections murataite is isotropic, transparent, color light yellow, n = 2.11 [2]. The fracture is conchoidal [1, 2]. H_M = 5 [2]. Microhardness HV (100 g load) = 782 to 870, mean 827 kg/mm^2, equivalent to H_M = 6.0 to 6.5. D(meas.) = 4.69 g/cm^3, D(calc.) = 4.64 g/cm^3 [1]; D(meas.) = 4.50 g/cm^3 [2].

References for 10.2.2.10:

[1] J. W. Adams, T. Botinelly, W. N. Sharp, K. Robinson (Am. Mineralogist **59** [1974] 172/6). – [2] A. M. Portnov, L. S. Dubakina, G. K. Krivokoneva (Dokl. Akad. Nauk SSSR **261** [1981] 741/4).

10.2.2.11 **Zirconolite** $\approx(Zr,Ca,RE,Fe^{2+},U,Th)_2(Ti,Nb,Fe^{3+})_2O_7$

For a mineral description see "Tantal" A 1, 1969, p. 209. After D. D. Hogarth (Am. Mineralogist **62** [1977] 403/10, 408) this mineral is synonymous with zirkelite. For a compilation of analyses and RE_2O_3 and Y_2O_3 contents of both minerals (39 analyses) see "Seltenerdelemente" A 4, 1979, pp. 36, 9.

10.2.3 Complex Oxides with Nb, Ta ± Ti

10.2.3.1 **Koppite** $(Ca,Ce)_2(Nb,Fe)_2O_6(O,OH,F)$

Koppite is a variety of pyrochlore; Fe_2O_3 found in the analysis of Brandenberger [1] is probably due to contamination. The name should be dropped [2]. For a mineral description see "Seltenerdelemente" A 1, 1938, p. 98; for koppite from the carbonatite in the Kaiserstuhl, southern Baden, West Germany, see [3]. For a compilation of analyses and of $(RE,Y)_2O_3$ content, mean and variance of 11 analyses, see "Seltenerdelemente" A 4, 1979, pp. 23, 9.

References for 10.2.3.1:

[1] E. Brandenberger (Z. Krist. **76** [1931] 322/34, 324, 328). – [2] D. D. Hogarth (Am. Mineralogist **62** [1977] 403/10, 406). – [3] W. Wimmenauer (Lapis **3** No. 3 [1978] 14/7).

10.2.3.2 **Brocenite** $CeNbO_4$

Brocenite is a Ce-analogue of fergusonite [1, 2], it should be called beta-fergusonite-(Ce) [3]. The name β-brocenite for the non-metamict equivalent [4] is unnecessary [2].

Occurrence and Paragenesis

Brocenite was first found in a skarn in contact with granitic rocks in northern China (special locality not given) associated with phlogopite, Ce-apatite, diopside, topaz, and calcite [4]. Another locality for brocenite are olivine-bearing phlogopite-calcite and phlogopite-dolomite-calcite carbonatites of the Chernigovsk zone of the Ukrainian Shield, where it occurs in anhedral grains as inclusions of micas and carbonates, at grain boundaries, and intergrown with magnetite and monazite [5], see also [1]; it is associated with pyrochlore, columbite, monazite, and baddeleyite [6].

Chemistry

For a compilation of brocenite analyses and $(RE,Y)_2O_3$ content, variance and mean of 3 analyses, see "Seltenerdelemente" A 4, 1979, pp. 16, 6. An analysis of brocenite of the USSR corresponds to the formula $(RE_{0.98}Ca_{0.06}U_{0.01}Th_{0.01})(Nb_{0.94}Ta_{0.01}Fe^{3+}_{0.03})O_4$ [1, 5], cf. [6]. The mineral contains as rare elements in %: Al 0.15, Cr 0.1, Mn 0.06, Pb 0.04, Zr 0.06, Sc 0.01, and Cu 0.004 [1, 5]. Distribution of rare earth elements in %:

Locality	La	Ce	Pr	Nd	Sm	Eu	Ref.
USSR	18.1	49.3	5.9	23.2	1.86	0.37	[1, 5]
China	24.0	42.7	7.2	21.5	2.69	0.49	[4]

Locality	Gd	Tb	Dy	Ho	Yb	Lu	Ref.
USSR	0.60	0.14	0.30	0.06	0.14	0.03	[1, 5]
China	1.04	—	0.36	—	—	—	[4]

In the Russian samples Er and Tm could not be found [1, 5].

Morphology and Crystal Structure

Brocenite is monoclinic [5], it forms pyramidal, euhedral crystals [4]. The mineral is X-ray amorphous, after heating the following unit cell parameters can be calculated:

Heating temperature	a in Å	b in Å	c in Å	β	Ref.
700°C	5.170	11.372	5.497	85°20′	[1, 5, 6]
950°C	5.174	11.384	5.517	85°40′	[5]
1000°C	5.19	11.34	5.48	84°57′	[4]

Brocenite is isostructural with fergusonite [4], see also p. 60.

Optical Properties

Luster is adamantine for a Russian specimen [1, 5, 6], vitreous to greasy for a Chinese sample [4]. The color is red [1] to dark red [5, 6], and red to reddish brown [4]; calcinated material is brown to dark brown [5]. The streak is weakly rose [5, 6]. Under the microscope brocenite is transparent or semitransparent and brownish red [1, 5, 6]; it is isotropic or anomalously birefringent [5] with $n = 2.155 \pm 0.005$ [1, 6], increasing after calcination to 700 and 950°C [5].

Specific bands of the IR spectra of the mineral in natural state are at 660, 810, and 450 cm^{-1}, and at 1620 and 3450 cm^{-1} [5], see also [1, 6]. After calcination the first three absorption bands become more distinct, due to recrystallization of the originally metamict mineral [5].

Other Physical Properties

Microhardness with load 100 g = 778 to 845 kg/mm^2 [1], mean 810 kg/mm^2, corresponds to $H_M \approx 6$ [5]. The fracture is conchoidal [1, 5, 6]. The mineral is brittle [5]. The density D(meas.) = 5.34 g/cm^3 (Chinese mineral) [4], 5.44 g/cm^3 (Russian mineral) [1, 5, 6]; D(calc.) = 5.99 g/cm^3 (heated sample) [5].

Dependence of specific magnetic susceptibility on temperature see [1].

Chemical Behavior and Differential Thermal Analysis

Brocenite is insoluble in acids [5, 6]. The thermal curve of the mineral shows two exothermic effects with maximum at 365 and 675°C, due to the stagewise transition from metamict to crystalline state. An endothermic effect at 100°C is caused by the loss of molecular water [1, 5, 6].

References for 10.2.3.2:

[1] E. Ya. Marchenko, A. I. Chashka, V. A. Khvostova, A. V. Bykova (Dokl. Akad. Nauk SSSR **225** [1975] 1168/70; Dokl. Earth Sci. Sect. **225** [1975] 131/3). – [2] G. Y. Chao (from M. Fleischer, G. Y. Chao, A. Kato, Am. Mineralogist **60** [1975] 485/9, 485). – [3] M. Fleischer (from M. Fleischer, G. Y. Chao, A. Kato, Am. Mineralogist **60** [1975] 485/9, 485). – [4] C.-T. Kuo, I.-H. Wang, H.-C. Wang, C.-K. Wang, H.-C. Hou (Ti Ch'iu Hua Hsueh No. 2 [1973] 86/92 from Am. Mineralogist **60** [1975] 485). – [5] A. I. Chashka, E. Ya. Marchenko, V. A. Khvostova, A. V. Bykova (Zap. Vses. Mineral. Obshchestva **105** [1976] 457/62, 457/8, 460/1).

[6] A. I. Chashka, E. Ya. Marchenko (Geol. Geofiz. **1976** No. 4, pp. 141/3).

10.2.3.3 **Fergusonite** $YNbO_4$

10.2.3.4 **Formanite** $YTaO_4$

For a mineral description of the two modifications fergusonit-α (tetragonal) and fergusonite-β (monoclinic), as well as of formanite, see "Tantal" A 1, 1969, pp. 175/82. For a compilation of analyses and variance and mean of Y_2O_3 and $(RE,Y)_2O_3$ of both modifications of fergusonite and of formanite, 61, 2, and 5 analyses, respectively, see "Seltenerdelemente" A 4, 1979, pp. 19/20, 6, 7.

10.2.3.5 **Ashanite** $(Mn,RE,U,Nb,Ta,Fe)_4O_8$

Ashanite was found in the central part of a pegmatite in albitized two-mica granite in the Altai Mountains, NW China. It is associated with Ta-rich columbite, Th-rich monazite, and ishikawaite. An analysis with 1.40% RE_2O_3 corresponds to the crystal chemical formula $(Nb_{1.61}Ta_{0.61}Fe_{0.66}Mn_{0.63}U_{0.51}RE_{0.07})O_{7.98}$. The mineral is the Nb-analogue of ixiolite.

Ashanite is orthorhombic. Cell parameters a = 5.869, b = 4.873, c = 5.216 Å, Z = 1, space group Pcan-D_{2h}^{14} (No. 60, standard setting Pbcn).

The luster is semimetallic to pitch-like, the mineral is dark in color, with brownish streak. Under the microscope it is brownish red, with pleochroism X light brownish red, Z dark brownish red. Optically biaxial, positive; $n_x = 2.31$, $n_y = 2.35$, $n_z = 2.40$, 2V = 70° to 75°. Orientation X = b, Y = a, Z = c. Reflectance at 580 nm = 10.8%. The IR spectrum shows a slight maximum at about 1000 cm^{-1}.

Microhardness measurements ranged from 570.1 to 851 kg/mm^2, this corresponds to $H_M = 5.8$ to 6.6. The fracture is conchoidal. Density D(meas.) = 6.61 ± 0.01 g/cm^3, D(calc.) = 6.60 g/cm^3.

Reference for 10.2.3.5:

Z. Rubo, T. Huixin, P. Zhizhong, M. Zhesheng, H. Fengming, J. Zebei (Kexue Tongbao **25** [1980] 510/4 from Am. Mineralogist **66** [1981] 217).

10.2.3.6 **Yttrotantalite** $(Y,U,Ca)(Ta,Fe^{3+})_2(O,OH)_6$

Information concerning Ta(Nb) content and mineral description is found in "Tantal" A 1, 1969, pp. 31, 173/4; compilation of analyses, formula, and Y_2O_3 and $(RE,Y)_2O_3$ contents, mean and variance of 6 analyses, are located in "Seltenerdelemente" A 4, 1979, pp. 35, 7.

10.2.3.7 **Samarskite** $(Y,U,Ca)(Nb,Fe^{3+})_2(O,OH)_6$

For a mineral description of samarskite and its varieties, among them ishikawaite, and for the contents of Nb(Ta,Ti) and of $(RE,Y)_2O_3$ in minerals of the samarskite group, see "Tantal" A 1, 1969, pp. 31, 166, 168/72. For mean and variance of Y_2O_3 and $(RE,Y)_2O_3$ in 10 and 72 samples, respectively, of this group, see "Seltenerdelemente" A 4, 1979, pp. 8, 29/30.

10.2.3.8 **Fersmite** $(Ca,Ce,Na)(Nb,Ti,Fe,Al)_2(O,OH)_6$

For a mineral description see "Tantal" A 1, 1969, pp. 152/8. For a compilation of analyses and $(RE,Y)_2O_3$ content, variance and mean of 6 samples, see "Seltenerdelemente" A 4, 1979, pp. 20, 9.

10.2.4 Complex Oxides with Nb, Ta, and Ti

10.2.4.1 Yttrobetafite $(Y,U,Ca)_{2-x}(Nb,Ta,Ti)_2(O,OH)_7$

For a compilation of analyses and for Y_2O_3 and $(RE,Y)_2O_3$ contents, variance and mean of 1 and 4 analyses, respectively, see "Seltenerdelemente" A 4, 1979, pp. 35, 8.

10.2.4.2 Schetelegite $(Ca,Y,Fe,Mn)_2(Ti,Ta,Nb,Sb,W)_2(O,OH)_7$

See "Seltenerdelemente" A 1, 1938, p. 100. The mineral may possibly belong to the betafite subgroup of the pyrochlore group, D. D. Hogarth (Am. Mineralogist **62** [1977] 403/10, 407).

10.2.4.3 Yttromicrolite $(Y,RE,Ca,Mn,Mg,U,Pb,Th)_{2-m}(Ta,Nb,Ti,Fe^{3+},Sn,W)_2(O,OH,F)_{1-n}$

Yttromicrolite occurs in a granite pegmatite near the Kårarfvet Mine, Falun, Sweden, associated with allanite, gadolinite, tantalite, albite, muscovite, garnet, microcline, and quartz; it was observed within the replacement unit of the pegmatite, and was earlier named hjelmite [1]; see "Seltenerdelemente" A 1, 1938, p. 92. It is classified as yttromicrolite by [1], corresponding to the classification of [2].

Calculation of one analysis gives the crystal chemical formula $(Y_{0.431}RE_{0.251}Ca_{0.251}Mn_{0.123}Mg_{0.062}U_{0.082}Pb_{0.021}Th_{0.013})(Ta_{1.036}Nb_{0.661}Fe^{3+}_{0.141}Sn_{0.131}Ti_{0.028}W_{0.003})(O_{0.072}OH_{0.934})$ which is idealized to the formula given in the heading [1].

The mineral occurs as dense masses or single crystals of octahedral habit with {111}, {001}, {010}, and rarely {210}. It has no observable cleavage. Cubic cell parameter a = 10.179(1) Å, V = 1054.7 $Å^3$, space group is $Fd3m-O_h^7$ (No. 227), which is consistent with the pyrochlore group [1].

Yttromicrolite is black, but appears faint green-black in thin fragments. The streak is greenish-black. The crystals are isotropic, occasionally displaying weak anomalous birefringence in nonmetamict material with amber, pale brown to brown zonal colors. Refractive index n = 2.26(2) to 2.34(2) [1].

Hardness H_M = 5 to 5.5. The fracture is granular to subconchoidal. D(meas.) = 5.565(5) to 6.412(5) g/cm³, it is variable due to the degree of metamictization; D(calc.) = 6.530 g/cm³ [1].

This type "yttromicrolite" has been discredited by the Commission on New Minerals and Mineral Names of the I.M.A., as Peacor et al. have shown it to be an amorphous mixture of calcium sulfate, tantalite, and heterogeneous microlite. Especially the determination of Y seems to be in error [3].

References for 10.2.4.3:

[1] W. W. Crook III (Am. Mineralogist **64** [1979] 890/2). – [2] D. D. Hogarth (Am. Mineralogist **62** [1977] 403/10, 405). – [3] D. R. Peacor, W. B. Simmons Jr., E. J. Essene, E. W. Heinrich (Am. Mineralogist **67** [1982] 156/69, 164/5).

10.2.4.4 Yttropyrochlore (Obruchevite) $(Y,Na,Ca)_2(Nb,Ti,Ta)_2O_6(O,OH,F)$

The name obruchevite was given to two different species; the cinnamon colored mineral was named samiresite [1]; the other, a black mineral, was renamed yttropyrochlore by Kupriyanova [2], see also [3].

For a mineral description, under the name obruchevite, see data for pyrochlore in "Tantal" A 1, 1969, pp. 191/203. For a compilation of chemical analyses and of Y_2O_3 and $(RE,Y)_2O_3$ contents, mean and variance of 6 analyses, see yttropyrochlore in "Seltenerdelemente" A 4, 1979, pp. 35, 8.

References for 10.2.4.4:

[1] C. A. Gorzhevskaya, G. A. Sidorenko (Tr. Mineral. Muzeya Akad. Nauk SSSR No. 19 [1969] 138/45, 144). – [2] I. I. Kupriyanova (from [3]). – [3] D. D. Hogarth (Am. Mineralogist **62** [1977] 403/10, 407).

10.2.4.5 Ceriopyrochlore (Marignacite) $(Ce,Ca,Na)_2(Nb,Ta,Ti)_2O_6(O,OH,F)$

Marignacite is ceriopyrochlore with $\Sigma Ce > \Sigma Y$, where ΣCe = La through Eu, ΣY = Y + (Gd through Lu), D. D. Hogarth (Am. Mineralogist **62** [1977] 403/10, 404/6). For the occurrence see "Tantal" A 1, 1969, pp. 72, 191. For a compilation of analyses and $(RE,Y)_2O_3$ content of 11 analyses, see "Seltenerdelemente" A 4, 1979, pp. 16/7, 8.

10.2.4.6 Pisekite $(Y,Ce,Ca,As,Fe,U)(Nb,Ti,Ta)(O,OH)_4$

See "Tantal" A 1, 1969, pp. 172/3. A new microprobe analysis of the type material from Pisék, southern Bohemia, gives $\Sigma Y_2O_3 = 9.3\%$ and $\Sigma Ce_2O_3 = 4.2\%$. The analysis corresponds to the formula

$$(Y_{0.24}Ce_{0.08}As_{0.25}Ca_{0.13}Fe_{0.09}Al_{0.03}Th_{0.02}Sn_{0.02}Mg_{0.01}Sc_{0.01})(Nb_{0.78}Ti_{0.13}Ta_{0.08}W_{0.02})(O_{3.75}OH_{0.25}).$$

0.21% Tb_2O_3, 0.94% Yb_2O_3, 0.13% Lu_2O_3, 1.58% La_2O_3, and 2.51% Ce_2O_3 are determined by neutron activation analysis.

After hydrothermal recrystallization of the core of greenish brown pisekite, the monoclinic cell has a = 6.570, b = 6.840, c = 6.217 Å, $\beta = 95°25'$, space group $P2_1/c\text{-}C^5_{2h}$ (No. 14, standard setting $P2_1/b$); Z = 4. D(meas.) after recrystallization = 5.89, D(calc.) = 5.84 g/cm³. The rims of pisekite crystals, recrystallized by heating, are composed of a mixture of phases. The DTA curve of pisekite has endothermic peaks at 160 and 420°C, due to loss of water, and an exothermic peak at 740°C, due to recrystallization. These data correspond to the DTG curve. An abrupt decrease of weight, partly corresponding to a weak endothermic peak on the DTA curve, is interpreted as loss of As_2O_3 [1].

Ramdohr and Strunz have proposed that pisekite is an isotropized monazite with Nb, Ta, Ti, U, Y, Yb, and Th [2].

References for 10.2.4.6:

[1] V. L. Bouška, J. Johan (Lithos **5** [1972] 93/103). – [2] P. Ramdohr, H. Strunz (Klockmanns Lehrbuch der Mineralogie, 16th Ed., Enke, Stuttgart 1978, pp. 1/877, 626).

10.2.4.7 Euxenite $(Y,Ce,Ca,U,Pb)(Nb,Ta,Ti)_2(O,OH)_6$

10.2.4.8 Tanteuxenite $(Y,Ce,Ca,U,Pb)(Ta,Nb,Ti)_2(O,OH)_6$

10.2.4.9 Polycrase $(Y,Ce,Ca,U,Th)(Ti,Nb,Ta)_2(O,OH)_6$

The three minerals differ by their contents of Nb, Ta, and Ti, each of which may be predominant. The contents of $(RE,Y)_2O_3$ of these three minerals are similar; for the means varying from 23.8% for 36 samples of euxenite to 24.9% for 35 samples of polycrease, and to 25.2% for 8 samples of tanteuxenite, see "Seltenerdelemente" A 4, 1979, pp. 7, 18/9, 28, 31. For contents of Ta, Nb, Ti, and Ca, U, Th, Na, K, Mg, Mn, Fe, Pb, and for a mineral description, see "Tantal" A 1, 1969, pp. 29, 151, 154/8.

10.2.4.10 Kobeite $(Y,U)(Ti,Zr,Nb,Ta)_2(O,OH)_6$

For a mineral description see "Tantal" A 1, 1969, p. 210. For $(RE,Y)_2O_3$ content, variance and mean of 3 analyses, see "Seltenerdelemente" A 4, 1979, pp. 7, 23.

10.2.4.11 Blomstrandine (Priorite) $(Y,Ce,Th,Ca,Na,U)(Ti,Nb,Ta)_2O_6$

For the content of Nb, Ta, Ti, and of U, Th, Na, K, Mg, Mn, Fe, Pb, and for a mineral description see "Tantal" A 1, 1969, pp. 29, 151, 163/5. For a compilation of analyses and for $(RE,Y)_2O_3$ content, mean and variance of 28 analyses, see "Seltenerdelemente A 4, 1979, pp. 14, 7.

10.2.4.12 Eschynite $(Ce,Th,Ca)(Ti,Nb,Ta)_2O_6$

For the content of Nb, Ta, and of Ca, RE_2O_3, U, Th, Na, K, Mg, Mn, Fe, Pb, and for a mineral description of eschynite and its varieties lyndochite and Ta-, Nb-, Al-, and Th-eschynite, see "Tantal" A 1, 1969, pp. 29, 151, 158. For a compilation of analyses and for contents of Y_2O_3 and $(RE,Y)_2O_3$, mean and variance of 58 analyses of these minerals, see "Seltenerdelemente" A 4, 1979, pp. 10, 7.

10.2.4.13 Polymignite $(Ce,La,Y,Th,Mn,Ca)(Ti,Zr,Nb,Ta)_2O_6$

For the Nb, Ta, Ti content see „Tantal" A 1, 1969, p. 38, for a mineral description see "Seltenerdelemente" A 1, 1938, p. 104, and also "Tantal" A 1, 1969, pp. 208, 209.

10.2.4.14 Vigezzite $(Ca,Ce)(Nb,Ta,Ti)_2O_6$

Occurrence, paragenesis, and chemistry. Vigezzite occurs as well-shaped crystals in cavities of an albitic rock near Orcesco, Val Vigezzo, northern Italy, associated with beryl, bavenite, pyrochlore, columbite, and fersmite [1, 2]; see also [3].

A microprobe analysis, as mean of 12 point analyses, gives a mean of 10.5% Ce_2O_3 and indicates the formula $(Ca_{0.82}Ce_{0.24})(Nb_{0.90}Ta_{0.62}Ti_{0.50})O_6$. Fe and Mn occur in traces [1, 2].

Crystal form and structure. Orthorhombic. Axial ratio after structural data a:b:c = 0.685:1:0.486. Vigezzite occurs in form of prismatic crystals elongated along the a axis and flattened on {010}. The crystals are terminated by the pinacoids {100} and strongly striated on {010}. A distinct cleavage is observed parallel to {100} [1], cf. [2].

Unit cell parameters a = 7.559, b = 11.028, c = 5.360 Å; the space group is Pmnb-D_{2h}^{16} (No. 62, standard setting Pnma) [1, 2]. Vigezzite belongs to the eschynite group, it represents a Ca analogue of eschynite rather than an Nb analogue of rynersonite, $Ca(Ta,Nb)_2O_6$ [1], as it is suspected by [4].

Physical properties. The luster is vitreous to adamantine and resinous to adamantine [2]. The mineral is transparent. The color is intense orange-yellow [1], or pale orange-yellow to orange-red and red-brown [2]. Under the microscope the mineral is completely transparent, yellow, it shows no pleochroism, but extreme dispersion. It is biaxial, positive, $n_x = 2.14 \pm 0.02$, $n_z = 2.315$, $2V_Z$ is large, the elongation is positive, X = c, Y = b, Z = a. A brilliant green fluorescence can be seen in a common quartz lamp; but no fluorescence in short- or long-wave ultraviolet radiation [1].

Vickers hardness with load 50 g VHN = 396 kg/mm^2 corresponds to $H_M = 4.5$ to 5. The fracture is conchoidal. The mineral is very brittle. D(calc.) = 5.54 g/cm^3 [1].

References for 10.2.4.14:

[1] S. Graeser, H. Schwander, H. Hänni, V. Mattioli (Mineral. Mag. **43** [1979] 459/62). – [2] V. Mattioli (Riv. Mineral. Ital. [2] **1** [1979] 5/13, 5/9; C.A. **92** [1980] No. 183753). – [3] C. M. Gramaccioli (Lapis **5** No. 7/8 [1980] 49/56, 54, 56). – [4] E. E. Foord, M. E. Mrose (Am. Mineralogist **63** [1978] 709/14, 713).

10.2.4.15 Sinicite $2MO \cdot 4M_2O_3 \cdot 10TiO_2 \cdot ThO_2 \cdot 3(Nb,Ta)_2O_5$

For two analyses and the formula, see [1]. Sinicite is a species of doubtful validity [2], see also [3]. For a compilation of analyses and $(RE,Y)_2O_3$ contents, variance and mean of 9 analyses, see "Seltenerdelemente" A 4, 1979, pp. 30, 7.

References for 10.2.4.15:

[1] H. Chen-Tsi, C. Chi-Chen (Kexue Tongbao No. 12 [1957] 378 from Zap. Vses. Mineral. Obshchestva **87** [1958] 479). – [2] M. H. Hey (Mineralog. Mag. **32** [1959/61] 981). – [3] F. V. Chukhrov, E. M. Bohnshtedt-Kupletskaya (Mineraly Spravochnik, Vol. 2, Pt. 3, Nauka, Moscow 1967, pp. 1/676, 361).

10.3 Hydroxides

10.3.1 Yttrotungstite $(Y,Ln,Ca,Mg)_2(W,Al,Si,Ti,Fe)_4(O,OH,H_2O)_{18}$

Occurrence and Chemistry

A mineral of the Kramat Pulai tin mine in Perak, Malaysia, has been described by Scrivenor and Shenton and named "thorotungstite" on account of an analysis showing considerable amounts of WO_3 and ThO_2 besides minor amounts of RE_2O_3 [1]. A reexamination of Beard [2] of the composition of this mineral, based on a new analysis by Chadwick, published only later by Bradford [3], showed the mineral to contain Y_2O_3 and RE_2O_3 in considerable amounts, but no ThO_2; Beard therefore renamed this mineral yttrotungstite [2].

Yttrotungstite of the named locality occurs at the base of an eluvial deposit overlying granite; it is probably formed as the result of decomposition of scheelite and/or wolframite [1] under tropic conditions. It forms thick fine-fibrous and spherolitic deposits, altered to yellow-greenish masses at the surface [4].

For $RE_2O_3 + Y_2O_3$ contents and for literature on analyses of yttrotungstite from Malaysia see "Seltenerdelemente" A 4, 1979, pp. 8 and 36. Chadwick's analysis as given by Bradford [3] leads to the ideal formula $(Y,Ln)_2W_4O_{18}$, where Ln means La and lanthanides. Considering diadochy of the elements found in the analysis the true formula, given above, results. Si, Al, Fe, Ti replace W, and the RE belong with Ca(Mg) [5, pp. 270, 267, 276]. Semenov et al. from their analysis derive $Y_2W_5O_{14}(OH)_8$ or simplified $YW_3O_9(OH)_3$ [4]. Sahama gives $YW_2O_7(OH)\cdot H_2O$ [8].

The RE distribution shows strong predominance of Y, and among the lanthanides, predominance of Ce and Nd, see also p. 68. The RE composition is given as follows:

						Σ(RE + Y) = 100									Ref.
Y	La	Ce	Pr	Nd	Sm	Eu	Gd	Tb	Dy	Ho	Er	Tm	Yb	Lu	
58.9	1.1	8.1	1.3	8.9	2.8	—	4.5	0.8	4.8	—	3.9	0.5	3.9	0.4	[6][1)]
48	3	10.8	1.7	8.6	3.1	0.3	4.2	1.2	6.6	1.6	4.8	—	5.2	1.0	[4]
49.5[2)]	3.5	13	1.3	8.5	3.2	—	4.8	—[2)]	5.2	1.3	3.5	0.3	5.2	0.2	[4]

[1)] At.%. – [2)] 49.5 = Y + Tb.

Crystal Forms, Structure, and Physical Properties

Monoclinic [5, p. 263]. Crystals with acicular habit; transverse and longitudinal cleavage [1]. Laths, flattened on {100}, elongated along [001], often bevelled on their long edges by {110} and sometimes terminated by {101}. Cleavage (010) and $(10\overline{1})$. Always twinned on {100} [5, pp. 263, 268].

Unit cell constants a = 6.954, b = 8.637, c = 5.771 Å, β = 104°56′, V = 334.8 A^3. Space group $P2_1m\text{-}C^2_{2h}$ (No. 11) [5, p. 263]. a = 5.774 ± 0.005, b = 8.643 ± 0.005, c = 6.957 ± 0.005 Å, β = 104°47′ ± 10′, V = 336 A^3 [7]. The crystal structure consists of WO_6 octahedra sharing non-opposite edges in zig-zag chains running parallel to [010]. Y is in approximately trigonal prismatic coordination between the chains, with the water molecule as a seventh neighbor at one prism face. Water is probably hydrogen bonded to chain oxygen atoms and becomes coordinated to the Y by a shear between chains away from strict close-packing of the oxygen atoms [5, pp. 275/7].

Optical character negative [1]; n_x = 1.89, n_y = 1.98, n_z = 2.02 [4]. Z = [010], X: [001] = 26°; $2V_x \approx 68°$ [5, pp. 268/9]. Pleochroism X pale yellow, Z dark yellow [4]. For IR spectra see [4], [5, p. 267].

Density D = 5.55 [1], 5.82 [4], and 5.96 g/cm³; variations probably due to admixture of kaolinite [5, p. 268].

A thermogravimetric curve shows weight loss to occur in two stages: about 0.4% is lost at temperatures up to 150°C, and the remaining 4.8% between 250 and 600°C, mainly above 400°C [5, p. 267], cf. [4].

References for 10.3.1:

[1] J. B. Scrivenor, J. C. Shenton (Am. J. Sci. [5] **13** [1927] 487/90). – [2] E. H. Beard (Colonial Geol. Mineral Resources **1** [1950] 50/1 from Mineral. Abstr. **11** [1950/52] 189). – [3] E. F. Bradford (Proc. 9th Pacific Sci. Congr. [Bangkok Meeting], Bangkok 1957 [1961], Sect. 12, pp. 378/98 from Mineral. Abstr. **16** [1963/64] 619/20). – [4] E. I. Semenov, Z. T. Kataeva, E. S. Rudnitskaya (Dokl. Akad. Nauk SSSR **163** [1965] 447/9). – [5] R. J. Davis, G. W. Smith (Mineral. Mag. **38** [1971] 261/85).

[6] J. R. Butler (Geochim. Cosmochim. Acta **12** [1957] 190/4, 192). – [7] T. G. Sahama, O. v. Knorring, M. Lehtinen (Bull. Geol. Soc. Finland No. 42 [1970] 223/8, 226). – [8] T. G. Sahama (Mineral. Rec. **12** [1981] 81/7, 86).

10.3.2 Cerotungstite $(RE,Ca)_2(W,Al,Fe)_4(O,OH)_{18}$

Cerotungstite was found in the tungsten mines Kirwa and Nyamulilo in the Kigezi district, SW Uganda, as radiated groups of shiny bladed crystals.

An analysis of this cerotungstite calculated on the basis of O + OH = 18.00 results in the formula given above.*) The relative amount of RE in cerotungstite differs from that of yttrotungstite, see p. 67, by the predominance of Ce-earth elements, particularly Ce and Nd, and by the minor role of Y in cerotungstite. Because of their similar properties yttrotungstite and cerotungstite are thought to represent varieties of the same species.

Monoclinic with {001}, {100}, {101}, {011}, and {110}. Cleavage perfect parallel to (100); intimate twinning on (001). Unit cell parameters $a = 5.874 \pm 0.005$, $b = 8.700 \pm 0.005$, $c = 7.070 \pm 0.005$ Å, $\beta = 105°27' \pm 05'$, $V = 348$ A^3. Space groups $P2_1\text{-}C_2^2$ (No. 4) or $P2_1/m\text{-}C_{2h}^2$ (No. 11), Z = 2.

Optical properties: Color orange-yellow and yellow with a slightly greenish tint under the microscope. No detectable pleochroism. $n'_x = 1.89 \pm 0.01$, $n'_y = 1.95 \pm 0.01$, $n'_z = 2.02 \pm 0.01$; Z parallel to crystallographic b axis. Straight extinction on (100) and on (010).

Hardness $H_M \approx 1$.

Reference for 10.3.2:

T. G. Sahama, O. v. Knorring, M. Lehtinen (Bull. Geol. Soc. Finland No. 42 [1970] 223/8).

10.4 Borates

Braitschite $7(Ca,Na_2)O \cdot RE_2O_3 \cdot 11(B_2O_3) \cdot 7H_2O$

Braitschite has been found in white to reddish pink nodules in an anhydrite rock, overlying the potash zone in the Cane Creek Mine, Grand County, Utah. Besides an amount of ~65% finely crystalline aggregate of braitschite the nodules contain anhydrite, dolomite, halite, small amounts of hematite, and a few crystals of chalcopyrite. Some of the hematite crystals are intimately intergrown with braitschite. The occurrence is the first known of a RE borate in marine evaporites [1, 2].

One analysis, after deduction of the impurities Si, Al, Fe, CO_2, and organic C, leads to the formula given above. RE distribution, $(RE,Y)_2O_3 = 100$, calculated from the analysis: Y 7.3, La 22.2, Ce 37.1, Pr 4.9, Nd 17.8, Sm 4.6, Eu 1.9, Gd 1.6, Tb 0.5, Dy 1.2, Ho 0.3, Er 0.4, Tm 0.1, Yb 0.1, and Lu 0.05.

Braitschite forms hexagonal plates with only prism and pinacoid faces. It is hexagonal, $a = 12.156 \pm 0.001$, $c = 7.377 \pm 0.005$ Å, Z = 1, $V = 944.02 \pm 0.13$ $Å^3$. The failure of absorption spectra in the visible region for Pr and Nd in the microspectroscope is probably related to properties of the crystal structure. The mineral is colorless to white, the luster is vitreous. Braitschite is uniaxial positive. $n_o = 1.646$, $n_e = 1.647 \pm 0.002$ each. D(meas.) = 2.903 ± 0.002 g/cm^3, D(calc.) = 2.837 g/cm^3. Insoluble in water from 25 to 100°C, easily soluble in dilute hydrochloric, nitric, and sulfuric acid [2].

References:

[1] O. B. Raup, A. J. Gude, H. L. Groves (U.S. Geol. Surv. Profess. Papers No. 575-C [1967] 38/41). – [2] O. B. Raup, A. J. Gude, E. J. Dwornik, F. Cuttitta, H. J. Rose (Am. Mineralogist **53** [1968] 1081/95).

---

*) Lateron expressed as $CeW_2O_7(OH) \cdot H_2O$, T. G. Sahama (Mineral. Rec. **12** [1981] 81/7, 86/7).

10.5 Carbonates

10.5.1 Anhydrous Carbonates

Under this heading those anhydrous carbonates are described that contain, besides the RE, the cations Na, Ca, Sr, and Ba. These are sahamalite, carbocernaite, burbankite, khanneshite, and ewaldite.

The existence of broad isomorphic series is proved for the minerals carbocernaite, see p. 71, and burbankite, see p. 78. These two minerals resemble each other in structure, composition and some other properties, although they belong to different symmetry systems [1].

The scarce occurrence of the minerals of this group is caused by their extreme instability, which is shown by the appearance of many pseudomorphic replacements, e.g., of burbankite by carbocernaite and ancylite, and further by strontianite plus bastnaesite [2, 3]. Under hypergene conditions carbocernaite is more stable than burbankite and more slowly decomposed than the latter mineral [4].

References for 10.5.1:

[1] A. G. Bulakh, V. V. Kondrat'eva, E. N. Baranova (Zap. Vses. Mineral. Obshchestva **90** [1961] 42/9, 49). – [2] T. B. Zdorik (Tr. Mineral. Muzeya Akad. Nauk SSSR No. 17 [1966] 60/75, 60, 73). – [3] T. B. Zdorik (Geol. Mestorozhd. Redk. Elem. Vses. Nauchn. Issled. Inst. Mineral. Syr'ya No. 30 [1966] 121/31, 130). – [4] Yu. L. Kapustin (Mineralogiya Karbonatitov, Nauka, Moscow 1971, pp. 1/288, 180).

10.5.1.1 Sahamalite $(Mg, Fe)(Ce, La, Nd)_2[CO_3]_4$

Occurrence and Paragenesis

The mineral was first found in the southwestern part of the Sulphide Queen deposit near Mountain Pass, San Bernardino County, California, in coarse-grained light brown to dark brown barite-dolomite rock. Very small crystals of sahamalite occur in subparallel radiating clusters of singly terminated prisms intimately associated with parisite and altered bastnaesite. Sahamalite also embays dolomite rhombs along cleavage traces. As a late mineral it is very fresh and is associated with quartz, hematite, pyrite, calcite, and ankerite [1, pp. 751/3], cf. [2 to 4]. A second occurrence of this very rare mineral is the volcanic vent of Kangankunde, Chilwa Island, Malawi, where sahamalite occurs in siderite-dolomite carbonatites, associated with synchysite, monazite, quartz, and strontianite [5].

Chemistry

One analysis of the mineral, showing 31.4% Ce_2O_3 and 27.8% $(La, Nd)_2O_3$, leads to the formula given above, corresponding to a general formula $AB_2[XO_3]_4$ with $A = Mg$ and Fe^{2+}, $B =$ Ce-earth group (principally Ce^{3+}, La^{3+}, and Nd^{3+}), and $X = C^{4+}$. Appreciable substitutions in A and B may be expected. Further elements found in a spectrographic analysis of sahamalite are Mn, Ba, and Si; Pb, Al, Ca, and Y; Cu and Sr; Ag and Be. The three elements in the first group were present at concentrations of the order of tenths of a percent by weight; for each succeeding group the concentration is an order of magnitude lower than the preceding. The distribution of the five spectrographically determined RE_2O_3, recalculated to 100%, is: Ce : La : Nd : Pr : Sm = 53 : 30 : 12 : 4 : 1 [1, pp. 748/9].

Crystal Forms and Structure

Monoclinic 2/m; axial ratio a:b:c = 0.3635:1:0.2857, $\beta = 106°45'$. Sahamalite forms rather richly developed, elongated prisms and plates. Fifteen forms were identified: b{010}, e{120}, m{110}, l{230}, f{011}, d{021}, $\sigma\{\bar{2}01\}$, $k\{\bar{1}12\}$, $q\{\bar{1}11\}$, $\gamma\{\bar{1}42\}$, $G\{\bar{3}23\}$, $a\{\bar{3}43\}$, $s\{\bar{1}31\}$, $n\{\bar{2}11\}$, $x\{\bar{2}51\}$. Vicinal forms, produced by etching and partial solution, are observed on some faces; the cleavage is poorly developed parallel to (010), traces of cleavage have been observed occasionally on grains lying on $(\bar{2}01)$ [1, pp. 743/6].

Unit cell parameters are a = 5.92 ± 0.02, b = 16.21 ± 0.05, c = 4.63 ± 0.02 Å, $\beta = 106°45'$, space group $P2_1/a\text{-}C_{2h}^5$ (No. 14, standard setting $P2_1b$), Z = 2, cell volume = 425.6 $Å^3$ [1, p. 743], cf. [6].

The available facts permit the coordination of various cations in sahamalite to be predicted from Pauling's electrostatic valency principle [1, pp. 750/1].

Physical Properties

Sahamalite is colorless and nonpleochroic in thin sections and in fragments. The optical character is negative with 2V = 57° (measured and calculated). Refractive indices are n_x = 1.679 ± 0.002, n_y = 1.776 ± 0.002, n_z = 1.807 ± 0.002, dispersion r < v. The optical orientation is $X \wedge a$, Y = b, $Z \wedge c$ = 29° [1, p. 746], cf. [7]*). The measured mean index of refraction $(n_x + n_y + n_z)/3$ = 1.754 agrees approximately with that calculated, by the rule of Gladstone and Dale, to be dK + 1 = 1.738 [8]. The application of this rule has already proved very useful in establishing the composition of sahamalite [1, pp. 749/50].

The density D(meas.) = 4.30 g/cm³, it is equal to that calculated for Mg:Fe = 85:15 [1, p. 743], cf. [2, 7].

The mineral is slightly more magnetic than the accompanying parisite and bastnaesite [1, p. 742].

Chemical Behavior

Sahamalite is more resistent to acids than most normal carbonate minerals, dolomite for example. It is insoluble in boiling aqueous tartaric acid and in boiling 1n hydrochloric acid, but soluble in boiling 3n hydrochloric acid [1, pp. 746, 748].

References for 10.5.1.1:

[1] H. W. Jaffe, R. Meyrowitz, H. T. Evans Jr. (Am. Mineralogist **38** [1953] 741/54). – [2] J. Murdoch, R. W. Webb (Bull. Calif. Div. Mines Geol. No. 173 [1956] 1/452, 289). – [3] E. I. Semenov (in: K. A. Vlasov, Geochemistry and Mineralogy of Rare Elements and Genetic Types of Their Deposits, Vol. 2: Mineralogy of Rare Elements, Israel Program Sci. Trans., Jerusalem 1966, pp. 249/328, 259/60 [Russian Original, Moscow 1964]). – [4] C. Hintze (Handbuch der Mineralogie, Suppl.-Vol. 2, De Gruyter, Berlin 1960, pp. 1/958, 339/41). – [5] M. S. Garson (1955 from [7]).

[6] H. Neumann, K. O. Bryn (Avhandl. Norske Videnskaps. Akad. Oslo I Mat. Naturw. Klasse **1958** No. 1, pp. 1/15, 6, 9). – [7] Yu. L. Kapustin (Mineralogiya Karbonatitov, Nauka, Moscow 1971, pp. 1/288, 183/4). – [8] H. W. Jaffe (Am. Mineralogist **41** [1956] 757/77, 764).

*) With incorrect n_z = 1.887 instead of 1.807.

10.5.1.2 **Carbocernaite** $(Na,Ca)(RE,Sr,Ca)[CO_3]_2$

10.5.1.2.1 Occurrences and Paragenesis

Carbocernaite is a rare mineral of late stage carbonatites, found in vugs in dolomitized burbankite-calcite veins together with chlorite, pyrite, marmatite, anatase, and quartz; it is developed by hydrothermal alteration of less stable burbankite, taking up its components [1].

The first find of the mineral (originally thought to be "ambatoarinite") see p. 187, in dolomite-calcite carbonatites of Vuoriyarvi, Kola, USSR, shows that the mineral is a member of one of the latest and least wide-spread formations of the carbonatites, intersecting older rocks as veins. It occurs in voids, associated with calcite, dolomite, and ankerite as main, and with sulfides as minor constituents. It is suggested to be of hydrothermal origin, formed at low temperature in the postmagmatic formations of the carbonatites [2, pp. 42/3, 49]. Two generations of carbocernaite occur at this locality: the first in dolomitic veins together with pyrite, anatase, and barite, the second formed by decomposition of the burbankite in burbankite-calcite veins together with chlorite and calcite in leaching hollows of burbankite [3, pp. 138, 140, 168/70]. The mineral was also found in a carbonatite from Eastern Siberia, associated with sulfides, monazite, and other minerals in the aegirine-dolomite and ankerite stage IV, replacing completely the burbankite of stage III, the crystals of which are preserved as pseudomorphs [4, pp. 61, 67, 73], [5, pp. 125, 129]. Sporadically occurring relics of carbocernaite in a veinlike body of late carbonatite from a multiple pluton in Siberia are replaced by barite and RE minerals [6]. Further occurrences of carbocernaite are: a Ba-rich variety in masses and stringers developed in barite-calcite metasomatite of the Phan Si Pan Range, North Vietnam, accompanied by small amounts of pyrrhotite, apatite, quartz, and secondary ancylite [7], and an Sr-rich variety occurring as distinct layers and clusters in an albite-biotite-calcite carbonatite on the east side of Sturgeon Lake, district of Thunder Bay, Ontario. Here the mineral is closely associated with purple fluorite, calcite, pyrite, quartz, barite, galena, and chalcopyrite, and with minor amounts of ancylite [8, p. 813]. Recently carbocernaite has been found in a rare earth ore of Inner Mongolia, China [13].

Finally two rare minerals of unknown composition from Namibia are thought to be carbocernaite according to X-ray analyses (powder data) and paragenesis. The first occurs as ragged patches in close association with biotite in a micaceous sövite from the Kalkfeld Complex, 210 km NW of Windhoek, the second, associated with ancylite, Mg-dolomite, calcite, pistomesite, and strontianite, in amphibole beforsite of Ondurakorume, 13 km NE of Kalkfeld [9, 10].

10.5.1.2.2 Chemistry

For a compilation of analyses and $(RE,Y)_2O_3$ content, mean and variance of 5 samples, see "Seltenerdelemente" A 4, 1979, pp. 16, 7.

Chemical **analyses** show considerable differences in the ΣRE_2O_3 and in the content of alkaline earths, especially in Sr and Ba, see [1]. In view of structural findings, which show all cations to occupy two crystallographically independent positions, A and B, the gross **formula** should be written as $(Na,Ca)(RE,Sr,Ca)[CO_3]_2$ or generally as $AB[CO_3]_2$, corresponding to a double salt similar to dolomite; see also structure p. 73. Thus the mineral from Vuoriyarvi corresponding to one analysis has the formula $(Ca_{0.37}Na_{0.23})(RE_{0.22}Sr_{0.16}Ba_{0.03})CO_3$ [2, pp. 47/8], that from Inner Mongolia has the formula $(Ca_{0.76}Na_{0.26})(Sr_{0.44}RE_{0.40}Ba_{0.10})[CO_3]_2$ [13]. The diadochic entrance of Na and RE follows the heterovalent scheme $Na^+ + RE^{3+} \rightarrow 2Ca^{2+}$ with additional capture of Sr^{2+} atoms, which are equidimensional with RE^{3+} atoms [2, p. 48]. This substitution is further discussed in [3]. For the carbocernaite from Eastern Siberia a modified formula $(Ca,Na)_3(RE,Sr,Ba)_2[CO_3]_5$ is given in [4, p. 67], [5, p. 125]. For the sample from Canada, which is rich in Sr and F, but contains no Ba or Na, the analysis corresponds to the formula $(Ca_{2.13}Sr_{1.24}RE_{0.74})[CO_3]_{4.0}F_{1.16}$ [8, pp. 812, 814]. Carbocernaite from Vietnam has more Ba and less Sr than the specimens from Kola and Eastern Siberia [7].

The Fe_2O_3 found in all analyses has been assumed to be present as an admixture, e.g., Fe-hydroxide [2, p. 47], or finely disseminated Fe-oxide [8, p. 815]. On the basis of thermal analysis, however, the presence of Fe^{3+} as admixed oxide, hydroxide, phosphate, or sulfate was judged unlikely [7].

The mode of occurrence of H_2O in the mineral is obscure. Possibly it is adsorbed and not incorporated into the structure [2, p. 48]. The presence of some F (0.03 wt%) in the North Vietnam sample could indicate the entry of hydroxyl groups, which are isomorphously replaced by F^- [7]. This seems to be supported by the occurrence of an appreciable amount of F^- and some Cl^- (3.86 and 0.09 wt%, respectively) in the mineral from Canada, which also suggests the presence of some $(OH)^-$; the stoichiometric subscript of the group (F,OH) might equal two in the formula [8, p. 815].

The **distribution of the RE** in carbocernaites, expressed as wt% of the oxides ($\Sigma RE \pm Y$) for the samples from Vuoriyarvi, Kola (Nos. 1 and 2) and those from Canada (No. 4) and from Inner Mongolia (No. 5), and as sum of the elements (ΣRE) for the sample from Eastern Siberia (No. 3) are given in the following table:

No.	La	Ce	Pr	Nd	Sm	Eu	Gd	Tb	Dy	Ho to Lu	Y	Ref.
1	42.2	42.2	—	8.5	4.25	—	—	—	—	—	2.9	[2, p. 47]
2	34	54	2.6	9	0.4	—	—	—	—	—	—	[1, 12]
3	44.5	44.5	—	7.6	0.73	—	0.71	—	—	—	—	[4, p. 74]
4	27.9	52.6*)	4.8	13.7	0.5	0.3	0.1	0.1	0.1	<0.3		[8, p. 814]
5	37.84	44.27	4.27	10.65	1.29	—	1.68	—	—	—	—	[13]

*) CeO_2.

The nearly twofold excess of CeO_2 over La_2O_3 in the mineral from Canada contrasts with the approximately equal amounts of Ce_2O_3 and La_2O_3 in the sample No. 1 from Kola [8, pp. 814/5] and No. 3 from Eastern Siberia.

10.5.1.2.3 Crystal Forms and Structure

Orthorhombic [2, pp. 43/4], [5, p. 125], [11]. The crystals from Kola in druses are small and tabular, but well developed with rhombic appearance [1, p. 179], [3]. They are slightly elongated, and sometimes pseudohexagonal. The habit is determined by the pinacoid {100}, subordinate pinacoids {010} and {001}, and prisms {021}, {540}, and {110}; truncating forms are {305} and {210}, see **Fig. 4** [2, pp. 43/4], [3]. The Siberian crystals are fine-lamellar; as pseudomorphs after burbankite they accept the forms of this mineral [4, p. 67], [5, p. 125]. Besides well developed isometric crystals, stringers and patches (Vietnam) [7], and lath-shaped grains (Canada) [8, pp. 812/3] are observed. Fine striations characterize the faces of the prism zone [2, p. 43].

For the unit cell parameters and the density of carbocernaites from several localities, see the following table:

Locality	Unit cell parameters			Density in g/cm³		Ref.
	a in Å	b in Å	c in Å	D(meas.)	D(calc.)	
Vuoriyarvi, Kola[1)]	6.40±0.01	7.28±0.02	5.22±0.01	3.53	3.58	[2, pp. 45, 48]
Vuoriyarvi, Kola[2)]	6.41	7.30	5.23	—	—	[11]
Vuoriyarvi, Kola	6.34	7.25	5.62	3.50	—	[1]
Eastern Siberia	6.39±0.03	7.31±0.05	5.18±0.03	3.4	—	[4, pp. 66/7]
Thunder Bay, Ontario	6.410	7.279	5.198	3.33 to 3.45	—	[8, pp. 817/8]
Inner Mongolia	5.214(4)	6.430(3)	7.301(12)	3.54	3.64	[13]

[1)] The parameters for the first sample from Kola, originally calculated in kX units [2, p. 45] are incorrectly given as Å units in [3] and moreover with incorrect $c_0 = 6.21$ instead of $c_0 = 5.21$ kX in [1]. – [2)] The unit cell parameters $a_0 = 6.241$, $b_0 = 7.30$, and $c_0 = 5.23$ Å ascribed to a sample from North Vietnam in [1] are according to [7] the data of [11] for a sample from Kola. The value for $a_0 = 6.241$, however, must be 6.41 Å.

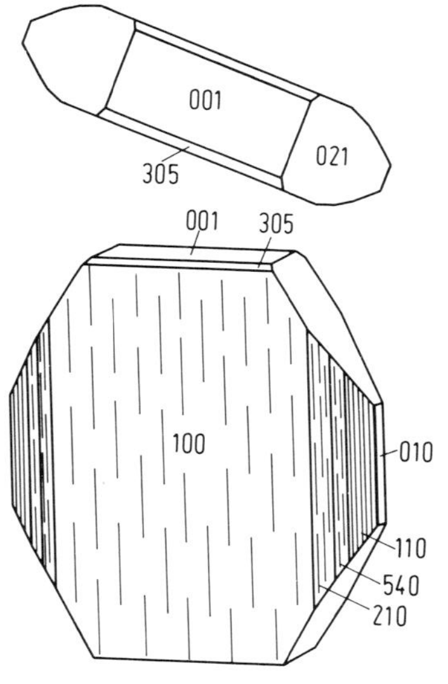

Fig. 4. Forms of a carbocernaite crystal of Kola from [2, p. 44].

For carbocernaite from Kola the axial ratio calculated from unit cell parameters is a:b:c = 0.8791:1:0.7166. The volume of the unit cell with Z = 4 is V = 243.54 Å³ [2, p. 45, 48], with Z = 2 it is V = 244.797 Å³ [13]. Of three possible space groups, C_{2v}^2-$Pb2_1m$ (No. 26, standard setting $Pmc2_1$) was chosen and subsequently confirmed [11, 13]. The space groups D_{2h}^5-Pcmm and C_{2v}^4-Pbm2 were ruled out as inconsistent with an observed strong piezoelectric effect and by examination of interatomic vector functions [11]. For structural similarity between carbocernaite and burbankite, see [2, p. 49].

The structure model obtained by means of two-dimensional Patterson- and Fourier-syntheses shows the atoms of Na, Ca, RE, and Sr to be placed as pairs in crystallographically independent twofold positions in the plane of symmetry. The character of the isomorphous replacement of the cations corresponds to the formulas with A and B cations; see p. 71. The cations of the group A have a coordination number of 7 and are placed in polyhedra which consist of distorted trigonal prisms with one centered side face. The cations of group B (among them RE) have the coordination number 10; the corresponding large polyhedra are not rectangular. Apart from these units the structure contains CO_3 triangles of 2 sorts with planes parallel to the c axis and angles of ~24° and 22°, respectively, with the (010) plane [11].

10.5.1.2.4 Physical Properties

The luster on crystal faces is strongly vitreous; fractures [2, p. 46] and grains [8, p. 813] are greasy. The crystals are sometimes greasy to vitreous, but aggregates are dull [4, p. 67]. The mineral is commonly colorless and transparent, but it is turbid white, yellowish, rose or brown if altered [2, p. 46]; in different localities the color is bright yellow [1, 3] to colorless [4, p. 67], pale and brownish yellow [7], pale pink to brick-red [8, p. 813], or brick-red [10]. A white to reddish turbidity is characteristic for the mineral from the Kalkfeld, Namibia [9, 10]. In transmitted light carbocernaite is colorless [2, p. 46] or transparent with marked pseudo-absorption [7].

Carbocernaite is biaxial negative [2, p. 46], [4, p. 67], [7], with an axial angle 2V between 52° and 56° [1]. The following table gives specified values of the refractive indices, of the axial angle 2V and of the dispersion, r to v, for carbocernaites from several localities:

Locality	Refractive indices n_x	n_y	n_z	2V	Dispersion	Ref.
Vuoriyarvi, Kola, sample I	1.570	—	1.700	54°	r>v	[1, 3]
sample II	1.569±0.002	1.679±0.002	1.708±0.002	52°	r>v	[2, p. 46]
Eastern Siberia	1.570±0.02	—	1.709±0.02	—	—	[4, p. 67]
Phan Si Pan Range, North Vietnam	1.566±0.003	1.679±0.003	1.708±0.003	56°	r>v	[7]
Kalkfeld, Namibia	—	—	1.70±0.01	—	—	[9]

The optical orientation is $n_x \| [010]$, $n_y \| [100]$, and $n_z \| [001]$ [2, p. 46]. The large double refraction (n_z-n_x = 0.139), the optical negativity, and the orientation of the indicatrix all are direct consequences of the structural characteristics of carbocernaite [11].

The mineral luminesces under UV illumination [2, p. 46].

The hardness $H_M \approx 3$; the mineral is brittle [2, p. 46] or fairly brittle [8, p. 813]. The measured density ranges are 3.50 to 3.60 g/cm³ [3] or 3.40 to 3.656 g/cm³ [1]; individual density values are given in the table on p. 73. The measured density of a carbocernaite from North Vietnam is 3.656 ± 0.005 g/cm³ [7].

10.5.1.2.5 Chemical Behavior and DTA

Carbocernaite readily dissolves in hydrochloric acid, even at considerable dilution [2, p. 46], [9], cf. [10].

The heating curve up to 1200°C of carbocernaite from Vuoriyarvi is characterized by two distinct endothermal effects, at 650 and 950°C, showing dissociation of the mineral in 2 stages. A minimum at 1200°C possibly corresponds to melting. There is also an unexplained small endothermal effect at 700°C [2, p. 46]. The somewhat deviating values for a second sample show two endothermal maxima, at 615 and 930°C, corresponding to the decomposition of the mineral to oxides of Ca, Sr, and Ba, and to the decomposition of Na_2CO_3. This is established by physical and structural data and spectral analyses [3], cf. [1].

In contrast, a thermal study up to 1200°C of carbocernaite from Vietnam shows endothermal dissociations at temperatures of about 650, 700, 900, and 1100°C with weight losses of 8, 0.5, 5, and >7%, respectively [7].

The thermal analyses (DTA and TGA) of the F-containing sample from Canada deviate considerably from those of the F-free samples: these curves show a strong endothermal reaction at 444°C (with weight loss of 1%) and three distinct endothermal peaks, at 565, 625, and 860°C (with corresponding weight losses of 4.1, 2.7, and 17.6%, respectively). Less marked endothermal reactions occur at approximately 700, 800, and 900°C. On heating to 1250°C a total weight loss of 25.9% was recorded. The strong exothermal peak at 444°C is difficult to account for by a change in the mineral itself; it is probably due to a reaction of small amounts of heavy liquid (for separation) remaining in the grains. The distinct peak at 860°C represents evolution of the gaseous phase and decomposition of the mineral [8, pp. 815/6].

References for 10.5.1.2:

[1] Yu. L. Kapustin (Mineralogiya Karbonatitov, Nauka, Moscow 1971, pp. 1/288, 170, 179/81). – [2] A. G. Bulakh, V. V. Kondrat'eva, E. N. Baranova (Zap. Vses. Mineral. Obshchestva **90** [1961] 42/9). – [3] Yu. L. Kapustin (in: E. I. Semenov, R. P. Tikhonenkova, Mineralogiya i Geneticheskie Osobennosti Shchelochnykh Massivov, Nauka, Moscow 1964, pp. 135/94, 168/70). – [4] T. B. Zdorik (Tr. Mineral. Muzeya Akad. Nauk SSSR No. 17 [1966] 60/75). – [5] T. B. Zdorik (Geol. Mestorozhd. Redk. Elem. Vses. Nauchn. Issled. Inst. Mineral. Syr'ya No. 30 [1966] 121/31).

[6] Yu. L. Kapustin (Dokl. Akad. Nauk SSSR **202** [1972] 422/5; Dokl. Earth Sci. Sect. **202** [1972] 116/9). – [7] A. G. Bulakh, E. P. Izokh (Dokl. Akad. Nauk SSSR **175** [1967] 175/7; Dokl. Earth Sci. Sect. **175** [1967] 118/20). – [8] D. C. Harris (Can. Mineralogist **11** [1971/73] 812/8). – [9] W. J. Verwoerd (S. Africa Geol. Surv. Ann. Rept. **2** [1963] 119/35, 124/5). – [10] W. J. Verwoerd (S. Africa Geol. Surv. Handbook **6** [1967] 1/452, 208, 224/6).

[11] A. A. Voronkov, N. G. Shumyatskaya, Yu. A. Pyatenko (Zh. Strukt. Khim. **7** [1966] 461/2; J. Struct. Chem. [USSR] **7** [1966] 436/7). – [12] Yu. L. Kapustin (Geokhimiya **1966** 1311/21; Geochem. Intern. **3** [1966] 1054/64, 1060). – [13] N. Shi, Z. S. Ma, Z. Peng (K'o Hsueh T'ung Pao **27** [1982] 42/5 [chinese]; C.A. **97** [1982] No. 41621).

10.5.1.3 Burbankite $(Na,Ca)_3(Ca,Sr,Ba,RE)_3[CO_3]_5$

10.5.1.3.1 Occurrence and Paragenesis

Burbankite as a rare mineral is known from several carbonatites, fenites, and metasomatites, it is typomorphic for late calcitic veins or for the first stage in the ankerite-dolomite veins, associated with pyrrhotite and forming solid masses or veinlets in calcite, often corroding the latter [1].

Burbankite was originally found in carbonatites in the Bearpaw Mountains, Montana, where it occurs together with dominant calcite and subordinate RE carbonates such as calkinsite, lanthanite, and ancylite in veins intersecting shonkinite [2, p. 1171]. In the Vuoriyarvi pluton, Kola, USSR, it is found in solid masses of irregular form or in veinlets in calcitic veins together with pyrrhotite and phlogopite, further in ankerite-dolomite veins associated with Fe-Mg carbonates, pyrite, sphalerite, phlogopite, and alkali hornblende [3, 4], cf. [5]. In East Siberian carbonatites, two generations of burbankites (I and II) are minerals of the third amphibole-calcite and amphibole-dolomite stage and of the fourth aegirine-dolomite to ankerite stage, respectively. They only exist in a narrow interval of temperatures and pH values; burbankite I is associated with Mn arfvedsonite and burbankite II with sulfides, monazite, and other minerals [6], [7, pp. 123, 130]. In the ultrabasic alkaline complex of Arbarastakh, Yakutia, the greatest amount of burbankite occurs in the youngest dolomite-ankerite formations of the carbonatites assembled with strontianite, witherite, sulfides, and other RE carbonates (bastnaesite, huanghoite) [8]. In another Siberian vein-like carbonatite body burbankite is enriched in the earliest of the late carbonatite stages occurring together with carbocernaite, florencite, several carbonates, sulfides, and ilmenite [9].

New finds of burbankite in alkaline rocks indicate, that the mineral is not only a mineral typomorphic of late carbonatitic rocks, but can be formed under a broader range of genetic conditions. The occurrence of burbankite in alkaline metasomatites of the western slope of the Southern Urals confirms this; it is here associated with neighborite a mineral of a late, strongly alkaline, metasomatically altered sodalite-analcite-cancrinite zone of the complex [10]. In the Khibina alkaline massif, Kola, burbankite occurs together with calcite and Mn calcite in a special type of carbonate veins crosscutting foyaites and monchiquites [11], in amounts up to 30% in some parts of the veins [12]; it is a widely distributed accessory mineral of these late carbonatites [13]. Sr-rich burbankite occurs in foyaites, rischorrites, and albitized parts of fenites, associated with fluorite, ilmenorutile, and zircon, but is also a mineral of the calcite veins. The foyaites contain 1346 to 2134 g/t burbankite [14]. In the Lovozero alkaline massif, Kola, burbankite occurs as a rare accessory mineral of metasomatic formations in the contact zone, together with ilmenite and ramsayite in veinlets of microclinite, which cut nepheline-feldspar fenites [15]. In apophyllitized varieties of charoitic rocks, a special type of alkaline rocks from Yakutia, Siberia, burbankite is evidently connected with a carbonatization process [16]. In nepheline syenites from the Mont St. Hilaire, Quebec, burbankite is embedded in analcite and natrolite, occurring as crystal molds or "hollow-crystals" in miaroles and pegmatitic dikes, commonly associated with microcline, aegirine, and minor amounts of calcite, siderite, and ancylite, partly altered and filled with the latter mineral [17], cf. [18] and a paragenesis of burbankite with albite and siderite in altered syenite pegmatites and silicate vugs [19]. In the nearby Miron quarry in Montreal, Canada, burbankite occupies small cavities in dolomite, analcite, and natrolite in dikes and sills of satellite bodies of the Monteregian intrusion; it is only seldom associated with ancylite [20]. Burbankite occurring in corroded relics in altered, carbonatized alkaline rocks (biotite melteigite) from the Goudini Volcano, Zeerust District, Transvaal, South Africa, is a further example of its intimate association with analcite and cancrinite besides aegirine, biotite, and subordinate chlorite, perowskite, magnetite, and calcite as other associated minerals [21, 22].

The unique sedimentary burbankite occurs as an authigenic mineral in the lacustrine Eocene deposits of the Green River Formation in Utah and Wyoming; it is a microscopic rarity accompanying barytocalcite [23]; never abundant, it is only locally developed, but fairly widespread [24], cf. [28]. In Utah it is associated with nahcolite, in Wyoming with trona [25]. The authigenic mineral is segregated in a highly alkaline lacustrine environment especially in Sweetwater Co., Wyoming, especially at the base of the Wilkins Peak Member of the Green River Formation [26, pp. 158/9], [27]. It is characterized by inclusions of organic matter, which are not evenly distributed, but adsorbed at the surface of the individual crystals, with higher concentrations of this "kerogen" in the darker pieces [26, p. 161].

For intergrowths of burbankite with ancylite, see p. 216.

In the Vuoriyarvi occurrence burkankite, probably formed under high hydrothermal conditions from alkali-rich solutions, replaces calcite along microfractures. Itself is at later stages readily replaced by ancylite, barite, and late calcite, under hypergene conditions by barite, calcite, and strontianite as well as ochreous products [3], cf. [4, 5]. In the Lovozero massif, Kola, burbankite replaces microcline and is itself replaced by fluorite, albite, and late, acicular aegirine [15]. In Siberia burbankite as the RE mineral, first formed at decreasing pH and temperature, is followed by a complex series of pseudomorphs containing bastnaesite, strontianite, and sometimes barite and ancylite [7, pp. 125/6]. In the Bearpaw Mountains it is altered to lanthanite and calkinsite [2, p. 1183], cf. [29], in South Africa it is probably derived from nepheline [21]. For replacement of burbankite by carbocernaite see p. 71, by ancylite see p. 216.

References for 10.5.1.3.1:

[1] Yu. L. Kapustin (Mineralogiya Karbonatitov, Nauka, Moscow 1971, pp. 1/288, 179). – [2] W. T. Pecora, J. H. Kerr (Am. Mineralogist **38** [1953] 1169/83). – [3] L. S. Borodin, Yu. L. Kapustin (Dokl. Akad. Nauk SSSR **147** [1962] 462/5; Dokl. Earth Sci. Sect. **147** [1962] 144/7). – [4] Yu. L. Kapustin (in: E. I. Semenov, R. P. Tikhonenkova, Mineralogiya i Geneticheskie Osobennosti Shchelochnykh Massivov, Nauka, Moscow 1964, pp. 135/94, 165/8). – [5] Yu. L. Kapustin (Geokhimiya **1966** 1311/21; Geochem. Intern. **3** [1966] 1054/64, 1056).

[6] T. B. Zdorik (Tr. Mineral. Muzeya Akad. Nauk SSSR No. 17 [1966] 60/75, 61, 74/5). – [7] T. B. Zdorik (Geol. Mestorozhd. Redk. Elem. Vses. Nauchn. Issled. Inst. Mineral. Syr'ya No. 30 [1966] 121/31, 122/3, 129/30). – [8] A. G. Zhabin, N. G. Shumyatskaya, N. S. Samsonova (Tr. Mineral. Muzeya Akad. Nauk SSSR No. 20 [1971] 202/4). – [9] Yu. L. Kapustin (Dokl. Akad. Nauk SSSR **202** [1972] 422/5; Dokl. Earth Sci. Sect. **202** [1972] 116/9). – [10] A. F. Efimov, E. M. Es'kova, Z. T. Kataeva (Tr. Mineral. Muzeya Akad. Nauk SSSR No. 19 [1969] 165/9, 165).

[11] Yu. M. Kirnarskii, L. V. Kozyreva (in: P. M. Tatarinov et al., Mineraly i Paragenezisy Mineralov Magmaticheskikh Metasomaticheskikh Gornykh Porod, Nauka, Leningrad 1974, pp. 64/73 from Ref. Zh. Geol. **1975** No. 2 V 318). – [12] O. B. Dudkin, F. V. Minakov (in: I. V. Bel'kov, Materialy po Mineralogii i Geokhimiya Shchelochnykh Kompleksov Porod Kol'skogo Poluostrova, Apatity 1975, pp. 159/74, 160, 168). – [13] O. B. Dudkin, L. I. Polezhaeva, F. V. Minakov (Veshchestvennyi Sostav Shchelochnykh Intrusivnykh Kompleksov Kol'skogo Poluostrova, Apatity 1981, pp. 73/81 from Ref. Zh. Geol. **1981** No. 12 V 179). – [14] R. P. Tikhonenkova (Nov. Dannye Mineral. Geokhim. Redk. Elem. **1974** 89/92 from Ref. Zh. Geol. **1975** No. 10 V 175; C.A. **87** [1977] No. 41935). – [15] R. P. Tikhonenkova, M. E. Kazakova (in: E. I. Semenov, R. P. Tikhonenkova, Mineralogiya i Geneticheskie Osobennosti Shchelochnykh Massivov, Nauka, Moscow 1964, pp. 40/4).

[16] K. A. Lazebnik, N. V. Zayakina (Byul. Nauchn. Tekhn. Inst. Yakutsk. Fil. Sib. Otd. Akad. Nauk SSSR **1979** 19/20 from Ref. Zh. Geol. **1980** No. 6 V 209). – [17] T. T. Chen, G. Y. Chao (Can. Mineralogist **12** [1973/74] 342/5, 342). – [18] G. Y. Chao, D. C. Harris, A. W. Hounslow, J. A. Mandarino, G. Perrault (Can. Mineralogist **9** [1967] 109/23, 112/3, 118). – [19] L. Marble, A. Regis (Rocks Minerals **54** [1979] 4/25, 11). – [20] A. P. Sabina (Geol. Surv. Paper Geol. Surv. Can. No. 78-1A [1978] 253/8, 253/4).

[21] W. J. Verwoerd (S. Africa Geol. Surv. Ann. Rept. **2** [1963] 119/35, 126/7). – [22] W. J. Verwoerd (S. Africa Geol. Surv. Handbook **6** [1967] 1/452, 127). – [23] C. Milton, J. J. Fahey (Am. J. Sci. A **258** [1960] 242/6). – [24] C. Milton (Univ. Wyoming Contrib. Geol. **10** [1971] 57/63, 60). – [25] C. Milton, H. P. Eugster (in: P. H. Abelson, Researches in Geochemistry, Wiley, New York 1959, pp. 118/50, 123/4, 136).

[26] J. Fitzpatrick, A. Pabst (Am. Mineralogist **62** [1977] 158/63). – [27] J. J. Fitzpatrick (Diss. Univ. California 1976, pp. 1/120 from Diss. Abstr. Intern. B **38** [1977] 557). – [28] C. Milton, B. Ingram, J. R. Clark, E. J. Dwornik (Am. Mineralogist **50** [1965] 593/612, 608). – [29] P. C. Sæbø (Norsk Geol. Tidsskr. **41** [1961] 311/7, 312).

10.5.1.3.2 Chemistry

Compilation of **analyses**, mean and variance of Y_2O_3 (3 samples) and $(RE,Y)_2O_3$ (13 samples), including 2 Sr-rich samples, see "Seltenerdelemente" A 4, 1979, pp. 16, 8. The maximum value of ΣRE_2O_3 of 21.72 wt% has been found in authigenic, sedimentary burbankite from Wyoming [1, pp. 159, 161].

The high variance of the cations Na, Ca, Sr, Ba, and RE in burbankites can be seen in the analyses of several localities as given by [1, p. 161], [2 to 4]. Besides the isovalent replacement between Ca, Sr, and Ba [5], a heterovalent replacement NaRE→2Ca [3, 6] or NaRE→2(Ca, Sr, Ba) is observed [5, 7]. This is verified by the similar slope of increasing amounts both of Na and RE, e.g., for a sample from the Urals [3]. The mutual diadochic relationship between the atoms in the structure and structural data show the presence of two nonequivalent groups of cations in burbankite; therefore the general **formula** is given as $A_3B_3[CO_3]_5$. In agreement with the crystal chemical character of the corresponding ions and with their mutual substitution in other natural compounds the most probable distribution of the cations is A = (Na, Ca), B = (Ca, Sr, Ba, RE); thus the final formula may be written as $(Na,Ca)_3(Ca,Sr,Ba,RE)_3[CO_3]_5$. The A group consists of 90% Na, while the probability for Na to occupy B positions is low. As the effective radius of the B positions (1.11 Å) corresponds to Sr^{2+} (1.12 Å), being close to the radius of Ce^{3+} (1.07 Å), the formula may be written more specifically in the form $Na_3(Sr_2Ce)[CO_3]_5$ [8]. The substitutions in burbankites of Ba for Sr up to 13.6 wt% and of Ca(Sr) by RE up to 15.1 wt% with predominance of Ce [9], correspond to 0.5 and 0.7 atoms per formula unit and indicate the ideal formula $(Na_2Ca)(Sr_2Ca)[CO_3]_5$ [10]. A comparison of published and new chemical analyses of various burbankites indicates the existence of a series Sr burbankite-burbankite-Ba burbankite*⁾, in which the Sr variety and the common burbankite are the most stable members; further the existence of a Ca-RE variety is proved while an RE burbankite is not possible. The whole group of minerals with burbankite structure and with a broad range of contents of Sr, Ba, RE, and Ca is characterized by strong predominance of Sr over the other elements in the "B position", a very instable ratio Sr:Ba varying from 4 to 17, and by 65 to 90% Na in the "A position" [17].

The synthesis of burbankites from RE-free materials gives a compound with the formula $Na_2Ca_2Sr_2[CO_3]_5$ with smaller parameters of the unit cell; the results of this experiments suggest, that the RE present in the natural burbankites are not essential for the mineral [10].

The following individual formulas given for burbankites from several localities show the broad range of diadochy between the cations in A and B positions and frequently a deficiency of CO_3 in relation to the cations, which may be compensated in some cases by H_2O. The calculated formula for burbankite from the Bearpaw Mountains, Montana, $(Na_{2.17}Ca_{1.61}Sr_{1.26}Ba_{0.60}RE_{0.36})[CO_3]_5$ shows a ratio cations:CO_3 = 6:5, which indicates a slight excess of total positive charge [11, p. 1174]. For burbankites from Vuoriyarvi, assuming 15 oxygen atoms (five CO_3) in the anion group, the following formulas are given: 1) burbankite from dolomite veins $Na_2(Ca_{1.2}Sr_{0.8}Ba_{0.6}RE_{0.7}Na_{0.7})[CO_3]_5$, 2) burbankite from calcite veins $Na_2(Ca_{1.4}Sr_{0.9}Ba_{0.5}RE_{0.6}Na_{0.6})[CO_3]_5$ [7], cf. [6, p. 167].

Reduced Ba contents are characteristic features of Siberian samples, with formula $Na_2(Ca_{1.4}Sr_{1.3}Ba_{0.2}RE_{0.9}Na_{0.6})[CO_3]_5$ for an Eastern Siberian Sr-rich mineral [12] and $(Na_{2.367}Ca_{0.487}K_{0.081})(Ca_{1.088}Sr_{1.020}Ba_{0.170}RE_{0.625}Fe^{3+}_{0.029}Mn_{0.004})[CO_3]_5$, or slightly idealized $(Na_{2.62}Ca_{0.38})(Ca_{0.19}Sr_{1.02}Ba_{0.17}RE_{0.62})[CO_3]_5$ for the mineral from Arbarastakh, Yakutia [13]. Other burbankites with low Ba and elevated Sr contents from Khibina, Kola, have the formulas $(Na_{5.2}Ca_{0.8})(Ca_{2.2}Sr_{2.2}RE_{1.4}Ba_{0.2})[CO_3]_{9.6}$ [16], or $Na_2(Ca_{1.4}Sr_{1.3}Ba_{0.1}RE_{0.8}Na_{0.4})[CO_3]_5$; this mineral is chemically similar to burbankite from the fenitized zone of Lovozero, Kola [14], for which the formula $(Na_{2.9}Sr_{1.5}Ca_{0.8}Ba_{0.1}Ce_{0.7})[CO_3]_{4.3}$ reveals a CO_2 deficiency [5]. The mineral from the Urals with its formula $(Na_{2.5}Ca_{0.5})(Ca_{0.39}Sr_{1.21}Ba_{0.44}RE_{0.62})[CO_3]_5 \cdot H_2O$ deviates from the theoretical formula, $Na_3(Ca,Sr,Ba,RE)_3[CO_3]_5$ by the existence of a considerable quantity of 3.94% H_2O and a deficit in cations. This is probably interrelated and can be explained by leaching of cations from the mineral simultaneously with the hydration. In burbankite from Vuoriyarvi

*⁾ Ba burbankite has recently been given the name khanneshite, cf. p. 85.

one finds up to 2.6% H_2O together with some deficit of cations, which evidently confirm this supposed process of hydration [3]. The Sr-rich and RE-poor burbankite from the Mont St. Hilaire, Quebec, recalculated on the basis of 5 oxygen atoms and assuming a stoichiometric amount of the nonanalyzed CO_2, has the formula $(Na_{1.78}Ca_{1.07})(Ca_{0.35}Sr_{2.07}Ba_{0.48}RE_{0.10})[CO_3]_5$ [10].

For the sedimentary burbankite from the Green River Formation, Wyoming, the recalculation of an analysis to 100% gives a formula close to the ideal formula $(Na_{5.0}Ca_{0.62})$ $(Ca_{1.48}Sr_{2.22}Ba_{0.56}RE_{1.74})$ $[CO_3]_{10}$, but with a slight deficiency of cations in A position (which is also observed for the sample from Mont St. Hilaire, see [10]); this cation deficiency may be due in part to vacancies on the Na site, in part due to counting errors on Na (in electron microprobe analysis) [1, p. 161].

The elements Mn and Fe in East Siberian material may be admixtures [12]. Spectral analysis showed that Mn, Fe, Al, Si, Zn, and Pb are present in burbankite from Vuoriyarvi [7].

Distribution of RE. Burbankites are generally characterized by a dominance of the Ce group of RE. Besides a Ce-La maximum, considerable amounts of Nd are present in several burbankites, e.g., of metasomatic and authigenic origin, see the following table, in which the relative amounts of RE, expressed as wt% of the oxides, are given.

Locality	Σ (RE $\pm$ Y)$_2$O$_3$ = 100								Ref.
	La	Ce	Pr	Nd	Sm	Gd	Eu	Y	
Kola, USSR	33	52	3.5	11	0.4	0.1	—	—[1]	[2]
	31.3	54.1	3.5	10.7	0.4	0.1	—	—[2]	[1, p. 161]
Lovozero, Kola	24	46	5.2	16	0.6	—	—	—	[5]
Urals, USSR	30.5	41.9	3.4	22.9	0.6	—	0.1	—	[3]
Eastern Siberia	43	42	3.2	10.7	0.8	—	—	—	[12, p. 74]
Yakutia, Siberia	26.2	48.0	3.1	22.2	0.5	—	—	—	[15]
	26.1	51.4	4.0	15.1	1.05	0.74	—	1.6[2]	[1, p. 161]
Montana, USA	25.9	51.4	4.1	15.5	1.1	0.9	—	1.1[2]	[11, p.1173]
	25.8	53.6	4.2	15.4	1.1	0.8	—	—	[18]
Wyoming, USA	19.3	44.8	8.2	23.9	2.9	0.0	—	1.0[2]	[1, p. 161]

[1] These values are recalculated from values incorrectly given in [7]. Further incorrect values, especially too low Nd values, are given in [5, 6, 15]. – [2] Recalculated to 100% from analyses.

The RE-poor burbankite from Mont St. Hilaire, Quebec, contains Ce_2O_3 2.12, Nd_2O_3 0.13, Dy_2O_3 0.08, and Yb_2O_3 0.10 wt%; La, Pr, Sm, and Gd are not determined [10]; the Ba-poor burbankite from Khibina, Kola, contains La_2O_3 24.2, CeO_2 46.0, Pr_2O_3 5.7, Nd_2O_3 15.4, Sm_2O_3 2.6, $(Eu,Gd)_2O_3$ 3.2, $(Tb,Y)_2O_3$ 1.8, Dy_2O_3 0.8, and Er_2O_3 0.3, $\Sigma = 100\%$ [16].

References for 10.5.1.3.2:

[1] J. Fitzpatrick, A. Pabst (Am. Mineralogist **62** [1977] 158/63). – [2] Yu. L. Kapustin (Mineralogiya Karbonatitov, Nauka, Moscow 1971, pp. 1/288, 170, 180). – [3] A. F. Efimov, E. M. Es'kova, Z. T. Kataeva (Tr. Mineral. Muzeya Akad. Nauk SSSR No. 19 [1969] 165/9). – [4] C. Milton, B. Ingram, J. R. Clark, E. J. Dwornik (Am. Mineralogist **50** [1965] 593/612, 609). – [5] R. P. Tikhonenkova, M. E. Kazakova (in: E. I. Semenov, R. P. Tikhonenkova, Mineralogiya i Geneticheskie Osobennosti Shchelochnykh Massivov, Nauka, Moscow 1964, pp. 40/4).

[6] Yu. L. Kapustin (in: E. I. Semenov, R. P. Tikhonenkova, Mineralogiya i Geneticheskie Osobennosti Shchelochnykh Massivov, Nauka, Moscow 1964, pp. 135/94, 167/8). – [7] L. S. Borodin, Yu. L. Kapustin (Dokl. Akad. Nauk SSSR **147** [1962] 462/5; Dokl. Earth Sci. Sect. **147** [1962] 144/7). – [8] A. A. Voronkov, N. G. Shumyatskaya (Kristallografiya **13** [1968] 246/52; Soviet Phys.-Cryst. **13** [1968] 192/6). – [9] A. S. Povarennykh (Crystal Chemical Classification of Minerals, Vol. 2, Plenum, New York 1972, pp. 459/766, 611). – [10] T. T. Chen, G. Y. Chao (Can. Mineralogist **12** [1973/74] 342/5).

[11] W. T. Pecora, J. H. Kerr (Am. Mineralogist **38** [1953] 1169/83). – [12] T. B. Zdorik (Tr. Mineral. Muzeya Akad. Nauk SSSR No. 17 [1966] 60/75, 63/4). – [13] A. G. Zhabin, N. G. Shumyatskaya, N. S. Samsonova (Tr. Mineral. Muzeya Akad. Nauk SSSR No. 20 [1971] 202/4). – [14] R. P. Tikhonenkova (Nov. Dannye Mineral. Geokhim. Redk. Elem. **1974** 89/92 from Ref. Zh. Geol. **1975** No. 10 V 175). – [15] E. I. Semenov (Mineralogiya Redkikh Zemel', Nauka, Moscow 1963, pp. 1/412, 92).

[16] O. B. Dudkin, L. I. Polezhaeva, F. V. Minakov (Veshchestv. Sostav Shcheloch. Intrusivn. Kompleksov Kol'sk. Poluostrova, Apatity 1981, pp. 73/81, 79; Ref. Zh. Geol. **1981** No. 12 V 179). – [17] R. P. Tikhonenkova, N. G. Shumyatskaya, M. E. Kazakova (Novye Dannye Mineral. Mineral. Metod. Issled., Nauka, Moscow 1977, pp. 3/11, 3, 5, 7; Ref. Zh. Geol. **1978** No. 5 V 248). – [18] E. I. Semenov, R. L. Barinskii (Geokhimiya **1958** 314/33; Geochemistry [USSR] **1958** 398/419, 400).

10.5.1.3.3 Crystal Forms and Crystal Structure

Burbankite is hexagonal [1, pp. 1172, 1174], [2, 3, 4]. The **forms** are hexagonal prisms, which usually terminate in simple pedions [5, pp. 162/3], [6, p. 63].

Burbankite occurs in coarse hexagonal crystals, with conic or prismatic forms, or in minute crystals forming rosettes and sheaf-like intergrowths [6, pp. 61/3] or fan-shaped accumulations [7]. Large elongated crystals of poor shape can take the form of hexagonal prisms with obtusely deformed faces; sometimes small hexagonal dipyramidal crystals with lesser evolved prism faces occur [2]. For coarse elongated crystals with hexagonal prismatic habit, see [8], for poorly developed crystals from ankerite-dolomite veins [3], for crystals with prismatic habit and hexagonal cross section in microclinites [9], and for elongated small crystals [10]. Small long-prismatic crystals with hexagonal or dihexagonal prismatic habit and shallow pyramidal termination are developed at Mont St. Hilaire, Quebec [11, pp. 342/3], where the slender crystals occur in radiating masses [4]. Similarly elongated single crystals with hexagonal prismatic habit, which are randomly oriented to bundles of fibers, forming microbotryoidal crusts of velvety texture, are authigenically developed in the Green River Formation, Wyoming [5, p. 163]; cf. fibrous spheroidal aggregates of microscopic size (Wyoming and Utah) [12], and radially diverging intergrowths of single prismatic crystals forming veins or veinlets in youngest dolomite-ankerite formations of carbonatites in Yakutia [13]. In the alkaline metasomatic complexes of the Urals burbankite forms individual crystals or more rarely monomineralic veinlets with granular aggregates [14].

Hairlike to acicular aggregates of burbankite occur in cavities and fine flaky encrustations on natrolite in igneous rocks near Montreal, Canada [15], and powdery segregations and accumulations of irregular grains in alkaline rocks of Kola [16]. Minute blebs of burbankite form, together with cancrinite, analcite, and calcite, an intimate intergrowth occurring as fibrous crusts in alkaline dikes of Goudini Volcano, West-Transvaal, South Africa [17]. Synthetic crystals resemble natural crystals with prismatic habit and shallow pyramidal termination [11, p. 344].

Burbankite crystals from Quebec are heavily striated parallel to c on prismatic faces [11, p. 343]. For vertical striation on deformed faces see [2], on dull faces [8]. Such striations have not been found on any crystal from the Green River Formation [5, p. 163].

The **cleavage** is prismatic and distinct [11, p. 343] or prismatic {100} distinct to imperfect [1, p. 1172], also distinct on (001) in crystals from ankerite-dolomite veins, and not present in crystals from calcite veins [2], see also one distinct cleavage [18], indistinct cleavage parallel to elongation [6, p. 63], indistinct cleavage [9] and no cleavage [14, 19]. Two partings are observed: one according to the basic pinacoid and one diagonal to the elongation (20° to 25°) of the crystals [6, p. 63].

The **unit cell constants** in Å for burbankites from several localities with considerable deviations in chemical composition, see formulas pp. 78/9, are compiled in the following table:

Locality	a in Å	c in Å	Ref.
Montana, USA	10.46±0.08[1)]	6.48±0.03[1)]	
	10.53±0.05[2)]	6.47±0.02[2)]	[1, p. 1174]
Mont St. Hilaire, Quebec, Canada	10.514(3)	6.520(2)	[11, p. 343]
Vuoriyarvi, Kola[3)]	10.41	6.48	[2, 8]
	10.49	6.42	[3]
	10.52±0.04	6.51±0.02	[20, p. 192], cf. [21]
Urals	10.57±0.03	6.43	[14]
East Siberia	10.45[4)]	6.39[4)]	[6, p. 64]
Yakutia, Siberia	10.49±0.03	6.40±0.02	[13]
Green River Formation, Wyoming	10.541(3)[5)]	6.510(2)	[5, p. 160]
Synthetic material[6)]	10.477(2)	6.456(3)	[11, p. 343]

[1)] Data from single crystal by Weissenberg photograph. – [2)] Data from X-ray powder pattern. – [3)] Three different samples. – [4)] In the original paper the values are given as kX; these values are incorrectly repeated as Å in [3]. – [5)] In the author abstract of [5, p. 158] and in [22] erroneously given as 10.451. – [6)] RE-free $Na_2Ca_2Sr_2[CO_3]_5$.

A direct dependence of the unit cell parameters from the Σ of atoms Ba and Sr is assumed [6, p. 64].

The centrosymmetric space group $P6_3mmc\text{-}D_{6h}^4$ (No. 194) previously assumed for burbankite by [2] and [1, p. 1176], and similarly the space group $P\bar{6}2c\text{-}D_{3h}^4$ (No. 190), are ruled out due to the strong piezoelectric effect of the mineral. This fact and the crystal analysis indicate $P6_3mc\text{-}C_{6v}^4$ (No. 186) as the most probable space group [20, p. 192], cf. [1, 21]; Z=2 [20, p. 193], [21, 23]. The cell volume calculated from cell dimensions is 621 Å^3 [1, p. 1174].

Burbankites from Goudini Volcano, West-Transvaal, South Africa [17, 18], and from Yakutia [19] are identified by X-ray patterns.

Structure description. Patterson projections and electron density distributions indicate the existence of 2 nonequivalent groups of cations A and B, as it was already shown on p. 78, in the burbankite structure. With Z=2, the cations are in 2 sets of 6 each per cell. The set with the A cations is 8-coordinated, the set with the B cations, among them the RE, has the coordination number 10. The CO_3 triangles are of 3 types; for a description of the structure, see [20, pp. 193/4]. For the existence of this structure the diadochic interrelationship between the large cations, at least in one of the groups A or B, is an indispensable condition [20, pp. 195/6].

References for 10.5.1.3.3:

[1] W. T. Pecora, J. H. Kerr (Am. Mineralogist **38** [1953] 1169/83). – [2] L. S. Borodin, Yu. L. Kapustin (Dokl. Akad. Nauk SSSR **147** [1962] 462/5; Dokl. Earth Sci. Sect. **147** [1962] 144/7). – [3] Yu. L. Kapustin (Mineralogiya Karbonatitov, Nauka, Moscow 1971, pp. 1/288, 179/80). – [4] L. Marble, A. Regis (Rocks Minerals **54** [1979] 4/25, 11). – [5] J. Fitzpatrick, A. Pabst (Am. Mineralogist **62** [1977] 158/63).

[6] T. B. Zdorik (Tr. Mineral. Muzeya Akad. Nauk SSSR No. 17 [1966] 60/75). – [7] T. B. Zdorik (Geol. Mestorozhd. Redk. Elem. Vses. Nauchn. Issled. Inst. Mineral. Syr'ya No. 30 [1966] 121/31, 123). – [8] Yu. L. Kapustin (in: E. I. Semenov, R. P. Tikhonenkova, Mineralogiya i Geneticheskie Osobennosti Shchelochnykh Massivov, Nauka, Moscow 1964, pp. 135/94, 167). – [9] R. P. Tikhonenkova, M. E. Kazakova (in: E. I. Semenov, R. P. Tikhonenkova, Mineralogiya i Geneticheskie Osobennosti Shchelochnykh Massivov, Nauka, Moscow 1964, pp. 40/4). – [10] O. B. Dudkin, F. V. Minakov (in: I. V. Bel'kov, Materialy po Mineralogii i Geokhimiya Shchelochnykh Kompleksov Porod Kol'skogo Poluostrova, Apatity 1975, pp. 159/74, 168).

[11] T. T. Chen, G. Y. Chao (Can. Mineralogist **12** [1973/74] 342/5). – [12] C. Milton, H. P. Eugster (in: P. H. Abelson, Researches in Geochemistry, Wiley, New York 1959, pp. 118/50, 136). – [13] A. G. Zhabin, N. G. Shumyatskaya, N. S. Samsonova (Tr. Mineral. Muzeya Akad. Nauk SSSR No. 20 [1971] 202/4). – [14] A. F. Efimov, E. M. Es'kova, Z. T. Kataeva (Tr. Mineral. Muzeya Akad. Nauk SSSR No. 19 [1969] 165/9, 165/6). – [15] A. P. Sabina (Geol. Surv. Paper Geol. Surv. Can. No. 78-1A [1978] 253/8, 254).

[16] R. P. Tikhonenkova (Nov. Dannye Mineral. Geokhim. Redk. Elem. **1974** 89/92). – [17] W. J. Verwoerd (S. Africa Geol. Surv. Handbook **6** [1967] 1/452, 127). – [18] W. J. Verwoerd (S. Africa Geol. Surv. Ann. Rept. **2** [1963] 119/35, 127). – [19] K. A. Lazebnik, N. V. Zayakina (Byul. Nauchn. Tekhn. Inst. Yakutsk. Fil. Sib. Otd. Akad. Nauk SSSR **1979** 19/20 from Ref. Zh. Geol. **1980** No. 6 V 209). – [20] A. A. Voronkov, N. G. Shumyatskaya (Kristallografiya **13** [1968] 246/52; Soviet Phys.-Cryst. **13** [1968] 192/6).

[21] A. A. Voronkov, N. G. Shumyatskaya, Yu. A. Pyatenko (Kristallografiya **12** [1967] 135; Soviet Phys.-Cryst. **12** [1967] 107). – [22] J. J. Fitzpatrick (Diss. Univ. California 1976, pp. 1/120 from Diss. Abstr. Intern. B **38** [1977] 557). – [23] H. Strunz, C. Tennyson (Mineralogische Tabellen, 6th Ed., Geest and Portig, Leipzig 1977, pp. 1/621, 240).

10.5.1.3.4 Optical Properties

The luster of burbankite is vitreous at most localities [1 to 5]; it is dull on deformed and vertically striated prism faces [6]. The mineral is transparent [7, 8], especially the grains in microclinite [3] and splinters of the mineral [9, p. 63], [2].

The color of burbankite is predominantly yellow with greenish or rose tints, sometimes colorless: Freshly broken, the mineral is pale yellow (Montana) [10, p. 1172], lemon-yellow (East Siberia) [9, pp. 61/3], [5], and yellowish [11] to light cinnamon (Khibina, Kola) [20]. Further colors given are: yellow, greenish yellow, or rose (Vuoriyarvi, Kola) [1, 5, 6]; yellow and yellowish rose (Lovozero, Kola) [3, 5], bright rose [12], colorless and pale yellow to pink [4], or pale to light yellow, but also brown due to goethite inclusions (Mont St. Hilaire, Quebec) [13]. The color is pale green and greenish yellow if fresh, but yellowish rose or reddish brown in decomposed varieties (Urals) [5]. Aggregates of the mineral are colorless, white or reddish, flaky incrustations are bluish white (Miron quarry, Montreal) [14], fibrous spheroidal aggregates of authigenic burbankite are pinkish (Utah/Wyoming) [15], crusts or seams have yellow to dark brown color, often varying markedly within a small fragment and sometimes with a colorless outer part (Wyoming) [16].

The powder of burbankite is grayish yellow [10, p. 1172]. In thin section the mineral is colorless [1, 3, 17], in splinters nearly colorless [9, p. 63].

The optical character of burbankite is constantly uniaxial negative [1, 5, 8], [9, p. 63], [10, p. 1172]. The mean index of refraction, calculated from the chemical composition and the D value after the law of Gladstone and Dale, n = 1.632, is in close agreement with the mean value n = 1.625, calculated from the measured indices of burbankite from Montana [10, p. 1174]. The birefringence is low, see ~0.012 [9, p. 63] and the following table; one burbankite from Yakutia, however, has a remarkably high value of 0.029 [8].

Refractive indices n_o and n_e and density values of analyzed burbankites from several localities are listed in the following table:

Locality	n_o	n_e	D (g/cm^3)	Ref.
Montana	1.627 ± 0.001	1.615 ± 0.001	3.50	[10, p. 1172]
Mont St. Hilaire	1.616(1)	1.597(1)	3.50*)	[4]
Vuoriyarvi	1.632 ± 0.002	1.620 ± 0.002	3.54	[1, 6, 7]
(3 samples)	1.635 ± 0.002	1.623 ± 0.002	3.58	[1, 6, 7]
	1.629	1.622	3.48	[7]
Lovozero	1.618 ± 0.002	1.606 ± 0.002	3.60**)	[3, 5]
Khibina	1.632	1.627	—	[11]
Ba-poor sample	1.628	1.616	3.43	[2]
Urals	1.629	1.621	3.50	[5]
East Siberia	1.629 ± 0.002	1.617 ± 0.002	3.58 to 3.60	[9, p. 65], [5]
Yakutia	1.627 to 1.628	1.616 ± 0.002	3.386, 3.393*)	[12]
(2 samples)	1.626	1.597	—	[8]
South Africa	—	1.605 ± 0.003	—	[17]
Synthetic $Na_2Ca_2Sr_2[CO_3]_5$	1.615(1)	1.596(1)	3.25*)	[4]

*) Calculated values. – **) In [7] incorrectly given as 3.50.

Higher refractive indices and D values of burbankite from Vuoriyarvi, Kola, compared with those of burbankite from Bearpaw Mountains, Montana, have been ascribed to the elevated RE content of the first sample, which contains less Sr and varies only slightly in the Ba content [1]. The association of low refractive indices with low RE content for the burbankite from Mont St. Hilaire, Quebec, confirms this observation [4]. The interdependence of refractive indices and main element contents in burbankites, however, seems to be very complex caused by the various cation substitutions [5], cf. [3]; reduced values of refractive indices are confirmed only for increasing Sr contents in the different burbankite varieties [19].

Cathodic luminescence was observed for burbankite from Lovozero, Kola [2].

10.5.1.3.5 Other Physical Properties

The hardness of burbankites from different occurrences $H_M \approx 3.5$*) (Montana) [10, p. 1172], cf. [7], ≈4 (Mont St. Hilaire) [4], 4.5 (Lovozero) [3] (and Eastern Siberia) [9, p. 63], and 4.5 to 5 (Vuoriyarvi) [1].

The extremely fragile burbankite aggregates of the Green River Formation seem to be much softer than the megascopic crystals [16].

The fracture of the mineral is shelly or conchoidal [7], generally uneven [5].

For density values D(meas.) of analyzed burbankites see table p. 83. The calculated density value of burbankite from Mont St. Hilaire D(calc.) = 3.50 g/cm^3 (see [4] in the table) is close to the measured values of burbankites from several other localities; the relative low values D(meas. and calc.) of a burbankite of Yakutia correspond to their low Ba content, cf. [12] in the table and formula p. 78.

The density values of burbankites with little variance in the Ba content are shown to increase with increasing RE content and decreasing Sr content [1, 3]. It must be considered, that the Ba content influences these values too [6]. Theoretically the D values of burbankites should be directly proportional to the Ba, Sr, and RE contents; deviations of this rule are probably due to different degrees of hydration of the burbankites from several occurrences [5].

Burbankite from Khibina is electromagnetic [2]. The piezoelectric effect of burbankite from Vuoriyarvi is strong [18].

References for 10.5.1.3.4 and 10.5.1.3.5:

[1] L. S. Borodin, Yu. L. Kapustin (Dokl. Akad. Nauk SSSR **147** [1962] 462/5; Dokl. Earth Sci. Sect. **147** [1962] 144/7). – [2] R. P. Tikhonenkova (Nov. Dannye Mineral. Geokhim. Redk. Elem. **1974** 89/92). – [3] R. P. Tikhonenkova, M. E. Kazakova (in: E. I. Semenov, R. P. Tikhonenkova, Mineralogiya i Geneticheskie Osobennosti Shchelochnykh Massivov, Nauka, Moscow 1964, pp. 40/4, 40/2). – [4] T. T. Chen, G. Y. Chao (Can. Mineralogist **12** [1973/74] 342/5). – [5] A. F. Efimov, E. M. Es'kova, Z. T. Kataeva (Tr. Mineral. Muzeya Akad. Nauk SSSR No. 19 [1969] 165/9, 166, 168).

[6] Yu. L. Kapustin (in: E. I. Semenov, R. P. Tikhonenkova, Mineralogiya i Geneticheskie Osobennosti Shchelochnykh Massivov, Nauka, Moscow 1964, pp. 135/94, 165/8). – [7] Yu. L. Kapustin (Mineralogiya Karbonatitov, Nauka, Moscow 1971, pp. 1/288, 179/80). – [8] K. A. Lazebnik, N. V. Zayakina (Byul. Nauchn. Tekhn. Inst. Yakutsk. Fil. Sib. Otd. Akad. Nauk SSSR **1979** 19/20 from Ref. Zh. Geol. **1980** No. 6 V 209). – [9] T. B. Zdorik (Tr. Mineral. Muzeya Akad. Nauk SSSR No. 17 [1966] 60/75). – [10] W. T. Pecora, J. H. Kerr (Am. Mineralogist **38** [1953] 1169/83).

[11] O. B. Dudkin, F. V. Minakov (in: I. V. Bel'kov, Materialy po Mineralogii i Geokhimiya Shchelochnykh Kompleksov Porod Kol'skogo Poluostrova, Apatity 1975, pp. 159/74, 168). – [12] A. G. Zhabin, N. G. Shumyatskaya, N. S. Samsonova (Tr. Mineral. Muzeya Akad. Nauk SSSR No. 20 [1971] 202/4). – [13] L. Marble, A. Regis (Rocks Minerals **54** [1979] 4/25, 11). – [14] A. P. Sabina (Geol. Surv. Paper Geol. Surv. Can. No. 78-1A [1978] 253/8). – [15] C. Milton, H. P. Eugster (in: P. H. Abelson, Researches in Geochemistry, Wiley, New York 1959, pp. 118/50, 136).

[16] J. Fitzpatrick, A. Pabst (Am. Mineralogist **62** [1977] 158/63, 159). – [17] W. J. Verwoerd (S. Africa Geol. Surv. Ann. Rept. **2** [1963] 119/35, 127). – [18] A. A. Voronkov, N. G. Shumyatskaya, Yu. A. Pyatenko (Kristallografiya **12** [1967] 135; Soviet Phys.-Cryst. **12** [1967] 107). – [19] R. P. Tikhonenkova, N. G. Shumyatskaya, M. E. Kazakova (Nov. Dannye Mineral. Mineral. Metod. Issled., Nauka, Moscow 1977, pp. 3/11, 10; Ref. Zh. Geol. **1978** No. 5 V 248). – [20] O. B. Dudkin, L. I. Polezhaeva, F. V. Minakov (Veshchestvennyi Sostav Shchelochnykh Intrusivnykh Kompleksov Kol'skogo Poluostrova, Apatity 1981, pp. 73/81, 79).

---

*) In [1] incorrectly given as 4.5.

10.5.1.3.6 Chemical Behavior and DTA

Burbankite dissolves with strong effervescence in cold, dilute hydrochloric acid [1 to 3], cf. [4, 5]; it is also readily soluble in nitric acid [6].

The thermic investigation of burbankites confirms the assumed mixed carbonate or double salt: constantly 2 endothermal peaks of the DTA curve correspond to the decomposition of the mineral in 2 stages to oxides of RE, Ca, Sr, and Ba, respectively; the final decomposition occurs above 900°C [1]. For burbankites from several occurrences the following temperature values of the 2 endothermal maxima are given:

Occurrence	Maxima in °C		Ref.
	T_1	T_2	
Vuoriyarvi	670	730	[1, 2, 7]
Lovozero	640	695	[8]
Urals	635	695	[9]
East Siberia	680	750	[6]
Yakutia*)	~640	~740	[10]

*) Values estimated from a diagram.

The burbankite sample from Yakutia (weight 96 mg) shows a loss of 6 mg at 800°C and 5 mg at 920°C, altogether 19 mg, both curves (TG and DTA) are similar to standard [10]. The heating curves for burbankites from Vuoriyarvi and from Urals are very similar [9]. Yellow-brown aggregates of the authigenic burbankite from Wyoming lose their color upon heating, due to destruction or removal of organic matter (kerogen), see p. 76. This burbankite remains unchanged up to 515°C and is converted to a mixture of new phases only after heating up to 615°C for 24 h, where new principal phases are recognized as $(Sr,Ba)CO_3$, BaO, and traces of calcite. The contrast to burbankites from other localities with decomposition peaks at higher temperatures, see table above, may be due to their rapid heating during DTA [11].

References for 10.5.1.3.6:

[1] L. S. Borodin, Yu. L. Kapustin (Dokl. Akad. Nauk SSSR **147** [1962] 462/5; Dokl. Earth Sci. Sect. **147** [1962] 144/7). – [2] Yu. L. Kapustin (in: E. I. Semenov, R. P. Tikhonenkova, Mineralogiya i Geneticheskie Osobennosti Shchelochnykh Massivov, Nauka, Moscow 1964, pp. 135/94, 167/8). – [3] W. J. Verwoerd (S. Africa Geol. Surv. Ann. Rept. **2** [1963] 119/35, 127). – [4] W. T. Pecora, J. H. Kerr (Am. Mineralogist **38** [1953] 1169/83, 1172). – [5] T. T. Chen, G. Y. Chao (Can. Mineralogist **12** [1973/74] 342/5, 343).

[6] T. B. Zdorik (Tr. Mineral. Muzeya Akad. Nauk SSSR No. 17 [1966] 60/75, 63/4). – [7] Yu. L. Kapustin (Mineralogiya Karbonatitov, Nauka, Moscow 1971, pp. 1/288, 179). – [8] R. P. Tikhonenkova, M. E. Kazakova (in: E. T. Semenov, R. P. Tikhonenkova, Mineralogiya i Geneticheskie Osobennosti Shchelochnykh Massivov, Nauka, Moscow 1964, pp. 40/4, 42). – [9] A. F. Efimov, E. M. Es'kova, Z. T. Kataeva (Tr. Mineral. Muzeya Akad. Nauk SSSR No. 19 [1969] 165/9, 166). – [10] A. G. Zhabin, N. G. Shumyatskaya, N. S. Samsonova (Tr. Mineral. Muzeya Akad. Nauk SSSR No. 20 [1971] 202/4).

[11] J. Fitzpatrick, A. Pabst (Am. Mineralogist **62** [1977] 158/63, 160).

10.5.1.4 Khanneshite $(Na,Ca)_3(Ba,Sr,RE,Ca)_3[CO_3]_5$

The mineral had already been found by Eremenko in 1974 in the carbonatite complex of Khanneshin, Afghanistan. It now is described as a new mineral in the burbankite series with the general formula $A_3B_3[CO_3]_5$ and is named khanneshite.

Khanneshite occurs impregnated in fine-grained, hydrothermally altered dolomite- and barite-containing dolomite-ankerite carbonatites of the schorl facies. The massive fine-grained aggregates of radial structure are closely intergrown with dolomite, calkinsite, carbocernaite, barite, and chlorite. Commonly the mineral is strongly altered and replaced by powdery secondary barite.

Two chemical analyses on contaminated material (about 80 wt% khanneshite together with mainly barite and some dolomite and chlorite), after deduction of the mechanical admixtures show quantitative ratios of the atomic amounts of the sum of cations to those of CO_2 to be 1.209 (close to the theoretical value of 1.2) and 1.232. The formulas calculated from these analyses based on 15 atoms oxygen are:

1) $(Na_{2.13}Ca_{0.87})(Ba_{1.82}Sr_{0.53}RE_{0.49}K_{0.15})C_{4.96}O_{15}\cdot 0.67\,H_2O$ and

2) $(Na_{2.02}Ca_{0.98})(Ba_{1.55}Sr_{0.60}RE_{0.51}Ca_{0.25}K_{0.12})C_{4.89}O_{15}\cdot 0.55\,H_2O$.

The water found, as in analyses of burbankites, is not in accordance with the structure of burbankite and indicates that the mineral is easily replaced by secondary products and is altered by hydration; without the content of water the formula corresponds to the general formula given above with Na > Ca and with Ba > Sr + Ca + RE*).

The relative amounts of the RE (ΣRE + Y = 100) in khanneshite are identified by 2 spectral analyses; they show the selective ceric character of the RE:

	La	Ce	Pr	Nd	Sm	Eu	Gd	Dy	Ho	Er	Tm	Yb	Y
1)	20.3	44.0	6.0	9.5	6.0	0.1	4.8	0.3	0.2	0.8	0.05	0.5	7.5
2)	20.9	51.2	4.7	7.3	4.7	0.1	3.7	0.4	0.1	0.9	0.56	0.2	5.8

The mineral is hexagonal, it forms elongated hexagonal-prismatic crystals, occurring in aggregates of radial structure. It has an indistinct cleavage along the elongation and a coarse parting in transversal direction.

The unit cell constants are $a = 10.65 \pm 0.01$, $c = 6.58 \pm 0.01$ Å, the axial ratio is $c/a = 0.618$. The space group is $P6_3mc\text{-}C_{6v}^4$ (No. 186). The unit cell is larger than those of burbankite and strontioburbankite, which is the consequence of the large dimensioned Ba ion.

The color of khanneshite is pale yellowish, nearly colorless; it is uniaxial with negative optical character. The refractive indices are $n_o = 1.623$ to 1.620, $n_e = 1.610$ to 1.608. The IR spectrum is characterized by absorption lines at 700, 730, 870, and 1060 cm^{-1}, and by broad zones with absorption maxima at 1340, 1390, and 1535 cm^{-1}; further an elevated absorption is observed at 3400 cm^{-1}.

The mineral has a low hardness and is brittle; the measured density is 3.8 to 3.9 g/cm^3 (by hydrostatic balance).

Khanneshite is easily soluble in hydrochloric acid. The DTA curve shows endothermal effects at 645 and 590°C, indicating somewhat lower temperatures of reaction compared with those of burbankites (cf. p. 85); an additional small effect at 360°C is connected with 1.5% loss of weight, probably the loss of water [1].

*) A Ba-rich mineral differing from strontioburbankite with (Sr > Ba, RE, Ca) by (Ba > Sr, Ca, RE) has been predicted to occur in late carbonatites or metasomatites, connected with them, by Tikhonenkova et al. [2], cf. [1].

References for 10.5.1.4:

[1] G. K. Eremenko, V. A. Bel'ko (Zap. Vses. Mineral. Obshchestva **111** [1982] 321/4). – [2] R. P. Tikhonenkova, N. G. Shumyatskaya, M. E. Kazakova (Nov. Dannye Mineral. Mineral. Metod. Issled., Nauka, Moscow 1977, pp. 3/11, 5, 10).

10.5.1.5 **Ewaldite** Ba(Ca, RE, Na, K)$[CO_3]_2$

Occurrence and Paragenesis

Ewaldite has first been found in the Green River Formation of Wyoming, at 4 localities in the Sweetwater Co., always intergrown with mackelveyite [1, p. 187], see also mackelveyite, p. 199. It is a local, extremely scarce mineral [2, p. 57]. A second occurrence is the nepheline syenite of Mont St. Hilaire, Quebec, where ewaldite occurs in pegmatite dikes, miarolitic cavities and interstices; it commonly forms syntactic intergrowths with mackelveyite and donnayite and is further associated with a great number of minerals [3]. In unaltered pegmatites, silicate vugs, and hornfels inclusions of this occurrence it is associated with labuntsovite and leucosphenite [4]. A third occurrence of ewaldite are ultra-agpaitic types of pegmatites in rischorrites and ijolite-urtites of the southwestern part of the Khibina alkali massif, Kola. The mineral is found in a pegmatitic rock from a drill core with dominant feldspar (in large blocks, cemented by fine-grained adularia) and some aegirine, pyrite, and a hisingerite type mineral [5]. Ewaldite is also described from precipitates, obtained by evaporation of aqueous inclusions from quartz, galena, and sphalerite, found in the deposit of Madzharovsk, East Rhodope Mountains, Bulgaria. For this mineral, determined by X-rays only, the formula $BaCa[CO_3]_2$ without any RE content was presented [6].

The ewaldite from the Green River Formation contains inclusions of hematite, pseudomorphous after magnetite, of material rich in sodium and low in calcium, and of apparently carbonaceous material [1, p. 187], cf. for hematite [7, 8]. The ewaldite from Kola is turbid due to dust-like inclusions of Fe-sulfides and probably of organic matter [5].

A polycrystal, showing the outer forms of mackelveyite but identified by X-ray diffraction as ewaldite, is a pseudomorph of ewaldite after mackelveyite [1, p. 199], see also p. 88.

Chemistry

Chemical analyses of very small amounts of impure crystals from Wyoming showed the mineral to be a carbonate. After reduction of impurities (Fe, Na) the formula of this ewaldite with Ba = 1 is given in correspondence with structure determination, see p. 88, as $Ba_{1.00}(Ca_{0.48}RE_{0.20}Na_{0.15}K_{0.11}Sr_{0.03}U_{0.02}\square_{0.01})[CO_3]_{2.00}$ [1, pp. 187/9]. It is simplified to Ba(Ca, RE, Na, K)$[CO_3]_2$ by [10]. For ewaldite from Mont St. Hilaire the formulas are similar, but include Y [3, 4]; for ewaldite from Kola the empirically derived formula is $Ba_{0.95}(Ca_{0.32}RE^{Y}_{0.26}RE^{Ce}_{0.05}Na_{0.25}Sr_{0.17})[CO_3]_2$; the essential difference of this mineral compared with that from Wyoming are elevated values of Sr and of RE of the Ce group, which are typomorph elements of this occurrence [5], cf. the RE composition below. For the ewaldite from Wyoming another cation distribution and an elevated content of RE is indicated by the formula $(Na_2Ca)(Ba_4RE_{1.7}U_{0.3})[CO_3]_9$ from [2, pp. 58, 60].

Ewaldite is an Y- and Y-earth-concentrating carbonate mineral [2, p. 60], [11]; the relative amounts of the RE ($\Sigma RE + Y$ in at.%) for the minerals from Wyoming (No. 1) and from Kola (No. 2) are:

No.	La	Ce	Nd	Sm	Eu	Gd	Dy	Ho	Er	Yb	Y	Ref.
1	1	—	—	—	1	6	5	1	7	6	73	[1, p. 189]
2	2.4	7.2	11.6	9.7	—	7.2	4.8	—	4.0	1.6	51.4	[5]

Crystal Forms and Crystal Structure

The symmetry is hexagonal [1, p. 191], [4, 10]; dihexagonal-pyramidal according to the space group, see below.

The axial ratio is c/a = 2.400 (goniometric), c/a = 2.419 (by X-rays). Ewaldite from Wyoming has been found with one exception of a pure crystal [8], only as polycrystals in syntactic intergrowth with mackelveyite [1, p. 187], [2, p. 60]. The crystal forms in these polycrystals referred to the ewaldite cell are indexed c{0002}, c′{000$\bar{2}$}, e{11$\bar{2}$4}, r{11$\overline{22}$} with the transformation matrix from mackelveyite to ewaldite ⅔ ⅓ 0 | $\overline{⅓}$ ⅓ 0 | 0 0 ⅔ [1, pp. 197/8]. The crystals from Kola are isometrical and have well developed faces of the hexagonal prism and of the pyramid {10$\bar{1}$1} [5]. Twins [8] and cleavage [9, p. 203] are not observed in crystals from Wyoming, a perfect cleavage after the basic pinacoid is, however, characteristic for ewaldite crystals from Kola [5]. The polycrystals show, regardless of their composition, tabular to pyramidal habit [1, p. 197]; they are hemimorphic [4], sometimes doubly terminated with "caps" of donnayite (see p. 198) and mackelveyite (see p. 199) at the Mont St. Hilaire, Quebec [3]. Dimorphism for ewaldite and mackelveyite is indicated by electron microprobe data, except for possibly different water contents. The transformation of mackelveyite to ewaldite is considered to be an ordering one: ewaldite, compared with mackelveyite, has the higher symmetry, the smaller cell, fewer disorder and thus a smaller free energy. The transformation must proceed in the direction mackelveyite to ewaldite with a composition surface between the 2 phases in a polycrystal not to be seen and completely irregular. The ordering may proceed randomly without affecting the perfection of the mutual orientation, and in particular it does not have to be planar and reticular (0001) as in the polycrystals of the bastnaesite-synchysite series [1, pp. 187, 199].

For the unit cell constants, number of formula units (Z), cell volume (V), and calculated density values of ewaldite, see the following table:

	a in Å	c in Å	Z	V in Å^3	D(calc.) in g/cm^3	Ref.
Wyoming	5.284 ± 7	12.78 ± 1	2	309.1	3.37	[1, p. 192]
	5.291 ± 0.006	12.751 ± 0.012	—	309.1 ± 4	—	[8]
Kola	5.30	12.80	—	—	—	[5]

By converting the chemical analysis to a structural formula with a total of 4 cations, which are grouped to their "effective ionic radius", the cell content can be written as $(Ba_{1.68}K_{0.16}Na_{0.12}Sr_{0.04})(Na_{0.92}Ca_{0.76}Y_{0.32})[CO_3]_{3.56}$. The ions in the first parenthesis with coordination number 9 have a weighted average radius of 1.38 Å, those in the second parenthesis with coordination number 6 have a weighted average radius of 0.99 Å [7]. The lattice parameters of ewaldite (a_e, c_e) are related to those of mackelveyite (a_m, c_m) as follows: $a_e = a_m/\sqrt{3}$, $c_e = 2c_m/3$, $V_e = 2V_m/9$ [8], cf. [1, pp. 192/3].

The space group is $P6_3mc$-C_{6v}^4 (No. 186) [1, p. 192], [9, p. 203], [10]. The 2 types of cations in ewaldite lie between carbonate groups parallel to the basal plane. Ba has coordination number 9 and Ca, partly substituted by RE and other cations, is 6-coordinated [9, pp. 209/11].

Physical Properties

Ewaldite has a dull luster and is semi-transparent [5]. Pure crystals from Wyoming are deep green [8] or bluish green [1, p. 192], the commonly occurring polycrystals of ewaldite and mackelveyite are dull light bluish green to dark green [8], bluish green [4], bluish green or light colored [1, pp. 186, 194]. Crystals from Kola are brick-red on the surface, but gray in the inner parts [5]. Under the microscope the polycrystals are green with darker colored inclusions [1, p. 187].

Ewaldite is uniaxial negative, the refractive indices of crystals from Kola are $n_o = 1.644 \pm 0.002$, $n_e = 1.542 \pm 0.002$ [5]; a fragment of a light-colored polycrystal from Wyoming with only a minimum amount of mackelveyite has $n_o = 1.646 \pm 3$, $n_e \leqq 1.572$; pleochroism: E pale yellowish green, O dark bluish green. The optical negative character and a high birefringence with maximum at least 0.074 suggest, that the carbonate groups should lie in or close to the (0001) plane in the ewaldite structure [1, p. 194].

The IR spectrum of an ewaldite crystal is in agreement with the expected structure [9, p. 209].

The ewaldite from Kola has $H_M \sim 3$, it is brittle and has a density D(meas.) = 3.42 g/cm^3 [5]. The density of ewaldite from Wyoming D(meas.) = 3.25 ± 5 g/cm^3 [1, p. 192], [9. p. 203], it is low compared with the calculated density D(calc.) = 3.37 g/cm^3; this is due to a considerable amount of less dense, sodium-rich and carbonaceous inclusions present in all the crystals. Both D values are the same in ewaldite and mackelveyite [1, pp. 191/3].

Ewaldite is readily soluble at room temperature in dilute hydrochloric acid with effervescence [5].

References for 10.5.1.5:

[1] G. Donnay, J. D. H. Donnay, M. H. Hey (Tschermaks Mineral. Petrogr. Mitt. [3] **15** [1971] 185/200). – [2] C. Milton (Univ. Wyoming Contrib. Geol. **10** [1971] 57/63). – [3] G. Y. Chao, P. R. Mainwaring, J. Baker (Can. Mineralogist **16** [1978] 335/40, 337/8). – [4] L. Marble, A. Regis (Rocks Minerals **54** [1979] 4/25, 11). – [5] A. P. Khomyakov, T. A. Kurova, G. N. Muravitskaya, A. T. Timchenko (Dokl. Akad. Nauk SSSR **262** [1982] 964/6).

[6] V. Dimov, V. Breskovska, M. Maleev (Geokhim. Mineral. Petrol. **13** [1980] 37/44, 39/42). – [7] G. Donnay, H. Preston (Carnegie Inst. Washington Yearbook **1968/69** 288/90). – [8] G. Donnay, J. D. H. Donnay (Carnegie Inst. Washington Yearbook **1969** 218/9). – [9] G. Donnay, H. Preston (Tschermaks Mineral. Petrogr. Mitt. [3] **15** [1971] 201/12). – [10] C. Hintze (Handbuch der Mineralogie, Suppl.-Vol. 4, de Gruyter, Berlin 1974, pp.1/373, 27, 263).

[11] J. Fitzpatrick, A. Pabst (Am. Mineralogist **62** [1977] 158/63, 161/2).

10.5.2 Anhydrous Carbonates with Additional Anions

Review

The anhydrous carbonates of RE with additional anions (F, OH) form two series proceeding from bastnaesite. They are characterized by a successive replacement of RE by Ca in the series bastnaesite – parisite – roentgenite – synchysite and of RE by Ba in the series bastnaesite – cordylite – huanghoite – "Ba-parisite" – zhonghuacerite. In the RE-Ca series two additional subseries exist, caused by a replacement of a RE of the Ce group by a RE of the Y group and by a replacement of F by OH.

RE-Ca Series

The very detailed structural investigations of Donnay and Donnay explained the close correlation between the members of the bastnaesite-synchysite series with regard to their chemistry, structure, and physical properties, and especially the correlation of physical properties with the successive replacement of RE by Ca [1, pp. 932/63].

The **structures** of the members of the series are characterized by hexagonal layers of RE-F parallel to (0001) alternating with CO_3 groups in bastnaesite and with layers of Ca and of CO_3 groups in parisite and synchysite; the CO_3 triangles are arranged parallel to the c axis [2]. This structure is essentially in agreement with the scheme given for the series including roentgenite by Donnay and Donnay. The 4 members of the series have very similar values of a′ of about 4.1 Å and of a = 7.11 to 7.18 Å; the 3 species containing Ca are characterized by Ce-F layers alternating with various amounts of Ca layers, and two types of CO_3 groups, arranged between two Ce-F layers and between Ce-F and Ca layers, respectively [1, pp. 935/6, 943/5]. These layers have been designated as follows:

	RE-F	Ca	CO_3 between RE-F	CO_3 between RE-F and Ca	Ref.
1)	A	C	B	D	[3]
2)	d	e	f	g	[4]

The sequence of these layers in the four minerals of the series can be seen in **Fig. 5**. (The layer sequences given in table 1 of [4] are, however, incorrect.)

All members of the series are characterized by 1 to 3 pseudoperiods in c direction; the most marked pseudoperiod c‴ decreases from bastnaesite to synchysite from 4.9 to 4.6 Å; a well-marked pseudoperiod c″ is characteristic for parisite and synchysite, and a less marked one c′ occurs in all members. The true c values, obtained from rotation photographs, correspond to the pseudoperiods as follows:

Bastnaesite	c = 2c‴	—	c′ = c‴
Parisite	c = 3c′	= 18c‴	c′ = 6c‴
Roentgenite	c = 3c′	= 15c‴	c′ = 5c‴
Synchysite	c = 3c′	= 12c‴	c′ = 4c‴

The pseudoperiods and the true period in c direction are given in Fig. 5. The number of cation layers contained in the pseudocells ac′, 4 layers (2 of Ce, 2 of Ca) for parisite, and 5 layers (3 of Ce, 2 of Ca) for roentgenite, confirms the chemical formulas extrapolated from c‴ plotted against mol% $CaCO_3$ [1, pp. 935/7, 943/5]; it furthermore indicates vaterite (hexagonal μ-$CaCO_3$) which is formed by replacement of all Ce-F layers by Ca layers to be the end member of the series [5], a fact, later confirmed by [6, 7], cf. [8]. The considerable collapse in the c direction of 0.33 Å between adjacent Ce-F and Ca layers of the Ca-containing members of the series, and of 0.63 Å between adjacent Ca layers of vaterite, suggests that the CO_3 groups remain vertical in vaterite, but require an offset in the vertical sequence [6].

Syntaxic intergrowths and polycrystals. The existence of parallel intergrowths between bastnaesite, parisite and synchysite according to their close structural relation (identical a axes, similar RE-F planes) had been already supposed by Oftedal [2], and has been confirmed for the bastnaesite-synchysite series inclusive roentgenite. This series is characterized by syntaxy in the extended meaning for chemically different substances; it occurs between the members of the series (without bastnaesite-synchysite), when the pseudoperiods c‴ differ by as much as 5.5%, but not when the difference reaches 7.0%. The a axes of the two intergrown species may have the same sense or opposite senses. Such edifices, composed of two or more structurally distinct species in syntaxic intergrowth and simulating a single crystal, have been named polycrystals. The frequent occurrence of polycrystals in the bastnaesite-synchysite series is caused by the inability of the system $CeFCO_3$-$CaCO_3$ to form solid solutions. The alternation of two species requires periodically and discontinuously changing conditions (especially in the pH of the solution) during the precipitation of the polycrystals [1, pp. 939/42].

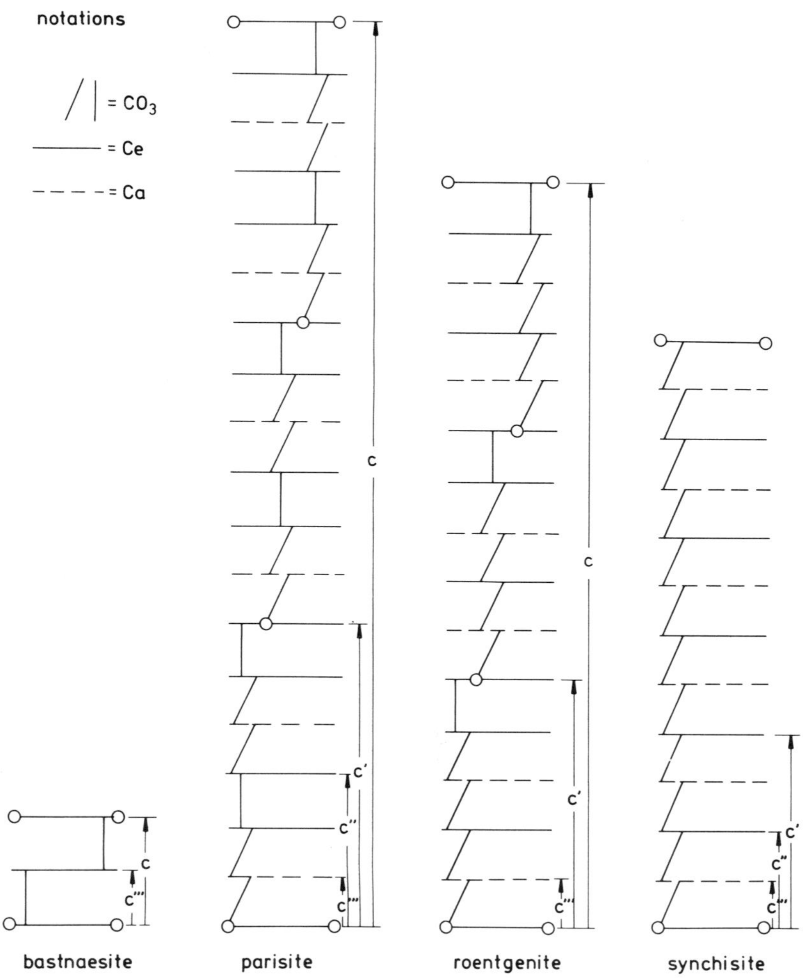

Fig. 5. Scheme of the RE fluorocarbonates from [1, p. 944]. The notations correspond to the letters of [3]: —— to A, – – – – to C, | to B, and / to D, cf. p. 90. The symbol ○ indicates the rhombohedral lattice points.

These polycrystals are the result of a superposition of platelets of bastnaesite (B) and of synchysite (S). The members of the series accordingly may be designated B, BS(parisite), BS_2(roentgenite) and S [7]. Besides these compounds three new mixed layer compounds were discovered by electron microscope and electron diffraction studies. They have the block symbols BS_4, B_3S_4, and B_3S_2 and are designated as block crystals [4]. The term block crystal has been used in a somewhat deviating sense for parisite, see pp. 146/7. For a review on ordered syntaxy and polytypism of the bastnaesite-synchysite series, see [3]. Examples for natural occurrences of syntaxic intergrowths and polycrystals of the B-S series, see pp. 119/20, 148, 155, and 168. Syntaxic intergrowths of BS, BS_2, BS_4, and S on a very fine scale indicate a perfect microsyntaxy in this series. Sequential faults (shown by electron microscope) and imperfect stacking sequences, which often prevent the determination of the exact layer sequence, lead to a variation in the ratio c'''/a' from specimen to specimen [4].

Physical properties. In the bastnaesite-synchysite series the refractive indices n_o and n_e decrease in a linear gradation from bastnaesite to synchysite and, extrapolated, to vaterite as the end member of the series [5, 6]. The decrease of n_o^2 and $1/n_e^2$ from bastnaesite to vaterite as a function of the volume fraction of vaterite in the minerals of this series is also observed [7]. The density D(calc.) decreases in the same direction [5].

The **chemical behavior** of the members of the series is characterized by a markedly increasing solubility in cold dilute acid from bastnaesite (insoluble) to parisite (slowly soluble), roentgenite (readily soluble) and synchysite (very quickly soluble); this permits the study of the physical nature of the intergrowths [1, p. 941].

Subseries of the RE-Ca Series with Y for Ce and OH for F

Yttrium analogues of the Ce-F carbonates described by several authors, see pp. 104/5, 141, and 159/60, always occur in fine-grained intergrowths or aggregates together with other minerals and generally appear to be restricted to late stage alteration products of monazite, xenotime, gadolinite, and other RE minerals [9]. Fine-grained reddish intergrowths of aggregates of yttrium fluorocarbonates often occur instead of well-developed polycrystals, which are common among the cerium analogues [10]. The existence of a series with members the bastnaesite-(Y)–parisite-(Y)–synchysite-(Y) analogous to the common series bastnaesite–synchysite with predominantly Ce-earth has been suggested by Mineev [11].

The transition from Ce- to Y members is connected with a reduction of the cell parameters, due to the decreasing ionic radii of the RE [12]; Y members may be indicated by a reduced parameter of a of about 4.01 Å, instead of 4.10 to 4.14 Å, for members with Ce-earth predominating [5], cf. tables on pp. 122, 148, and 169. A common property of all members with Y-earth is a red-brown color, which is possibly caused by their contents of Er and Nd [12]. The identification of minerals of the common bastnaesite-synchysite series by merely optical methods is lessened due to the overlapping of the optical properties of the Ce- and Y series [13].

The existence of a series of RE hydroxylcarbonates $RE[(OH)|CO_3]$–$CaRE[(OH)|(CO_3)_2]$–$CaCO_3$, analogous to the RE-F series, is proposed from mineral analyses by Mineev [14] and is further indicated by some hydroxylbastnaesites, which have been found as late stage or alteration products, see pp. 97, 103, and a hydroxylparisite in a carbonatite, see p. 140. Synthetic hexagonal $Ln(OH)CO_3$ of B type (bastnaesite type) under the electron microscope shows the same lamellar form as natural bastnaesite [15]. Small variations in the unit cell dimensions and in the physical properties of the members of the bastnaesite–synchysite series have been supposed to depend on the ratio $(OH)^-:F^-$ in these minerals [16].

RE-Ba Series

The existence of a second complete series originating from bastnaesite with $BaCO_3$-β (the hexagonal modification of witherite) as end member and the intermediate members cordylite $RE_2Ba[F_2|(CO_3)_3]$ and huanghoite $REBa[F|(CO_3)_2]$ has been supposed after the find of huanghoite [17], and has been confirmed by the find of "barium-parisite" $RE_2Ba_3[F_2|(CO_3)_5]$ [18], and of zhonghuacerite $REBa_2[F|(CO_3)_3]$ [19]. All these minerals found in China are considered to be complex salts (or double salts) of $RE[F|(CO_3)]$ and $BaCO_3$ with ratios between the two components in simple integers [18], cf. [17]. Cordylite, huanghoite, and "barium-parisite" have a very low content of yttrium-earth [18].

The unit cell data are characterized by relatively constant values of a, see for cordylite and huanghoite [8], for "barium-parisite" [18], and for zhonghuacerite [19]. The presence of Ba in the lattice of cordylite and huanghoite causes a change in the orientation of a large fraction of the CO_3 groups which become perpendicular to the c axis; this leads to a marked change in the c parameter [20]. The physical properties of the minerals of this series are not characterized by a linear relationship to the chemistry as it is the case in the RE-Ca series; the refractive indices n_o and n_e and the density D correspond to curves, which are apparently due to the Ba content of the minerals [17]. The optical negative character of cordylite and huanghoite, see pp. 179 and 183, is ascribed to the arrangement of CO_3 triangles perpendicular to the c axis [20], cf. [2].

References:

[1] G. Donnay, J. D. H. Donnay (Am. Mineralogist **38** [1953] 932/63). – [2] I. Oftedal (Z. Krist. **79** [1931] 437/64, 456, 461/4). – [3] P. E. Caro (J. Solid State Chem. **6** [1973] 396/401, 398/400). – [4] J. van Landuyt, S. Amelinckx (Am. Mineralogist **60** [1975] 351/8). – [5] E. I. Semenov (Tr. Inst. Mineral. Geokhim. Kristallokhim. Redk. Elem. No. 2 [1959] 181/6).

[6] J. D. C. McConnell (Mineral. Mag. **32** [1960] 535/44, 540/2). – [7] J. D. H. Donnay, G. Donnay (Bull. Soc. Franc. Mineral. Crist. **84** [1961] 25/9). – [8] E. I. Semenov (in: K. A. Vlasov, Geochemistry and Mineralogy of Rare Elements and Genetic Types of Their Deposits, Vol. 2: Mineralogy of Rare Elements, Israel Program Sci. Trans., Jerusalem 1966, pp. 249/328, 264/5, 276 [Russian Original, Moscow 1964]). – [9] A. A. Levinson, R. A. Borup (Am. Mineralogist **47** [1962] 337/43, 337). – [10] W. J. Verwoerd (S. Africa Geol. Surv. Ann. Rept. **2** [1963] 119/35, 127).

[11] D. A. Mineev (in: L. N. Ovchinnikov, N. A. Solodov, Mestorozhdeniya Litofil'nykh Redkikh Metallov, Nedra, Moscow 1980, pp. 178/84, 181/2). – [12] N. I. Aleksandrova, I. V. Maksimova, N. M. Svirskaya (Izv. Vysshikh Uchebn. Zavedenii Geol. Razved. **20** No. 10 [1977] 80/7, 81/2). – [13] F. Emiliani, G. Gandolfi (Mineral. Petrogr. Acta **11** [1965] 123/31, 127). – [14] D. A. Mineev (Geokhimiya Apogranitov i Redkometal'nykh Metasomatitov Severo-Zapadnogo Tarbagataya, Moscow 1968, pp. 1/183, 97). – [15] H. Dexpert, E. Antic-Fidancev, J. P. Coutures, P. Caro (J. Cryst. Spectrosc. Res. **12** No. 2 [1982] 129/42, 141).

[16] P. C. Sæbø, H. Neumann (Norsk Geol. Tidsskr. **41** [1961] 247/54, 249). – [17] P.-S. Chang [Chzhan, Pei-Shan] (Sci. Sinica [Peking] **11** [1962] 251/8, 252/3, 256/7 [russ.]; C.A. **57** [1962] 10799; Ref. Zh. Geol. **1963** No. 2 V 145). – [18] H.-C. Wang, H.-F. Su, C.-T. Kuo, S.-P. Li, C.-K. Wang (Ti Ch'iu Hua Hsueh **1973** No. 1, pp. 31/8; Geochimica [Peking] **1** [1973/74] 30/7, 30/1, 35/6). – [19] P.-S. Zhang, K.-J. Tao (Sci. Geol. Sinica [Ti Chih K'o Hsueh] **1981** 195/6; Mineral. Abstr. **33** [1982] 82 M/0707). – [20] A. S. Povarennykh (Crystal Chemical Classification of Minerals, Vol. 1, pp. 1/458, 181/2, Vol. 2, pp. 459/766, 616, translated from Russian by J. E. S. Bradley, Plenum, New York 1972).

10.5.2.1 Bastnaesite Group $(RE,Y,Th)[(F,OH)|CO_3]$

10.5.2.1.1 Name and Varieties

In a system of nomenclature for RE minerals Levinson has proposed to use the name bastnaesite as the group name, while minerals of this group with one rare earth element predominating over the other RE are characterized by the predominant RE as, e.g., bastnaesite-(Ce) or bastnaesite-(La); bastnaesite-(Ce) is usually the most widely distributed species [1]. This species is mostly called bastnaesite without notation, see pp. 94/9. Semenov uses the names La-bastnaesite, Nd-bastnaesite, and Y-bastnaesite [2] or yttrium-bastnaesite [3]. The last one, however, is really an intermediate member of the Ce-Yb series, in which Y has not been determined [4].

A mineral with $Th \gg RE$ has been named thorbastnaesite [5], cf. [6], see pp. 104, 109. Minerals with a relative high content of water (H_2O^+ or hydroxyl groups) and relative low content of fluorine have been designated as hydroxylbastnaesite [7, 8]. Besides these varieties

Povarennykh has enumerated La-hydroxylbastnaesite, Ca-yttrobastnaesite, Ce-yttrobastnaesite, and Dy-yttrobastnaesite [9].

Mineral names, that had been given to minerals which are identical or similar to bastnaesite, are beiyinite from Beiyin Obo, Inner Mongolia [10, 11], and buszite [12], cf. [13].

Kyshtymite (kis(c)htimite) from Kyshtym, Urals, deviates somewhat from bastnaesite by its chemical and physical properties, especially by a low F content (⅓ to ½ that of bastnaesite) and an elevated H_2O content; it is denoted to be a variety of bastnaesite with hydroxyl groups instead of fluorine [14], cf. [15]. Consequently, it is a member of the series bastnaesite–hydroxylbastnaesite [8], which further is characterized by relative high La content [13, 14]. The original name "kyshtymo-parisite" [11] has been shown to be inappropriate for this mineral, as it contains no Ca [16, 17].

Weibyeite is a pseudomorph after ancylite, consisting largely of bastnaesite together with some relics of ancylite and some octahedral-shaped bastnaesite-ancylite pseudomorphs after zircon [18], cf. [13].

Two types of the hydroxycarbonate $RE(OH)CO_3$ have been synthesized, one of them, the B type, is isostructural with bastnaesite [19, 20]. For the A type see p. 213.

References for 10.5.2.1.1:

[1] A. A. Levinson (Am. Mineralogist **51** [1966] 152/8, 153/5). – [2] E. I. Semenov (in: R. P. Tikhonenkova, E. I. Semenov, Mineralogiya Pegmatitov i Gidrotermalitov Shchelochnykh Massivov, Nauka, Moscow 1967, pp. 1/168, 36/40). – [3] E. I. Semenov (in: K. A. Vlasov, Geochemistry and Mineralogy of Rare Elements and Genetic Types of Their Deposits, Vol. 2: Mineralogy of Rare Elements, Israel Program Sci. Trans., Jerusalem 1966, pp. 249/328, 266 [Russian Original, Moscow 1964]). – [4] D. A. Mineev, T. I. Lavrishcheva, A. V. Bykova (Zap. Vses. Mineral. Obshchestva **99** [1970] 328/32, 328 FN). – [5] A. S. Pavlenko, L. P. Orlova, M. V. Akhmanova, K. I. Tobelko (Zap. Vses. Mineral. Obshchestva **94** [1965] 105/13, 111).

[6] M. Fleischer (Am. Mineralogist **50** [1965] 1504/8, 1505). – [7] A. S. Kirillov (Dokl. Akad. Nauk SSSR **159** [1964] 1048/50; Dokl. Earth Sci. Sect. **159** [1964] 93/5). – [8] A. S. Kirillov (Zap. Vses. Mineral. Obshchestva **95** [1966] 51/9, 53, 55, 57). – [9] A. S. Povarennykh (Crystal Chemical Classification of Minerals, Vol. 2, translated from Russian by J. E. S. Bradley, Plenum, New York 1972, pp. 459/766, 614). – [10] T. L. Ho (Bull. Geol. Soc. China [Peking] **14** [1935] 279/82).

[11] C. Palache, H. Berman, C. Frondel (The System of Mineralogy, 7th Ed., Vol. 2, Wiley, New York 1951, pp. 1/1124, 289, 291). – [12] J. Goñi, C. Guillemin (Bull. Soc. Franc. Mineral. Crist. **76** [1953] 124/9). – [13] H. Strunz, C. Tennyson (Mineralogische Tabellen, 6th Ed., Geest and Portig, Leipzig 1977, pp. 1/621, 243, 512, 587). – [14] N. V. Svyazhin (Tr. Inst. Geol. Akad. Nauk SSSR Ural'sk. Filial No. 70 [1965] 249/52). – [15] T. L. Sverdrup, K. Ø. Bryn, P. C. Sæbø (Norsk Geol. Tidsskr. **39** [1959] 237/47, 246).

[16] A. Lacroix (Bull. Soc. Franc. Mineral. **35** [1912] 108/13, 113). – [17] V. Silberminz [V. A. Zil'bermintz] (Compt. Rend. Acad. Sci. USSR A **1929** 55/60, 55/6). – [18] P. C. Sæbø (Norsk Geol. Tidsskr. **43** [1963] 441/3). – [19] J. Sawyer, P. Caro, L. Eyring (Rev. Chim. Minerale **10** [1973] 93/104, 93/4). – [20] J. A. K. Tareen, T. R. N. Kutty, I. Mohamed (Indian Mineralogist **21** [1980] 43/8, 43, 46).

10.5.2.1.2 Occurrence and Paragenesis

10.5.2.1.2.1 Review

Bastnaesite, formerly known from a few occurrences only, has been considered to be most frequently a contact-metamorphic mineral [1]; later, however, it has also been found in granite and syenite pegmatites, in veins together with other carbonate minerals, fluorite, and barite, and in alluvial deposits [2]. In a review of 27 occurrences of bastnaesite in 1964, a genetically broader range of formations was visible: the most numerous occurrences (10) are magmatic or probably magmatic, including pegmatites and aplites, 3 are from carbonatites and 3 from contact metasomatic occurrences, 5 from hydrothermal occurrences, and 6 bastnaesites are secondary alteration products of allanite (4) and tysonite (2) in granites or pegmatites [3, pp. 235/7]. According to this compilation bastnaesite is apparently formed in a wide range of p-T conditions and is predominantly associated with other RE minerals [4], cf. [5]. Bastnaesite is one of the most common RE minerals, characteristic of deposits of the miaskite alkali series and of hydrothermal deposits related to alkali granosyenites. Occasionally it occurs in alkali granite pegmatites and as an accessory mineral in granites. Bastnaesite is generally a late hydrothermal and supergene mineral replacing other Ce minerals; a most unusual occurrence is supergene bastnaesite in the weathering crust of one of the alkali massifs in Siberia [6].

10.5.2.1.2.2 Magmatic Rocks

Granites and Granite Pegmatites

Bastnaesite is probably a primary accessory mineral in the coarse-grained Redstone granite from Westerly, Rhode Island [7], and in biotite granite of the Iisori massif, Tarbagatai, East Kazakhstan [8, p. 16]; it is a secondary mineral in the granite of the Elberton area, Georgia [9]. As "black bastnaesite" it has been found in the magnetic part of the heavy mineral fraction of the granite from Honkamäki, Finland [10].

In the **anatectic granitoids** of the Mecsek Mountains, South Hungary, bastnaesite of secondary origin (with RE mobilized at granitization) is a frequent mineral in: 1) porphyroblastic microcline granodiorite, 2) fine-grained veined or aplitic granites, and 3) veins of biotite amphibolite (here with a marked Y content); the minerals of the bastnaesite group are assumed to be more abundant in rock types influenced by carbonate solutions, which dissolve RE from allanite [11], see p. 103.

In **alkali granites** bastnaesite occurs in the ekerite from the Oslo region, Norway, together with many fluorides [12], in riebeckite granite from the Mount Rosa, and in the alkali-rich granite from the Pikes Peak, Colorado [13]. In riebeckite granitoides bastnaesite is assumed to be a mineral of the "carbonate stage" [14], cf. metasomatic formations pp. 97/8.

Bastnaesite has been recorded from several **granite pegmatites**; where it often is a secondary alteration product of other RE minerals (see pp. 103/5): it occurs in Black Cloud pegmatite near Lake George and in St. Peter's Dome pegmatite, further in small amounts in the pegmatites of the Silver Plume granite near Jamestown, Colorado [15], in the latter associated with cerite [16], and in pegmatite of the Mount Rosa area, El Paso and Teller Co., Colorado [17]. In the Scrub Oaks mine, Morris Co., New Jersey, bastnaesite, intimately associated with leucoxene, occurs in form of spots together with xenotime and synchysite-(Y) (doverite) in the pegmatite, but more abundantly in the coarse-grained magnetite ore; as small irregular colloidal or microcrystalline masses bastnaesite encloses aggregates of synchysite-(Y) (doverite) and chevkinite [18], cf. [19, 20], for a description of this deposit, see "Rare Earth Elements" A 8, 1984, Chapter 11.7.2.3.2. Further bastnaesite has been found in some pegmatites of the region of Ambatofinandrahana, Madagascar [21], where it is associated with biotite and monazite, and has been formed at a relatively high temperature of about 800°C [22], see also "Rare Earth Elements" A 8, 1984, Chapter 11.7.4.3. Bastnaesite has been found in grains together with synchysite-(Y) (doverite) in a radioactive pegmatitic to aplitic zone adjacent to alkali granite of the Jabal Sayid complex, Saudi Arabia [23]. A fine-grained bastnaesite occurs together with aeschynite-priorite and ilmenite in an allanite-monazite-quartz rock near the confluence of the rivers Lunsemfwa and Masano in Zambia [24].

The widespread occurrence of minerals of the bastnaesite-hydroxylbastnaesite group together with allanite in many granites and granite-pegmatites is clearly related to a progressive development of bastnaesite from allanite by a postmagmatic, hydrothermal or by a supergene process of "bastnaesitization" [25]; bastnaesite is more wide-spread especially as a secondary mineral in pegmatites containing allanite than previously supposed [26]. For such secondary occurrences of bastnaesite in some granites and in numerous pegmatites from the USA, Japan, Norway, and the USSR, see pp. 103/4.

Alkali Syenites and Related Rocks

Bastnaesite occurs in the nordmarkite of Flaen near Grorud, Oslo region, Norway, together with helvine and a great number of other minerals [27] (this occurrence has been designated "Grefsen syenite" by [28]). It has been found in nepheline syenite pegmatites of the Langesund-fjord, of Langangen, of Sandefjord, and of Vesterøya, all in South Norway [29]. Together with other RE carbonates it occurs in the veins and vugs of the nepheline syenite from the Mont St. Hilaire, Quebec [30], especially on aegirine in carbonatite vugs in pegmatitic dikes of this occurrence [31]. It has been further observed together with parisite in a syenite pegmatite near Marathon, Ontario [32], in a nepheline syenite pegmatite of the Pichekhol' massif, Eastern Tuva [33], and together with albite, aegirine, and/or acmite, besides neptunite and rare Th minerals as the latest formations in green veins, which surround the Ilímaussaq alkali intrusion in South Greenland [4]. Bastnaesite is also present in alkali dike rocks of the Augusta County, Virginia [34], and, close intergrown with parisite and cordylite, in unspecified alkali rocks of the East Pamir [35]. Some bastnaesites related to quartz- and granosyenites and to K-alkalic basaltoid rocks are of hydrothermal origin, see pp. 98/9; this is also valid for most of the bastnaesite varieties occurring together with carbonatitic rocks, see in the following section.

Carbonatites

Bastnaesite is rare in carbonatites of the hypabyssal type (Kola, Siberia), and frequent only in carbonatite stocks of the surface-near subvolcanic type (Tundulu, Chilwa Lake, Kangan-kunde in East Africa) and in independent dikes of late carbonatites (Karonge, Burundi, and Mountain Pass, California); it is common in ankerite-dolomite veins, where it occurs together with parisite, fluorite, quartz, thorite, and monazite in form of impregnations or micro-veinlets [36], cf. bastnaesite together with parisite, strontianite, and sulfides in typical hydrothermal-metasomatic late carbonatites [37, p. 244/6]. Large quantities of bastnaesite occur in the Sulphide Queen carbonate body and in numerous smaller veins and mineralized shear zones in the Mountain Pass district, California, see "Rare Earth Elements" A 8, 1984, Chapter 11.7.2.3.3.2. In this locality the bastnaesite is fine-grained and interlocked with many of the associated minerals [38], e.g., with barite [39] and with sahamalite [40], see p. 69; it is especially intimately intergrown with calcite [41]. Bastnaesite is only rarely replaced by younger minerals [2], see p. 106.

Most of the bastnaesites in carbonatites are characterized by their occurrence in **late stage veins** of predominantly dolomitic to ankeritic, sometimes also sideritic, or late calcitic compositions. Bastnaesites of this type occur in Th-bearing veins in the Powderhorn district, Gunnison Co., Colorado [42], cf. [43], [3, p. 237], and closely associated with ancylite, silicates, apatite, and monazite, further with sulfates, sulfides, and oxides, among them Nb-bearing rutile (ilmenorutile), in a probably metasomatic association resembling carbonatite in the northern Lemhi Co., Idaho [44]; for a description of these two deposits, see "Rare Earth Elements" A 8, 1984, Chapters 11.7.2.3.3.2 and 11.7.2.3.3.1, respectively. Further bastnaesites are known from dolomite-ankerite veins on the Prince of Wales Island, Salmon Bay, Alaska [45], in the Cerro Manomo carbonatite complex, eastern Bolivia [46], cf. for Cerro Manomo also "Rare Earth Elements" A 8, 1984, Chapter 11.7.3.2, in dolomite carbonatite of the Kovdor massif, and in ankerite veins from Vuoriyarvi, Kola [47]; in the latter occurrence bastnaesite appears in form of hydroxylbastnaesite [48], which indicates a deficit in fluorine at its formation [49]. The common bastnaesite with fluorine occurs together with strontianite in the fourth ankeritic stage, and the hydroxylbastnaesite occurs together with barite in the fifth, late calcitic stage of an East Siberian carbonatite massif; both minerals are pseudomorphs, hydroxylbastnaesite also forms veinlets [50, 51]. Small amounts of bastnaesite, besides parisite as the dominating RE carbonate, are typical of Fe-poor dolomite-ankerite carbonatites, formed at low temperature (200 to 300°C) in East Sayan, Siberia [49], as the youngest minerals, besides burbankite, huanghoite, witherite, strontianite, magnetite, and sulfides; they are characteristic of dolomite-ankerite formations in the carbonatite complex from Arbarastakh, Yakutia, Siberia [52]. Further bastnaesite, besides monazite and goyazite, occurs in late stage carbonatites of the Wigu Hill, Tansania, and, besides monazite and florencite, in the carbonatite of Kangankunde, Malawi [53], in the latter occurrence in mottled carbonatitic rocks [54]. In the carbonatite of the Tundula area, southern Malawi, bastnaesite is a mineral of ankeritic association [3, p. 237]; it has been also found in dark manganiferous carbonatites, in apatite-sövite, and more frequently in less pure apatite rocks than in pure apatite rocks; further in pure beforsite of the Tundulu Ring complex and in beforsite veins from the Nathace Hill southwest of Tundulu [55]. In the carbonatite of the Chilwa Island in southern Malawi bastnaesite occurs in dark brown sideritic carbonatite and in druses of another rock of this type [56]. A brecciated carbonatite dike ore body of Karonge, Burundi, is cemented by quartz and bastnaesite [37, p. 267].

Metasomatic Rocks

The average concentration of bastnaesite increases from 5 to 7 ppm in normal granites to 67 to 900 ppm in albitized and fluoritized varieties up to as much as 2000 ppm in rapakivi granites from Karelia [25]. Bastnaesite is present in biotite-microcline apogranites, riebeckite-apogranites, quartz-feldspar-metasomatites and -pegmatoids of alkali massifs of the Northwest Tarbagatai, East Kazakhstan [8, pp. 15/6, 25, 28/9, 34/5], it occurs in an albite-aegirine metasomatite of the zentral zone of the alkali complex Burpala, Northern Pribaikalia, Siberia [57], and in albitites of Siberia, where it occurs in aggregates with quartz, fluorite, microcline, and RE minerals [58]. Bastnaesite further occurs in albitized pegmatites and in hydrothermally altered, disturbed pegmatites of the Transbaikal region, Siberia [59], in albitized aegirine-augite syenites of the Vishnevye Gory, South Urals [60], and in strongly albitized and cancrinitized varieties of alkali rocks (biotite miaskites and miaskite pegmatites) and other syenitic rocks from Borsuksai, Mugodzhary, northwestern Kazakhstan [61]. In metasomatic albitites of the European part of the USSR, that consist of apogranites and apodiabases (= aegirine-amphibole-albite-metasomatites), bastnaesite occurs as secondary mineral after britholite and allanite in two or three varieties [62, 63], cf. [64]. Further bastnaesite occurs in radioactive microcline bodies in a metasomatically altered quartzmonzonite in the southern Caballo Mountains, Sierra Co., New Mexico [65], and is disseminated together with U minerals, fluorite, micas, and pyrite in a (contact-)metasomatically influenced trachytic rock of the Rexspar property, British Columbia, Canada [66].

In **fenites** (= fenitized exocontact rocks) of a miaskite intrusion in the Vishnevye Gory, South Urals, bastnaesite is typomorphic (besides allanite) for 1) the aegirine-augite-microcline-fenite-aureoles and 2) the postmagmatic pneumatolytic-hydrothermal stage of carbonatization with development of calcite veinlets in the fenites [67]; it occurs in quartz-feldspar veins, in silicified miaskite pegmatites, and in other fenitized country rocks, where in ovoidal aggregates a core of bastnaesite is surrounded by several RE minerals in concentric shells [68]; in quartz lenses of the aegirine-augite fenites the bastnaesite is intimately associated with rhabdophane, which occurs on cleavage planes of the bastnaesite [69]. Bastnaesite is further concentrated as an accessory mineral in carbonatized and albitized nepheline-bearing fenites of the massif Soustova south of the Khibina pluton, Kola [70], and occurs as a minor accessory mineral in fenitized perthosites (=syenitic rocks of metasomatic origin) in the alkali complex of Bagbe, Gola Forests, SE Sierra Leone [71]. For hydroxyl-containing kyshtymite, thorbastnaesite, and bastnaesite-(Y) as secondary minerals, see pp. 103/5.

At the type locality of bastnaesite, the Mg metasomatic cerite deposit of Bastnäs, Sweden, this mineral replaces tysonite in the cerite ore, often along cleavage cracks [72], cf. "Seltenerdelemente" A 1, 1938, p. 57. Later it has been found at three additional localities in the neighboring Norberg district (Östanmossa, Södra Hackspik, and Stripåsen) and is assumed to be also related to the widespread Mg metasomatism [73]. Further bastnaesite occurs in pneumatolytic-hydrothermal metasomatites of Siberia, of the Ukraine, and of the Krasnoyarskii Krai [74].

Bastnaesite together with tourmaline, beryl, graphite, sulfides, and other minerals has been found in biotite-albite-phlogopite **greisen** of the USSR [75], and in the pneumatolytic quartz-bastnaesite-tourmaline-fluorite paragenesis in the granite of San Martin, province San Luis, Argentina [76]. At several localities in the Potgietersrus tin fields, Transvaal, South Africa, bastnaesite occurs in syngenetic quartz-tourmaline ore bodies or lodes in the Bobbejaanskop granite and in the pegmatitic Lease granite. It is one of the latest (mesothermal) minerals and markedly younger than the quartz-tourmaline bodies, where it frequently fills fissures and fractures; it is also younger than the associated minerals cassiterite, tysonite, parisite, and pyrite; sometimes it is accompanied by a Ca-Y silicate (kainosite?), by other Y silicates and RE minerals, by wolframite, scheelite, and many other sulfides, among them especially molybdenite [77].

Hydrothermal Formations

A late hydrothermal character of bastnaesites is predominantly confirmed for those in deposits related to alkali-granosyenites, and for those in hydrothermal quartz- and quartz-barite veins [74], and for bastnaesites in epithermal quartz-barite-fluorite-sulfide veins [3, pp. 233/5], cf. [4].

In the quartz-barite veins of the Karonge area, Burundi, bastnaesite, associated mainly with barite and sulfides, occurs in irregular fine quartz veinlets in brecciae and schists [78] (cf. [74]), and it is distributed on phosphate minerals, such as monazite and rhabdophane, on goethite, and on silicates [79], see also "Rare Earth Elements" A 8, 1984, Chapter 11.7.4.3. In the epithermal fluorite-copper-bastnaesite deposits (veins and breccia fillings) of the Gallinas Mountains, New Mexico, bastnaesite is an ubiquitous accessory mineral and occurs in nearly every one of the deposits, associated with quartz, barite, fluorite, and the sulfides [3, pp. 227/34], or associated with baritocalcite, calcite, Fe oxides and hydroxides [1]. In the Red Cloud mines it is accompanied by a fine-grained purple fluorite matrix [80], cf. for this and other mines of this district [81] and "Rare Earth Elements" A 8, 1984, Chapter 11.7.2.3.3.2. Further occurrences of this type are: bastnaesite ("beiyinite") associated with magnetite, pyrite, and barite in fluorite veins in the iron ore deposit of Beiyin Obo, Inner Mongolia [82], a Th-containing bastnaesite associated with fluorite, calcite, barite, and numerous other minerals in veins of Beylikahir, West-Anatolia, Turkey [83], bastnaesite in a late stage hydrothermal Pb-Zn mineralization in quartz veins with fluorite, barite, and sulfides, cementing brecciated material in granites of the North Rhodopes, Bulgaria [84], and bastnaesite in a hydrothermal-metasomatic quartz-fluorite formation in an altered syenite in the southeastern Ukraine, where it replaces chevkinite [85], see also p. 104.

In hydrothermal iron ores bastnaesites are present in sideritic and hematitic associations related to quartz syenite in the western Tannu-Ola, Tuva, Siberia [86], cf. the same associations with bastnaesite related to a granosyenite of an unspecified locality in the USSR [87]. In these associations bastnaesite generally occurs together with fluorite in the sideritic type and is mostly enriched in the hematitic type of these hydrothermal ore formations, which have been designated as carbonate-fluorite-hematite metasomatites and are found, related to iron-ore bodies, in Siberia and Central Asia [37, pp. 271, 274, 276]. For another hydrothermal bastnaesite occurring together with huanghoite, other RE minerals and fluorite in Inner Mongolia, see p. 181. At Morro do Ferro in the Pocos de Caldas district, Minas Gerais, Brazil, bastnaesite occurs together with allanite in a U-Th-containing zircon-baddeleyite ("caldasite") ore; the mineralization with RE and Th follows the emplacement of magnetite [88]. In a U-Th-vein system of the Bokan Mountains, southeast Alaska, bastnaesite is common in some veins besides five other RE minerals and numerous other minerals [89].

In the North Baikal alkali province bastnaesite and parisite are minerals in zeolite-calcite veinlets of a hydrothermal carbonate association, which is related to rocks of ultrabasic to syenitic composition [90, 91]. In a hydrothermal formation related to K-alkaline basaltoid rocks of the Verkhnedunkel'dykskii massif, South Pamir, bastnaesite, parisite, and cordylite are accessory minerals [92]. Further bastnaesite occurs together with synchysite in calcite veinlets and joints cutting dark gray shale and siltstone within the pyrite zone of the basal Nonesuch shale, northern Michigan [93].

In Alpine fissure veins homogeneous bastnaesite crystals have been found only at the Ovi di Scimfuss, Gotthard, Switzerland [94], cf. [95]; in fissures in the southeastern Gotthard massif and at some other localities in Switzerland bastnaesite occurs in syntaxic intergrowths with clearly dominating synchysite, see p. 168.

10.5.2.1.2.3 Sedimentary and Metamorphic Rocks

In the **weathering zone** of an alkali massif in the USSR, bastnaesite occurs in fine-dispersed argillaceous masses, preferably in ferrihalloysite, associated with limonite, pyrolusite, and other supergene minerals [96]; in the kaoline fraction on a weathered gneiss, but not in the soil above this gneiss at Ekiti, Ibadan, Nigeria, bastnaesite probably is a weathering product of allanite [97]. The species bastnaesite-(Nd) occurs together with other supergene RE minerals especially phosphates) in the weathering crust of a metamorphic rock (restite with different Ti minerals) in Kazakhstan [98]. Further bastnaesites occur in the weathering crusts of granites and in **placers:** in some buried coastal placers of the Ukraine the bastnaesite concentration exceeds 4000 to 8000 ppm [25]; in pebbles in gold placers from the confluence of the Mochaline-log and the Borzovka river, Kyschtym district, Urals, bastnaesite (originally described as kyshtymite) is constantly accompanied by allanite, cerite, lessingite, and other minerals [99].

In the lower part of the **bauxite** profile in the San Giovanni Rotondo deposit, Italy, secondary bastnaesite fillings occur in fissures or micropores [100]; on the footwall limestone of the Marmara bauxite deposit in Greece a mineral, filling fissures and micropores in a brindleyite (= nimesite) coating, has been determined to be bastnaesite-(La), formed by descending solutions [101], and accompanied by bastnaesite-(Ce), collected only a few meters apart [102].

In **metamorphic rocks** bastnaesite has been found in dolomite geodes in a talc quarry from Trimouns, Luzenac, Département Ariège, France [103], and hydroxylbastnaesite occurs in coarse-crystalline calcite associated with stillwellite, monazite, garnet, titanite, pyrite, K-feldspar, and quartz in veins and lenses in sugary diopside-calcite rock enclosed in marble at Desmont Mining Corporation property near Wilberforce, Ontario [104].

In deep **ocean water** a crystal of bastnaesite has been identified by its high refractivity [105].

References for 10.5.2.1.2.1 to 10.5.2.1.2.3:

[1] J. J. Glass, R. G. Smalley (Am. Mineralogist **30** [1945] 601/15, 604/8, 613). – [2] J. C. Olson, D. R. Shawe, L. C. Pray, W. N. Sharp (U.S. Geol. Surv. Profess. Paper No. 261 [1954] 1/75, 34/6, 41). – [3] R. M. Perhac, E. W. Heinrich (Econ. Geol. **59** [1964] 226/39). – [4] J. Hansen (Medd. Groenland **181** No. 8 [1968] 1/47, 19, 31/2). – [5] L. B. Zubkov, N. I. Galadzheva, A. B. Chernyakovskii (Dokl. Akad. Nauk Belorussk. SSR **16** No. 7 [1972] 646/9).

[6] E. I. Semenov (in: K. A. Vlasov, Geochemistry and Mineralogy of Rare Elements and Genetic Types of Their Deposits, Vol. 2: Mineralogy of Rare Elements, Israel Program Sci. Trans., Jerusalem 1966, pp. 249/328, 269 [Russian Original, Moscow 1964]). – [7] W. L. Smith, E. A. Cisney (Am. Mineralogist **41** [1956] 76/81, 77). – [8] D. A. Mineev (Geokhimiya Apogranitov i Redkometalnykh Metasomatitov Severo-Zapadnogo Tarbagataya, Moscow 1968, pp. 1/183). – [9] L. T. Silver, M. Grunenfelder (Geol. Soc. Am. Bull. **68** [1957] 1796). – [10] V. Marmo, V. Hoffrén, K. Hytönen, P. Kallio, O. Lindholm, J. Siivola (Bull. Comm. Geol. Finlande No. 221 [1966] 1/34, 17/8).

[11] G. Panto (Acta Geol. Acad. Sci. Hung. **19** [1975] 59/93, 62/3, 77, 88). – [12] G. Raade, J. Haug (Mineral. Rec. **11** [1980] 83/91, 83). – [13] E. B. Gross, E. W. Heinrich (Am. Mineralogist **50** [1965] 1273/95, 1279/81). – [14] Yu. L. Kapustin (in: L. N. Ovchinnikov, N. A. Solodov, Mestorozhdeniya Litofil'nykh Redkikh Metallov, Nedra, Moscow 1980, pp. 208/25, 211/2, 217). – [15] C. V. Haynes (in: S. M. Del Rio, Mineral Resources of Colorado First Sequel, Mineral Resources Board, Denver, Colo., 1960, pp. 370/85, 376/7).

[16] E. N. Goddard, J. J. Glass (Am. Mineralogist **25** [1940] 381/404, 385/8, 401). – [17] E. B. Gross, E. W. Heinrich (Am. Mineralogist **51** [1966] 299/323, 307, 312/3). – [18] H. Klemic, A. V. Heyl, A. R. Taylor, J. Stone (U.S. Geol. Surv. Bull. No. 1082-B [1959] 29/60, 39, 46/7, 52). – [19] W. L. Smith, J. Stone, D. D. Riska, H. Levine (Science **122** [1955] 31). – [20] W. L. Smith, J. Stone, D. R. Ross, H. Levine (Am. Mineralogist **45** [1960] 92/8, 92/4).

[21] T. G. Murdock (U.S. Bur. Mines Inform. Circ. No. 8196 [1963] 1/147, 102). – [22] J. Behier (Rep. Malgache Ann. Geol. Madagascar No. 29 [1960] 1/78, 18). – [23] A. R. A. Turkistany, K. Koyama (IAG [Inst. Appl. Geol.] Bull. **3** [1980] 99/104, 99/100). – [24] S. Vrána (Rec. Geol. Surv. Zambia **12** [1972] 91/4). – [25] D. A. Mineev, K. I. Rozanov, N. V. Smirnova, T. I. Matrosova (Dokl. Akad. Nauk SSSR **210** [1973] 925/8; Dokl. Earth Sci. Sect. **210** [1973] 149/52).

[26] T. L. Sverdrup, K. Ø. Bryn, P. C. Sæbø (Norsk Geol. Tidsskr. **39** [1959] 237/47, 237, 247). – [27] I. Oftedal, C. Sæbø (Norsk Geol. Tidsskr. **43** [1963] 405/9). – [28] G. Raade (Mineral. Rec. **3** No. 1 [1972] 7/11). – [29] G. Raade, J. Haug, R. Kristiansen, A. O. Larsen (Lapis **5** No. 10 [1980] 22/8, 24, 26). – [30] G. Y. Chao, D. C. Harris, A. W. Hounslow, J. A. Mandarino, G. Perrault (Can. Mineralogist **9** [1967] 109/23, 112/3).

[31] L. Marble, A. Regis (Rocks Minerals **54** [1979] 4/25, 6/7, 11). – [32] J. Mandarino, W. M. Tovell (Can. Mineralogist **7** [1963] 819). – [33] Yu. L. Kapustin (Geokhimiya **1982** 533/40, 540). – [34] Anonymous (U.S. Geol. Surv. Profess. Paper No. 450-A [1962] A-1/A-257, A-86). – [35] E. A. Dmitriev (Dokl. Akad. Nauk Tadzh. SSR No. 3 [1961] 29/31).

[36] Yu. L. Kapustin (Mineralogiya Karbonatitov, Moscow 1971, pp. 1/288, 184/5). – [37] Yu. L. Kapustin (in: L. N. Ovchinnikov, N. A. Solodov, Mestorozhdeniya Litofil'nykh Redkikh Metallov, Nedra, Moscow 1980, pp. 243/85). – [38] V. E. Shaw (U.S. Bur. Mines Rept. Invest. No. 5474 [1959] 1/12, 2). – [39] J. B. Zadra, A. L. Engel, E. S. Shedd, A. C. Rice (U.S. Bur. Mines Rept. Invest. No. 4919 [1952] 1/15, 1, 7). – [40] H. W. Jaffe, R. Meyrowitz, H. T. Evans (Am. Mineralogist **38** [1953] 741/54, 752).

[41] H. W. Jaffe (Geol. Soc. Am. Bull. **66** [1955] 1247/56, 1255/6). – [42] J. C. Olson, S. R. Wallace (U.S. Geol. Surv. Bull. No. 1027-O [1956] 693/723, 702). – [43] D. C. Hedlund, J. C. Olson

(U.S. Geol. Surv. Profess. Paper No. 424-B [1961] 283/6). – [44] A. L. Anderson (Econ. Geol. **55** [1960] 1179/201, 1184, 1189, 1196). – [45] J. R. Houston, R. G. Bates, R. S. Velikanje, H. Wedow (U.S. Geol. Surv. Bull. No. 1058-A [1958] 1/31, 10).

[46] C. J. N. Fletcher, J. D. Appleton, B. C. Webb, I. R. Basham (Trans. Inst. Mining Met. B **90** [1981] 37/50, 40/1). – [47] Yu. L. Kapustin (in: K. A. Vlasov, Mineralogiya i Geneticheskie Osobennosti Shchelochnykh Massivov, Moscow 1964, pp. 135/94, 170). – [48] A. S. Kirillov (Zap. Vses. Mineral. Obshchestva **95** [1966] 51/9, 51, 58). – [49] M. Ya. Somina, A. G. Bulakh (Mineral. Geokhim. **1972** No. 4, pp. 64/73, 64/5, 73). – [50] T. B. Zdorik (Tr. Mineral. Muzeya Akad. Nauk SSSR No. 17 [1966] 60/75, 61, 68/70).

[51] T. B. Zdorik (Geol. Mestorozhd. Redk. Elem. Vses. Nauchn. Issled. Inst. Mineral'n. Syr'ya No. 30 [1966] 121/31, 125/7). – [52] A. G. Zhabin, N. G. Shumyatskaya, N. S. Samsonova (Tr. Mineral. Muzeya Akad. Nauk SSSR No. 20 [1971] 202/4). – [53] D. McKie (Mineral. Mag. **33** [1962] 281/97, 282, 295/6). – [54] D. N. Holt (Malawi Geol. Surv. Dept. Bull. **20** [1965] 1/130, 14; Mineral. Abstr. **17** [1965] 369). – [55] M. S. Garson (Nyasaland Geol. Surv. Dept. Mem. No. 2 [1962] 1/248, 108/9, 167).

[56] M. S. Garson, W. C. Smith (Nyasaland Geol. Surv. Dept. Mem. No. 1 [1958] 1/127, 51/3). – [57] A. P. Khomyakov (Izv. Akad. Nauk SSSR Ser. Geol. **1972** No. 1, pp. 94/101, 98). – [58] I. I. Kupriyanova (Tr. Mineral. Muzeya Akad. Nauk SSSR No. 18 [1968] 206/9). – [59] A. G. Mironov, A. M. Gofman (Tr. Geol. Inst. Buryat. Fil. Sib. Otd. Akad. Nauk SSSR No. 23 [1980] 64/73, 65, 71/2). – [60] E. B. Khalezova, I. I. Nazarenko (Tr. Inst. Mineral. Geokhim. Kristallokhim. Redk. Elem. Akad. Nauk SSSR No. 2 [1959] 99/101).

[61] V. A. Cheverdin (Tr. Kaz. Nauchn. Issled. Inst. Mineral'n. Syr'ya No. 2 [1960] 89/91). – [62] L. V. Ganzeeva [Ganzeyeva], A. A. Ganzeev [Ganzeyev] (Geokhimiya **1975** 74/80; Geochem. Intern. **12** [1975] 47/52, 47, 49). – [63] L. V. Ganzeeva (Redk. Elem. Syr'e Ekon. No. 7 [1972] 112/24, 112, 114, 118). – [64] N. A. Solodov, K. D. Subbotin (in: L. N. Ovchinnikov, N. A. Solodov, Mestorozhdeniya Litofil'nykh Redkikh Metallov, Nedra, Moscow 1980, pp. 233/43, 238/9). – [65] M. H. Staatz, J. W. Adams, N. M. Conklin (U.S. Geol. Surv. Profess. Paper No. 525-D [1965] D48/D51).

[66] A. H. Lang (Can. Mining J. **77** No. 6 [1956] 71/6). – [67] A. G. Zhabin, G. N. Mukhitdinov, M. E. Kazakova (Tr. Inst. Mineral. Geokhim. Kristallokhim. Redk. Elem. Akad. Nauk SSSR No. 4 [1960] 51/73, 54/6). – [68] A. G. Zhabin, N. V. Svyazhin (Tr. Inst. Mineral. Geokhim. Kristallokhim. Redk. Elem. No. 9 [1962] 55/66, 56/7, 62/3). – [69] E. B. Khalezova, D. N. Knyazeva (Tr. Mineral. Muzeya Akad. Nauk SSSR No. 12 [1961] 235/8). – [70] R. P. Tikhonenkova, E. D. Osokin (in: K. A. Vlasov, Mineralogiya i Geneticheskie Osobennosti Shchelochnykh Massivov, Moscow 1964, pp. 123/34, 130/1).

[71] N. W. Wilson (Sierra Leone Geol. Surv. Bull. No. 4 [1965] 1/102, 71, 88/91; Mineral. Abstr. **17** [1966] 650/1). – [72] P. Geijer (Sveriges Geol. Unders. Arsbok C **14** No. 6 [1920/21] 1/24, 7/9). – [73] P. Geijer (Arkiv Mineral. Geol. **3** [1962/65] 99/105, 99/100). – [74] N. I. Aleksandrova, I. V. Maksimova, N. M. Svirskaya (Izv. Vysshikh Uchebn. Zavedenii Geol. Razved **20** No. 10 [1977] 80/7, 81). – [75] N. A. Solodov, K. D. Subbotin (in: L. N. Ovchinnikov, N. A. Solodov, Mestorozhdeniya Litofil'nykh Redkikh Metallov, Nedra, Moscow 1980, pp. 158/78, 176).

[76] F. Ahlfeld, V. Angelelli (Univ. Nacl. Tucuman Inst. Geol. Mineria Publ. No. 458 [1948] 1/304, 156). – [77] C. A. Strauss (S. Africa Geol. Surv. Mem. **46** [1954] 1/241, 36, 84, 107/20, 126, 157/8, 164/5). – [78] J. Thoreau, B. Aderca, L. van Wambeke (Acad. Roy. Sci. Coloniales [Brussels] Classe Sci. Nat. Med. [2] **4** No. 3 [1958] 684/715, 689/91, 694, 699). – [79] B.-M. Aderca, R. van Tassel (Acad. Roy. Sci. Outre Mer Classe Sci. Nat. Med. [2] **18** No. 5 [1971] 1/117, 83, 86). – [80] R. S. De Mark (Mineral. Rec. **11** [1980] 69/72).

[81] J. H. Soulé (U.S. Bur. Mines Rept. Invest. No. 3854 [1946] 1/25, 5, 21). – [82] T. L. Ho (Bull. Geol. Soc. China [Peking] **14** [1935] 279/82). – [83] N. Çagatay (Turk. Jeol. Kurumu Bul. **24** No. 2 [1981] 59/65, 59, 63 [English on p. 59]; C.A. **96** [1982] No. 220682). – [84] R. Marinova (Rudoobraz. Protsesi Mineralni Nakhodishta **14/15** [1981] 132/6 [Bulgarian, German on p. 136]; C.A. **97** [1982] No. 58728). – [85] E. Ya. Marchenko, O. I. Chashka, E. P. Gurova (Dopov. Akad. Nauk Ukr. RSR Ser. B **30** No. 1 [1968] 69/72).

[86] A. P. Khomyakov (Geokhimiya **1964** 85/8; Geochem. Intern. **1** [1964] 40/3). – [87] A. P. Khomyakov (in: Mineralogiya i Genetecheskie Osobennosti Shchelochnykh Massivov, Moscow 1964, pp. 56/82, 56/9, 77). – [88] H. Wedow (U.S. Geol. Surv. Profess. Paper No. 424-D [1961] D214/D216). – [89] M. H. Staatz (Econ. Geol. **73** [1978] 512/23, 516/7). – [90] V. V. Gerasimovskii (Tr. Mineral. Muzeya Akad. Nauk SSSR No. 15 [1964] 194/202, 194).

[91] V. V. Gerasimovskii (Vestn. Mosk. Univ. Ser. IV Geol. **19** No. 6 [1964] 43/9, 43). – [92] A. S. Lomteva, E. A. Dmitriev (Dokl. Akad. Nauk Tadzh. SSR **19** No. 7 [1976] 53/6). – [93] W. S. White, J. C. Wright (Econ. Geol. **61** [1966] 1171/90, 1178). – [94] S. Graeser, H. A. Stalder (Schweizer Strahler **3** No. 7 [1974] 265/77, 265). – [95] M. Gramaccioli (Mineral. Rec. **8** [1977] 287/93, 288).

[96] E. I. Semenov, A. P. Khomyakov, A. V. Bykova (Tr. Mineral. Muzeya Akad. Nauk SSSR No. 11 [1961] 202/4). – [97] J. C. Hughes, G. Brown (Clay Minerals **12** [1977] 319/29, 324). – [98] E. K. Podporina, A. P. Niyazov, M. D. Brylin (in: V. V. Burkov, Ekzogennye Mestorozhdeniya Redkikh Elementov, Moscow 1980, pp. 1/134, 21, 23). – [99] V. Silberminz [V. A. Zilbermintz] (Compt. Rend. Acad. Sci. USSR A **1929** 55/60, 55/6, 59). – [100] G. Bárdossy, G. Pantó (Congr. ICSOBA, Nice 1973, Vol. 3, pp. 47/53, 48, 50, 52).

[101] Z. Maksimović, G. Pantó (4th Intern. Congr. Study Bauxites Alumina Alum., Preprints, Athens 1978, Vol. 2, pp. 540/52, 545/6). – [102] Z. Maksimović, G. Pantó (Bull. Acad. Serbe Sci. Arts Classe Sci. Nat. Math. Sci. Nat. **72** No. 20 [1980] 35/42, 35/6, 39). – [103] R. Pierrot (Bull. Soc. Franc. Mineral. Crist. **97** [1974] 521/4). – [104] A. P. Sabina (Geol. Surv. Paper Geol. Surv. Can. No. 78-1A [1978] 253/8). – [105] J. Jedwab (Earth Planet. Sci. Letters **49** [1980] 551/64, 554).

10.5.2.1.2.4 Inclusions and Admixtures

High amounts of SiO_2, Al_2O_3, Fe_2O_3, and P_2O_5 in bastnaesites, particularly in those of supergene origin, indicate the presence of admixtures or impurities [1]. In a bastnaesite from a Siberian carbonatite contents of SiO_2, Al_2O_3, and K_2O indicate a contamination by muscovite, of MgO (and partly CaO) by dolomite, of Fe_2O_3 by hematite and of P_2O_5 by monazite [2]. Fine-fibrous bastnaesite in apodiabasic rocks in the European USSR is intergrown with aegirine, alkali-amphibole, titanite, and carbonates [3], cf. inclusions of aegirine needles and of albite, microcline, and hematite in bastnaesite of the "green veins" in alkali rocks of South Greenland [4]. A bastnaesite from a Pb-Zn deposit in Bulgaria, which contains Si, Al, Ca, Mg, Ba, and Fe, is intimately associated with fluorite, ankerite, and barite [5], a supergene bastnaesite from the weathering zone of an alkali massif in the USSR contains Fe_2O_3, Al_2O_3, and SiO_2 due to mechanical admixture of limonite and ferrihalloysite [6]. The reddish and yellowish bastnaesite varieties with several percent Fe in Karonge, Burundi, contain inclusions of goethite [7], and the bastnaesite from a radioactive RE deposit of the Scrub Oaks mine, New Jersey, contains dustlike inclusions of pyrite and a high amount of Ti, which may be within the structure or due to an intimately intermixed component such as leucoxene; the dark-colored bastnaesite is further intermixed with xenotime, hematite, chevkinite, and synchysite-(Y) (doverite) [8], cf. TiO_2 in hydroxylbastnaesite from Vuoriyarvi, Kola, apparently due to admixtures p. 103.

Elevated amounts of P in bastnaesite from Karonge, Burundi, are due to monazite [9] and are not a constituent of the lattice, as it has been earlier assumed [7], cf. minor amounts of P in bastnaesite, which is intimately associated with monazite in the Redstone granite, Westerly, Rhode Island [10], and P caused by monazite in a bastnaesite from a Siberian carbonatite [2]. The metasomatic bastnaesite in diabase from the Russian Platform contains 1.66 wt% P_2O_5 apparently in micro-inclusions of apatite [11], and the supergene bastnaesite from an alkali massif in the USSR has 1.8 wt% P_2O_5 in rhabdophane [6]. Elevated contents of Gd, Dy, and Y together with Be, Si, and Fe indicate gadolinite, and 3.6 wt% Ca a parisite component in bastnaesite from the Russian Platform [11].

The presence of SiO_2, TiO_2, Fe_2O_3, MgO, CaO, BaO, SO_3, and P_2O_5 in a hydroxylbastnaesite from Vuoriyarvi, Kola, is obviously related to mechanical admixtures of quartz, anatase, hydrogoethite, dolomite, barite, and probably rhabdophane [12, 13]. In kyshtymite from the Urals no inclusions of minerals but sometimes gaseous-fluid inclusions have been observed on fissures [14].

Contents of SiO_2, Al_2O_3, and Fe_2O_3 in thorbastnaesite from East Siberia have been ascribed to admixtures of Fe hydroxides, albite, and quartz [15].

For the intergrowths of bastnaesite with accompanying minerals at the Mountain Pass, see p. 96.

10.5.2.1.2.5 Alteration

Minerals of the bastnaesite group have often been formed as secondary products by the alteration of various RE minerals. Frequently bastnaesite replaces allanite, sometimes Ce minerals such as cerite, tysonite, chevkinite, but also fluorite [1]. Bastnaesite-(Y) or bastnaesite-(Ce-Y) have been found merely as secondary minerals, pseudomorphously replacing gagarinite, gadolinite, yttrium-britholite, and tysonite [16].

Replacement of Other Minerals by Bastnaesite

Allanite. The alteration of allanite to bastnaesite has been observed in many granitic and pegmatitic rocks [17]; it often has been ascribed to weathering [18]. "Bastnaesitization" takes place "in situ" under endogene conditions and includes the formation of minerals of the bastnaesite group, or more generally, the formation of pseudomorphs after allanite, which are composed of an aggregate of fine-grained bastnaesite, magnetite, and fluorite, and sometimes also kaolinite and thorite [20], cf. [19]. The "bastnaesitization" is effected by mineralizing solutions with deuteric character [22, 23], and with increasing acidity (due to increasing activity of CO_2 and F) [20, 21]; the presence of fluorine contradicts a formation by weathering [24].

The study of many specimens of altered accessory allanites in granites and granite pegmatites from various regions shows, that minerals of the bastnaesite-hydroxylbastnaesite group usually are the first products of a postmagmatic hydrothermal or supergene alteration of allanite; the development of bastnaesite from allanite is a progressive process with the possibility of formation of hydroxylbastnaesite, when the solutions are characterized by a deficit of fluorine [25]. An example for the formation of hydroxylbastnaesite from allanite in the first of two stages of alteration is the occurrence in radioactive granitic rocks of the Canadian Shield [26]. Secondary bastnaesites with a replacement of F by OH have been found in metasomatically fenitized granite pegmatites of the South Urals (kyshtymite) [14], and in microcline-quartz pegmatite from Karasugawa, Fukushima Prefecture, Japan [27]. Compared with allanite the bastnaesite is characterized by a higher ΣRE and by a variable change in the RE distribution with a tendency to increasing relative concentrations in intermediate RE (Sm to Ho), in accordance with the acid trend of the replacement reaction, and to loss in Ce, La, and heavy RE (Er to Lu) in an alkaline environment [25], cf. [21].

The pseudomorphs after allanite contain bastnaesite intimately intermixed with several other minerals, that are defined by the original composition and by the conditions of alteration of allanite. An acid trend is characterized by bastnaesite and kaolinite as endproducts, an alkalic trend is characterized by bastnaesite, thorite, and montmorillonite in the first stage of alteration [26], cf. "Rare Earth Elements" A 8, 1984, Chapter 10.10.6.10.2.3.1, and kaolinite and fluorite as indicators of the chemistry and temperature of the solutions [28, p. 89].

Bastnaesite as an alteration product of allanite has been observed in the granites of Elberton, Northeast Georgia [22], and of the Pikes Peak, Colorado [23], in granitic rocks at the Rough Rock Lake, Ontario [17], in granitoid intrusions of the Kyzylkum, Uzbekistan, USSR [29], in granitoid rocks from eastern Kazakhstan [30], and in metasomatic rocks of the European part of the USSR [3, 31]. Bastnaesite, pseudomorphous after allanite, is also a frequent mineral in pegmatites; some examples of the numerous occurrences are: the syenitic RE pegmatite of the Adirondack Mountain area, Essex Co., New York [32], the pegmatite on the Constitution Drive near Allentown, Lehigh Co., Pennsylvania [33], and the Rutherford pegmatites, Amelia Co., Virginia [24, 34]; it occurs further in pegmatites of two other counties in Virginia [35, 36], in pegmatites of Texas [37], Ontario [17], Manitoba [38], southern and northern Norway [39, 40], southwestern Finland [41], and from the South Urals [20], cf. [28, p. 89] and p. 96, and from eastern Kazakhstan [28, pp. 86/9].

Chevkinite. Bastnaesite together with Ti minerals in pseudomorphs after chevkinite in the pegmatite of Ambatofinandrahana, Madagascar, has been formed by solutions containing F and CO_2 [42], cf. formation of bastnaesite from chevkinite by carbonate halide metasomatism with decreasing acidity in the temperature range 330 to 320°C in a hydrothermal quartz-fluorite formation in the Ukrainian Crystalline Shield [43]. Chevkinite is also replaced by bastnaesite in the Vishnevye Gory, Urals [20], and in the radioactive RE deposit at the Scrub Oaks Mine, Morris Co., New Jersey [8], and, possibly, in nodules in the pegmatite of Southeastern Rappahannock Co., Virginia [35]. See also "Rare Earth Elements" A 8, 1984, Chapter 10.10.6.6.2.

Zircon, thorite, and cerite. Bastnaesite occurs together with ancylite in a few pseudomorphs after zircon in nepheline-syenite-pegmatites on the island Øvre Arø, Langesundfjord, South Norway [44], cf. [45]; bastnaesite has also been found together with many rare fluorides pseudomorphous after zircon in cavities in ekerite near Lake Gjerdingen, Nordmarka, Oslo Region, Norway [46].

"Ferrithorite" is supposed to be the original mineral, from which thorbastnaesite has been formed by replacement, the latter mineral has been found adjacent to relics of ferrithorite in microcline albitites in nepheline syenite from Eastern Siberia [15], cf. [47].

Bastnaesite as an alteration product of cerite has been found in Bastnäs at Riddarhyttan, Västmanland, Sweden [48], in coarse cerite from the Jamestown district, Boulder Co., Colorado [49], and in cerite from the Mountain Pass, California [50].

Gadolinite. As a weathering product bastnaesite-(Y) forms pseudomorphs after gadolinite in granitepegmatites [1].

Monazite, britholite, and abukumalite. Bastnaesite somewhat corrodes monazite in the Nb-bearing rutile deposit in the Northern Lemhi Co., Idaho [51], and clearly replaces monazite in pegmatitic veins in the northwestern Tarbagatai, eastern Kazakhstan [28, p. 97]. In the pegmatite of Mostad, Aust-Agder area, southern Norway, monazite is closely associated with bastnaesite and apparently an alteration product of it [40].

Bastnaesite replaces and forms complete pseudomorphs after britholite in amphibole-albitites (=apodiabasic rocks) of the Russian Platform [31], cf. [3]. Bastnaesite-(Y) has been found as weathering product of britholite-(Y) (and thalenite) in pegmatites of alkali granites from Kola, USSR, and is, probably, the supergene alteration product of another britholite-(Y) from the USSR [52].

Bastnaesite-(Y) is further developed as one of the alteration products of abukumalite in granitoids of the northwestern USSR [53].

RE oxides and fluorides. Pseudomorphs of bastnaesite after pyrochlore occur in vugs of ekerite near the Lake Gjerdingen, Nordmarka, Oslo region, Norway [46].

Bastnaesite replaces tysonite along cleavage planes in the cerite ore of Bastnäs at Ryddarhyttan, Västmanland, Sweden [48], and tysonite from the tin deposit at the Mutue Fides farm, Potgietersrus district, Transvaal, South Africa [54], cf. p. 15. The replacement of tysonite by bastnaesite for the occurrence Pikes Peak, El Paso Co., Colorado [55, 56, 57], is questioned by [58].

Minerals of the group bastnaesite–bastnaesite-(Y) have replaced gagarinite in the pegmatites and in the quartz-microcline metasomatites of a riebeckite-granite massif in eastern Kazakhstan [21], which was later specified as the Akzhailyau massif in northwestern Tarbagatay, where bastnaesite-(Y) was formed together with fluorite as an alteration product of gagarinite in a low temperature, alkali-free association; it is characterized by a more basic RE composition than the gagarinite. The alteration corresponds to the reaction:

$$NaCaRE_YF_6 \xrightarrow[-F^-]{+CO_3^{2-}} RE_{Ce\ to\ Y}CO_3F + CaF_2 + NaF$$ [28, pp. 95/6]

or, somewhat deviating to:

$$Na(Y,Ca,Na)_2F_6 \xrightarrow[-F^-]{+CO_3^{2-}} YCO_3F + CaF_2 + Na^+$$ [16].

RE carbonates. Bastnaesite has replaced burbankite together with strontianite in the ankeritic stage and together with calcite in a late calcitic stage, both in the form of pseudomorphs after burbankite, in an East Siberian carbonatite massif [59]; the second generation of bastnaesite of this occurrence has a reduced F and an increased OH content, and thus is very close to hydroxylbastnaesite [60]. Hydroxylbastnaesite as a late stage hydrothermal mineral has also replaced burbankite in the carbonatite from Vuoriyarvi, Kola [13].

The pseudomorphous development of bastnaesite after parisite is often readily discernible [1], e.g., in hydrothermal veinlets of an alkali massif in the USSR and in hydrothermal veins in an alkali syenite and on samples from an alkali pegmatite of Narsarsuk, southern Greenland [6].

Bastnaesites in pseudomorphs after cordylite and huanghoite have also been observed [1]; huanghoite of a late veinlike carbonatite in a Siberian pluton has been replaced by bastnaesite in close association with barite and quartz [61].

Ancylite crystals in druses in the Grefsen syenite (= nordmarkite) at Bonkall, Grorud district, Oslo region, Norway, have been completely altered by bastnaesite [62], cf. [63]. Bastnaesite pseudomorphous after ancylite is the dominant mineral in the "weibyeite" from the island Øvre Arø in the Langesundfjord, southern Norway [44], cf. p. 104.

Hydroxylbastnaesite, as the last mineral in the alteration series burbankite → carbocernaite → ancylite → hydroxylbastnaesite, replaces ancylite and other RE carbonates in the carbonatite from Vuoriyarvi, Kola, USSR [13].

Minerals without RE. Bastnaesite replaces the outer parts of calcite and zeolites (stilbite and heulandite) in veinlets, which occur in alkali granitoids in the northern Baikal area, USSR [64, 65], and locally replaces albite-oligoclase and anatase in the radioactive RE deposit of the Scrub Oaks Mine, Morris Co., New Jersey [8]. The hydroxyl-containing kyshtymite forms metasomatically on feldspar or allanite in fenitized granite-pegmatites in the South Urals [14].

Replacement of Bastnaesite by Other Minerals

Bastnaesite is replaced by several minerals, most often by allanite, e.g., in calcite veins in the northern Baikal area, USSR [64, 65], in albitites of Siberia [66], and in the latest stage of carbonatites in East Siberia [60], by monazite in Karonge, Burundi [9], and by rhabdophane in the USSR [21]. A replacement of bastnaesite by calcite has been observed in the Nb-bearing rutile deposits in the northern Lemhi Co., Idaho [51].

Under surface conditions, especially hydration with loss of F, bastnaesite alters to rhabdophane and to lanthanite (see p. 191); at oxidation of Ce and loss of trivalent lanthanides cerianite is formed [1].

The kyshtymite from the Urals is replaced by cerite [14].

References for 10.5.2.1.2.4 and 10.5.2.1.2.5:

[1] E. I. Semenov (in: K. A. Vlasov, Geochemistry and Mineralogy of Rare Elements and Genetic Types of Their Deposits, Vol. 2: Geochemistry of Rare Elements, Israel Program Sci. Trans., Jerusalem 1966, pp. 249/328, 266, 269 [Russian Original, Moscow 1964]). – [2] M. Ya. Somina, A. G. Bulakh (Mineral. Geokhim. **1972** No. 4, pp. 64/73, 67). – [3] L. V. Ganzeeva (Redk. Elem. Syr'e Ekon. No. 7 [1972] 112/24, 118/9). – [4] J. Hansen (Medd. Groenland **181** No. 8 [1968] 1/47, 20, 28). – [5] R. Marinova (Rudoobraz. Protsesi Mineralni Nakhodishta **14/15** [1981] 132/6, 133 [Bulgarian, German on p. 136]; C.A. **97** [1982] No. 58728).

[6] E. I. Semenov, A. P. Khomyakov, A. V. Bykova (Tr. Mineral. Muzeya Akad. Nauk SSSR No. 11 [1961] 202/4). – [7] J. Thoreau, B. Aderca, L. van Wambeke (Acad. Roy. Sci. Coloniales [Brussels] Classe Sci. Nat. Med. [2] **4** No. 3 [1958] 684/715, 708/10). – [8] H. Klemic, A. V. Heyl, A. R. Taylor, J. Stone (U.S. Geol. Surv. Bull. No. 1082-B [1959] 29/60, 47, 52/3). – [9] B.-M. Aderca, R. van Tassel (Acad. Roy. Sci. Outre Mer Classe Sci. Nat. Med. [2] **18** No. 5 [1971] 1/117, 85/6). – [10] W. L. Smith, E. A. Cisney (Am. Mineralogist **41** [1956] 76/81, 78/9).

[11] L. B. Zubkov, N. I. Galadzheva, A. B. Chernyakovskii (Dokl. Akad. Nauk Belorussk. SSR **16** No. 7 [1972] 646/9). – [12] A. S. Kirillov (Dokl. Akad. Nauk SSSR **159** [1964] 1048/50; Dokl. Earth Sci. Sect. **159** [1964] 93/5). – [13] A. S. Kirillov (Zap. Vses. Mineral. Obshchestva **95** [1966] 51/9, 51, 53, 58). – [14] N. V. Svyazhin (Tr. Inst. Geol. Akad. Nauk SSSR Ural'sk. Filial No. 70 [1965] 249/52). – [15] A. S. Pavlenko, L. P. Orlova, M. V. Akhmanova, K. I. Tobelko (Zap. Vses. Mineral. Obshchestva **94** [1965] 105/13, 109, 111/2).

[16] D. A. Mineev, T. I. Lavrishcheva, A. V. Bykova (Zap. Vses. Mineral. Obshchestva **99** [1970] 328/32). – [17] P. Černý, I. Černá (Can. Mineralogist **11** [1971/73] 541/3). – [18] J. C. Hughes, G. Brown (Clay Minerals **12** [1977] 319/29, 319). – [19] V. I. Chernov, O. F. Krol', A. V. Stepanov (Geol. Geokhim. Mineral. Mestorozhd. Redk. Elem. Kaz. **1966** 138/44, 139, 142/3). – [20] D. A. Mineev, B. A. Makarochkin, A. G. Zhabin (Geokhimiya **1962** 590/7; Geochemistry [USSR] **1962** 684/93, 685/6, 691).

[21] D. A. Mineev [Mineyev] (Geokhimiya **1963** 1082/100; Geochemistry [USSR] **1963** 1129/49, 1142, 1146/7). – [22] L. T. Silver, M. Grunenfelder (Geol. Soc. Am. Bull. **68** [1957] 1796). – [23] J. W. Adams, E. J. Young (U.S. Geol. Surv. Profess. Paper No. 424-C [1961] 292/4). – [24] W. D. Riesmeyer (Virginia J. Sci. **18** [1967] 188). – [25] D. A. Mineev, K. I. Rozanov, N. V. Smirnova, T. I. Matrosova (Dokl. Akad. Nauk SSSR **210** [1973] 925/8; Dokl. Earth Sci. Sect. **210** [1973] 149/52).

[26] A. L. Littlejohn (Geol. Surv. Paper Geol. Surv. Can. No. 81-1B [1981] 95/104, 95, 100, 102/3). – [27] K. Sakurai, H. Wakita, A. Kato, K. Nagashima (Bull. Chem. Soc. Japan **42** [1969] 2725/8). – [28] D. A. Mineev (Geokhimiya Apogranitov i Redkometal'nykh Metasomatitov Severo-Zapadnogo Tarbagataya, Moscow 1968, pp. 1/183). – [29] Kh. R. Rakhmatullaev, P. L.

Gertman, E. P. Nikolaeva (Zap. Uzbek. Otd. Vses. Mineral. Obshchestva Akad. Nauk Uz. SSR No. 32 [1979] 35/7; C.A. **92** [1980] No. 200998). – [30] G. N. Gogel' (Izv. Akad. Nauk Kaz. SSR Ser. Geol. **33** No. 4 [1976] 58/64, 59).

[31] L. V. Ganzeeva [Ganzeyeva], A. A. Ganzeev [Ganzeyev] (Geokhimiya **1975** 74/80; Geochem. Intern. **12** No. 1 [1975] 47/52, 49). – [32] E. B. Rowley (Rocks Minerals **37** [1962] 341/7, 453/60, 346/7). – [33] R. G. Wells (Rocks Minerals **53** [1978] No. 3, pp. 108/11). – [34] R. S. Mitchell (Am. Mineralogist **53** [1968] 2102). – [35] C. R. Halladay (Virginia Div. Mineral. Resour. Publ. No. 7 [1978] 87/9).

[36] R. S. Mitchell, G. E. Redline (Rocks Minerals **55** [1980] 245/9). – [37] W. W. Crook (Mineral. Rec. **8** [1977] 88/90). – [38] P. Černý, N. A. Bristol (Can. Mineralogist **11** [1971/73] 560/3). – [39] Th. L. Sverdrup (Norg. Geol. Unders. No. 211 [1959] 124/96, 156, 178). – [40] T. L. Sverdrup, K. Ø. Bryn, P. C. Sæbø (Norsk Geol. Tidsskr. **39** [1959] 237/47, 237/9, 243/4).

[41] A. Vorma, P. Ojanperä, V. Hoffrén, J. Siivola, A. Löfgren (Bull. Comm. Geol. Finlande No. 222 [1966] 241/74, 245, 248). – [42] A. Lacroix (Bull. Soc. Franc. Mineral. **38** [1915] 106/25, 115/6, 118). – [43] E. Ya. Marchenko, O. I. Chashka, E. P. Gurova (Dopov. Akad. Nauk Ukr. RSR Ser. B **30** No. 1 [1968] 69/72; C.A. **68** [1968] No. 71136). – [44] P. C. Sæbø (Norsk Geol. Tidsskr. **43** [1963] 441/3). – [45] G. Raade, J. Haug, R. Kristiansen, A. O. Larsen (Lapis **5** No. 10 [1980] 22/8, 28).

[46] G. Raade, J. Haug (Mineral. Rec. **11** [1980] 83/91, 83, 88). – [47] M. Fleischer (Am. Mineralogist **50** [1965] 1504/8, 1505). – [48] P. Geijer (Sveriges Geol. Unders. Arsbok C **14** No. 6 [1920/21] 1/24, 9, 16, 22). – [49] E. B. Eckel (U.S. Geol. Surv. Bull. No. 1114 [1961] 1/399, 66). – [50] K. J. Murata, H. J. Rose Jr., M. K. Carron, J. J. Glass (Geochim. Cosmochim. Acta **11** [1957] 141/61, 157).

[51] A. L. Anderson (Econ. Geol. **55** [1960] 1179/201, 1189). – [52] E. I. Semenov (Mineralogiya Redkikh Zemel', Moscow 1963, pp. 1/412, 102). – [53] A. Ya. Lunts (Mineralogiya Geokhimiya i Genezis Redkozemel'nykh Pegmatitov Shchelochnykh Granitov Severo-Zapada SSSR, Nedra, Moscow 1972, pp. 1/176, 141). – [54] J. G. D. Steyn (Rep. S. Africa Dept. Mines Geol. Surv. Bull. No. 35 [1961] 1/31, 19/21). – [55] O. D. Allen, W. J. Comstock (Am. J. Sci. [3] **19** [1880] 390/3).

[56] S. A. Northrop (Minerals of New Mexico, Albuquerque 1959, pp. 1/665, 135). – [57] W. F. Hillebrand (Am. J. Sci. **157** [1899] 51/7, 51). – [58] R. Köchlin (Tschermaks Mineral. Petrogr. Mitt. [2] **31** [1911/12] 525/36, 528). – [59] T. B. Zdorik (Geol. Mestorozhd. Redk. Elem. Vses. Nauchn. Issled. Inst. Mineral'n. Syr'ya No. 30 [1966] 121/31, 125/7). – [60] T. B. Zdorik (Tr. Mineral. Muzeya Akad. Nauk SSSR No. 17 [1966] 60/75, 68, 70, 73/5).

[61] Yu. L. Kapustin (Dokl. Akad. Nauk SSSR **202** [1972] 422/5; Dokl. Earth Sci. Sect. **202** [1972] 116/9). – [62] I. Oftedal, P. C. Sæbø (Norsk Geol. Tidsskr. **45** [1965] 171/5). – [63] G. Raade (Mineral. Rec. **3** No. 1 [1972] 7/11). – [64] V. V. Gerasimovskii (Tr. Mineral. Muzeya Akad. Nauk SSSR No. 15 [1964] 194/202, 196). – [65] V. V. Gerasimovskii (Vestn. Mosk. Univ. Ser. IV Geol. **19** No. 6 [1964] 43/9, 43).

[66] I. I. Kupriyanova (Tr. Mineral. Muzeya Akad. Nauk SSSR No. 18 [1968] 206/9).

10.5.2.1.3 Chemistry

For a compilation of **analyses** and $(RE,Y)_2O_3$ contents, mean and variance of 37 samples of bastnaesite and some of its varieties, see "Seltenerdelemente" A 4, 1979, pp. 13/4, 5.

According to the analyses of minerals of the bastnaesite group, besides the most common species, cerfluorbastnaesite $Ce[F|CO_3]$, there exist several more or less complete diadochic series. They are caused by a complex diadochic replacement of the cations (RE with each other and RE by Y, Ca, and Th) and by a replacement of the anions F and OH, as it is expressed in the formulas.

The **formulas** most frequently given for bastnaesites are: for bastnaesite-(Ce) $Ce[CO_3]F$ [1, 2], $(Ce,La)[CO_3]F$ [3], [4, p. 446], $(Ce,La,Nd)[CO_3]F$ [2]. For bastnaesite-(La) with dominance of La over Ce the formula is $(La,Ce)[CO_3]F$ [1]; the formulas $Y[CO_3]F$ or $(Ce,Y)[CO_3]F$ have been given to bastnaesite with Y-earth dominating or accompanying Ce-earth [5]; in the formula $(Y,Ce,Dy)[CO_3]F$ the symbol Y represents yttrium in addition to the lanthanides Tb to Lu, and Ce represents the light lanthanides [1]. For a bastnaesite from the Vishnevye Gory, South Urals, the analysis corresponds to $(Ce,La,Pr)[CO_3]F$ [6], for a bastnaesite of the Marmara bauxite deposit, Greece, the dominance of La and a small amount of Ca lead to the formula $(La,Nd,Ce...)_{1.0}Ca_{0.06}[CO_3]_{0.93}F_{0.70}$ [7, p. 37]. The existence of hydroxyl groups in some bastnaesites is indicated by formulas such as $Ce[CO_3](OH,F)$ [1] or $(Ce,La,Nd)[CO_3](F,OH)$ [4, p. 447], [2], simplified to $RE[CO_3]F \rightarrow RE[CO_3](F,OH)$ for the whole series from common bastnaesite to hydroxylbastnaesite by [8, p. 55], cf. [61]. For an intermediate member of this series (kyshtymite from the South Urals, see p. 94), the formula $(Ce,La,Nd,Pr)[CO_3](F,OH)$ has been proposed [9]. The analysis of the hydroxylbastnaesite from Vuoriyarvi, Kola, (after deduction of many admixed elements) has been calculated as $(RE_{0.98}Ca_{0.01}Th_{0.01}Na_{0.01})[CO_3](OH_{0.87}F_{0.14})$ [10]. The analysis of a secondary, hydroxyl-containing bastnaesite-(Y), replacing gagarinite of Kazakhstan, leads to $(RE_{Y_{0.90}}Ca_{0.09}Th_{0.01})[CO_3](F_{0.73}OH_{0.19}) \cdot 0.5H_2O$ [5]. For thorbastnaesite, the calculated formula is $ThCe_{0.2}Ca_{0.8}[CO_3]_2F_2 \cdot 3H_2O$ [11, pp. 110/1] or simplified to $ThCeCa[CO_3]_2F_2 \cdot 3H_2O$ [4, p. 450]. Assuming Al and Pb are components of the structure, the formula is $(Th_{0.98}Ca_{0.65}RE_{0.28}Pb_{0.02}Al_{0.02})(C_{1.74}Al_{0.26})O_6[F_{1.77}(OH)_{0.18}O_{0.05}] \cdot nH_2O$ [11, p. 111].

The existence of a series bastnaesite-(Ce) → bastnaesite-(Ce, Y) → bastnaesite-(Y), in which Ce and Y represent RE plus Y, is proposed by [5]. The mutual diadochic replacement of cations of the RE and Y, mostly of Ce by La, Nd, Pr, and Y [12], has been thought to be limited to two or three maxima such as bastnaesite-(La), -(Ce), and -(Y) [13]. According to a great number of more recent analyses the variance in the RE composition is essentially broader and bastnaesites with maxima at Nd and Dy, e.g., have been found, see Table 2, pp. 114/5.

The replacement of RE by Y in bastnaesites is confirmed by high relative amounts of Y of about 11 to about 40 wt% from the total amount of RE plus Y in 5 samples, which are all pseudomorphous after the more or less Y-containing minerals gagarinite, gadolinite, britholite, chevkinite, and ferrithorite, see Table 2, pp. 114/5; it is further confirmed by smaller amounts of Y_2O_3 given in wt% in a few analyses, e.g., in bastnaesite in bauxite of the Marmara deposit, Greece, 2.3 [7, p. 38], in some bastnaesites from albitites of the USSR 1.7 [14], 1.53 and 0.97 [15], and in bastnaesite from a green vein in alkali rocks from South Greenland about 1 (spectroscopically determined) [16]; further in thorbastnaesite 1.73 [11, p. 110]. Very low amounts below 1 wt% Y_2O_3 have been found in 2 samples of bastnaesite from Burundi with 0.17 each [17], cf. <0.5 [18, pp. 707/8], and in bastnaesite from the Redstone granite, Rhode Island, between 0.5 and 0.1 [19], while in bastnaesite from an amphibolite of the Mecsek Mountains, Hungary, a considerable amount of Y has been found qualitatively [20].

The diadochic replacement between RE and other cations in bastnaesites can be seen from the formulas. For the replacement of the RE by Ca, Th, and Na in a hydroxylbastnaesite the following balance of valences has been proposed: $RE^{3+} + Ca^{2+} \leftarrow Th^{4+} + Na^+$; $RE^{3+} + Na^+ \leftarrow 2Ca^{2+}$; $2RE^{3+} \leftarrow Ca^{2+} + Th^{4+}$; $3RE^{3+} \leftarrow 2Th^{4+} + Na^+$ [8, p. 55].

Only a small replacement of Ce by Th and Na has been determined in bastnaesite from carbonatites of the East Sayan [21], and a replacement of RE by Th and Ca has been found in bastnaesite after allanite from the Rutherford pegmatite, Virginia [22]. In thorbastnaesite the replacement follows the scheme $Ca^{2+}Th^{4+} \rightarrow 2Ce^{3+}$ [4, p. 444]. In bastnaesite-(Y) the Y is replaced by Ca and RE of both groups (Ce to Yb) [12]. The existence of Ca in bastnaesites from the Russian Platform, however, combined with reduced contents of RE and F, may indicate a parisite component in the mineral [23].

For elevated Th contents in bastnaesites from various localities see, e.g., (in wt% ThO_2): 2.8 in supergene bastnaesite of Siberia [24], 5.36 in a bastnaesite of an ankerite-barite vein in carbonatite, East Siberia [25, 26], 25.60 in an intermediate member of the bastnaesite group from Beylikahir, Turkey [27], up to 46.79 in thorbastnaesite (with Th$\gg$RE) from East Siberia [11, pp. 109/10]. Contents <1% ThO_2 have been found in various bastnaesites, see, e.g., 0.93 [21], 0.39 [19], 0.16 [22], 0.09 [14], in a hydroxylbastnaesite 0.53 [10], in kyshtymite 0.70 [9], and in bastnaesite-(Y) 0.72 [5]. For the Th contents in bastnaesites, besides a replacement of Th for RE, an admixture of Th-containing minerals must be considered, e.g., uranoan thorianite in bastnaesite of the Redstone granite, Rhode Island [19], thorogummite associated with bastnaesite (with 0.68% Th and 0.30% U) in granite pegmatite of the Transbaikalia region, USSR [28, p. 72], and tiny inclusions of monazite in bastnaesite with 0.02% ThO_2 of Karonge, Burundi [18, p. 709], see p. 102. The amount of 2.09% ThO_2 in a pseudomorph after allanite from Kazakhstan, completely calculated for bastnaesite, corresponds to 5.70% ThO_2 in the mineral [29].

The diadochy between F and OH can be seen from the formulas p. 108. The existence of OH ions in bastnaesite has been confirmed by IR spectra, see p. 131. For the relations between cell size and physical properties with the replacement of F by OH, see pp. 123, 129, 133. Various recent analyses have shown that the theoretical value of F = 8.67 wt% for the end member $(Ce_{0.5}La_{0.5})[CO_3]F$ is not reached in several natural bastnaesites, and that the content of H_2O^+ approaches the theoretical value of the pure endmember $(Ce_{0.5}La_{0.5})[CO_3](OH) = 4.16$ wt%. This is the case in the following hydroxylbastnaesites (% = wt%): metasomatic bastnaesite from the European part of the USSR with 2.43% F and 4.10% H_2O^+ [30], bastnaesite from East Yakutia with 0.22% F and 4.00% H_2O^+, which is near to the hydroxyl-endmember [2], and bastnaesite from Vuoriyarvi, Kola, with 1.21% F and 3.68% H_2O^+, corresponding to 87% hydroxylbastnaesite and 14% fluorbastnaesite, see formula p. 108 [8, p. 55], [10]. Partial replacement of F by OH is further indicated for bastnaesites pseudomorphous after allanite from Japan by 2.1% H_2O^+ [31], and from Kazakhstan with 1.07% H_2O^+ and hydroxyl groups confirmed by the IR spectrum [29]. For further bastnaesites with reduced F content see, e.g. [23, 24, 32, 33] and for bastnaesite-(Y) [34, p. 102]. A high amount of 10.9% H_2O in supergene bastnaesite-(Nd) may indicate a strong hydration [35], as it is also assumed for thorbastnaesite with 9.04% H_2O^+ besides 6.87% F; for this mineral only an insignificant amount of hydroxyl groups is supposed to replace the F ions [11, pp. 110/1] while the presence of molecular water, see formulas p. 108, is confirmed by IR spectra [4, pp. 450/1]. In bastnaesite-(La) a deficiency of CO_3 ions is not explicable [7, p. 37]. Chlorine has not been observed in bastnaesites [2].

Besides the elements given in the formulas, see p. 108, and those which are caused by admixtures, see pp. 102/3, chemical or spectrographic analyses of bastnaesites show the presence of several trace elements, the most frequent of which are Ba, Be, Cu, Sr, Sn, Zr, and Mn: bastnaesite from the Birthday area, Mountain Pass, California, contains Ba, Sr, and Cu [32], bastnaesite from the Russian Platform Ba, Sr, Be, Cu, Sn, Mn, Zr, Co, Nb, Ta, Cd, and Ga [23], other bastnaesites of the USSR Be [33], Ba, Be, and Mn [21], Ba, Be, Sn, and Zr [28], and Ba, Mn, and Mo [36, p. 67]; hydroxylbastnaesite from Vuoriyarvi, Kola, has Ba, Be, Zr, SO_3, and P_2O_5 [8, pp. 53/4], [10] (partly due to mechanical admixtures, see p. 103).

RE composition. All bastnaesites, except bastnaesites-(Y), are strongly Ce-selective minerals and show a large increase in light lanthanides [1], often with enrichment of La [37]. The Ce selectivity is in accordance with the elevenfold coordination in bastnaesite [38]. Examples for this dominance of the RE of the Ce-group over those of the Y-group are numerous for bastnaesites of nearly all genetic types, see Table 2, pp. 112/5, and "Seltenerdelemente" A 4,

1979, p. 52. Besides the common fluorbastnaesite the Ce selectivity is also shown by the hydroxylbastnaesite [2], by the strongly Ce-selective kyshtymite [9], and by thorbastnaesite with its low content of $(RE, Y)_2O_3$ and anomalously low content of Ce itself [11, p. 109]. The predominance of the Ce group is reflected by the values of $\Sigma(RE \pm Y)$, e.g., $\Sigma La + Ce + Pr$ (in at.%) = 81 to 92 for bastnaesites from several localities, among them three from the Mountain Pass, California, with 81 to 91 [39, pp. 149, 151/3, 157], cf. for the latter occurrence $\Sigma La + Ce + Nd = 80$ to 85 and for Karonge, Burundi, $\Sigma La + Ce + Nd + Pr = 90$ to 95 [40], and 96.6 [17]. For hydroxylbastnaesite from Vuoriyarvi, Kola, ΣCe-earth is 95.6 and ΣY-earth 4.4% [8, p. 59].

The $\Sigma La + Ce + Pr$ (in at.%) is more elevated for bastnaesites as replacing minerals relative to that of the replaced minerals, e.g., cerite [39, p. 157], cf. a more marked "Ce selectivity" in bastnaesite pseudomorphous after allanite from Kazakhstan, but a less marked one in pseudomorphs from the South Urals [41, p. 89]. An enrichment of light lanthanides has also been observed in bastnaesite-(Y), which has been developed from gagarinite by a bastnaesitization process [41, pp. 95/6].

The order $Ce > La > Nd$ is the most typical one in bastnaesites from magmatic-metasomatic to hydrothermal inclusive alkali-rich and carbonatitic occurrences. A marked predominance of Ce above other light lanthanides, e.g., $Ce \gg La \gg Nd$ has been observed for bastnaesites from granite at Honkamäki, Finland [42], or $Ce \gg La > Nd$ in bastnaesite from a metasomatite in East Kazakhstan [41, pp. 97, 128], and in one of the supergene bastnaesites in bauxite from the Marmara deposit, Greece, which shows the highest amount of Ce relative to the other RE [7, pp. 39/40], cf. $Ce > La > Nd$ and $Ce > La > Pr$ for secondary bastnaesites in bauxites from Allauch, France, and from San Giovanni Rotondo, Italy [43]. For bastnaesite from the Redstone granite, Westerly, Rhode Island, the order of the Ce-earth metals is $Ce > La \geqslant Nd > Pr \geqslant Sm > Eu$, the order of the Y-earth metals is $Gd > Dy \geqslant Er \geqslant Lu \geqslant Tb > Ho > Yb$ [19], for a sample pseudomorphous after allanite in a pegmatite of Kangasala, southwestern Finland, however, the order $Ce > Nd > La$ has been observed [44]. In bastnaesites of the carbonate-hematite veins of the Western Tannu Ola, Tuva, Siberia, the concentration of the Ce-earth La to Eu increases from the sideritic to the hematitic association with changing order from $Ce > Nd > La > Pr > Sm$ to $Ce > La > Nd > Pr > Sm$ [45], [36, pp. 79/80].

La is relatively enriched close to Ce values in bastnaesites especially from alkalic and carbonatitic rocks [34], cf. $La \approx Ce$ in two of three bastnaesites from the Mountain Pass, California, and in bastnaesite from the Red Cloud Mine, Gallinas, New Mexico [39, p. 151], and $La = Ce \gg Nd > Pr$ in kyshtymite from the Urals, which is more enriched in La than most of the bastnaesites from the USA [9]. $La > Ce$ has been found in a bastnaesite from a Siberian carbonatite [34], and in one from a hydrothermal deposit in Bulgaria [46]; $La > Nd \geqslant Ce > Pr$ in some bastnaesites in bauxite from Marmara, Greece, which occur very close to Ce-rich bastnaesites [47], [7, pp. 38/40], cf. above, and in a supergene bastnaesite in a soil sample from Ibadan, Nigeria [48]. A high La value relative to the other RE has been further observed in a bastnaesite from a carbonatite of East Africa [49]. For the supergene bastnaesite-(Nd) from Kazakhstan the order is $Nd > La > Sm = Gd > Pr > Dy > Eu$; an Eu content of 2.2 to 3.8% and the extreme low content of 0.4% Ce from the sum of the RE + Y are unique [35].

For six bastnaesites-(Y) (yttrobastnaesites) the order of the RE is characterized by $Dy > Gd > Er (\geqslant Sm \gtreqless Nd)$, $Ce \lesseqgtr Nd$, $Nd \lesseqgtr Yb$; ratios $Dy/Yb > 2$ show a "Dy type" of RE enrichment, which is characteristic for metasomatites [41, pp. 126/9]; in another bastnaesite from a metasomatite of the USSR equal amounts of Ce and Y have been observed [50]. For thorbastnaesite from USSR the order is $Y > La > Ce > Nd > Dy$ [11, p. 112].

Table 1, p. 111, shows the distribution of the RE in at.% together with some parameters (ratios between RE, and the $\Sigma La + Ce + Pr$) of minerals of the bastnaesite group. It can be seen,

Table 1
At.% of La and Lanthanides in Bastnaesites from Various Rocks from [1].

Genetic group	La	Ce	Pr	Nd	Sm	Eu	Gd	Tb	Dy	Ho	Er	Tm	Yb	Lu	$\frac{100\,Y}{Y+RE}$	Σ(La + Ce + Pr)	La to Nd	La/Nd
Common bastnaesites																		
Granites and granite pegmatites (7)[1]	25.0	44.3	5.2	15.2	2.6	0.1	2.0	0.4	2.2	0.5	1.4	0.1	0.9	0.1	23.0(3)[2]	74.5	89.7	1.64
Hydrothermal occurrences (21)	28.8	48.1	5.0	15.7	1.4	0.1	0.5	0.1	0.1	0	0.1	0	0.1	0	5.1(5)	81.9	97.6	1.83
Alkaline rocks (13)	34.4	49.3	4.1	10.2	1.0	—	0.6	—	0.3	—	0.1	—	—	—	2.1(2)	87.7	98.0	3.37
Carbonatites (15)[3]	33.5	49.1	3.6	12.5	1.0	0	0.2	0	0.1	0	0	0	0	—	1.3(3)	86.2	98.7	2.67
Bastnaesite-(Y)																		
Granites and granite pegmatites (9)	4.3	14.0	3.1	11.5	9.8	0.4	15.3	2.5	17.6	3.0	9.9	0.9	6.6	1.1	35.3(3)	21.4	32.9	0.37

[1] Number in parentheses indicates the number of analyses. – [2] Number of analyses, in which Y is reported. – [3] Including bastnaesite-(La) and hydroxylbastnaesite.

Table 2
Distribution of Rare Earth Elements in Bastnaesites*).

No.	Locality	La	Ce	Pr	Nd	Sm	Eu	Gd	Tb	Dy	Ho	Er	Tm	Yb	Lu	Y	Ref.
	Common bastnaesites from magmatites (1 to 6) and metasomatites (7 to 10)																
1	Pikes Peak, Colorado	25.4	53.3	5.4	14.6	1.0	—	—	—	—	—	—	—	—	—	0.3	[37]
2	Madagascar	34	53	3.8	8.5	0.7	—	0.1	—	—	—	—	—	—	—	—	[51]
3	Vishnevye Gory, South Urals	45.7	45.9	2.0	6.4	—	—	—	—	—	—	—	—	—	—	—	[52]
		39	50	3.1	7.7	0.2	—	—	—	—	—	—	—	—	—	—	[51]
		32.1	54.4	3.3	9.8	0.3	0.1		0.1		—	—	—	—	—	—	[53]
4	Narsarsuk, South Greenland	28	51	4.8	14	1.0	0.1	0.5	—	0.3	—	—	—	—	—	—	[37]
5	Borsuksai, NW Kazakhstan	30.9	51.6	6.2	10.3	1.0	—	—	—	—	—	—	—	—	—	—	[33]
6	Eastern Tuva	32	53	3	9	1.1	—	1.3	—	0.4	—	0.2	—	—	—	—	[54]
7	Siberia	36.7	42.8	6.1	12.2	1.2	—	0.4	—	tr.	—	—	—	—	—	0.5	[14]
*8		43.9	38.4	3.0	10.0	1.2	—	1.2	—	—	—	—	—	—	—	2.3	[14]
9	Burpala, North Baikalia	39.3	53.0	1.0	6.4	0.3	—	—	—	—	—	—	—	—	—	—	[55]
*10	Tarbagatai, Kazakhstan	26.1	41.8	7.4	17.4	3.0	—	1.4	0.9	0.5	0.1	0.2	0.1	1.0	0.2	—	[41, p. 125]
	Common bastnaesites from hydrothermal occurrences																
11	Gallinas, New Mexico	42.9	45.6[1)]	3.9[1)]	7.1	0.5	—	—	—	—	—	—	—	—	—	0.1	[39]
12	Karonge, Burundi	29.4	48.8	4.4	15.4	1.3	—	0.5	—	—	—	—	—	—	—	0.2	[37]
		26.6	51.3[1)]	4.3[1)]	14.6	2.1	0.3	0.5	0.05	0.08	0.01	0.01	<0.01	<0.01	—	0.3	[17]

		31.4	48.0	4.2	14.7	1.0	0.1	0.4	—	0.1	—	—	—	—	—	—	[52]
		34.0	48.5	1.8	13.0	1.0	0.2	1.0	—	—	—	tr.	—	—	—	0.4	[56]
13	Mongolia	27	54	4.1	14	0.3	—	0.4	—	—	—	—	—	—	—	—	[37]
14	Tuva, Siberia	25	50	5.6	18	0.9	—	0.5	—	—	—	—	—	—	—	—	[51]
15	Tuva, Siberia[2)]	17	54	6.5	18	1.9	0.4	0.6	0.05	0.4	0.1	0.3	—	0.05	—	—	[45]
	[3)]	27	50	4.5	17	0.8	0.1	0.4	—	0.1	—	0.1	—	—	—	—	[45]
*16	Vishnevye Gory, South Urals	35	53.7	3.2	8	0.1	—	—	—	—	—	—	—	—	—	—	[57]
17	Russian Platform	29	49[1)]	4.7[1)]	14.5	1.5	—	0.5	—	0.2	—	—	—	—	—	0.6	[23]
*18	North Baikalia	23.4	54.7	5.0	15.6	1.3	—	—	—	—	—	—	—	—	—	—	[58]
	Carbonatites[4)]																
19	Mountain Pass, California																
	Sulphide Queen Body	26.1	50.5	4.9	15.9	1.8	—	0.9	—	—	—	—	—	—	—	0.2	[39]
		25.7	49.8[1)]	4.8[1)]	15.6	1.8	—	0.9	—	—	—	—	—	—	—	0.2	[51]
20	Birthday Claim	29.6	50.3[1)]	4.4[1)]	14.3	1.3	—	—	—	—	—	—	—	—	—	—	[32]
		42.1	44.6[1)]	3.5[1)]	8.4	0.8	—	—	—	—	—	—	—	—	—	—	[39]
21	Average ore	28.6	55.8	4.1	10.5	0.7	—	0.3[5)]	—	—	—	—	—	—	—	—	[59]
22	Typical ore	34	50	4	11	0.5	0.2	0.1	—	—	—	0.2[6)]	—	—	—	—	[60]
23		32.3	49.0	4.5	13.1	0.5	0.1	0.2	—	0.3	—	—	—	—	—	—	[56]
24	Vuoriyarvi, Kola	37.8	52.8	5.0		2.4	—	2.0[7)]	—	—	—	—	—	—	—	—	[61]
25		37.9	52.9	4.8		2.2	—	—	—	—	—	—	—	—	—	0.04	[10]
26		22	49	4	24	1	—	—	—	—	—	—	—	—	—	—	[62]
27	Kovdor, Kola	23	46	5	25	1	—	—	—	—	—	—	—	—	—	—	[62, 63]
28	East Sayan	42	40	3.3	14	1.2	—	—	—	—	—	—	—	—	—	—	[34]
*29		42.3	39.6	3.3	13.6	1.2	—	—	—	—	—	—	—	—	—	—	[64]

Table 2 (continued)

No.	Locality	La	Ce	Pr	Nd	Sm	Eu	Gd	Tb	Dy	Ho	Er	Tm	Yb	Lu	Y	Ref.
*30	East Yakutia	36.0	50.4	3.3	10.3	—	—	—	—	—	—	—	—	—	—	—	[2]
*31	Eastern Siberia	39.3	45	4.1	11.6	—	—	—	—	—	—	—	—	—	—	—	[25]
	Bauxites (32 to 34) and a weathering crust (35)																
*32	Marmara, Greece	30.1	27.3	6.8	27.7	3.9	—	3.2	—	1.0	—	—	—	—	—	—	[47]
**33		30.2	23.7	6.8	28.5	3.9	—	3.2	—	0.9	—	—	—	—	—	2.8	[7, p. 38]
**34		11.7	72.2	1.6	7.5	1.8	—	1.7	—	0.4	—	0.2	—	—	—	2.9	[7, p. 38]
35	Kazakhstan, USSR	13.6	0.4	7.8	43.0	12.0	2.8	12.0	1.1	3.8	0.7	1.7	0.3	0.7	0.1	—	[35]
	Pseudomorphs[8)]																
36	Siberia, USSR	39	45	3.3	11	0.8	0.1	0.5	0.1	0.3	—	0.1	—	—	—	—	[24, 37]
37	European USSR	28[9)]	44	4	23	0.6	—	0.5	—	0.1	—	—	—	—	—	—	[34, 37]
38	Mongolia	22	51	5.6	19	1.4	0.1	0.9	0.1	0.2	—	—	—	—	—	—	[34]
39		24.4	52.3	4.2	17.5	1.6	—	—	—	—	—	—	—	—	—	—	[2]
*40	Tarbagatai, Kazakhstan	27.2	54.4	5.9	10.4	1.5	—	0.5	—	0.1	—	—	—	—	—	—	[41]
41	Russian Platform	28.6	53.1	2.7	11.8	1.6	0.1	1.4	—	0.4	0.1	0.1	0.04	0.05	—	—	[43]
*42	Ilmen Mountains, Urals	36.7	43.3	4.7	6.5	2.6	0.2	1.1	—	1.9	0.2	1.7	0.2	1.3	0.2	—	[65]
*43	Tarbagatai, Kazakhstan	31.1	51.8	4.2	9.3	1.8	—	1.1	—	0.5	—	0.3	—	—	—	—	[41, 29]
		28.4	48.9	5.9	13.9	1.8	—	0.7	—	0.3	—	—	—	—	—	—	[41, 29]
*44	Ukraine	23	40	3	18	2.4	0.3	1.2	0.1	0.6	—	0.2	—	—	—	11.2	[66]
*45	Vishnevye Gory, South Urals	34.1	53.0	3.1	8.8	0.6	—	0.3	—	0.1	—	—	—	—	—	—	[65]

*46	Tarbagatai, Kazakhstan	1.4	7.0	1.8	6.2	5.3	0.6	6.8	1.6	11.0	2.6	7.5	1.1	5.3	1.7	40.1	[5]
		4.7	13.6	2.6	6.6	10.5	0.6	16.3	2.8	20.6	3.7	10.5	0.9	5.8	0.7	—	[5]
		2.8	9.2	3.3	10.9	9.8	0.2	21.0	3.1	19.9	3.7	10.0	0.7	4.6	0.9	—	[41, p.126]
		4.6	9.2	0.2	9.0	5.0	—	15.0	2.5	21.2	4.2	14.2	1.7	10.4	2.9	—	[41, p. 126]
		2.5	7.6	2.5	7.2	7.8	0.2	18.0	2.7	21.5	4.5	13.5	1.2	9.4	1.4	—	[41, p. 126]
		4.9	14.1	3.2	7.2	11.1	0.5	16.2	3.5	20.4	—	11.1	0.9	6.0	0.9	—	[41, p. 126]
		2.3	7.3	2.5	7.3	8.2	0.2	18.6	2.5	20.9	4.4	13.8	1.3	9.7	1.3	—	[41, p. 126]
		4.8	19.1	2.9	14.3	3.8	—	9.5	1.0	11.9	2.9	11.8	1.3	16.7	—	—	[56]
*47	European USSR	34.2	43.4	3.9	12.9	2.3	0.3	1.2	—	—	0.1	0.2	<0.1	<0.1	—	1.63	[30]
48	Russian Platform	31.3	47.8	3.3	13.7	1.8	0.2	1.0	0.1	0.1	0.1	0.1	<0.1	<0.1	—	—	[15]
49	Kola, USSR	7.3	22.0	4.2	12.0	4.6	0.5	3.5	0.8	5.6	1.2	4.0	0.6	3.5	0.6	(30)[10]	[34]
		6.4	20	2.9	14	7.2	—	8.9	1.4	12	1.2	5.2	1.2	5.2	—	14	[34]
50	Mongolia	2.8	14	3.5	16	12	0.7	9	1.4	6	1.4	3	0.4	2.5	0.4	(25)[10]	[34]
51	East Siberia	21.3	17.2	4.6	16.4	3.4	—	3.4	—	4.3	1.6	2.8	—	2.0	—	23.2	[11]

*) For a key of the symbols used see p. 116; for references see pp. 116/7.

1) CeO_2 and Pr_6O_{11}. – 2) Sideritic association. – 3) Hematitic association. – 4) Nos. 19 to 23 and 26 to 29 are common bastnaesites, Nos. 24, 25, 30, and 31 are hydroxylbastnaesites. – 5) Eu + Gd. – 6) Tb to Lu and Y. – 7) Eu, Gd, Tb, Dy, and Y. – 8) Common bastnaesite pseudomorphs after: parisite (36 and 37), huanghoite (38 and probably 39), monazite (40), allanite (41 to 43), chevkinite (44 and 45), and britholite (47 and 48). Pseudomorphs of bastnaesite-(Y) after gagarinite (46), of bastnaesite-(Ce-Y) after britholite-(Y) (49) and gadolinite (50), and thorbastnaesite after ferrithorite (51). – 9) Erroneously given as 22 in [34]. – 10) Calculated as difference against 100.

that there is a marked increase in light lanthanides in the common bastnaesites from granites and granite pegmatites through hydrothermal and alkaline to carbonatitic samples [1].

In Table 2, pp. 112/5, the composition of RE ± Y of bastnaesite and its varieties from various rocks and of different genesis is given as $(RE \pm Y)_2O_3 = 100$, as RE ± Y = 100 (numbers with one asterisk), or as at.% (numbers with two asterisks); tr = traces.

References for 10.5.2.1.3:

[1] M. Fleischer (Can. Mineralogist **16** [1978] 361/3). – [2] I. V. Aleksandrova, V. I. Ivanov, L. A. Sin'kova (Zap. Vses. Mineral. Obshchestva **94** [1965] 323/6). – [3] C. Palache, H. Berman, C. Frondel (The System of Mineralogy, Vol. 2, 7th Ed., Wiley, New York 1951, pp. 1/1124, 289/90). – [4] M. V. Akhmanova, L. P. Orlova (Geokhimiya **1966** 571/8; Geochem. Intern. **3** [1966] 444/51). – [5] D. A. Mineev, T. I. Lavrishcheva, A. V. Bykova (Zap. Vses. Mineral. Obshchestva **99** [1970] 328/32).

[6] E. V. Khalezova, I. I. Nazarenko (Tr. Inst. Mineral. Geokhim. Kristallokhim. Redk. Elem. Vopr. Mineral. Geokhim. Genezisa Mestorozhd. Redk. Elem. No. 2 [1959] 99/101). – [7] Z. Maksimović, G. Pantó (Bull. Acad. Serbe Sci. Arts Classe Sci. Nat. Math. Sci. Nat. **72** No. 20 [1980] 35/42). – [8] A. S. Kirillov (Zap. Vses. Mineral. Obshchestva **95** [1966] 51/9). – [9] N. V. Svyazhin (Tr. Inst. Geol. Akad. Nauk SSSR Ural'sk. Filial No. 70 [1965] 249/52). – [10] A. S. Kirillov (Dokl. Akad. Nauk SSSR **159** [1964] 1048/50; Dokl. Earth Sci. Sect. **159** [1964] 93/5).

[11] A. S. Pavlenko, L. P. Orlova, M. V. Akhmanova, K. I. Tobelko (Zap. Vses. Mineral. Obshchestva **94** [1965] 105/13). – [12] A. S. Povarennykh (Crystal Chemical Classification of Minerals, Vol. 2, translated from Russian by J. E. S. Bradley, Plenum, New York 1972, pp. 459/766, 614). – [13] E. I. Semenov (in: Mineralogiya Pegmatitov i Gidrotermalitov Shchelochnykh Massivov, Moscow 1967, pp. 1/168, 36/40). – [14] I. I. Kupriyanova (Tr. Mineral. Muzeya Akad. Nauk SSSR No. 18 [1968] 206/9). – [15] L. V. Ganzeeva [Ganzeyeva], A. A. Ganzeev [Ganzeyev] (Geokhimiya **1975** 74/80; Geochem. Intern. **12** No. 1 [1975] 47/52).

[16] J. Hansen (Medd. Groenland **181** No. 8 [1968] 1/47, 31). – [17] W. Ooghe, F. Verbeek (Anal. Chim. Acta **73** [1974] 87/95, 92). – [18] J. Thoreau, B. Aderca, L. van Wambeke (Acad. Roy. Sci. Coloniales [Brussels] Classe Sci. Nat. Med. [2] **4** No. 3 [1958] 684/715). – [19] W. L. Smith, E. A. Cisney (Am. Mineralogist **41** [1956] 76/81, 78/9). – [20] G. Pantó (Acta Geol. Acad. Sci. Hung. **19** [1975] 59/93, 88).

[21] M. Ya. Somina, A. G. Bulakh (Mineral. Geokhim. [Leningrad] No. 4 [1972] 64/73, 66/7). – [22] W. D. Riesmeyer (Virginia J. Sci. **18** [1967] 188). – [23] L. B. Zubkov, N. I. Galadzheva, A. B. Chernyakovskii (Dokl. Akad. Nauk Belorussk. SSR **16** [1972] 646/9). – [24] E. I. Semenov, A. P. Khomyakov, A. V. Bykova (Tr. Mineral. Muzeya Akad. Nauk SSSR No. 11 [1961] 202/4). – [25] T. B. Zdorik (Tr. Mineral. Muzeya Akad. Nauk SSSR No. 17 [1966] 60/75, 72, 74).

[26] V. S. Gaidukova, T. B. Zdorik (Geol. Mestorozhd. Redk. Elem. No. 17 [1962] 86/117, 113; Ref. Zh. Geol. **1963** No. 10 V 121). – [27] N. Çagatay (Turk. Jeol. Kurumu Bul. **24** No. 2 [1981] 59/65, 63 [Turkish with English Abstr.]; C.A. **96** [1982] No. 220682). – [28] A. G. Mironov, A. G. Gofman (Tr. Geol. Inst. Buryat. Filial Sib. Otd. Akad. Nauk SSSR No. 23 [1980] 64/73). – [29] V. I. Chernov, O. F. Krol', A. V. Stepanov (Geol. Geokhim. Mineral. Mestorozhd. Redk. Elem. Kaz. **1966** 138/44, 141 and table after p. 140). – [30] L. V. Ganzeeva (Redk. Elem. Syr'e Ekon. No. 7 [1972] 112/24, 119).

[31] K. Sakurai, H. Wakita, A. Kato, K. Nagashima (Bull. Chem. Soc. Japan **42** [1969] 2725/8). – [32] J. C. Olson, D. R. Shawe, L. C. Pray, W. N. Sharp (U.S. Geol. Surv. Profess. Paper No. 261 [1954] 1/75, 35). – [33] V. A. Cheverdin (Tr. Kaz. Nauchn. Issled. Inst. Mineral'n. Syr'ya No. 2 [1960] 89/91). – [34] E. I. Semenov (Mineralogiya Redkikh Zemel', Moscow 1963, pp. 1/412, 98/100, 102). – [35] E. K. Podporina, A. P. Niyazov, M. D. Brylin (in: V. V. Burkov, Ekzogennye

Mestorozhdeniya Redkikh Elementov Sb. Nauchn. Statei Akad. Nauk SSSR, Moscow 1980, pp. 1/134, 24/6).

[36] A. P. Khomyakov (in: K. A. Vlasov, Mineralogiya i Geneticheskie Osobennosti Shchelochnykh Massivov, Moscow 1964, pp. 56/82). – [37] E. I. Semenov (in: K. A. Vlasov, Geochemistry and Mineralogy of Rare Elements and Genetic Types of Their Deposits, Vol. 2: Mineralogy of Rare Elements, Israel Program Sci. Trans., Jerusalem 1966, pp. 249/328, 266/9 [Russian Original, Moscow 1964]). – [38] M. Fleischer (Geochim. Cosmochim. Acta **29** [1965] 755/72, 766/8). – [39] K. J. Murata, H. J. Rose, M. K. Carron, J. J. Glass (Geochim. Cosmochim. Acta **11** [1957] 141/61). – [40] Yu. L. Kapustin (in: L. N. Ovchinnikov, N. A. Solodov, Mestorozhdeniya Litofil'nykh Redkikh Metallov, Nedra, Moscow 1980, pp. 243/85).

[41] D. A. Mineev (Geokhimiya Apogranitov i Redkometal'nykh Metasomatitov Severo-Zapadnogo Tarbagataya, Moscow 1968, pp. 1/183). – [42] V. Marmo, V. Hoffrén, K. Hytönen, P. Kallio, O. Lindholm, J. Siivola (Bull. Comm. Geol. Finlande No. 221 [1966] 1/34, 25). – [43] G. Bardossy, G. Pantó (Congr. ICSOBA, Nice 1973, Vol. 3, pp. 47/53, 49/50). – [44] A. Vorma, P. Ojanperä, V. Hoffrén, J. Siivola, A. Löfgren (Bull. Comm. Geol. Finlande No. 222 [1966] 241/74, 248). – [45] A. P. Khomyakov (Geokhimiya **1964** 85/8; Geochem. Intern. **1** [1964] 40/3).

[46] R. Marinova (Rudoobraz. Protsesi Mineralni Nakhodishta **14/15** [1981] 132/6, 133 [Bulgarian, German on p. 136]; C.A. **97** [1982] No. 58728). – [47] Z. Maksimović, G. Pantó (4th Intern. Congr. Study Bauxites Alumina Alum. Preprints, Athens 1978, Vol. 2, pp. 540/52, 548/50; Mineral. Abstr. **31** [1980] No. 80-2783). – [48] J. C. Hughes, G. Brown (Clay Minerals **12** [1977] 319/29, 327). – [49] D. McKie (Mineral. Mag. **33** [1962] 281/97, 285). – [50] D. A. Mineev (in: L. N. Ovchinnikov, N. A. Solodov, Mestorozhdeniya Litofil'nykh Redkikh Metallov, Nedra, Moscow 1980, pp. 178/84, 182).

[51] E. I. Semenov, R. L. Barinskii (Geokhimiya **1958** 314/33; Geochemistry **1958** 398/419, 400). – [52] E. I Semenov, A. P. Khomyakov (Diagnost. Diagnost. Svoistva Mineral. **1981** 88/93, 90). – [53] A. G. Zhabin, N. V. Svyazhin (Tr. Inst. Mineral. Geokhim. Kristallokhim. Redk. Elem. Akad. Nauk SSSR No. 9 [1962] 55/66, 62/3). – [54] Yu. L. Kapustin (Geokhimiya **1982** 523/40, 538). – [55] A. P. Khomyakov (Izv. Akad. Nauk SSSR Ser. Geol. **1972** No. 1, pp. 94/101, 98).

[56] N. I. Aleksandrova, I. V. Maksimova, N. M. Svirskaya (Izv. Vysshikh Uchebn. Zavedenii Geol. Razved. **20** No. 10 [1977] 80/7, 84). – [57] E. B. Khalezova, D. N. Knyazeva (Tr. Mineral. Muzeya Akad. Nauk SSSR No. 12 [1961] 235/8). – [58] V. V. Gerasimovskii (Tr. Mineral. Muzeya Akad. Nauk SSSR No. 15 [1964] 194/202, 195). – [59] D. R. Maneval (Mineral. Ind. **31** No. 8 [1962] 1, 3/5, 8, 3). – [60] J. G. Cannon (Mining Eng. **21** No. 1 [1969] 53/4).

[61] A. A. Kukharenko, M. P. Orlova, A. G. Bulakh et al. (Kaledonskii Kompleks Ultraosnovnykh Shchelochnykh Porod i Karbonatitov Kol'skogo Poluostrova i Severnoi Karelii, Nedra, Moscow 1965, pp. 1/772, 534). – [62] Yu. L. Kapustin (Geokhimiya **1966** 1311/21; Geochem. Intern. **3** [1966] 1054/64, 1060). – [63] Yu. L. Kapustin (Mineralogiya Karbonatitov, Moscow 1971, pp. 1/288, 170). – [64] E. E. Vainshtein, L. K. Pozharitskaya, N. V. Turanskaya (Geokhimiya **1961** No. 11, pp. 1031/4; Geochemistry [USSR] **1961** 1151/4). – [65] D. A. Mineev, B. A. Makarochkin, A. G. Zhabin (Geokhimiya **1962** 590/7; Geochemistry [USSR] **1962** 684/93, 689).

[66] E. Ya. Marchenko, O. I. Chashka, E. P. Gurova (Dopov. Akad. Nauk Ukr. RSR B **30** No. 1 [1968] 69/72).

10.5.2.1.4 Crystal Forms and Intergrowths

Crystal forms. Bastnaesite is hexagonal ditrigonal-dipyramidal [1, 2, 4], [3, p. 34] as are thorbastnaesite (according to its space group) [5, p. 106] and a synthetically formed bastnaesite, $CeFCO_3$ [6]; for synthetic $Ln[CO_3](OH)$ only the hexagonal system is confirmed [7]. A hexagonal exterior symmetry is confirmed for hydroxylbastnaesite [8], and for bastnaesite from East Sayan; for the latter the crystal class is questioned [9, p. 66].

The morphological axial ratios $c/a = 1.3598$ [1] and 1.369 [10] correspond to those determined from unit cell data, see pp. 122/3.

The most frequently observed forms for common bastnaesite are the pinacoid c$\{0001\}$, the prisms a$\{2\bar{1}\bar{1}0\}$, and m$\{10\bar{1}0\}$, and three hexagonal bipyramids q$\{10\bar{1}1\}$, p$\{10\bar{1}2\}$, and t$\{10\bar{1}3\}$ [10, pp. 950/1]. Besides c and m crystals from the Pikes Peak, Colorado, have a$\{11\bar{2}0\}$, p$\{10\bar{1}1\}$, q$\{20\bar{2}1\}$, and s$\{11\bar{2}1\}$ [11, p. 529]. Nearly the same forms (c, m, q, p, t, and a, but s$\{11\bar{2}2\}$) have been observed with crystals from East Sayan, where two of three different types of development show prisms and all up to four pyramids besides the dominant pinacoid [9, pp. 65/7], see **Fig. 6**. Crystals from North Baikalia show only c and m [12, 13], those from South Africa ("buszite") c, m, a$\{11\bar{2}0\}$, h$\{21\bar{3}0\}$, and s$\{11\bar{2}2\}$ [4], and the hydroxylbastnaesite from Vuoriyarvi, Kola, besides c and a, has the hexagonal dipyramids $\{11\bar{2}2\}$, $\{10\bar{1}1\}$, and $\{20\bar{2}1\}$ and the dihexagonal-bipyramid $\{12\bar{3}9\}$ [8], or, somewhat deviating, the pyramids $\{10\bar{1}2\}$, $\{10\bar{1}3\}$, and $\{20\bar{2}3\}$ besides c and m on two types of crystals [14, p. 52], see Fig. 6. The figure shows the narrow bands or sometimes total absence of the prism faces and the conspicuous pyramids, cf. [15, p. 605].

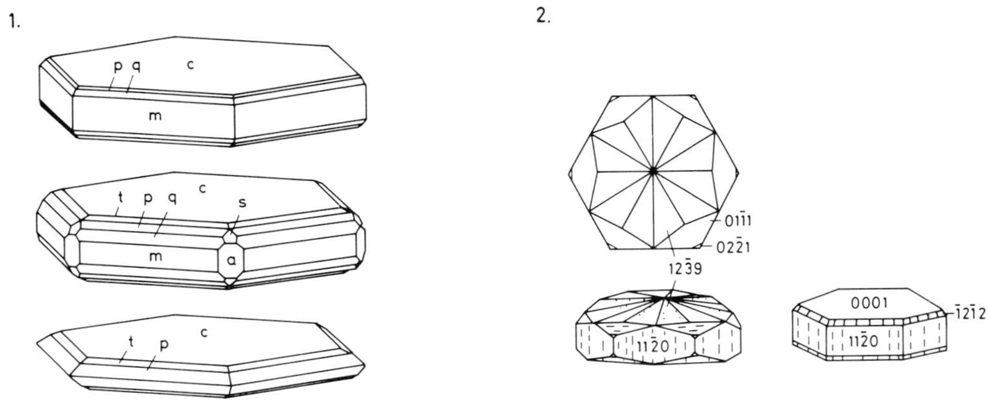

Fig. 6. Forms of bastnaesite crystals.
1. Common bastnaesite of East Sayan from [9, p. 67].
2. Hydroxylbastnaesite of Vuoriyarvi, Kola, from [8].

Habit. Most bastnaesite crystals are tabular on c$\{0001\}$ [1], [10, pp. 950/1], [16, 17], see for crystals from California [3, p. 34], Madagascar [18, 19], Bulgaria [20], and the Urals [21], and for large tabular crystals, single and in groups, from Karonge, Burundi [22]. A prismatic habit with pyramidal faces has been observed on crystals from New Mexico [23], from Colombia [10, pp. 950/1] and from Colorado, Siberia, and South Africa, see above. Often the crystals are tabular to lamellar [24], fine lamellar [9, p. 65], or occur in very thin laminae, as in Alpine fissures [25]. Coarse euhedral to subhedral tabular crystals, flattened to the base, but also tiny grains (in druses) have been observed in tin granites of South Africa [26]. Recrystallized bastnaesite is also coarse tabular [27] or barrel-shaped [28 to 30], as are also pseudomorphs after allanite [31]; a pseudomorph after zircon forms flat bipyramidal crystals [32]. Fibrous crystals occur in another pseudomorph after allanite [33], and fibers and needles in apatite rock in East Africa [34, p. 99]. Further bastnaesite occurs in small grains [35] of sandy to platy form [36], sugary form [37], in anhedral and sometimes rectangular form [38], or irregular form [39].

In thin section crystals with hexagonal outline have been observed from the Mountain Pass, California [3, pp. 45, 47], from Bastnäs, Sweden [40], from the Russian Platform [41], from the North Baikal area, Siberia [12], and from Alpine fissure veins in Switzerland [42].

Aggregates of bastnaesite crystals are common and have been observed for prismatic crystals at Bastnäs, Sweden [40], for platy crystals in New Mexico, and for tabular crystals in Burundi [15, pp. 605, 608]. Acicular crystals aggregate to sheaves and felted masses in Bolivia [43], to sheaves, tufts, and stellate or radiating groups in East Africa [34, pp. 97/8, 103, 108], to spherical radiating groups in Quebec [44], to rosettes in Kola [45], and to fan-shaped or star-like aggregates in Siberia [46]; pearly flakes are aggregated in druses in carbonatitic rocks in Malawi [47], cf. [48]. Prismatic crystals of bastnaesite in granite from Rhode Island consist of minute crystallites [49], strontianite-bastnaesite-pseudomorphs from Siberia contain granular to radiated or sheaflike aggregated bastnaesite [28]. A supergene bastnaesite, consisting of fibers, elongated parallel to [0001] and aggregated to crossed, tangled, or partly radiating groups, shows a hexagonal outline under the electron microscope [52]. Further bastnaesite occurs in anhedral granular masses [1], cf. [50, 51].

Single crystals of hydroxylbastnaesite from Vuoriyarvi, Kola, are tabular, somewhat short-columnar, rarely needle-shaped; they form accumulations, nodular aggregates, and brush-shaped forms [14]. Bastnaesite(-Y) forms large hexahedral crystals (up to 8 cm in diameter) with a very typical, pyramidal termination; under the microscope it consists of an aggregate of elongated grains, which frequently shows an irregular ragged form [53]. Thorbastnaesite consists of microindividuals, which are subparallel arranged to cryptocrystalline aggregates; under the electron microscope little balls are visible, sometimes also fine lamellae with hexagonal outline [5, p. 105].

The synthetic compound $Ln(OH)CO_3$ of B-type in high resolution electron microscopy clearly shows the same lamellar nature as natural bastnaesite (see figures in [54]) [7].

A horizontal **striation** occurs on crystals from New Mexico with rough edges [15, p. 605]; deep horizontal grooves on bastnaesite crystals are due to oscillatory combination of $\{10\bar{1}0\}$ and $\{10\bar{1}1\}$ and resemble a pile of thin plates [1, 50]. The crystal faces of bastnaesite are often rough [11, p. 529], sometimes uneven with fine pores [41].

Twins have only been observed for hydroxylbastnaesite from Vuoriyarvi, Kola, where penetration twins $\{10\bar{1}0\}$ with intergrowths at $\{10\bar{1}2\}$ are characteristic and form triplet crystals and subparallel intergrowths of many individuals [8, 14]. For twins with composition plane $\{0001\}$ formed after burbankite in cavities at this locality, see [45].

A prismatic **cleavage** $\{10\bar{1}0\}$ (formerly supposed to be a distinct one [11, p. 531]), is rather indistinct [1, 16]; it is poorly developed and rarely observed on crystals from New Mexico [15, p. 605], [23], and also present on crystals from the Mountain Pass, California [3, p. 34]. A basal cleavage $\{0001\}$ is sometimes observed [26], designated as well-developed [30] or even perfect [55], or brilliant [22], it may be a distinct to perfect **parting** $\{0001\}$ of secondary origin (cf. [1]) and can easily be mistaken as a cleavage [23]. This secondary basal parting has been observed on small short-prismatic crystals from New Mexico, which easily can be split to thin laminae [50], giving them an appearance of mica [56]; it develops with increasing weathering, which causes a weaker bond between the laminae [15, p. 605]. For hydroxylbastnaesite an imperfect prismatic cleavage $\{11\bar{2}0\}$ has been observed [8, 14].

Syntaxic Intergrowths. Examination of a great number of specimens shows, that single crystals of bastnaesite are rare and most specimens are polycrystals: extended lamellar intergrowths ("syntaxy") had been at first observed between bastnaesite and other members of the series bastnaesite-synchysite, e.g., parisite and roentgenite, but not between the two end members bastnaesite and synchysite. The observed intergrowths are characterized by common c and a axes of the two individuals, the a axes may have the same sense or opposite senses [10, pp. 939/41], the composition plane of such bidimensional syntaxic intergrowths is the basis (0001) [57]. The existence of such intergrowths of tabular subindividuals of two RE-F carbonates is confirmed for bastnaesite and parisite in "block crystals" in a carbonatite of the East Sayan, USSR [9, p. 65], for bastnaesite with synchysite in Switzerland in Alpine fissure veins in the Grimsel granite [58] and at the Piz Blas [42], cf. [59], and in a carbonatite of the Tundulu complex, southern Malawi [34, pp. 103, 109]. More recent determinations on a mixed layer compound (bastnaesite from Narsarsuk, Greenland) shows intimate syntaxy not only between bastnaesite (B), roentgenite (BS_2), parisite (BS), and synchysite (S), but also for three additional members with block symbols BS_4, B_3S_4, and B_3S_2; the compounds have been identified by diffraction techniques and by electron microscopic images, which confirm a perfect microsyntaxy on a very fine scale of about 100 Å [54, pp. 352/5], see also p. 91.

At the Mountain Pass, California, bastnaesite (locally perched on cerite in cavities) is intimately intergrown with cerite in parallel growths with a preferred orientation of the hexagonal c axis tending to lie normal to the crystal surface of cerite [60]; at Bastnäs, Sweden, bastnaesite is intergrown with cerite, but also replaces this mineral [40, pp. 16, 20/2].

Bastnaesite often forms oriented intergrowths and alteration pseudomorphs after tysonite, with the crystal axes of the two minerals oriented in parallel positions [1], cf. pp. 15, 18. For bastnaesite associated with tysonite of the Pikes Peak, Colorado [61], [15, p. 609], the intimate parallel intergrowth and the same orientation of these minerals indicate oriented intergrowths as more probable than formerly assumed formation of bastnaesite from tysonite [11, pp. 527/31]; Oftedal, however, inspite of the observed intergrowth of these two minerals (X-ray investigation), supposed bastnaesite to be an alteration product of tysonite [62], cf. [10, p. 942].

References for 10.5.2.1.4:

[1] C. Palache, H. Berman, C. Frondel (The System of Mineralogy, 7th Ed., Vol. 2, Wiley, New York 1951, pp. 1/1124, 289). – [2] H. Strunz, C. Tennyson (Mineralogische Tabellen, 6th Ed., Geest and Portig, Leipzig 1977, pp. 1/621, 242). – [3] J. C. Olson, D. R. Shawe, L. C. Pray, W. N. Sharp (U.S. Geol. Surv. Profess. Paper No. 261 [1954] 1/75). – [4] J. Goñi, C. Guillemin (Bull. Soc. Franc. Mineral. **76** [1953] 124/9). – [5] A. S. Pavlenko, L. P. Orlova, M. V. Akhmanova, K. I. Tobelko (Zap. Vses. Mineral. Obshchestva **94** [1965] 105/13).

[6] G. B. Jansen, G. B. Magin Jr., B. Levin (Am. Mineralogist **44** [1959] 180/1). – [7] H. Dexpert, E. Antic-Fidancev, J. P. Coutures, P. Caro (J. Cryst. Spectrosc. Res. **12** [1982] 129/42, 129, 141). – [8] A. S. Kirillov (Dokl. Akad. Nauk SSSR **159** [1964] 1048/50; Dokl. Earth Sci. Sect. **159** [1964] 93/5). – [9] M. Ya. Somina, A. G. Bulakh (Mineral. Geokhim. **1972** No. 4, pp. 64/73). – [10] G. Donnay, J. D. H. Donnay (Am. Mineralogist **38** [1953] 932/63).

[11] R. Köchlin (Tschermaks Mineral. Petrogr. Mitt. [2] **31** [1911/12] 525/36). – [12] V. V. Gerasimovskii (Tr. Mineral. Muzeya Akad. Nauk SSSR No. 15 [1964] 194/202, 194). – [13] V. V. Gerasimovskii (Vestn. Mosk. Univ. Ser. IV Geol. **19** No. 6 [1964] 43/9, 47). – [14] A. S. Kirillov (Zap. Vses. Mineral. Obshchestva **95** [1966] 51/9, 51/2). – [15] J. J. Glass, R. G. Smalley (Am. Mineralogist **30** [1945] 601/15).

[16] E. I. Semenov (in: K. A. Vlasov, Geochemistry and Mineralogy of Rare Elements and Genetic Types of Their Deposits, Vol. 2: Mineralogy of Rare Elements, Israel Program Sci. Trans., Jerusalem 1966, pp. 249/328 [Russian Original, Moscow 1964]). – [17] A. L. Anderson (Econ. Geol. **55** [1960] 1179/201, 1189). – [18] A. Lacroix (Bull. Soc. Franc. Mineral. **38** [1915] 106/25, 265/71, 110). – [19] J. Behier (Rep. Malgache Ann. Geol. Madagascar No. 29 [1960] 1/78, 18). – [20] R. Marinova (Rudoobraz. Protsesi Mineralni Nakhodishta **14/15** [1981] 132/6, 133 [Bulgarian, German on p. 136]; C.A. **97** [1982] No. 58728).

[21] A. G. Zhabin, G. N. Mukhitdinov, M. E. Kazakova (Tr. Inst. Mineral. Geokhim. Kristallokhim. Redk. Elem. **1960** No. 4, pp. 51/73, 69; C.A. **56** [1962] 11269). – [22] B.-M. Aderca, R. van Tassel (Acad. Roy. Sci. Outre Mer Classe Sci. Nat. Med. [2] **18** No. 5 [1971] 1/117, 83). – [23] R. M. Perhac, E. W. Heinrich (Econ. Geol. **59** [1964] 226/39, 232). – [24] I. I. Kupriyanova (Tr. Mineral. Muzeya Akad. Nauk SSSR No. 18 [1968] 206/9). – [25] A. Kipfer (Schweizer Strahler **3** No. 4 [1973] 133/69, 157).

[26] C. A. Strauss (S. Africa Geol. Surv. Mem. **46** [1954] 1/241, 158). – [27] Yu. L. Kapustin (Mineralogiya Karbonatitov, Moscow 1971, pp. 1/288, 184). – [28] T. B. Zdorik (Tr. Mineral. Muzeya Akad. Nauk SSSR No. 17 [1966] 60/75, 68/9). – [29] T. B. Zdorik (Geol. Mestorozhd. Redk. Elem. Vses. Nauchn. Issled. Inst. Mineral'n. Syr'ya No. 30 [1966] 121/31, 127). – [30] T. L. Sverdrup, K. Ø. Bryn, P. C. Sæbø (Norsk Geol. Tidsskr. **39** [1959] 237/47, 240).

[31] V. I. Chernov, O. F. Krol', A. V. Stepanov (in: Geol. Geokhim. Mineral. Mestorozhd. Redk. Elem. Kaz. **1966** 138/44, 139). – [32] G. Raade, J. Haug (Mineral. Rec. **11** [1980] 83/91, 88). – [33] J. W. Adams, E. J. Young (U.S. Geol. Surv. Profess. Paper No. 424-C [1961] 292/4). – [34] M. S. Garson (Nyasaland Geol. Surv. Dept. Mem. No. 2 [1962] 1/248). – [35] G. B. Griswold (New Mexico Bur. Mines Mineral Resour. Bull. No. 67 [1959] 1/117, 71).

[36] J. H. Soulé (U.S. Bur. Mines Rept. Invest. No. 3854 [1964] 1/25, 5). – [37] N. W. Wilson (Sierra Leone Geol. Surv. Bull. No. 4 [1965] 1/102, 92; Mineral. Abstr. **17** [1966] 650/1). – [38] V. A. Cheverdin (Tr. Kaz. Nauchn. Issled. Inst. Mineral'n. Syr'ya No. 2 [1960] 89/91). – [39] D. A. Mineev (Geokhimiya Apogranitov i Redkometal'nykh Metasomatitov Severo-Zapadnogo Tarbagataya, Moscow 1968, pp. 1/183, 97). – [40] P. Geijer (Sveriges Geol. Unders. Arsbok C **14** No. 6 [1920/21] 1/24, 20/1).

[41] L. B. Zubkov, N. I. Galadzheva, A. B. Chernyakovskii (Dokl. Akad. Nauk Belorussk. SSR **16** No. 7 [1972] 646/9). – [42] S. Graeser, H. A. Stalder (Schweizer Strahler **3** No. 7 [1974] 265/77, 265). – [43] C. J. N. Fletcher, J. D. Appleton, B. C. Webb, I. R. Basham (Trans. Inst. Mining Met. B **90** [1981] 37/50, 40/1). – [44] L. Marble, A. Regis (Rocks Minerals **54** [1979] 4/25, 11). – [45] Yu. L. Kapustin (in: K. A. Vlasov, Mineralogiya i Geneticheskie Osobennosti Shchelochnykh Massivov, Moscow 1964, pp. 135/94, 171).

[46] A. P. Khomyakov (in: K. A. Vlasov, Mineralogiya i Geneticheskie Osobennosti Shchelochnykh Massivov, Moscow 1964, pp. 56/82, 59, 67). – [47] M. S. Garson, W. C. Smith (Nyasaland Geol. Surv. Dept. Mem. No. 1 [1958] 1/127, 37). – [48] D. N. Holt (Malawi Geol. Surv. Dept. Bull. **20** [1965] 1/130, 14; Mineral. Abstr. **17** [1965] 369). – [49] W. L. Smith, E. A. Cisney (Am. Mineralogist **41** [1956] 76/81, 78). – [50] S. A. Northrop (Minerals of New Mexico, Albuquerque 1959, pp. 1/665, 135/6).

[51] J. Murdoch, R. W. Webb (Bull. Calif. Div. Mines Geol. No. 173 [1956] 1/452, 71). – [52] E. I. Semenov, A. P. Khomyakov, A. V. Bykova (Tr. Mineral. Muzeya Akad. Nauk SSSR No. 11 [1961] 202/4). – [53] D. A. Mineev, T. I. Lavrisheva, A. V. Bykova (Zap. Vses. Mineral. Obshchestva **99** [1970] 328/32). – [54] J. van Landuyt, S. Amelinckx (Am. Mineralogist **60** [1975] 351/8, 355). – [55] J. G. D. Steyn (Rep. S. Africa Dept. Mines Geol. Surv. Bull. No. **35** [1961] 1/31, 21).

[56] R. S. DeMark (Mineral. Rec. **11** [1980] 69/72). – [57] P. E. Caro (J. Solid State Chem. **6** [1973] 396/401, 398). – [58] Y. Iitaka, H. A. Stalder (Schweiz. Mineral. Petrogr. Mitt. **41** [1961] 485/8). – [59] Anonymous (Mineralienfreund **14** [1976] 115). – [60] J. J. Glass, H. T. Evans, M. K. Carron, F. A. Hildebrand (Am. Mineralogist **43** [1958] 460/75, 461/2, 468).

[61] O. D. Allen, W. J. Comstock (Am. J. Sci. [3] **19** [1880] 390/3). – [62] I. Oftedal (Norsk Geol. Tidsskr. **12** [1931] 459/67, 464/7).

10.5.2.1.5 Crystal Structure

The unit cell constants and the axial ratios (if given by the authors) of minerals of the bastnaesite group from several localities and with considerable deviations in their chemical composition are given in the following table: Nos. 1 to 10 are common bastnaesites from: granites (1 and 2), a green vein in an alkali rock (3), pegmatites (4 and 5), hydrothermal veins (6 and 7), carbonatites (8 to 10); Nos. 11 to 13 are hydroxylbastnaesites from carbonatites, formed by dissolution of ancylite (11) or pseudomorphous after burbankite (12 and 13); Nos. 14 to 20 are common bastnaesites pseudomorphous after allanite (14 to 18), ancylite (19), and monazite (20); No. 21 is bastnaesite-(Y) pseudomorphous after gagarinite; No. 22 is bastnaesite-(Ce-Y) after gadolinite; No. 23 is a sedimentary formed bastnaesite-(La) with La > Nd > Ce; No. 24 is "kyshtymite" with La ≈ Ce and an increased content of water; No. 25 is thorbastnaesite, probably formed from ferrian thorite; No. 26 is a synthetic bastnaesite with Ce only.

No.	Locality	a in Å	c in Å	c/a	Ref.
1	Cheyenne Canyon, Colorado	7.094 ± 0.007	9.718 ± 0.008	1.370	[1, pp. 463, 468]
2	Stove Mountain, near Pikes Peak, Colorado	7.12	9.80	—	[2]
3	Ilímaussaq, southern Greenland	7.120 ± 0.007	9.77 ± 0.02	1.372	[3]
4	Saudi Arabia	7.10	9.76	—	[4]
5	Madagascar	7.153 ± 0.003	9.814 ± 0.004	—	[5]
6	Grimsel area, Switzerland	7.08	9.74	—	[6]
7	Inner Mongolia	7.05 ± 0.01	9.79 ± 0.01	1.388	[7]
8	Mountain Pass, California	7.129	9.774	—	[8]
9	Mountain Pass, California	7.23	9.98	—	[9]
10	Eastern Sayan, Siberia	7.13 ± 0.02	9.67 ± 0.02	—	[10]
11	Vuoriyarvi, Kola, USSR	7.23 ± 0.02	9.98 ± 0.05	1.380	[11, pp. 52/3]
12	Eastern Siberia	7.11	9.88	—	[12]
13	Eastern Yakutia	7.21 ± 0.01	9.92 ± 0.01	1.376	[7]
14	Amelia Co., Virginia	≈7.07	≈9.72	—	[13]
15	Rough Rock Lake, western Ontario	7.08	9.74	—	[14]
16	Rømteland, Norway	7.068	9.779	1.383	[15]
17	Lapplaegret, Norway	7.033	9.497	1.350	[15]
18	Fukushima, Japan	7.10	9.75	—	[16]
19	Langesundfjord, Norway	7.084	9.800	1.383	[15]
20	Mostad, Norway	7.068	9.779	1.383	[15]
21	Tarbagatai, East Kazakhstan	6.87*)	9.48	1.380	[17]
22	Mongolia	6.95	9.56	1.376	[18, 19]
23	Ibadan, Nigeria	7.056	9.730	—	[20]
24	South Urals, USSR	7.095	7.95	1.121	[21]
25	East Siberia	6.99 ± 0.02	9.71 ± 0.02	—	[22]
26	—	7.24	9.92	—	[2]

*) The incorrect value for a = 6.57 Å in the text instead of a = 6.87 Å in the table of [17] has been cited for bastnaesite-(Y) by [23] and [24]; it is also the cause of the too high axial ratio c/a = 1.441 for this species [23].

The mean values for single crystals from several localities are a = 7.162 ± 0.022 Å, c = 9.787 ± 0.029 Å, c/a = 1.367 [25, pp. 935/6]. The parameters of the unit cell of hydroxylbastnaesite are increased relative to those of fluorbastnaesite [7], this is due to the ionic radius of OH^- = 1.40 Å > F^- = 1.36 Å [15]; the parameters of synthetic bastnaesite may be increased by a high content of La [19].

The cell volume V (in $Å^3$) calculated from single crystal data is V = 217.4 × 2 = 434.8 $Å^3$ [25, p. 936], cf. for a crystal from Greenland V = 428.96 $Å^3$ [3]. Corresponding to the unit cell parameters the cell size of hydroxylbastnaesite is somewhat enlarged [15]; it increases with increasing hydroxyl ions to the theoretical value of 458 $Å^3$ for bastnaesite with 100% OH^- ions [11, pp. 56, 58].

The unit cell parameters of bastnaesite-(Y) are considerably lower than those of common bastnaesite and bastnaesite-(Ce-Y) [23, 19], cf. [17]. Bastnaesite from the Tysfjord, Norway, which is an alteration product of allanite, has lattice constants a = 7.02 Å and c = 9.72 Å, intermediate between bastnaesite and bastnaesite-(Y), indicating an intermediate composition of the RE [19].

For thorbastnaesite the somewhat reduced unit cell constants are due to the smaller ionic radius of Th^{4+} (1.02 Å), compared with that of, e.g., Ce^{3+} (1.07 Å) [22].

The axial ratio seems to be relatively constant ≈1.38 for all bastnaesite varieties [19].

The space group for minerals of the bastnaesite group is $P\bar{6}2c$-D_{3h}^4 (No. 190)*) [24], cf. for common bastnaesites [25, p. 951], [1, p. 468], and for hydroxylbastnaesite [7]; it also includes bastnaesites-(Y) [19, 24], synthetic bastnaesite [2], and thorbastnaesite [22].

The number of formula units per unit cell for all bastnaesite varieties Z = 6 [1, p. 463], [25, p. 936], [19, 24]; for thorbastnaesite, however, Z = 3 [22].

Thorbastnaesite [22] and hydroxylbastnaesite [28] are isostructural with common fluorbastnaesite; from two synthesized polymorphic RE hydroxycarbonates $RE[CO_3](OH)$ only one, that of the "B type", with vertically oriented CO_3 groups, is isostructural with bastnaesite [5].

The unit cell consists of regularly constructed RE-F layers arranged vertically to the c axis and alternating with CO_3 layers, in which equilateral planar CO_3 groups are arranged vertically to these layers (parallel to the c axis), see **Fig. 7**, p. 124, and [1, pp. 462, 467, 469]. RE and F ions alternate at the vertices of regular hexagons; the orientation of the carbonate groups parallel to the cell edges causes strong vertical bonds between RE and O(1), which account for the observed lack of cleavage (0001) [25, p. 949]. The bastnaesite structure is a typical insular structure with Ce and F atoms and CO_3 groups oriented along the principal axis (= c axis); the planes of the CO_3 groups are, however, turned around this axis and are nearly mutually perpendicular [24]. The vertically oriented CO_3 groups hold together the hexagonal layers of composition $[(RE)F]_n^{2n+}$, whose arrangement alone can be described by a hexagonal subcell with a = 4.13 Å, c = 4.98 Å; it is, however, the orientation of the carbonate groups within the sheets which determines the final symmetry [5, p. 102]. The entrance of hydroxyl groups into the bastnaesite structure may take place layer by layer, and layers with hydroxyl groups may alternate with layers filled with fluorine atoms; the hydroxyl groups, however, are assumed to cause a change in the local symmetry of the carbonate ions, as their polarizing influence generates distorted CO_3 groups [28]. For the atomic positions and interatomic distances of RE, F, C, O(1), and O(2) in bastnaesite see [25, pp. 947/8].

Bastnaesite as an end member of the series bastnaesite-synchysite shows close structural relations to the other members of this series, see pp. 90/1.

*) The space group $C\bar{6}c2$-D_{3h}^2 (No. 188) has been erroneously given to bastnaesite by [26], as has been indicated by [25, p. 951].

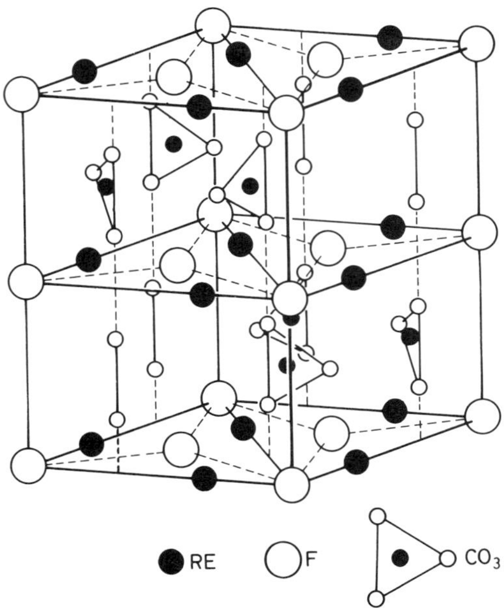

Fig. 7. Unit cell of bastnaesite from [1, p. 466].

The absence of metamictization in most bastnaesites has been explained by the difference in electronegativities between the participating cations, that is 230 kcal/g-atom in this mineral [27]. An incipient metamictization is, however, assumed to be the cause for diffuse X-ray patterns of some Norwegian bastnaesites [15].

References for 10.5.2.1.5:

[1] I. Oftedal (Z. Krist. **78** [1931] 462/9). – [2] G. J. Jansen, G. B. Magin Jr., B. Levin (Am. Mineralogist **44** [1959] 180/1). – [3] J. Hansen (Medd. Groenland **181** No. 8 [1968] 1/43, 31). – [4] A. R. A. Turkistany, K. Koyama (IAG [Inst. Appl. Geol.] Bull. **3** [1980] 99/104, 103). – [5] J. Sawyer, P. Caro, L. Eyring (Rev. Chim. Minerale **10** [1973] 93/104, 95, 102/3).

[6] Y. Iitaka, H. A. Stalder (Schweiz. Mineral. Petrogr. Mitt. **41** [1961] 485/8). – [7] I. V. Aleksandrov, V. I. Ivanov, L. A. Sin'kova (Zap. Vses. Mineral. Obshchestva **94** [1965] 323/6). – [8] J. J. Glass, H. T. Evans, M. K. Carron, F. A. Hildebrand (Am. Mineralogist **43** [1958] 460/75, 469). – [9] Yu. L. Kapustin (Mineralogiya Karbonatitov, Moscow 1971, pp. 1/288, 185). – [10] M. Ya. Somina, A. G. Bulakh (Mineral. Geokhim. **1972** No. 4, pp. 64/73, 66).

[11] A. S. Kirillov (Zap. Vses. Mineral. Obshchestva **95** [1966] 51/9). – [12] T. B. Zdorik (Tr. Mineral. Muzeya Akad. Nauk SSSR No. 17 [1966] 60/75, 72). – [13] W. D. Riesmeyer (Virginia J. Sci. **18** [1967] 188). – [14] P. Černý, I. Černá (Can. Mineralogist **11** [1971/73] 541/3). – [15] T. L. Sverdrup, K. Ø., Bryn, P. C. Sæbø (Norsk Geol. Tidsskr. **39** [1959] 237/47, 244/6).

[16] K. Sakurai, H. Wakita, A. Kato, K. Nagashima (Bull. Chem. Soc. Japan **42** [1969] 2725/8). – [17] D. A. Mineev (Geokhimiya Apogranitov i Redkometal'nykh Metasomatitov Severo-Zapadnogo Tarbagataya, Moscow 1968, pp. 1/183, 95). – [18] E. I. Semenov (Mineralogiya Redkikh Zemel', Moscow 1963, pp. 1/412, 100/2). – [19] E. I. Semenov (in: K. A. Vlasov, Geochemistry and Mineralogy of Rare Elements and Genetic Types of Their Deposits, Vol. 2: Mineralogy of Rare Elements, Israel Program Sci. Trans., Jerusalem 1966, pp. 249/328, 266/70 [Russian Original, Moscow 1964]). – [20] J. C. Hughes, G. Brown (Clay Minerals **12** [1977] 319/29, 324, 327).

[21] N. V. Svyazhin (Tr. Inst. Geol. Akad. Nauk SSSR Ural'sk. Filial No. 70 [1965] 249/52). – [22] A. S. Pavlenko, L. P. Orlova, M. V. Akhmanova, K. I. Tobelko (Zap. Vses. Mineral. Obshchestva **94** [1965] 105/13, 106, 110). – [23] D. A. Mineev, T. I. Lavrishcheva, A. V. Bykova (Zap. Vses. Mineral. Obshchestva **99** [1970] 328/32). – [24] A. S. Povarennykh (Crystal Chemical Classification of Minerals, Vol. 2, translated from Russian by J. E. S. Bradley, Plenum, New York 1972, pp. 459/766, 613/4). – [25] G. Donnay, J. D. H. Donnay (Am. Mineralogist **38** [1953] 932/63).

[26] C. Palache, H. Berman, C. Frondel (The System of Mineralogy, 7th Ed., Vol. 2, Wiley, New York 1951, pp. 1/1124, 289). – [27] A. S. Povarennykh (Zap. Vses. Mineral. Obshchestva **85** [1956] 593/7). – [28] M. V. Akhmanova, L. P. Orlova (Geokhimiya **1966** 571/8; Geochem. Intern. **3** [1966] 444/51, 444, 447, 449).

10.5.2.1.6 Optical Properties

The **luster** of common bastnaesite is vitreous to greasy [1 to 3] up to adamantine, but sometimes only waxy to resinous [4], resinous especially on fracture surfaces [5]; sometimes greasy to dull [6], dull [7], or silky [8]. On parting planes the luster is pearly [1, 4, 9], or nacreous [10], on grains in pseudomorphs weakly greasy [11]. Granular aggregates of hydroxylbastnaesite show waxy to resinous luster [12].

The common bastnaesite is **transparent** to translucent [1, 13], semitransparent [14, 15] to subtranslucent [16], sometimes opaque, except for the finest fragments [17], translucent only in very fine powder [11], sometimes not translucent even in finest splinters due to fine-fibrous aggregated material [18]. Hydroxylbastnaesite is totally transparent [19, 20].

Some bastnaesites show a **zonal structure:** At Mountain Pass, California, crystals with an irregular clear core are surrounded by alternating zones of bastnaesite with or without many dusty inclusions [5]; in Greenland an opaque core, pigmented by inclusions of silicates is surrounded by a transparent colorless to grayish border zone [21]; in Norway crystals with a quite dull, yellowish center have a glassy clear rim zone with pearly gray luster and lower refractive indices than the inner zone [22]. Other crystals with clear outer parts have a dull intermediate zone [23], show zonar turbidity [24], or are strongly banded with minute white, buff, gray, and reddish stripes [25]. In hydroxylbastnaesite from Vuoriyarvi, Kola, iron hydroxide is sometimes zonally distributed with yellowish to brown colors in thin section [20].

Common fluorbastnaesites and hydroxylbastnaesites show predominantly yellowish to brownish **colors** [1 to 3], sometimes they are colorless, white to gray, or reddish to reddish brown, and rarely rose to red or green; in some localities (Mountain Pass, California; Tundulu, Malawi; Kola; Bulgaria) the colors are very different, see table pp. 126/9. Bastnaesite-(Y) seems to be characterized by red colors and altered bastnaesite by green color, see in the following and table pp. 128/9. Dark colored bastnaesites are impure and intimately mixed with other minerals [25], several reddish to reddish brown bastnaesites have inclusions of iron oxides or hydroxides, see in the following.

The dominating colors of common bastnaesites are yellow to brown or reddish brown in various tints, sometimes also white, pink, greenish, and gray tints have been observed, see table pp. 126/9. The observed colors are: yellowish [24], waxy-yellow [26], honey-yellow to brownish yellow [4], dark yellow [27], light tan [28], pearly white to tan [8], and amber [29]; this color has also been observed for grains under the binocular microscope [30, 31]; further colors are yellow-brown to dark brown [32], yellow to red or brown [33], and brown [23], especially in clusters [34]. The common bastnaesite is rarely orange [35], green [33], or black (in aggregates)

[36]; a common bastnaesite from a perthosite is pale yellow [37], that from a dolomite geode in talc is yellowish white [38], and that from bauxite is white [39]. Hydroxylbastnaesite is brownish yellow to dark brown [12]; for yellowish to brown tints, but also colorless hydroxylbastnaesite, see No. 31, while bastnaesites-(Y) show reddish tints, see Nos. 35 and 36 in table p. 129, and cf. pale purple color for this species [40].

Pseudomorphs of bastnaesite after cerite are nearly colorless to pale yellow [41], those after britholite are light colored [42], after synchysite-(Y) (doverite) buff or white, after xenotime buff, and after chevkinite gray due to different composition of the replaced Y-containing minerals [25]; pseudomorphs after tysonite are brown [9], deep brown [29], or reddish brown [16]. Such reddish brown or reddish colors, which are due to staining by Fe (+Mn) [43] in form of iron oxides [16], or especially in form of goethite (α-FeOOH) [7], have frequently been observed for bastnaesite formed after allanite [12, 44, 45], cf. earthy brown bastnaesite in an allanite-monazite-quartz rock [46]; some bastnaesites formed after allanite are, however, white or blue-white (in thin crusts) or waxy-gray (in botryoidal crusts) [47].

Altered bastnaesite is green [7] or has a reddish rim [48].

In transmitted light (thin section) bastnaesites are colorless [49, 50], nearly colorless [51], to very pale yellow [13, 16], yellowish [6], or with a reddish touch [48]; the powder from black aggregates is also colorless [36]. Fine lamellas of bastnaesite from Burundi are yellow-brown to green [7], and the bastnaesites of the Scrub Oaks mine, New Jersey, show under the microscope widely varying colors from white to pale gray, buff, pink, reddish brown, waxy dark gray, and chocolate-brown; the younger bastnaesite is commonly white, the lightest colored bastnaesites appear to have been deposited as fillings and did not replace older RE minerals [25].

The pleochroism is very weak [52, 53] with E>O [5] in light yellow to yellowish [21], with O pale yellow, E yellowish [30, 31], or O colorless, E pale yellow [51], cf. [16].

In the following table the optical properties (color, refractive indices n_o and n_e) of minerals of the bastnaesite group are compiled together with the values for the measured density. For most of these bastnaesites a positive optical character has been determined; for some very fine crystalline samples it remains undetermined. The samples are arranged according to their genesis. Mother rocks of common bastnaesite (if not specified) and its varieties are: Nos. 1 to 5 granites and granite pegmatites; No. 6 an Sn deposit; Nos. 7 to 14 alkaline rocks and alkali metasomatites inclusive albitites and albitized rocks; No. 15 Mg metasomatite; Nos. 16 and 19 quartz-fluorite veins, No. 17 a quartz-barite vein, No. 18 a lead-zinc mineralization in quartz veins, No. 20 an Alpine fissure vein; Nos. 21 to 26 carbonatites (No. 22 hydroxylbastnaesite); Nos. 27 to 30 are pseudomorphs after allanite; Nos. 31 and 32 are hydroxylbastnaesites after ancylite and after several RE carbonates; No. 33 is a supergene, weathered, and hydrated bastnaesite, No. 34 an Y-containing bastnaesite pseudomorphous after chevkinite, No. 35 a bastnaesite-(Ce-Y) pseudomorphous after gadolinite, No. 36 a bastnaesite-(Y) pseudomorphous after gagarinite, and No. 37 thorbastnaesite pseudomorphous after ferrian thorite; it is cryptocrystalline with high contents of Th and molecular water; Nos. 38 and 39 are synthetic fluorbastnaesite and hydroxylbastnaesite.

No.	Locality	Color	Refractive indices n_o	n_e	D(meas.) g/cm^3	Ref.
1	Pikes Peak, Colorado	reddish brown	1.717	1.818	—	[16]
2	St. Peters Dome, Colorado	—	1.719	1.820	—	[54]
3	Jamestown, Colorado	reddish brown	1.716	1.817	—	[55]

No.	Locality	Color	Refractive indices n_o	n_e	D(meas.) g/cm^3	Ref.
4	Madagascar	reddish to yellowish brown	1.717	1.818	—	[16]
			1.719	1.820	—	[54]
5	Khan, Namibia	brown-red	1.718	1.815	5.12 ±0.02	[52]
6	Transvaal, South Africa	brown	1.716	1.816	4.965 ±0.005	[9]
7	Nunasarnaq, South Greenland	colorless to weakly pale yellow	1.722	1.823	—	[21]
			1.718	1.818	—	[56]
8	Langesundfjord, South Norway	gray-yellow to brown	1.691 to 1.700	1.780 to 1.790	—	[22]
9	South Urals, USSR	reddish brown	1.722	1.811	5.0254	[10]
		brownish yellow or light brown	1.718	1.788 to 1.792	5.01 to 5.025	[57]
		dark honey-yellow to reddish brown	1.714	1.798	4.85	[49]
10	Russian Platform	grayish beige, cinnamon to gray	1.720	1.818	4.9	[14]
11	European USSR	bright yellowish	1.720 ±0.002	>1.780	3.89	[18]
12	Kazakhstan	pale yellow to rose	1.716	1.816	4.99	[6]
13	Siberia	canary-yellow	1.714	>1.780	4.93	[50]
14	North Baikalia, USSR	yellowish brown to brown	1.720 ±0.002	1.820 ±0.004	4.86 ±0.02	[30]
			1.720	1.814	4.86 ±0.03	[31]
15	Bastnäs, Sweden	honey-yellow	1.7220	1.8253	—	[16]
			1.7225	1.8242	—	[58]
16	Gallinas, New Mexico	honey-yellow	1.718	1.819	4.99	[16]
17	Karonge, Burundi	honey-yellow to reddish brown	1.722	1.823	≈5	[16]
18	Rhodopes, Bulgaria	wax-yellow, light cream, pale pink	1.730	>1.780	5.05	[51]
19	Beiyin Obo, Inner Mongolia	greenish yellow	1.7169	1.7910	≈4.829	[53]
20	Grimsel, Switzerland	honey-yellow	1.718 ±0.002	1.818 ±0.003	—	[59]

No.	Locality	Color	Refractive indices n_o	n_e	D(meas.) g/cm³	Ref.
21	Mountain Pass, California					
	I Birthday area	yellowish	1.722	1.823	≈5	[5]
	II Vein No. 6 (interstitial)	light gray*)	1.721	1.822	—	[41]
	III from cavities	pale yellow to colorless	1.718	1.818	—	[41]
22	Soustova, Kola, USSR	cinamon with reddish touch	1.720	>1.780	4.1	[48]
23	Kola, USSR	dark yellow	1.728	1.821	4.94	[27]
24	East Sayan, Siberia	honey-yellow	1.712 ± 0.002	1.814 ± 0.002	4.91 ± 0.03	[15]
25	Tuva, Siberia					
	I sideritic	lemon to	1.719	—	4.8	
	II hematitic	honey-yellow	1.717	—	4.96	[60]
26	Tundulu, southern Malawi	(pale) yellow, yellow-brown, brownish, greenish-yellow	1.719	1.820	—	[61]
27	Laplaegret, North Norway	yellow-red	1.691 to 1.700	—	—	[22]
28	Urals, USSR	brownish yellow	1.710 to 1.723	1.820	4.7	[62]
		—	1.711	1.812	—	[54]
		greenish yellow	1.723	1.825	4.746	[63]
		greenish yellow	1.7225	1.8242	4.93	[63]
29	Kazakhstan, USSR	colorless with yellowish touch*)	1.722	1.780	—	[64]
30	Karasugawa, Fukushima Prefecture, Japan	—	1.72 ± 0.01	1.82 ± 0.01	—	[65]
31	Vuoriyarvi, Kola, USSR	waxy-yellow, dark brown, colorless	1.760 ± 0.003	1.870 ± 0.006	4.745	[19, 20]
32	Yakutia, USSR	yellowish	1.753 ± 0.002	—	—	[66]
33	USSR	white, yellow, brown	≈1.66 ± 0.01	≈1.76 ± 0.01	4.27	[67]
34	Ukraine	pale brownish green	1.785	1.827	≈4.47	[68]

No.	Locality	Color	Refractive indices n_o	n_e	D(meas.) g/cm^3	Ref.
35	Mongolia	reddish yellow	—	≈1.70	3.3	[69, 70]
36	Kazakhstan	brick-red to carmine	—	≈1.66 to 1.67	3.9 to 4.0	[70]
37	Siberia	brown	1.670 to 1.678		4.04	[71]
38	—	—	≈1.71	≈1.82	—	[72]
39	—	—	1.757	1.864	—	[66]

*) Color in transmitted light.

A comparison of bastnaesites of different genesis (from high temperature postmagmatic formations, e.g., albitites, and from formations of intermediate temperature such as hydrothermal veins) shows relative constant physical and chemical properties [50]; the optical data of bastnaesites from six localities indicate a nearly constant composition for the mineral [16]. Recent investigations, however, indicate for the series common bastnaesite (with F : O = 92 : 4) to hydroxylbastnaesite (with OH = 100) relatively regular increasing refractive indices n_o from 1.719 to 1.757, and n_e from 1.820 to 1.864; the highest values, however, have been found for the species from Vuoriyarvi, Kola (No. 31), with F : OH = 14 : 87 [20], and in hydroxylbastnaesite from East Yakutia (No. 32). An entering of Th and of RE of the yttrium group into the bastnaesite lattice possibly also elevates the refractive indices and density value [20]. The replacement of Ce by Y in the series bastnaesite – bastnaesite-(Ce-Y) – bastnaesite-(Y) causes a decrease of n_e from 1.825 to 1.66 or 1.67 [70], the hydration of bastnaesite apparently causes a similar effect [67], see No. 33 and for thorbastnaesite No. 37 with 2 to 3 molecules H_2O in the formula.

The mean refractive index $n_m = (2n_o + n_e)/3 = 1.7515$ deviates by 0.012 units from the calculated value $dK + 1 = 1.763$, as obtained by the rule of Gladstone and Dale [73].

References for 10.5.2.1.6:

[1] C. Palache, H. Berman, C. Frondel (The System of Mineralogy, 7th Ed., Vol. 2, Wiley, New York 1951, pp. 1/1124, 289/90). – [2] J. Murdoch, R. W. Webb (Bull. Calif. Div. Mines Geol. No. 173 [1956] 1/452, 71). – [3] E. I. Semenov (in: K. A. Vlasov, Geochemistry and Mineralogy of Rare Elements and Genetic Types of Their Deposits, Vol. 2: Mineralogy of Rare Elements, Israel Program Sci. Trans., Jerusalem 1966, pp. 249/328, 266 [Russian Original, Moscow 1964]). – [4] S. A. Northrop (Minerals of New Mexico, Albuquerque 1959, pp. 1/665, 135). – [5] J. C. Olson, D. R. Shawe, L. C. Pray, W. N. Sharp (U.S. Geol. Surv. Profess. Paper No. 261 [1954] 1/75, 34/5, 43, 45).

[6] D. A. Mineev (Geokhimiya Apogranitov i Redkometal'nykh Metasomatitov Severo-Zapadnogo Tarbagataya, Moscow 1968, pp. 1/183, 97). – [7] J. Thoreau, B. Aderca, L. van Wambeke (Acad. Roy. Sci. Coloniales [Brussels] Classe Sci. Nat. Med. [2] **4** No. 3 [1958] 684/715, 702, 708). – [8] L. Marble, A. Regis (Rocks Minerals **54** [1979] 4/25, 11). – [9] J. G. D. Steyn (Rep. S. Africa Dept. Mines Geol. Surv. Bull. No. 35 [1961] 1/31, 21). – [10] E. B. Khalezova, I. I. Nazarenko (Tr. Inst. Mineral. Geokhim. Kristallokhim. Redk. Elem. Vopr. Geokhim. Genezisa Mestorozhd. Redk. Elem. No. 2 [1959] 99/101).

[11] Kh. R. Rakhmatullaev, P. L. Gertman, E. P. Nikolaeva (Zap. Uzbek. Otd. Vses. Mineral. Obshchestva Akad. Nauk Uz.SSR No. 32 [1979] 35/7; C.A. **92** [1980] No. 200998). – [12] A. P. Sabina (Geol. Surv. Paper Geol. Surv. Can. No. 78-1A [1978] 253/8, 254). – [13] R. M. Perhac,

E. W. Heinrich (Econ. Geol. **59** [1964] 226/39, 232). – [14] L. B. Zubkov, N. I. Galadzheva, A. B. Chernyakovskii (Dokl. Akad. Nauk Belorussk. SSR **16** No. 7 [1972] 646/9). – [15] M. Ya. Semina, A. G. Bulakh (Mineral. Geokhim. **1972** No. 4, pp. 64/73).

[16] J. J. Glass, R. G. Smalley (Am. Mineralogist **30** [1945] 601/15, 605, 608/10). – [17] J. Mandarino, W. M. Tovell (Can. Mineralogist **7** [1963] 819). – [18] L. V. Ganzeeva (Redk. Elem. Syr'e Ekon. No. 7 [1972] 112/24, 118/9). – [19] A. S. Kirillov (Dokl. Akad. Nauk SSSR **159** [1964] 1048/50; Dokl. Earth Sci. Sect. **159** [1964] 93/5). – [20] A. S. Kirillov (Zap. Vses. Mineral. Obshchestva **95** [1966] 51/9, 52, 56/7).

[21] J. Hansen (Medd. Groenland **181** No. 8 [1968] 1/47, 20, 28, 30). – [22] T. L. Sverdrup, K. Ø. Bryn, P. C. Sæbø (Norsk Geol. Tidsskr. **39** [1959] 237/47, 240/1, 245/6). – [23] S. Graeser, H. A. Stalder (Schweizer Strahler **3** No. 7 [1974] 265/77, 265/6). – [24] Anonymous (Mineralienfreund **14** [1976] 115). – [25] H. Klemic, A. V. Heyl, A. R. Taylor, J. Stone (U.S. Geol. Surv. Bull. No. 1082-B [1959] 29/60, 52).

[26] R. S. DeMark (Mineral. Rec. **11** [1980] 69/72). – [27] Yu. L. Kapustin (Mineralogiya Karbonatitov, Moscow 1971, pp. 1/288, 184/5). – [28] M. H. Staatz (Econ. Geol. **73** [1978] 512/23, 517). – [29] E. B. Eckel (U.S. Geol. Surv. Bull. No. 1114 [1961] 1/399, 66). – [30] V. V. Gerasimovskii (Tr. Mineral. Muzeya Akad. Nauk SSSR No. 15 [1964] 194/202, 194).

[31] V. V. Gerasimovskii (Vestn. Mosk. Univ. Ser. IV Geol. **19** No. 6 [1964] 43/9, 47). – [32] R. Köchlin (Tschermaks Mineral. Petrogr. Mitt. [2] **31** [1911/12] 525/36, 525). – [33] B.-M. Aderca, R. van Tassel (Acad. Roy. Sci. Outre Mer Classe Sci. Nat. Med. [2] **18** No. 5 [1971] 1/117, 86). – [34] C. J. N. Fletcher, J. D. Appleton, B. C. Webb, I. R. Basham (Trans. Inst. Mining Met. B **90** [1981] 37/50, 40). – [35] E. B. Gross, E. W. Heinrich (Am. Mineralogist **51** [1966] 299/323, 312).

[36] V. Marmo, V. Hoffrén, K. Hytönen, P. Kallio, O. Lindholm, J. Siivola (Bull. Comm. Geol. Finlande No. 221 [1966] 1/34, 18). – [37] N. W. Wilson (Sierra Leone Geol. Surv. Bull. No. 4 [1965] 1/102, 92). – [38] R. Pierrot (Bull. Soc. Franc. Mineral. Crist. **97** [1974] 521/4). – [39] Z. Maksimović, G. Pantó (Bull. Acad. Serbe Sci. Arts Classe Sci. Nat. Math. Sci. Nat. **72** No. 20 [1980] 35/42, 36). – [40] M. S. Garson, N. Bradshaw, S. Rattawong (Intern. Tin Council 2nd Tech. Conf., Bangkok 1969, pp. 327/39, 338).

[41] J. J. Glass, H. T. Evans, M. K. Carron, F. A. Hildebrand (Am. Mineralogist **43** [1958] 460/75, 462/3). – [42] L. V. Ganzeeva [Ganzeyeva], A. A. Ganzeev [Ganzeyev] (Geokhimiya **1975** 74/80; Geochem. Intern. **12** No. 1 [1975] 47/52, 49). – [43] J. B. Zadra, A. L. Engel, E. S. Shedd, A. C. Rice (U.S. Bur. Mines Rept. Invest. No. 4919 [1952] 1/15, 1, 3, 9). – [44] D. A. Mineev, K. I. Rozanov, N. V. Smirnova, T. I. Matrosova (Dokl. Akad. Nauk SSSR **210** [1973] 925/8; Dokl. Earth Sci. Sect. **210** [1973] 149/52). – [45] A. Vorma, P. Ojanperä, V. Hoffrén, J. Siivola, A. Löfgren (Bull. Comm. Geol. Finlande No. 222 [1966] 241/74, 248).

[46] S. Vrana (Rec. Geol. Surv. Zambia **12** [1972] 91/4). – [47] R. S. Mitchell, G. E. Redline (Rocks Minerals **55** [1980] 245/9). – [48] R. P. Tikhonenkova, E. D. Osokin (in: K. A. Vlasov, Mineralogiya i Geneticheskie Osobennosti Shchelochnykh Massivov, Moscow 1964, pp. 123/34, 130). – [49] V. A. Cheverdin (Tr. Kaz. Nauchn. Issled. Inst. Mineral'n. Syr'ya No. 2 [1960] 89/91). – [50] I. I. Kupriyanova (Tr. Mineral. Muzeya Akad. Nauk SSSR No. 18 [1968] 206/9).

[51] R. Marinova (Rudoobraz. Protsesi Mineralni Nakhodishta **14/15** [1981] 132/6, 153 [Bulgarian, German on p. 136]; C.A. **97** [1982] No. 58728). – [52] J. Goñi, C. Guillemin (Bull. Soc. Franc. Mineral. **76** [1953] 124/9, 126). – [53] T. L. Ho (Bull. Geol. Soc. China [Peking] **14** [1935] 279/82). – [54] C. W. Beck (Am. Mineralogist **35** [1950] 985/1013, 998). – [55] E. N. Goddard, J. J. Glass (Am. Mineralogist **25** [1940] 381/404, 401).

[56] G. Donnay, J. D. H. Donnay (Am. Mineralogist **38** [1953] 932/63, 960). – [57] A. G. Zhabin, G. N. Mukhitdinov, M. E. Kazakova (Tr. Inst. Mineral. Geokhim. Kristallokhim. Redk. Elem. **1960**

No. 4, pp. 51/73, 69). – [58] P. Geijer (Sveriges Geol. Unders. Arsbok C **14** No. 6 [1920/21] 1/24, 10/1). – [59] Y. Iitaka, H. A. Stalder (Schweiz. Mineral. Petrogr. Mitt. **41** [1961] 485/8). – [60] A. P. Khomyakov (in: K. A. Vlasov, Mineralogiya i Geneticheskie Osobennosti Shchelochnykh Massivov, Moscow 1964, pp. 56/82, 67).

[61] M. S. Garson (Nyasaland Geol. Surv. Dept. Mem. No. 2 [1962] 1/248, 95/8, 108/9). – [62] N. V. Svyazhin (Tr. Inst. Geol. Akad. Nauk SSSR Ural'sk. Filial No. 70 [1965] 249/52). – [63] V. Silberminz [V. A. Zil'bermintz] (Compt. Rend. Acad. Sci. USSR A **1929** 55/60, 59/60). – [64] V. I. Chernov, O. F. Krol', A. V. Stepanov (Geol. Geokhim. Mineral. Mestorozhd. Redk. Elem. Kaz. **1966** 138/44, 139). – [65] K. Sakurai, H. Wakita, A. Kato, K. Nagashima (Bull. Chem. Soc. Japan **42** [1969] 2725/8).

[66] I. V. Aleksandrov, V. I. Ivanov, L. A. Sin'kova (Zap. Vses. Mineral. Obshchestva **94** [1965] 323/6). – [67] E. I. Semenov, A. P. Khomyakov, A. V. Bykova (Tr. Mineral. Muzeya Akad. Nauk SSSR No. 11 [1961] 202/4). – [68] E. Ya. Marchenko, O. I. Chashka, E. P. Gurova (Dopov. Akad. Nauk Ukr.RSR B **30** No. 1 [1968] 69/72). – [69] E. I. Semenov (Mineralogiya Redkikh Zemel', Moscow 1963, pp. 1/412, 101/2). – [70] D. A. Mineev, T. I. Lavrishcheva, A. V. Bykova (Zap. Vses. Mineral. Obshchestva **99** [1970] 328/32).

[71] A. S. Pavlenko, L. P. Orlova, M. V. Akhmanova, K. I. Tobelko (Zap. Vses. Mineral. Obshchestva **94** [1965] 105/13). – [72] G. J. Jansen, G. B. Magin Jr., B. Levin (Am. Mineralogist **44** [1959] 180/1). – [73] H. W. Jaffe (Am. Mineralogist **41** [1956] 757/77, 764).

10.5.2.1.7 IR Spectra and Luminescence

IR spectra of bastnaesites are basically characterized by the vibrations of the CO_3^{2-} group. Strong to medium intensities of these bands at about 1450, 870, 730, and 680 cm^{-1} have been found for bastnaesite from Japan [1], cf. distinct bands at 1443 and 875 cm^{-1}, and minor developed bands at 1086 and 728 cm^{-1} for bastnaesite from Bulgaria [2], and the same bands, but 868 instead of 875 cm^{-1} for ν_2 (recalculated from μ values) for bastnaesite from Zaire [3]. For a sedimentary bastnaesite from Nigeria the presence of CO_3 groups has also been confirmed by an IR spectrum [4].

The IR spectra of hydroxylbastnaesites show a much greater number of absorption bands than those of fluorbastnaesites due to the existence of hydroxyl groups; this is proved by a comparison with malachite [5, pp. 444, 447/9]. For natural bastnaesite from East Yakutia and for synthesized hydroxylbastnaesite the existence of hydroxyl groups is confirmed by the absorption bands in the range of 3470 to 3620 cm^{-1} (valence oscillation of O-H), which occur besides a splitting of the bands of the CO_3^{2-} groups at 1400 to 1500 cm^{-1} (valence oscillation) and in the interval between 700 and 800 cm^{-1} (deformation oscillation) [6]. A limited replacement of fluorine ions by hydroxyl ions in bastnaesite from Karasugawa, Japan, is confirmed by absorption bands at 3470 and 3630 cm^{-1} and a less marked splitting of the spectrum due to a lower water content than in hydroxylbastnaesite; a band at 1660 cm^{-1} is due to molecular water, another at 1020 cm^{-1} is ascribed to impurities such as SiO_2 and Al_2O_3 [1]. In a pseudomorph after allanite from Kazakhstan which contains bastnaesite together with illite and magnetite, absorption bands of the anions hydroxyl (2008 cm^{-1}), CO_3 (754, 870, and 1456 cm^{-1}), and SiO_4 (1020 cm^{-1}) have been observed [7].

For the B type of the synthetic compound $Ln(OH)CO_3$, see p. 94, two well defined double bands of the same intensity at 3465 to 3485 and at 3615 to 3635 cm^{-1} (OH stretching vibration) indicate the presence of two different hydroxyl sites with equivalent concentration [8]. The IR spectra of bastnaesite from Bulgaria [2] and of the F-rich bastnaesite from Inner Mongolia [6] give no reference for hydroxyl groups.

The IR spectrum of thorbastnaesite deviates from those of common bastnaesite by the occurrence of a diffuse band between 3000 and 3600 cm^{-1} with a maximum at 3450 cm^{-1}, which is characteristic for molecular water; the valence oscillation of hydroxyl groups seems to be absent, but may be masked by the broad bands of molecular water falling in this range; the splitting of the strong band between 1400 and 1500 cm^{-1} may be caused by the different cations Ce^{3+} and Th^{4+} surrounded by the CO_3 groups in thorbastnaesite; α radiation may change the length of the covalent bonding C-O [9]. The similarity of the IR spectra of natural thorbastnaesite and of heated bastnaesite and parisite may be explained by the α decay of the Th atoms, which causes locally structural changes in thorbastnaesite as it does heating in the other two minerals: the CO_3 groups are irreversibly dissociated and CO_2 evolves [5, pp. 450/1].

The absorption spectrum of bastnaesite in visible light (400 to 900 nm) is characterized by a weak band at 13350 cm^{-1}, an intermediate band at 19000 cm^{-1}, and a very strongly split band at 17250 cm^{-1}; the splitting is caused by different symmetry around Nd^{3+} cations [10].

An intense bottle-green fluorescence in UV light has been observed for bastnaesite from Sella, Lucendro, Switzerland, and grass-green to leek-green colors for two bastnaesites from pegmatites (not localized) [11]. For several other bastnaesites no fluorescence could be observed under long wave (27397 cm^{-1}) and short wave (39417 cm^{-1}) radiation [12]. In the thermo-luminescence diagram of bastnaesite from South Greenland a glow curve up to 400°C shows a peak between 150 and 400°C with a maximum at about 275°C; the peak is suggestive for a radiogenic origin (reorganization of trapped electrons, cf. DTA, p. 135) [13]. For kyshtymite from the Urals no luminescence is observed; this is explainable by the possible entrance of Fe into the lattice, which apparently causes the extinction of a potential luminescence [14].

References for 10.5.2.1.7:

[1] K. Sakurai, H. Wakita, A. Kato, K. Nagashima (Bull. Chem. Soc. Japan **42** [1969] 2725/8). – [2] R. Marinova (Rudoobraz. Protsesi Mineralni Nakhodishta **14/15** [1981] 132/6 [Bulgarian, German Abstr. on p. 136]; C.A. **97** [1982] No. 58728). – [3] H. H. Adler, P. F. Kerr (Am. Mineralogist **48** [1963] 839/53, 843, 848/9, 853). – [4] J. C. Hughes, G. Brown (Clay Minerals **12** [1977] 319/29, 326/7). – [5] M. V. Akhmanova, L. P. Orlova (Geokhimiya **1966** 571/8; Geochem. Intern. **3** [1966] 444/51).

[6] I. V. Aleksandrov, V. I. Ivanov, L. A. Sin'kova (Zap. Vses. Mineral. Obshchestva **94** [1965] 323/6). – [7] V. I. Chernov, O. F. Krol', A. V. Stepanov (Geol. Geokhim. Mineral. Mestorozhd. Redk. Elem. Kaz. **1966** 138/44, 141; C.A. **73** [1970] No. 79336). – [8] H. Dexpert, E. Antic-Fidancev, J. P. Coutures, P. Caro (J. Cryst. Spectrosc. Res. **12** [1982] 129/42, 130). – [9] A. S. Pavlenko, L. P. Orlova, M. V. Akhmanova, K. I. Tobelko (Zap. Vses. Mineral. Obshchestva **94** [1965] 105/13, 107/9, 111). – [10] A. N. Platonov, O. A. Krasil'shchikova, A. N. Tarashchan (Konst. Svoistva Mineral. **9** [1975] 58/64, 59/61).

[11] A. Kipfer (Schweizer Strahler **3** No. 9 [1975] 370/84, 377). – [12] W. R. Foster (Am. Mineralogist **34** [1949] 830/4). – [13] J. Hansen (Medd. Groenland **181** No. 8 [1968] 1/43, 37). – [14] N. V. Svyazhin (Tr. Inst. Geol. Akad. Nauk SSSR Ural'sk. Filial No. 70 [1965] 249/52).

10.5.2.1.8 Other Physical Properties

Hardness. The hardness of common fluorbastnaesites $H_M = 4$ to 4.5 [1 to 4]; $H_M = 4.5$, e.g., for bastnaesite from the Mountain Pass, California [5], for supergene bastnaesites [6, 7], and for bastnaesite from Inner Mongolia [8], 4.45 for bastnaesite from Kazakhstan [9], and 5 for bastnaesite ("buszite") from Southwest-Africa [10]. The microhardness of bastnaesites amounts to 520 to 590 kg/mm^2, corresponding to Mohs' hardness of 5.0 to 5.2 [11]; more exact values obtained on bastnaesite from the North Baikal area are 519 kg/mm^2 parallel to and 619 kg/mm^2 vertical to the c axis, corresponding to Mohs' hardness of 5.4 and 5.7 [12, 13]. A general range $H_M = 5.5$ to 6.0 is given by [14].

For hydroxylbastnaesite $H_M \approx 4$ [15, 16], for kyshtymite = 4 to 5 [17], for bastnaesite-(Y) ≈ 6 [14].

Density. The density values clearly depend on the chemistry and on the alteration of bastnaesites. The measured densities D(meas.) of common fluor- and of hydroxylbastnaesites show with values between <4 and about 5.20 g/cm^3, see table pp. 127/9, a broader range than those given with 4.9 to 5.2 [1, 2], 4.9 to 5.1 [14], or 4.723 to 5.025 g/cm^3 [4, 11]; this is probably caused by the inclusion of altered material [1]. The value D = 4.99 g/cm^3 (mean of three fragments), determined for bastnaesite from Gallinas, New Mexico [18], and for bastnaesite from the Pikes Peak, Colorado, has been considered the best representative of the mineral [1]. Similar values are D(meas.) = 4.95 and 5.018 for bastnaesite of the Red Cloud claim, Gallinas [19], 5.00 for bastnaesite from Madagascar [20], and 4.82 ± 0.02 to 5.10 ± 0.1 g/cm^3 for bastnaesites from Karonge, Burundi [21]. Lower density values are characteristic for polycrystals with D(meas.) = 4.78 ± 0.1 g/cm^3 [22], for supergene bastnaesite from a weathering zone (see table p. 128, No. 33), and for the strongly hydrated bastnaesite-(Nd) with 10.9 wt% H_2O and D(meas.) = 3.08 g/cm^3 [23]. Bastnaesite-(Y) has D(meas.) = 4.0 g/cm^3 [14]; for bastnaesites with higher contents of Y and hydroxyl groups, see Nos. 34 to 36 in table pp. 128/9. In the series bastnaesite – hydroxylbastnaesite the measured density decreases with decreasing ratio F : OH [16]. The large difference between D(meas.) = 4.04 g/cm^3 and D(calc.) = 5.70 g/cm^3 of thorbastnaesite can be explained by the molecular and hygroscopic water in the mineral [24]. D(calc.) of $Ce[CO_3]F$ = 5.12 g/cm^3 [1], 5.02 g/cm^3 according to new unit cell constants from single crystals [22].

Fracture. The fracture of bastnaesite is uneven [1, 4, 9, 18], or irregular [5, 12, 25]. Bastnaesites are brittle [1, 4, 12, 26]; crystals formed by alteration from allanite in Uzbekistan, USSR, are soft and friable [27], those from Tundulu, Malawi, extremely friable [28].

Magnetism. According to the strong paramagnetism of the RE ions and to the presence of Fe and Mn, the magnetic susceptibility χ is very characteristic for RE minerals, e.g., for bastnaesite; it increases relatively regularly from members with ceric to those with yttric composition [29]. Bastnaesite from the granite of Westerly, Rhode Island, is magnetic at 0.7 A in the prismatic variety, and at 0.5 A in the aggregated variety [30]. Bastnaesite can be separated from other accessory minerals at a current of ~0.7 to 1.3 A in the Cook Magnetic Separator at 20° slope and 10° tilt [31].

Piezoelectricity. Bastnaesite of the Cheyenne Canyon, Colorado, has been shown to have a strong piezoelectricity [32].

Radioactivity. Bastnaesites are frequently radioactive [2], caused by contents of <1 wt% Th and U [33]. Radioactivity due to Th has been observed for bastnaesites from the Mountain Pass, California [5], from Burundi (weakly) [34], and from the North Baikal area, Siberia, where this mineral shows intense β radiation [12, 13].

References for 10.5.2.1.8:

[1] C. Palache, H. Berman, C. Frondel (The System of Mineralogy, 7th Ed., Vol. 2, Wiley, New York 1951, pp. 1/1124, 289, 291). – [2] S. A. Northrop (Minerals of New Mexico, Albuquerque 1959, pp. 1/665, 135/6). – [3] J. Murdoch, R. W. Webb (Bull. Calif. Div. Mines Geol. No. 173 [1956] 1/452, 71). – [4] E. I. Semenov (in: K. A. Vlasov, Geochemistry and Mineralogy of Rare Elements and Genetic Types of Their Deposits, Vol. 2: Mineralogy of Rare Elements, Israel Program Sci. Trans., Jerusalem 1966, pp. 249/328, 266 [Russian Original, Moscow 1964]). – [5] J. C. Olson, D. R. Shawe, L. C. Pray, W. N. Sharp (U.S. Geol. Surv. Profess. Paper No. 261 [1954] 1/75, 34, 39).

[6] E. I. Semenov, A. P. Khomyakov, A. V. Bykova (Tr. Mineral. Muzeya Akad. Nauk SSSR No. 11 [1961] 202/4). – [7] V. Silberminz [V. A. Zil'bermintz] (Compt. Rend. Acad. Sci. USSR A **1929** 55/60, 60). – [8] T. L. Ho (Bull. Geol. Soc. China [Peking] **14** [1935] 279/82). – [9] D. A. Mineev (Geokhimiya Apogranitov i Redkometal'nykh Metasomatitov Severo-Zapadnogo Tarbagataya, Moscow 1968, pp. 1/183, 97). – [10] J. Goñi, C. Guillemin (Bull. Soc. Franc. Mineral. **76** [1953] 124/9).

[11] E. I. Semenov (Mineralogiya Redkikh Zemel', Moscow 1963, pp. 1/412, 97). – [12] V. V. Gerasimovskii (Tr. Mineral. Muzeya Akad. Nauk SSSR No. 15 [1964] 194/202, 194/5). – [13] V. V. Gerasimovskii (Vestn. Mosk. Univ. Ser. IV Geol. **19** No. 6 [1964] 43/9, 47). – [14] A. S. Povarennykh (Crystal Chemical Classification of Minerals, Vol. 2, translated from Russian by J. E. S. Bradley, Plenum, New York 1972, pp. 459/766, 613). – [15] A. S. Kirillov (Dokl. Akad. Nauk SSSR **159** [1964] 1048/50; Dokl. Earth Sci. Sect. **159** [1964] 93/5).

[16] A. S. Kirillov (Zap. Vses. Mineral. Obshchestva **95** [1966] 51/9, 56/7). – [17] N. V. Svyazhin (Tr. Inst. Geol. Akad. Nauk SSSR Ural'sk. Filial No. 70 [1965] 249/52). – [18] J. J. Glass, R. G. Smalley (Am. Mineralogist **30** [1945] 601/15, 605/6). – [19] J. H. Soulé (U.S. Bur. Mines Rept. Invest. No. 3854 [1946] 1/25, 5). – [20] J. Behier (Rep. Malgache Ann. Geol. Madagascar No. 29 [1960] 1/78, 18).

[21] B.-M. Aderca, R. van Tassel (Acad. Roy. Sci. Outre Mer Classe Sci. Nat. Med. [2] **18** No. 5 [1971] 1/117, 83/5). – [22] G. Donnay, J. D. H. Donnay (Am. Mineralogist **38** [1953] 932/63, 936). – [23] E. K. Podporina, A. P. Niyazov, M. D. Brylin (in: V. V. Burkov, Ekzogennye Mestorozhdeniya Redkikh Elementov, Moscow 1980, pp. 21/6, 25; Ref. Zh. Geol. **1981** No. 6 Zh 39). – [24] A. S. Pavlenko, L. P. Orlova, M. V. Akhmanova, K. I. Tobelko (Zap. Vses. Mineral. Obshchestva **94** [1965] 105/13, 106). – [25] V. A. Cheverdin (Tr. Kaz. Nauchn. Issled. Inst. Mineral'n. Syr'ya No. 2 [1960] 89/91).

[26] R. Marinova (Rudoobraz. Protsesi Mineralni Nakhodishta **14/15** [1981] 132/6 [Bulgarian, German Abstr. on p. 136]; C.A. **97** [1982] No. 58728). – [27] Kh. Rakhmatullaev, P. L. Gertman, E. P. Nikolaeva (Zap. Uzbek. Otd. Vses. Mineral. Obshchestva Akad. Nauk Uz.SSR No. 32 [1979] 35/7; C.A. **92** [1980] No. 200998). – [28] M. S. Garson (Nyasaland Geol. Surv. Dept. Mem. No. 2 [1962] 1/248, 108). – [29] E. I. Semenov, A. P. Khomyakov (Diagnost. Diagnost. Svoistva Mineral. **1981** 88/93, 88/9; Ref. Zh. Geol. **1982** No. 3 V 187). – [30] W. L. Smith, E. A. Cisney (Am. Mineralogist **41** [1956] 76/81, 78).

[31] N. W. Wilson (Sierra Leone Geol. Surv. Bull. No. 4 [1965] 1/102, 71/3). – [32] I. Oftedal (Z. Krist. **78** [1931] 462/9, 465). – [33] E. W. Heinrich (Mineralogy and Geology of Radioactive Raw Materials, McGraw-Hill, New York 1958, pp. 1/654, 127). – [34] J. Thoreau, B. Aderca, L. van Wambeke (Acad. Roy. Sci. Coloniales [Brussels] Classe Sci. Nat. Med. [2] **4** No. 3 [1958] 684/715, 713).

10.5.2.1.9 Chemical Behavior and Thermal Analyses

Bastnaesite is easily soluble in all mineral acids [1], in strong hot acids [2], cf. [3]; especially in concentrated hydrochloric and sulfuric acid [4]. It is easily soluble in diluted hydrochloric acid (1:20), but less easily in concentrated acid (1:1) [5]. An effervescence due to development of CO_2 has been observed in boiling concentrated hydrochloric acid and in nitric acid [6], but also in strong sulfuric acid [7] together with evolution of hydrofluoric acid [8, 9].

Hydroxylbastnaesite dissolves in cold hydrochloric and sulfuric acid [10, 11], the intermediate member kyshtymite dissolves best of various mineral acids in diluted hydrochloric acid at room temperature; development of chlorine may indicate the presence of Ce^{4+} [12].

DTA- and TG curves. The DTA curves of all bastnaesite varieties are characterized by a marked endothermal effect in the range of 410 to 625°C due to dissociation with development of CO_2 (and F) and formation of oxyfluorides and sesquioxides of RE (see p. 136); a second more diffuse and less marked endothermal effect at lower temperatures (up to 350°C) apparently is due to loss of absorbed water; an exothermal effect at about 650°C, due to oxidation of Ce^{3+} to Ce^{4+} occurs in some samples.

The main thermal effects of bastnaesite and its varieties from various localities are compiled in the following table. Nos. 1 to 11 are common fluorbastnaesites (No. 3 originally named tysonite, has been identified to be bastnaesite by DTA curve; No. 5 probably contains a parisite component; No. 8 is a supergene bastnaesite); Nos. 12 and 13 are hydroxylbastnaesites; No. 14 is bastnaesite-(Y) pseudomorphous after gagarinite and No. 15 thorbastnaesite after ferri-thorite. An asterisk means values taken from curves.

Thermal Effects of Minerals of the Bastnaesite Group.

No.	Locality	endothermal effects		exothermal effect	Ref.
		loss of water	dissociation	oxidation of $Ce^{3+} \rightarrow Ce^{4+}$	
1	Ampangabe, Madagascar	350°C	625°C	650°C	[15]
2	St. Peters Dome, El Paso Co., Colorado	—	620	655	[15]
3	Unspecified, Colorado	—	625	650	[15]
4	Nunasarnaq, South Greenland	<200	—	—[1)]	[17]
5	Russian Platform	—	570	—[2)]	[5]
6	Siberia, USSR	—	590[3)]	631	[22]
7	East Sayan, Siberia	—	*520[4)]	—	[18]
8	Siberia, USSR	—	*470[5)]	—	[19]
9	Mongolia	—	*580[6)]	—	[4]
10	Karonge, Burundi	—	480	—	[21]
11	Khan, Southwest Africa	≈100	460	—	[16]
12	Vuoriyarvi, Kola, USSR	—	580[7)]	—	[10, 11]
13	South Urals, USSR	—	*580[8)]	650	[15]
14	Kazakhstan, USSR	≈150	480	—	[20], cf. [23]
15	East Siberia, USSR	≤350[9)]	410	—	[24]

[1)] An exothermal effect in the range of 250 to 400°C is ascribed to a reorganization of trapped electrons. – [2)] A peak at ~900°C is ascribed to recrystallization with formation of new compounds. – [3)] Effects not explicitly explained. – [4)] Interval ~400 to ~600°C. – [5)] Interval 350 to 500°C. – [6)] Interval 420 to 600°C. – [7)] Interval 420 to 580°C, a second weak endothermal effect occurs between 770 and 800°C. – [8)] Interval 420 to 580°C. – [9)] Diffuse range due to loss of water.

The DTA curves of fluorbastnaesite of Inner Mongolia and of hydroxylbastnaesite of eastern Yakutia are in general identical; both show the characteristic endothermal effect between 400 and 530°C with a maximum between 490 and 495°C [13]. In contrast, among heating curves of different varieties of bastnaesites from carbonatites those of the varieties with reduced F^- and elevated OH^- content have an endothermal peak which is displaced to a lower temperature range [14].

The TG curves of bastnaesite from the East Sayan show the main loss of weight, corresponding very well with their content of CO_2, in the temperature range of the large endothermal effect of the DTA curves between about 400 and 600°C [18], cf. the main loss of weight in the same interval in supergene bastnaesite from Siberia [19], in hydroxylbastnaesite (kyshtymite) from the South Urals [12], and in bastnaesite-(Y) from Kazakhstan [20]. The total weight loss of bastnaesite from East Sayan is 17.2% upon heating at 800°C and 17.7% at 1100°C [18], that of the supergene bastnaesite is 23.26 wt% $CO_2 + H_2O$ upon heating up to 1000°C [19]; bastnaesite-(Y) shows a loss of 4.3 wt% of absorbed water between 160 and 200°C, and a loss of <1 wt% water (which probably is present as OH groups) at a higher temperature [20]. The loss of weight of hydroxylbastnaesite of Kola is given as 22.8 wt% [11].

Heating products. By means of thermogravimetric, X-ray, and chemical analysis it was shown that fluorbastnaesite at 480°C changes into an oxyfluoride (CeOF) with a = 5.54 Å; upon further heating fluorine escapes and at 600°C there appear sesquioxides of RE, that still increase upon heating up to 1000°C. Even at this temperature there still exist oxyfluorides; the fluorine content of 9.39% remains constant from room temperature up to about 500°C, it decreases to 6.76% at 800 and to 1.26% at 1000°C [21]. A mixture of fluorite and oxides, being formed as calcination product of bastnaesite from the East Sayan, is, however, inert up to 1000°C [18]. Calcination of fluorbastnaesite results in the formation of a cubic phase which is a solid solution of lanthanide sesquioxides in CeO_2 ($CeLnO_{3.5}$) with a = 5.55 Å, while the cubic phase of RE oxides forming upon calcination of bastnaesite-(Y) has a = 5.46 Å [4].

References for 10.5.2.1.9:

[1] M. S. Garson (Nyasaland Geol. Surv. Dept. Mem. No. 2 [1962] 1/248, 108/9). – [2] C. Palache, H. Berman, C. Frondel (The System of Mineralogy, 7th Ed., Vol. 2, Wiley, New York 1951, pp. 1/1124, 290). – [3] V. A. Cheverdin (Tr. Kaz. Nauchn. Issled. Inst. Mineral'n. Syr'ya No. 2 [1960] 89/91). – [4] E. I. Semenov (in: K. A. Vlasov, Geochemistry and Mineralogy of Rare Elements and Genetic Types of Their Deposits, Vol. 2: Mineralogy of Rare Elements, Israel Program Sci. Trans., Jerusalem 1966, pp. 249/328, 266/9 [Russian Original, Moscow 1964]). – [5] L. B. Zubkov, N. I. Galadzheva, A. B. Chernyakovskii (Dokl. Akad. Nauk Belorussk.SSR **16** No. 7 [1972] 646/9).

[6] Kh. R. Rakhmatullaev, P. L. Gertman, E. P. Nikolaeva (Zap. Uzbek. Otd. Vses. Mineral. Obshchestva Akad. Nauk Uz.SSR No. 32 [1979] 35/7; C.A. **92** [1980] No. 200998). – [7] E. B. Khalezova, I. I. Nazarenko (Tr. Inst. Mineral. Geokhim. Kristallokhim. Redk. Elem. Akad. Nauk SSSR No. 2 [1959] 99/101; C.A. **1960** 19318). – [8] J. J. Glass, R. G. Smalley (Am. Mineralogist **30** [1945] 601/15, 605, 609). – [9] J. C. Olson, D. R. Shawe, L. C. Pray, W. N. Sharp (U.S. Geol. Surv. Profess. Paper No. 261 [1954] 1/75, 35). – [10] A. S. Kirillov (Dokl. Akad. Nauk SSSR **159** [1964] 1048/50; Dokl. Earth Sci. Sect. **159** [1964] 93/5).

[11] A. S. Kirillov (Zap. Vses. Mineral. Obshchestva **95** [1966] 51/9, 52/3). – [12] N. V. Svyazhin (Tr. Inst. Geol. Akad. Nauk SSSR Ural'sk. Filial No. 70 [1965] 249/52). – [13] I. V. Aleksandrov, V. I. Ivanov, L. A. Sin'kova (Zap. Vses. Mineral. Obshchestva **94** [1965] 323/6). – [14] Yu. L. Kapustin (Mineralogiya Karbonatitov, Moscow 1971, pp. 1/288, 185). – [15] C. W. Beck (Am. Mineralogist **35** [1950] 985/1013, 998/1000, 1009).

[16] J. Goñi, C. Guillemin (Bull. Soc. Franc. Mineral. **76** [1953] 124/9). – [17] J. Hansen (Medd. Groenland **181** No. 8 [1968] 1/47, 29, 36). – [18] M. Ya. Somina, A. G. Bulakh (Mineral. Geokhim. **1972** No. 4, pp. 64/73, 66/7). – [19] E. I. Semenov, A. P. Khomyakov, A. V. Bykova (Tr. Mineral. Muzeya Akad. Nauk SSSR No. 11 [1961] 202/4). – [20] D. A. Mineev, T. I. Lavrishcheva, A. V. Bykova (Zap. Vses. Mineral. Obshchestva **99** [1970] 328/32).

[21] N. I. Aleksandrova, I. V. Maksimova, V. N. Svirskaya (Izv. Vysshikh Uchebn. Zavedenii Geol. Razved. **20** No. 10 [1977] 80/7, 85/6). – [22] I. I. Kupriyanova (Tr. Mineral. Muzeya Akad. Nauk SSSR No. 18 [1968] 206/9). – [23] D. A. Mineev (Geokhimiya Apogranitov i Redkometal'-nykh Metasomatitov Severo-Zapadnogo Tarbagataya, Moscow 1968, pp. 1/183, 96). – [24] A. S. Pavlenko, L. P. Orlova, M. A. Akhmanova, K. I. Tobelko (Zap. Vses. Mineral. Obshchestva **94** [1965] 105/13, 106).

10.5.2.2 Parisite $Ca(RE,Y)_2[(F,OH)_2|(CO_3)_3]$

The common parisite is generally characterized by a predominance of the Ce-earths, mostly with Ce selectivity, see Table 3, p. 145. A La variety, La-parisite or parisite-(La), has been found by [1]. A mineral characterized by a significant content of RE of the yttrium group has been named yttroparisite [2, 3] or parisite-(Y) with Y representing yttrium and the heavy lanthanides [4]. This mineral has been little studied, especially as regards to its chemical composition; it might be a fluorocarbonate deviating from parisite [5], cf. [6]. The mineral has not been confirmed [6]; the name, however, has been used in several papers, see, e.g., p. 140, and there exist parisites with significant absolute, see p. 142, and relative amounts of yttrium as well as detectable contents of Y-earth, see Table 3, p. 145. The existence of an hydroxylparisite is indicated by a mineral with a F-deficit [7], see p. 143.

A second phase occurring in parallel intergrowths in two parisites has been shown to be a rare polytypic species, very probably synchysite [8]. The three polytypes designated as parisite α, β, and γ [9], cf. [10], have been also revised to three species: Parisite α is synchysite, parisite β is roentgenite, and only parisite γ is the real parisite [11]. The existence of a structurally different second, more disordered form of parisite has been identified by [11] and was confirmed by [12], see p. 148.

Some parisites from the Swiss Alps, see p. 139, and of the USA [8] have been presumed to be synchysites. This is valid also for the mineral originally named parisite from Greenland. In spite of some similarities between parisite and synchysite both minerals can clearly be distinguished by several chemical and physical properties such as Ca content, crystal form, cleavage, density, and refractive indices [13]. In contrast the physical properties of both minerals from northern Pribaikalia are considered to be identical [14].

References:

[1] E. I. Semenov (in: R. P. Tikhonenkova, E. I. Semenov, Mineralogiya Pegmatitov i Gidrotermalitov Shchelochnykh Massivov, Nauka, Moscow 1967, pp. 1/168, 36/40). – [2] E. I. Nefedov (Dokl. Akad. Nauk SSSR **32** [1941] 361/4). – [3] C. Palache, H. Berman, C. Frondel (The System of Mineralogy, 7th Ed., Vol. 2, Wiley, New York 1951, pp. 1/1124, 282, 284). – [4] M. Fleischer (Can. Mineralogist **16** [1978] 361/3). – [5] E. I. Semenov (in: K. A. Vlasov, Geochemistry and Mineralogy of Rare Elements and Genetic Types of Their Deposits, Vol. 2: Mineralogy of Rare Elements, Israel Program Sci. Trans., Jerusalem 1966, pp. 249/328, 272 [Russian Original, Moscow 1964]).

[6] E. I. Semenov (Tr. Inst. Mineral. Geokhim. Redk. Elem. Akad. Nauk SSSR No. 2 [1959] 181/6, 186). – [7] V. A. Basina, E. K. Vasil'ev, A. A. Kashaev, A. V. Sharova (Vopr. Mineral. Gorn.

Porod Rud Vost. Sibiri **1972** 148/52, 151; Ref. Zh. Geol. **1973** No. 9 V 324). – [8] I. Oftedal (Z. Krist. **79** [1931] 437/64, 442/4, 462). – [9] H. Ungemach (Z. Krist. **91** [1935] 1/22, 10). – [10] N. W. Thibault (Am. Mineralogist **29** [1944] 249/78, 327/62, 359).

[11] G. Donnay, J. D. H. Donnay (Am. Mineralogist **38** [1953] 932/63, 938/9). – [12] M. Ya. Somina, A. G. Bulakh (Mineral. Geokhim. **1972** No. 4, pp. 64/73, 70/2). – [13] O. B. Bøggild (Medd. Groenland **149** No. 3 [1953] 1/442, 157/8). – [14] V. V. Gerasimovskii (Tr. Mineral. Muzeya Akad. Nauk SSSR No. 15 [1964] 194/202, 200).

10.5.2.2.1 Occurrence, Paragenesis, and Alteration

Occurrence and Paragenesis

Parisite occurs in granites and their pegmatites, in alkali rocks and alkali metasomatites, in greisen and other pneumatolytic to hydrothermal rocks, and very often in late carbonatites according to many recent investigations. In these rocks parisite is a late mineral, which is often observed in veinlets, where it is closely and predominantly associated with calcite, quartz, and fluorite, and sometimes also with sulfides.

According to a compilation of Semenov, parisite like bastnaesite is most characteristic for hydrothermal deposits genetically related to alkali syenites and granites (inclusive some carbonatite occurrences), where it is associated with calcite, fluorite, and other minerals. It further occurs as an accessory mineral in metasomatically altered nepheline and alkali syenites [1, 2].

Magmatites and metasomatites. A notable occurrence of parisite in the USA are pegmatite pipes in the alkali granite of Quincy, Massachusetts, where it is associated with silicates and Ti minerals [3]; it further occurs in giant quartz crystals in the zonal pegmatite of the Snowbird mine, Montana, where it extends into adjoining calcite, ankerite, and locally into fluorite (see p. 140) [4], and in granite pegmatites at Halvorsröd, south Norway, and at Hundholmen, north Norway [6], cf. [3, 7]. In veinlets in riebeckite granite at Montorfano, Baveno region, Italy [3], parisite is embedded in masses of chlorite, and surrounded by zeolites, fluorite, and quartz [5].

Parisite has been found in the nepheline syenites of Mont St. Hilaire, Quebec [8], and in nepheline syenite pegmatites of Marathon, Ontario [9, 10], and of the Langesundfjord, south Norway [3, 11]; this locality and others in south Norway are, however, doubted by [12]. For parisite together with ancylite in differentiates of the alkali peridotite of the Oka complex, Quebec, see [13].

Parisite occurs together with other RE minerals in coarse-grained albitites (riebeckite-albite metasomatites) of the USSR; it is a mineral of a late carbonate stage of these granitoids [14]. In the North Baikal area, parisite as an early crystallized idiomorphic mineral together with fluorite, occurs in barite-ankerite veinlets and in albitites [15]; in the Burek Kundus syenite-alkali granite massif, eastern Tuva, Siberia, parisite has been autometasomatically formed in albitized leucocratic syenites [16]. In the Dzhilisusk massif at the upper Khodzha Achkan river, Alai range, Kirgizia, parisite is an accessory mineral predominantly in the colored zones of liebeneritized nepheline syenites, but not in unaltered nepheline syenites; its occurrence is apparently connected with autopneumatolytic processes [17]. In the cassiterite-bearing veins at Irish Creek, Virginia, parisite occurs associated with other carbonates and fluorite in the younger group of minerals produced by greisenization [18], cf. [19].

Hydrothermal occurrences. In the Potgietersrus Tin Fields, Transvaal, South Africa, parisite is fairly common in pockets or pods in the mineralized stanniferous Bobbejaanskop granite, especially in some pipes on Zaaiplaats 236; it occurs together with bastnaesite, allanite, calcite, quartz, fluorite, sulfides, and scheelite, and closely associated with cassiterite and molybdenite; it crystallized during the mesothermal stage after cassiterite and before tysonite and barite [20]. In the alkali hornblende granite of Petrovo-Gnutovo at Zhdanov (Mariupol), Sea of Azov, Ukraine, parisite occurs together with quartz, sulfides, and other minerals in fluorite-calcite veins of hydrothermal genesis [21], cf. [22]; parisite comprises about 8 to 10% of the total mass of the vein and is closely intergrown with calcite [21], cf. [23]. In a brecciated zone in granosyenites with four hydrothermal stages at Tannu Ola, Tuva, Siberia, parisite is a mineral of the first, calcitic stage, where, followed by two generations of bastnaesite (see p. 99), it is uniformly distributed in ankerite calcite aggregates [24], cf. [25, 26].

Parisite reported from Alpine fissure veins in the Val Nalps, Swiss Alps, is synchysite [27]; the only known occurrence of parisite in the Alps is on the Ochsenfeld in the Binnatal, Switzerland [28], cf. [29, 30].

Parisite together with bastnaesite and cordylite is an accessory mineral of late hydrothermal formations of K-alkaline basaltic rocks of the southern Pamir, USSR [31], see also p. 99.

The type locality of parisite is the emerald deposit in the Muzo district, about 100 km north of Bogota, Colombia. The mineral here is associated with pyrite, marcasite, cerian dolomite, albite, calcite, quartz, gypsum, and emerald in veinlets and pockets in intensely folded carbonaceous shale beds [3]; it only occurs in the emerald-bearing calcite-quartz veins [32], especially as inclusions in emerald [33]. The deposit is typical for low temperature Be deposits, RE concentrations in parisite are the result of leaching and of redistribution of RE from the slates [34].

Carbonatites. Parisite has been found as a typomorphic accessory mineral especially in late dolomite-ankerite carbonatites of several localities in USA, USSR, and Africa; it is generally associated with chlorite, quartz, fluorite, sphalerite, pyrite, and hematite. Parisite forms impregnations in these rocks or occurs in druses [35], and is accompanied by bastnaesite and celestite [36]. In Siberia parisite is exclusively contained in ankeritic carbonatites and reaches up to 10% in some parts of these rocks, but occurs in considerably smaller amounts in the metasomatic rocks; accumulations or concretions of parisite crystals are common in vugs in the ankerite carbonatite, the interspaces between the crystals are filled with sphalerite or quartz [37]. The deposition of parisite together with other RE carbonates is generally ascribed to the third, ankerite-dolomite stage of carbonatite formation, before bastnaesite in the fourth stage [35]; parisite occurs, however, in the fourth stage of carbonatite formation in Eastern Siberia [38], and at the River Tuliok, Khibina massif, Kola, even in the fifth stage of carbonatite formation [39]. Parisite is thought to be a hydrothermal mineral in carbonatites on Kola and in Siberia [40], and also in the ankerite stage of a carbonatite massif in the Eastern Sayan Mountains; there it was found only in carbonatites which are almost totally replaced by ankerite and in ankerite fillings in the calcitic carbonatite, and was deposited together with fluorite and sphalerite after monazite [41]. Parisite in this massif dominates over bastnaesite and ancylite, it is deposited from hydrothermal solutions at low temperature of 200 to 300°C (according to fluid inclusions) and with increasing activity of F; it sometimes occurs in the form of "block crystals" (see pp. 147/9) with bastnaesite, closely associated with sulfides and further with Th-free monazite, florencite, sericite, quartz, and cleiophane [42, pp. 64/5, 67, 72/3]. In these late carbonatite stages parisite replaces several other RE minerals besides ankerite and albite, see p. 140. In the Mountain Pass RE deposit, California (see Rare Earth Elements A 8, 1984, Chapter 11.7.2.3.3.3), only small amounts of parisite have been locally identified in carbonate rocks in and near the Sulphide Queen Body and in a single granite sample [43], cf. [44]. Parisite here occurs intimately associated with altered bastnaesite and with sahamalite in light to dark brown barite-dolomite rock; it was formed accompanied by quartz, after bastnaesite and before sahamalite consisting with a decreasing availability of fluorine in the hydrothermal solutions [45]; see for paragenesis also p. 96.

In late carbonatitic rocks of the Tundulu complex, Malawi (cf. "Rare Earth Elements" A 8, 1984, Chapter 11.7.4.3), parisite especially occurs in apatite-poor bastnaesite-carbonatite; it is accompanied by bastnaesite and is sometimes lamellar intergrown with synchysite; these intergrowths are similar to those of bastnaesite and parisite from the Mountain Pass [46]. In an apatite breccia of Glenover, Transvaal, South Africa, parisite, followed by monazite, variscite, and metavariscite, has been deposited in drusy cavities after quartz, specularite, fluorite, and ankerite [47].

In the Snowbird fluorite deposit, Montana, parisite occurs in two generations in a quartz-carbonate body, accompanied by fluorite and minor amounts of pyrite and gersdorffite; the first generation is interpreted to be deposited from a concentrated fluid resembling a carbonatite magma, the second generation from aqueous solutions [48]. In the Salmon Bay area in the northern part of the Prince of Wales Island, Alaska, small amounts of parisite occur besides bastnaesite in essentially nonradioactive carbonate veins [49], which cut graywacke and have some similarity to the veins at the Mountain Pass; the parisite is especially abundant in one of the veins and generally a late stage mineral along post-vein fractures in the carbonate vein filling [50]. For parisite and the minerals associated with it in a late veinlike carbonatite body of a multiple pluton in Siberia, see p. 97.

Further parisite has been found in a dolomite geode of a talc quarry at Trimouns, Luzenac, Departement Ariège, France [51]. Radioactive parisite together with quartz occurs in highly altered, metamorphic limestone from the Whalen mine in the Nixon Fork mining district, western central Alaska [52], see also "Rare Earth Elements" A 8, 1984, Chapter 11.7.2.3.3.1.

The variety hydroxylparisite, associated with monazite, occurs in a carbonatite of Eastern Siberia [53]. The variety parisite-(Y) (yttroparisite) occurs as inclusions in violet octahedral fluorite crystals which themselves are included in albitized orthoclase, in a granite pegmatite of the Adun-Cholon district, Transbaikalia [54], cf. [2, 3]. An yttrium containing parisite (yttrian parisite) has further been found together with fluorite, zircon, and pyrite in a granite pegmatite at Shenyang (Mukden), Manchuria, northeastern China [3]. In a hydrothermal deposit of Central Asia parisite-(Y) together with bastnaesite-(Y) and synchysite-(Y) is a typical mineral of metasomatic biotite-sericite-chlorite rocks and of quartz-chlorite and quartz metasomatites [56].

Inclusions

Inclusions of Fe hydroxides have been observed in parisite of Siberia [37], and together with gaseous-fluid inclusions, on microfissures or on cleavage planes in parisite of East Sayan [42, p. 72]. Minute inclusions of thorite occur in parisite from northern Pribaikalia [15].

Replacements

Parisite replaces several minerals, especially in carbonatites: it is among the replacing products of burbankite in carbonatite occurrences in Siberia [35] and of Kola [40]; together with bastnaesite it replaces burbankite and ancylite in the East Sayan [42, pp. 67, 73], and apparently huanghoite in Siberia [57]. According to microprobe studies on the Khibina sample synchysite is replaced from the edge by parisite [39]. In the Sulfide Queen Body, Mountain Pass, California, parisite hydrothermally replaces bastnaesite and dolomite [45]. In the nepheline syenite pegmatite at Marathon, Ontario, parisite in the form of small crystals surrounds large crystals of bastnaesite [9]. Replacement of ankerite and albite by parisite has been observed in an unspecified carbonatite of USSR [37]. As pseudomorphs after allanite, parisite occurs in granitoids of the Gissar range, Pamir, USSR [1, 2], and, possibly, in occurrences of Kazakhstan [58].

Parisite-(Y) in two generations, partly together with tysonite, replaces monazite, xenotime, and yttrofluorite [55].

Parisite, in turn, is hydrothermally replaced by bastnaesite in alkali rocks of South Greenland [59] and, in its outer parts, in the alkali rocks at the Mont St. Hilaire, Quebec [8].

In a carbonatite massif of the East Sayan, parisite is almost totally replaced by ankerite [41], and in alkali granitoids of the North Baikal area, it is corroded by calcite, hematite, xenotime, and taeniolite [15]. Hydroxylparisite from carbonatites of eastern Siberia is in its marginal parts corroded by monazite [53].

Replacement of parisite by bastnaesite under surface conditions has been observed in the weathering crust of a nonspecified Russian alkali massif [59], cf. [2].

References for 10.5.2.2.1:

[1] E. I. Semenov (in: K. A. Vlasov, Geochemistry and Mineralogy of Rare Elements and Genetic Types of Their Deposits, Vol. 2: Mineralogy of Rare Elements, Israel Program Sci. Trans., Jerusalem 1966, pp. 249/328, 272 [Russian Original, Moscow 1964]). – [2] E. I. Semenov (Mineralogiya Redkikh Zemel', Moscow 1963, pp. 1/412, 103/5). – [3] C. Palache, H. Berman, C. Frondel (The System of Mineralogy, 7th Ed., Vol. 2, Wiley, New York 1951, pp. 1/1124, 284/5). – [4] R. Lasmanis (Mineral. Rec. **8** [1977] 83/6). – [5] A. Lacroix (Bull. Soc. Franc. Mineral. **38** [1915] 106/25, 123).

[6] C. Hintze (Handbuch der Mineralogie, Vol. 1, Pt. 3, de Gruyter, Berlin 1930, pp. 2675/4565, 3430). – [7] I. Oftedal (Z. Krist. **79** [1931] 437/64, 438). – [8] L. Marble, A. Regis (Rocks Minerals **54** [1979] 4/25, 13). – [9] J. Mandarino, W. M. Tovell (Can. Mineralogist **7** [1963] 819). – [10] V. Anderson (Mineral. Rec. **13** [1982] 107/12, 107/8).

[11] T. L. Sverdrup, K. Ø. Bryn, P. C. Sæbø (Norsk Geol. Tidsskr. **39** [1959] 237/47, 241). – [12] G. Raade, J. Haug, R. Kristiansen, A. O. Larsen (Lapis **5** No. 10 [1980] 22/8, 24, 26). – [13] D. P. Gold (Intern. Mineral. Assoc. 4th Gen. Meeting, New Delhi 1964; Papers Proc. Mineral. Soc. India IMA Vol. **1964** 109/25, 116, 124). – [14] Yu. L. Kapustin (in: L. N. Ovchinnikov, N. A. Solodov, Mestorozhdeniya Litofil'nykh Redkikh Metallov, Nedra, Moscow 1980, pp. 208/25, 225/33, 217, 227). – [15] V. V. Gerasimovskii (Tr. Mineral. Muzeya Akad. Nauk SSSR No. 15 [1964] 194/202, 198/201).

[16] A. S. Pavlenko, E. E. Vainshtein, N. V. Turanskaya (Geokhimiya **1959** 291/309; Geochemistry [USSR] **1959** 357/80, 358/60, 365/6). – [17] B. I. Omel'yanenko, N. A. Sirotinina (Mater. Geol. Rudn. Mestorozhd. Petrogr. Mineral. Geokhim. Akad. Nauk SSSR **1959** 414/22, 421/2). – [18] J. J. Glass, A. H. Koschmann, J. S. Vhay (Econ. Geol. **53** [1958] 65/84, 66, 71/4). – [19] C. V. Haynes (in: S. M. Del Rio, Mineral Resources of Colorado First Sequel, Mineral Resour. Board, Denver, Colo., 1960, pp. 370/85, 378). – [20] C. A. Strauss (S. Africa Geol. Surv. Mem. **46** [1954] 1/241, 61, 98, 117, 154, 157).

[21] V. I. Kuzmenko (Dopov. Akad. Nauk Ukr.RSR No. 3 [1940] 35/40 [engl. 38/40, 38/9]). – [22] V. P. Kuts (Dopov. Akad. Nauk Ukr.RSR B **1971** 892/5; Ref. Zh. Geol. **1972** 1 V 36). – [23] M. I. Ivantishin, O. A. Koulskaya, G. Y. Gornyi, G. D. Eliseeva (Geochem. Intern. **1** [1964] 372/87, 375, 378 [Transl. from: Tr. Inst. Geol. Sci. Akad. Nauk Ukr.RSR Ser. Petrogr. Mineral. Geokhim. No. 21 [1959]). – [24] A. P. Khomyakov (in: K. A. Vlasov, Mineralogiya i Geneticheskie Osobennosti Shchelochnykh Massivov, Moscow 1964, pp. 56/82, 68, 77, 81). – [25] A. P. Khomyakov (Geokhimiya **1964** 85/8; Geochem. Intern. **1** [1964] 40/3).

[26] Yu. L. Kapustin (in: L. N. Ovchinnikov, N. A. Solodov, Mestorozhd. Litofil'nykh Redkikh Metallov, Nedra, Moscow 1980, pp. 243/85, 271, 276). – [27] R. L. Parker, E. Brandenberger (Schweiz. Mineral. Petrogr. Mitt. **26** [1946] 12/8, 12/3). – [28] S. Graeser (Schweiz. Mineral.

Petrogr. Mitt. **45** [1965] 597/795, 773/5). – [29] A. Kipfer (Schweiz. Strahler **3** No. 4 [1973] 133/69, 157). – [30] M. Gramaccioli (Mineral. Rec. **8** [1977] 287/93, 288).

[31] A. S. Lomteva, E. A. Dmitriev (Dokl. Akad. Nauk Tadzh.SSR **19** No. 7 [1976] 53/6; C.A. **86** [1977] No. 75942). – [32] R. Wokittel (Colomb. Inst. Geol. Nacl. Compil. Estud. Geol. Oficiales Colombia **10** [1960] 1/393, 178; C.A. **58** [1963] 10004). – [33] J. Walton (Gemmologist **19** [1950] 123/5 from Mineral. Abstr. **11** [1951] 289). – [34] A. A. Beus, D. A. Mineev (Geol. Rudn. Mestorozhd. **1974** No. 4, pp. 18/30, 22/3, 28/9). – [35] Yu. L. Kapustin (Mineralogiya Karbonatitov, Moscow 1971, pp. 1/288, 78, 184).

[36] Yu. L. Kapustin (Geokhimiya **1966** 1311/21; Geochem. Intern. **3** [1966] 1054/64, 1055). – [37] V. S. Gaidukova, T. B. Zdorik (Geol. Mestorozhd. Redk. Elem. No. 17 [1962] 86/117, 111/2; Ref. Zh. Geol. **1963** No. 10 V 121). – [38] T. B. Zdorik (Geol. Mestorozhd. Redk. Elem. Vses. Nauchn. Issled. Inst. Mineral'n. Syr'ya No. 30 [1966] 121/31, 130). – [39] O. B. Dudkin, L. I. Polezhaeva, F. V. Minakov (Veshchestvennyi Sostav Shchelochnykh Intrusivnykh Kompleksov Kol'skogo Poluostrova, Apatity 1981, pp. 73/81, 78, 80; Ref. Zh. Geol. **1981** No. 12 V 179). – [40] Yu. L. Kapustin (in: K. A. Vlasov, Mineralogiya i Geneticheskie Osobennosti Shchelochnykh Massivov, Moscow 1964, pp. 135/94, 170/1).

[41] E. E. Vainshtein, L. K. Pozharitskaya, N. V. Turanskaya (Geokhimiya **1961** 1031/4; Geochemistry [USSR] **1961** 1151/4). – [42] M. Ya. Somina, A. G. Bulakh (Mineral. Geokhim. **1972** No. 4, pp. 64/73). – [43] J. C. Olson, D. R. Shawe, L. C. Pray, W. N. Sharp (U.S. Geol. Surv. Profess. Paper No. 261 [1954] 1/75, 36). – [44] L. A. Wright, R. M. Stewart, T. E. Gay, G. C. Hazenbush (Calif. J. Mines Geol. **49** [1953] 49/192, 125). – [45] H. W. Jaffe, R. Meyrowitz, H. T. Evans (Am. Mineralogist **38** [1953] 741/54, 751/3).

[46] M. S. Garson (Nyasaland Geol. Surv. Dept. Mem. No. 2 [1962] 1/248, 103, 108/9). – [47] W. J. Verwoerd (S. Africa Geol. Surv. Handbook **6** [1967] 1/452, 63). – [48] S. E. Clabaugh, C. R. Sewell (Geol. Soc. Am. Spec. Paper No. 76 [1963] 268/9). – [49] J. R. Houston, R. S. Velikanje, S. G. Bates, H. Wedow Jr. (U.S. Geol. Surv. Circ. No. 248 [1953] 6/13, 9/10). – [50] J. R. Houston, R. G. Bates, R. S. Velikanje, H. Wedow (U.S. Geol. Surv. Bull. No. 1058-A [1958] 1/31, 10, 12).

[51] Y. Moëlo, J.-P. Choutier, E. Melon (from R. Pierrot, Bull. Soc. Franc. Mineral. Crist. **97** [1974] 521/4, 523). – [52] M. G. White, J. M. Stevens (U.S. Geol. Surv. Circ. No. 279 [1953] 1/19, 10, 12). – [53] V. A. Basina, E. K. Vasil'ev, A. A. Kashaev, A. V. Sharova (Vopr. Mineral. Gorn. Porod Rud Vost. Sibiri **1972** 148/52; Ref. Zh. Geol. **1973** No. 9 V 324). – [54] E. I. Nefedov (Dokl. Akad. Nauk SSSR **32** [1941] 361/4). – [55] V. N. Zuev, A. V. Kosterin (Tr. Inst. Mineral. Geokhim. Kristallokhim. Redk. Elem. No. 3 [1959] 151/3).

[56] D. A. Mineev (in: L. N. Ovchinnikov, N. A. Solodov, Mestorozhdeniya Litofil'nykh Redkikh Metallov, Nedra, Moscow 1980, pp. 178/84, 181/2). – [57] Yu. L. Kapustin (Dokl. Akad. Nauk SSSR **202** [1972] 422/5; Dokl. Earth Sci. Sect. **202** [1972] 116/9). – [58] V. I. Chernov, O. F. Krol', A. V. Stepanov (Geol. Geokhim. Mineral. Mestorozhd. Redk. Elem. Kaz. **1966** 138/44, 139/41). – [59] E. I. Semenov, A. P. Khomyakov, A. V. Bykova (Tr. Mineral. Muzeya Akad. Nauk SSSR No. 11 [1961] 202/4).

10.5.2.2.2 Chemistry

For a compilation of analyses and $(RE,Y)_2O_3$ contents, mean and variance of 13 samples, including one sample of parisite-(Y), see "Seltenerdelemente" A 4, 1979, pp. 27, 5. A high Y content, 7.86% Y_2O_3, is given for an yttrian parisite from China, see [1],

The formulas of parisite derived from these analyses are: $Ce_2Ca[F_2|(CO_3)_3]$ [2 to 6], $(Ce,La)_2Ca[F_2|(CO_3)_3]$ [1, 7, 8] or in the oxide form $2CeFCO_3 \cdot CaCO_3$ [9, 10] and $2YFCO_3 \cdot CaCO_3$ for parisite-(Y) [10].

Parisite from East Sayan with 62.00% RE_2O_3 and 10.25% CaO corresponds to the standard composition [11], cf. theoretical values 60.89% RE_2O_3 and 10.44% CaO [1, 3, 12]. This is also the case for CaO values of parisites from Montana, Massachusetts, and Colombia [1]. The high values of 25.25% CaO and 29.80% CO_2 in Ukrainian parisite have been explained by the presence of closely intergrown calcite [13], cf. [3].

The crystalchemical formula of a parisite from Eastern Siberia with a deficit in fluorine, presumed to be balanced by H_2O (after exclusion of P and the part of RE connected with P) is $(Ce_{1.04}La_{0.88})(Ca_{0.62}Fe^{3+}_{0.02}Th_{0.02}Nb_{0.04})(CO_3)_3(F_{0.31}OH_{1.10})$; this formula is close to that of parisite [14]. The high OH content indicates a hydroxylparisite, cf. hydroxylbastnaesite, p. 108.

Minor elements in parisite are: SrO with 0.8 and BaO with 1.5% (besides only 7.02% CaO) in parisite from the complex core of the Khibina massif, Kola [15]; ThO_2 with 0.60 besides 0.85% Nb_2O_5 in parisite from Eastern Siberia (see formula above) [14]. Besides 1.7% Th, parisite of the Snowbird mine, Montana, contains 57.4 ppm U [16], parisite from the Alai range, Kirgizia, has 0.5 to 1% Th besides 0 to 0.3% U [17], and a sample from the Whalen mine, Alaska, has 20 ppm U [19]. Ukrainian parisites have traces of Th besides 0.21% MgO and MnO each [13] or (two samples) 0.94 and 1.00% MgO, 0.96 and 1.80% Fe_2O_3, 0.03 and 0.05% TiO_2 [12]. Trace elements are Sr, Ba, Mg, Mn, and Pb in parisite from Vuorijarvi, Kola [6], and Be, Al, Mg, Zr, Si in parisite from the Alai range, Kirgizia [17].

A content of 8.2% P_2O_5 in the parisite with OH > F, see formula above, from a carbonatite from Eastern Siberia is probably caused by monazite [14].

RE composition. All parisites, except parisite-(Y), are strongly Ce-selective minerals and show large relative amounts of light lanthanides [8]. Examples of this dominance of the RE of the Ce group over the RE of the Y group are numerous parisites of different genetic types, see Table 3, p. 145. The Ce selectivity is characteristic for parisites from granitic and alkalic rocks [3, 4], and especially for parisite from the Ukraine with $RE_{Ce} : RE_Y = 82$ [12], but also for parisite crystals in a dolomite geode in talc from France, with Ce, La, and Nd dominating, and Pr, Gd, Sm, and Dy subordinate [20]; it is most pronounced for parisites from carbonatites, see table p. 144. The Ce selectivity of parisite is less pronounced than that of bastnaesite at localities, where both minerals occur together, e.g., in the North Baikal area [21], and in Tannu Ola, Tuva, where parisite shows a higher content of RE of the Y group than bastnaesite [22, 23]. The typical order in parisites is Ce > La > Nd > Pr > Sm > Gd, see Table 3, p. 145. Parisite from alkali granites, Azov region, shows Ce > La > Nd > Pr [24] or Nd > La > Pr > Sm > Gd > Y [25]. An extremely high ratio Ce/Nd = 6.0 and ratios La/Nd = 4.3 and Pr/Nd = 0.48 have been found for parisite of eastern Tuva [26]. A La-dominated parisite is mentioned by [27]. Higher La/Ce ratios in parisite seem to be related to higher H_2O contents [28].

A remarkable portion of RE in a paragenesis may enter parisite: Parisite of ankerite carbonatites, East Sayan Mountains, concentrates about 90% of the total RE and >80% of Σ(La + Ce + Nd) [29], and parisite in calcitic ores of eastern Tuva concentrates 40% of total RE_2O_3, 40% of Ce-group rare earth elements (La to Eu) and 15% of Y-group RE (Gd to Lu) [22]. Parisite of the calcitic to dolomitic, emerald-bearing veins of Colombia contains >60% RE [30].

Parisite-(Y) from Adun-Cholon, Transbaikal region, shows significant relative values of Y and of the elements of the Y-earths [31]; moderate values of these elements have been found in parisites from the North Baikal region and from Muzo, Colombia, see Table 3, p. 145.

In the following table the averages of the RE composition are given in at.% together with Σ(La + Ce + Pr) and the ratios between Y and RE and La and Nd for parisites of different genetic associations; it shows the large increase in light lanthanides from granite and hydrothermal associations through alkaline rocks to carbonatites. The rocks are (number of samples in parentheses): No. 1 granite (1); No. 2 hydrothermal associations (7); No. 3 alkali rocks (5), No. 4 carbonatites (7) [8]:

No.	La	Ce	Pr	Nd	Sm	Eu	Gd	Tb	Dy	Ho	Er
1	19.3	49.1	7.3	16.7	3.5	0.2	1.9	0.1	0.4	—	0.5
2	23.6	45.7	5.8	18.7	2.6	0.2	1.5	0.2	0.9	0.1	0.2
3	31.3	47.8	5.0	14.1	1.0	0.1	0.6	0	0.1	—	—
4	32.1	51.4	4.1	11.3	0.9	—	0.2	—	—	—	—

No.	Tm	Yb	Lu	$\frac{100\,Y}{Y+RE}$	Σ(La + Ce + Pr)	La/Nd
1	—	1.0	—	—	75.7	1.16
2	—	0.5	—	8.65(4)*)	75.1	1.26
3	—	—	—	—	84.1	2.22
4	—	—	—	0.25(2)*)	87.6	2.84

*) The number in parentheses indicates the number of analyses in which Y is reported.

In Table 3, p. 145, the composition of RE ± Y of parisites and one parisite-(Y) from various rocks and of different genesis is given as $(RE \pm Y)_2O_3 = 100$ or RE ± Y = 100 (numbers with an asterisk). The numbers in the table refer to: Nos. 1 to 3 alkali rocks (2 and 3 albitized); Nos. 4 to 9 hydrothermal associations, mostly from alkali rocks (8 and 9 sideritic); No. 10 a secondary hydrothermal formation; Nos. 11 to 16 carbonatites (13 to 16 ankeritic); No. 17 parisite-(Y) from granite pegmatite.

References for 10.5.2.2.2:

[1] C. Palache, H. Berman, C. Frondel (The System of Mineralogy, 7th Ed., Vol. 2, Wiley, New York 1951, pp. 1/1124, 282, 284). – [2] H. Strunz, C. Tennyson (Mineralogische Tabellen, 6th Ed., Geest and Portig, Leipzig 1977, pp. 1/621, 242). – [3] E. I. Semenov (in: K. A. Vlasov, Geochemistry and Mineralogy of Rare Elements and Genetic Types of Their Deposits, Vol. 2: Mineralogy of Rare Elements, Israel Program Sci. Trans., Jerusalem 1966, pp. 249/328, 270/1 [Russian Original, Moscow 1964]). – [4] E. I. Semenov (Mineralogiya Redkikh Zemel', Moscow 1963, pp. 1/412, 103/4). – [5] Yu. L. Kapustin (Mineralogiya Karbonatitov, Moscow 1971, pp. 1/288, 170, 184).

[6] Yu. L. Kapustin (in: K. A. Vlasov, Mineralogiya i Geneticheskie Osobennosti Shchelochnykh Massivov, Moscow 1964, pp. 135/94, 170). – [7] J. Murdoch, R. W. Webb (Bull. Calif. Div. Mines Geol. No. 173 [1956] 1/452, 245). – [8] M. Fleischer (Can. Mineralogist **16** [1978] 361/3). – [9] C. W. Beck (Am. Mineralogist **35** [1950] 985/1013, 1000). – [10] E. I. Semenov (Tr. Inst. Mineral. Geokhim. Redkikh Elem. Akad. Nauk SSSR No. 2 [1959] 181/6, 184).

[11] M. Ya. Somina, A. G. Bulakh (Mineral. Geokhim. **1972** No. 4, pp. 64/73, 72). – [12] V. P. Kuts (Dopov. Akad. Nauk Ukr. RSR B **1971** 892/5). – [13] V. I. Kuzmenko (Dopov. Akad. Nauk Ukr. RSR No. 3 [1940] 35/40, 39 [Engl. 38/40]). – [14] V. I. Basina, E. K. Vasil'ev, A. A. Kashaev, A. V. Sharova (Vopr. Mineral. Gorn. Porod Rud Vost. Sibiri **1972** 148/52, 151/2; Ref. Zh. Geol. **1973** No. 9 V 324). – [15] O. B. Dudkin, L. I. Polezhaeva, F. V. Minakov (Veshchestvennyi Sostav Shchelochnykh Intrusivnykh Kompleksov Kol'skogo Poluostrova, Apatity 1981, pp. 73/81, 76; Ref. Zh. Geol. **1981** No. 12 V 179).

[16] R. Lasmanis (Mineral. Rec. **8** [1977] 83/6). – [17] B. I. Omel'yanenko, N. A. Sirotinina (Mater. Geol. Rudn. Mestorozhd. Petrogr. Mineral. Geokhim. Akad. Nauk SSSR **1959** 414/22,

Table 3

Distribution of Rare Earth Elements in Parisites.

No.	Locality	La	Ce	Pr	Nd	Sm	Eu	Gd	Tb	Dy	Ho	Er	Tm	Yb	Lu	Y	Ref.
1	South Greenland	29	45	5.5	15	3.2	0.1	2.1	—	—	—	—	—	—	—	—	[3]
2	Eastern Tuva, Siberia	36.7	51.1	3.7	8.5	—	—	—	—	—	—	—	—	—	—	—	[3]
*3	Burpala massif, northern Baikal area	34.0	51.3	3.1	10.4	0.6	—	0.3	—	—	—	—	—	—	—	—	[32]
4	Zhdanov (Mariupol), Ukraine	29.1	51	3.8	14.6	1.5	—	—	—	—	—	—	—	—	—	—	[33, 34]
5		28.4	46.5	4.9	16.5	1.2	—	1.4	—	—	—	—	—	0.2	—	1.0	[12]
*6	Northern Baikal area	18.7	35.0	3.9	16.2	3.6	—	2.7	0.8	2.9	0.7	1.0	—	0.7	—	13.5	[21]
*7	Tannu Ola, Tuva	14	42	7.5	25	5	0.5	3.4	0.4	1	0.2	0.3	—	1	0.2	—	[22, 23]
*8	Yenisey district, Siberia	23.3	46.6	5.1	21.3	1.6	0.7	0.9	0.2	0.3	—	—	—	—	—	—	[35]
9	Mongolia	29	55	3.2	10	0.3	—	0.6	—	0.2	—	0.4	—	0.3	—	—	[3, 4]
*10	Muzo, Colombia	23.0	43.0	6.5	18.0	2.5	—	1.1	—	0.3	—	0.1	—	0.2	—	5.0	
		24.1	48.6	5.3	17.2	2.2	—	1.3	—	0.2	—	0.1	—	1.0	—	—	[30]
11	Mountain Pass, California	31.1	48.8	4.3	14.2	1.4	—	—	—	—	—	—	—	—	—	0.2	[36]
12	Vuoriyarvi, Kola	33	50	4	11	1	—	0.8	—	0.2	—	—	—	—	—	—	[37]
13	East Sayan, Siberia	35	51	3	10	1	—	—	—	—	—	—	—	—	—	—	[5,37]
14		29.6	55.2	4.5	9.9	0.3	—	0.3[1)]	—	—	—	—	—	—	—	0.2	[11]
*15	[2)]	35.3	51.1	3.4	9.3	0.9	—	—	—	—	—	—	—	—	—	—	[29]
*16	[2)]	35.8	52.0	3.2	8.5	0.4	—	—	—	—	—	—	—	—	—	—	[29]
*17	Adun-Cholon, Transbaikalia	9.7	21.7	12.1	2.2	3.5	—	4.0	—	4.9	1.2	2.8	—	1.7	—	36.2	[38]

[1)] Gd + Eu. – [2)] Recalculated from values relative to Nd = 1.

421). – [18] V. S. Gaidukova, T. B. Zdorik (Geol. Mestorozhd. Redk. Elem. No. 17 [1962] 86/117, 113; Ref. Zh. Geol. **1963** No. 10 V 121). – [19] M. G. White, J. M. Stevens (U.S. Geol. Surv. Circ. No. 279 [1953] 1/19, 12, 19). – [20] Y. Moëlo, J.-P. Choutier, E. Melon (from R. Pierrot, Bull. Soc. Franc. Mineral. Crist. **97** [1974] 521/4, 523).

[21] V. V. Gerasimovskii (Tr. Mineral. Muzeya Akad. Nauk SSSR No. 15 [1964] 194/202, 200/1). – [22] A. P. Khomyakov (in: K. A. Vlasov, Mineralogiya i Geneticheskie Osobennosti Shchelochnykh Massivov, Moscow 1964, pp. 56/82, 68, 78/80). – [23] A. P. Khomyakov (Geokhimiya **1964** 85/8; Geochem. Intern. **1** [1964] 40/3). – [24] I. B. Borovskii, V. I. Gerasimovskii (Dokl. Akad. Nauk SSSR **49** [1945] 353/6). – [25] M. I. Ivantishin, O. A. Koulskaya, G. Y. Gornyi, G. D.Eliseeva (Geochem. Intern. **1** [1964] 372/87, 387 [Transl. from: Tr. Inst. Geol. Sci. Akad. Nauk Ukr.RSR Ser. Petrogr. Mineral. Geokhim. No. 21 [1959]]).

[26] A. S. Pavlenko, E. E. Vainshtein, N. V. Turanskaya (Geokhimiya **1959** 291/309; Geochemistry [USSR] **1959** 357/80, 358/9, 365). – [27] E. I. Semenov (in: R. P. Tikhonenkova, E. I. Semenov, Mineralogiya Pegmatitov i Gidrotermalitov Shchelochnykh Massivov, Nauka, Moscow 1967, pp. 1/168, 36/40, 38). – [28] N. V. Svyazhin (Tr. Inst. Geol. Akad. Nauk SSSR Ural'sk. Filial No. 70 [1965] 249/52). – [29] E. E. Vainshtein, L. K. Pozharitskaya, N. V. Turanskaya (Geokhimiya **1961** 1031/4; Geochemistry [USSR] **1961** 1151/4). – [30] A. A. Beus, D. A. Mineev (Geol. Rudn. Mestorozhd. **1974** No. 4, pp. 18/30, 25/6).

[31] E. I. Nefedov (Dokl. Akad. Nauk SSSR **32** [1941] 361/4). – [32] A. P. Khomyakov (Izv. Akad. Nauk SSSR Ser. Geol. **1972** No. 1, pp. 94/101, 98). – [33] E. I. Semenov, R. L. Barinskii (Geokhimiya **1958** 314/33; Geochemistry [USSR] **1958** 398/419, 400). – [34] E. I. Semenov, A. P. Khomyakov (Diagnost. Diagnost. Svoistva Mineral. **1981** 88/93, 90). – [35] E. I. Semenov, A. P. Khomyakov, A. V. Bykova (Tr. Mineral. Muzeya Akad. Nauk SSSR No. 11 [1961] 202/4).

[36] K. J. Murata, H. J. Rose Jr., M. K. Carron, J. J. Glass (Geochim. Cosmochim. Acta **11** [1957] 141/61, 148). – [37] Yu. L. Kapustin (Geokhimiya **1966** 1311/21; Geochem. Intern. **3** [1966] 1054/64, 1060). – [38] D. I. Ryabtchikov, M. M. Senyavin, Y. S. Sklyarenko (Proc. 2nd Intern. Conf. Peaceful Uses At. Energy, Geneva 1958, Vol. 4, pp. 333/40, 336).

10.5.2.2.3 Crystal Forms and Crystal Structure

Morphology

Symmetry and forms. Parisite is trigonal-pyramidal [1] in accordance with its space group, see p. 148. The morphological features of the parisite crystals of East Sayan are determined by the appearance of two opposing phenomena: 1) The outer symmetry appears to be higher than theoretically admissible, it approximates to a hexagonal symmetry, probably caused by a development of two twinning systems (see p. 147). 2) The real symmetry of the "block crystals" (see p. 149) is triclinic, probably on account of an irregular interchange in the borders of every block layer of ordered and disordered individuals which lay perpendicular to the c axis [2, pp. 69/72]. For parisite-(Y) (yttroparisite) a hexagonal symmetry has been assumed [3].

The baffling morphology of parisite is a long known problem [4, p. 933], cf. [5]. It is confirmed by later investigations on parisites from East Sayan, which show a very complex character of most of the crystals due to the superposition of the two opposite effects mentioned above [2, p. 72].

Crystal forms of parisite given by older authors are compiled in the handbook of Dana [5], cf. [6], and together with figures of crystals from various localities in the handbook of Hintze [7]. The crystal forms given by Palache and Warren of a crystal of Quincy, Massachusetts [8], and the forms of crystals of Muzo, Colombia, and Narsarsuk, Greenland, are indexed according to

½ c/a (= ½ × 11.722 to ½ × 11.807, see p. 148) and compiled by Donnay and Donnay. Frequent forms in the new setting with ½ c/a are {10$\bar{1}$1}, {22$\bar{4}$3}, {01$\bar{1}$2}, {20$\bar{2}$5}, {0003}, {11$\bar{2}$3} [4, p. 952]. For similar considerations of the relation between morphological and X-ray axial ratio as well as for forms of parisite in the new setting cf. [9]. The four types of crystals found in the East Sayan all show a pseudopinacoid {0001}, the prisms {01$\bar{1}$0} and {11$\bar{2}$0}, and numerous (up to five) inclined faces of pseudodipyramids [2, pp. 69/70]. For a discussion of the morphology of parisite and its relation to that of synchysite, see [10].

Habit. The habit of parisite enormously varies due to oscillatory combination of the steep rhombohedrons or pyramids with rudimentary or absent prism faces [8, 5], especially due to the different development of the five pseudodipyramids [2, pp. 67/70]. There exist long prismatic or elongated [11, 12], tabular to short columnar, prismatic, or sometimes corund-shaped crystals [2, pp. 67/9], and also barrel-shaped forms, if the crystal is cut by the pinacoid {0001} [13], cf. [14, 15, 16]. Barrel-shaped crystals are also found of OH-rich parisite [17]. Acute double hexagonal pyramids are common and elongated crystals sometimes have locally enlarged portions or scepter-like terminations [5]. For parisites from Russian carbonatites a marked hexagonal tabular habit is due to weakly developed prism faces [18] and lamellar forms are due to a combination of a dominating pinacoid with sparingly developed trigonal bipyramids [19], cf. prisms and plates at the Mountain pass, California [20], and hexagonal plates (in thin section) in carbonatite of Malawi [21]. Flat laths elongated parallel to the base occur in tin ore bodies in Transvaal [22], skeletal crystals besides short prismatic crystals with an imperfect or no hexagonal outline at the Mont St. Hilaire, Quebec [23], and elongated, sometimes distorted crystals in the Snowbird Mine, Montana [24].

Parisite in a pseudomorph after burbankite sometimes has a hexagonal prismatic, mostly an irregular form [25]; parisite after allanite, however, forms fine-grained accumulations [26]. Very often granular and irregular forms are observed for parisite [6], cf. grains (in greisen) [27], fine grains (in autopneumatolytic association) [28], and little groups of crystals (in a dolomitic geode) [15]. Flattened prismatic parisite crystals of a Siberian deposit are intergrown with each other [19]; parisites of northern Pribaikalia are intergrown at angles of 120° [13].

Parisite-(Y) occasionally forms well-developed hexagonal prisms, terminated by a pinacoid, but predominantly it occurs in irregular forms [3].

Striation and twinning. The lateral faces of parisite commonly are striated or deeply grooved [5, 13]. The horizontal striation, often accompanied by more or less deep furrows, originally has been ascribed to polysynthetic twinning [10]. The striation, however, is caused by salient and re-entrant angles formed by a multiple and irregular alternation of small faces of various dipyramids [2, pp. 69/70], or it is due to an alternating combination of dominant and subordinate pyramids [16]. Sometimes the striation is caused by an alternate development of a prism and a steep dipyramid [11].

Parisite from Narsarsuk, Greenland, is twinned by reticular merohedry; the twinned crystal consists of at least three horizontal slices intercalated in the main crystal, the composition plane is planar, parallel to the base [4, p. 952]. In contrast to their morphological appearance the inner symmetry of parisite "block crystals" from the East Sayan may be explained by twinning by two laws: one series of twins is formed by rotation of 60° around the axis a_3, in the other series the twinning plane is (0001), but the twinning axis is [01$\bar{1}$0] and the composition plane of the individuals is the Ca net plane. At twinning after these two laws the optical axes of the individuals coincide [2, p. 69]. For the hydroxyl-containing parisite from Eastern Siberia besides polycrystalline aggregates twins have been observed on barrel-shaped crystals [17].

Cleavage and parting. The cleavage often found in parisites parallel to the base (0001) [29] is perfect [13, 22], good [11, 16], or only distinct; a perfect parting parallel to the base has been observed [6]; it may be due to parallel intergrowths of parisite with other members of the bastnaesite series in lamellar form [30, p. 464]. Cleavage or parting can also be caused by alteration [5].

Syntaxic intergrowths. In thin sections lamellar intergrowths between parisite and bastnaesite have been observed in the northern part of the Sulphide Queen Body, Mountain Pass, California [31], and in block crystals in carbonatite of the East Sayan, USSR [2, p. 65], cf. p. 139; parisite occurs intergrown with synchysite in the Tundulu carbonatite complex, southern Malawi [21], and with roentgenite in a polycrystal from Narsarsuk, southern Greenland [4, p. 957]. Intergrowths of parisite with bastnaesite, tysonite (chlorite, chert, and sericite) are described from tin ore bodies in Transvaal, South Africa [22]. For general explanations of intergrowths in polycrystals of the bastnaesite series, see pp. 90/1.

Crystal Structure

Cell parameters. The unit cell constants and the axial ratios c/a of parisites from several localities and of parisite-(Y) from Siberia are given in the following table; Nos. 1 to 7 are common parisites from granite pegmatites (1 to 3), a secondary hydrothermal association (4), and carbonatites (5 to 7); No. 8 is a hydroxyl-containing parisite from a carbonatite; No. 9 a parisite-(Y) from a granite pegmatite. Values from the subcell (a′c‴), as given by the authors, are recalculated to the real unit cell ac: $a = a'\sqrt{3}$, $c = 18c'''$.

No.	Locality	a in Å	c in Å	Ref.
1	Quincy, Massachusetts	7.176[1] ± 0.3%	84.11 ± 0.3%	
2		7.124[2] ± 0.3%	84.11 ± 0.3%	[4, pp. 936, 952]
3		7.09[3]	83.88[3]	[30, p. 438]
4	Muzo district, Colombia	7.09[4]	83.7[4]	[30, p. 438]
5	East Sayan, USSR	7.07 ± 0.01	83.58 ± 0.06	[2, p. 72]
6	Siberia	7.10	88.92[5]	
7		7.16	88.11[6]	[18]
8	Eastern Siberia	7.15	84.6	[17]
9	Adun-Cholon, Transbaikalia	6.94[7]	80.44[7]	[3]

[1] Ordered structure, c/a = 11.722. – [2] Disordered structure, c/a = 11.807. – [3] $a' = 4.094 \pm 0.003$, $c''' = 4.66 \pm 0.01$ Å. – [4] $a' = 4.094 \pm 0.005$, $c''' = 4.65 \pm 0.01$ Å. – [5] $2c''' = 9.88$ Å. – [6] $2c''' = 9.79$ Å. – [7] $a' = 4.008 \pm 0.03$, $c''' = 4.469 \pm 0.04$ Å.

For parisite from East Sayan (No. 5), with a value c corresponding nearly totally with the common value, the somewhat lower value a indicates a disorder in the structure [2, p. 72] (cf. the lower value in No. 2 against No. 1). The smaller unit cell data of parisite-(Y) are apparently due to the Y-earth composition of the RE [6]. For the pseudoperiods in c direction ($c = 3c' = 6c'' = 18c'''$) of parisite compared with the other members of the bastnaesite series, see p. 90 and Fig. 5, p. 91.

The space group of parisite is R3-C_3^4 (No. 146) with Z = 18 [4, p. 936], cf. [6]. The formerly supposed space groups and Z = 6 correspond to a subcell (a c′) with a volume of ⅓ of the true cell (ac) and c′ = ⅓ c [30, p. 454]. The cell volume $V = 208.4 \times 18$ Å^3 for the ordered and 205.4×18 Å^3 for the disordered form of parisite [4, p. 936].

Structure description. The structure of parisite is characterized by clearly separated layers of RE-F and of Ca, arranged vertically to the c axis; the distance between two RE layers is 4.86 Å, between an RE-F and a Ca layer it is 4.55 Å. The CO_3 groups (triangles) lie parallel to the c axis between the RE-F and Ca layers [30, pp. 450/3, 463]. The Ce and F ions alternate at the vertices of regular hexagons and the Ca ions are placed above Ce ions in the alternating layers [4, p. 946]. The structure may be considered to consist of varying proportions of layers of bastnaesite and of synchysite [32], cf. the schematic representation of the crystal structure of parisite in Fig. 5, p. 91.

At Quincy, Massachusetts, besides the ordered, a more disordered second type of parisite occurs with the same c parameter, but a deviating parameter of a (see Nos. 1 and 2 in the table p. 148). The disordered structure results from irregularities in the sequence of the Ca- and RE-F layers, the decreased value of a of this species may be ascribed to a slight displacement of the F and Ce with respect to each other, so they no longer lie in one horizontal plane. This species probably crystallized at elevated temperature [4, pp. 938/9, 946]. The existence of a more disordered species has been confirmed for parisite crystals from the East Sayan, USSR: within one "block crystal", individuals with regular and irregular structure alternate perpendicularly to the c axis and lower the symmetry to triclinic [2, pp. 70/2], cf. p. 146.

The observed disordered parisite species can readily be shown by electron microscopy lattice images; from electron diffraction data it can be concluded, that parisite occurs in two rhombohedral polytypes with basic repeats of 14 Å (c = 42 Å) and 28 Å (c = 28 Å) in c direction [32].

References for 10.5.2.2.3:

[1] H. Strunz, C. Tennyson (Mineralogische Tabellen, 6th Ed., Geest and Portig, Leipzig 1977, pp. 1/621, 242). – [2] M. Ya. Somina, A. G. Bulakh (Mineral. Geokhim. **1972** No. 4, pp. 64/73). – [3] E. I. Nefedov (Dokl. Akad. Nauk SSSR **32** [1941] 361/4). – [4] G. Donnay, J. D. H. Donnay (Am. Mineralogist **38** [1953] 932/63). – [5] C. Palache, H. Berman, C. Frondel (The System of Mineralogy, 7th Ed., Vol. 2, Wiley, New York 1951, pp. 1/1124, 282/3, 285).

[6] E. I. Semenov (in: K. A. Vlasov, Geochemistry and Mineralogy of Rare Elements and Genetic Types of Their Deposits, Vol. 2: Mineralogy of Rare Elements, Israel Program Sci. Trans., Jerusalem 1966, pp. 249/328, 270/2 [Russian Original, Moscow 1964]). – [7] C. Hintze (Handbuch der Mineralogie, Vol. 1, Pt. 3, de Gruyter, Berlin 1930, pp. 2675/4565, 3419/21, 3431/8). – [8] C. Palache, C. H. Warren (Am. J. Sci. [4] **31** [1911] 533/57, 535/40). – [9] R. L. Parker, E. Brandenberger (Schweiz. Mineral. Petrogr. Mitt. **26** [1946] 12/8, 15/6). – [10] H. Ungemach (Z. Krist. **91** [1935] 1/22, 6/18).

[11] S. Graeser (Schweiz. Mineral. Petrogr. Mitt. **45** [1965] 597/795, 774). – [12] J. Murdoch, R. W. Webb (Bull. Calif. Div. Mines Geol. No. 173 [1956] 1/452, 245). – [13] V. V. Gerasimovskii (Tr. Mineral. Muzeya Akad. Nauk SSSR No. 15 [1964] 194/202, 199/200). – [14] A. P. Khomyakov (in: K. A. Vlasov, Mineralogiya i Geneticheskie Osobennosti Shchelochnykh Massivov, Moscow 1964, pp. 56/82, 68). – [15] N. I. Aleksandrova, I. V. Maksimova, N. M. Svirskaya (Izv. Vysshikh Uchebn. Zavedenii Geol. Razved. **20** No. 10 [1977] 80/7, 82).

[16] Y. Moëlo, J.-P. Choutier, E. Melon (from R. Pierrot, Bull. Soc. Franc. Mineral. Crist. **97** [1974] 521/4, 523). – [17] V. A. Basina, E. K. Vasil'ev, A. A. Kashaev, A. V. Sharova (Vopr. Mineral. Gorn. Porod Rud Vost. Sibiri **1972** 148/52; Ref. Zh. Geol. **1973** No. 9 V 324). – [18] Yu. L. Kapustin (Mineralogiya Karbonatitov, Moscow 1971, pp. 1/288, 184/5). – [19] V. S. Gaidukova, T. B. Zdorik (Geol. Mestorozhd. Redk. Elem. No. 17 [1962] 86/117, 111/2; Ref. Zh. Geol. **1963** No. 10 V 121). – [20] H. W. Jaffe, R. Meyrowitz, H. T. Evans (Am. Mineralogist. **38** [1953] 741/54, 752).

[21] M. S. Garson (Nyasaland Geol. Surv. Dept. Mem. No. 2 [1962] 1/248, 103, 109). – [22] C. A. Strauss (S. Africa Geol. Surv. Mem. **46** [1954] 1/241, 157). – [23] L. Marble, A. Regis (Rocks

Minerals **54** [1979] 4/25, 13). – [24] R. Lasmanis (Mineral. Rec. **8** [1977] 83/6). – [25] Yu. L. Kapustin (in: K. A. Vlasov, Mineralogiya i Geneticheskie Osobennosti Shchelochnykh Massivov, Moscow 1964, pp. 135/94, 170).

[26] V. I. Chernov, O. F. Krol', A. V. Stepanov (Geol. Geokhim. Mineral. Mestorozhd. Redk. Elem. Kaz. **1966** 138/44, 139). – [27] J. J. Glass, A. H. Koschmann, J. S. Vhay (Econ. Geol. **53** [1958] 65/84, 66). – [28] B. I. Omel'yanenko, N. A. Sirotinina (Mater. Geol. Rudn. Mestorozhd. Petrogr. Mineral. Geokhim. Akad. Nauk SSSR **1959** 414/22, 421). – [29] O. B. Bøggild (Medd. Groenland **149** No. 3 [1953] 1/442, 157). – [30] I. Oftedal (Z. Krist. **79** [1931] 437/64).

[31] J. C. Olson, D. R. Shawe, L. C. Pray, W. N. Sharp (U.S. Geol. Surv. Profess. Paper No. 261 [1954] 1/75, fig. after p. 36). – [32] J. van Landuyt, S. Amelinckx (Am. Mineralogist **60** [1975] 351/8, 351, 356/7).

10.5.2.2.4 Optical Properties

The **luster** of parisite is vitreous to resinous [1 to 3], cf. vitreous [4] and resinous [5]; it also may be greasy [6], silky [7], or pearly on parting planes [3]. On well developed crystals all faces including the base are brilliant [8].

Parisite is transparent to translucent [1, 3], cf. [9, 10]; colorless crystals are clear, brownish crystals turbid [11]. Transparent parisite is often turbid due to inclusions [12].

Color. Parisites predominantly show yellowish to brown(ish) colors [1], cf. [3, 13 to 15], but sometimes they are pink to reddish and rarely greenish or gray, see table p. 151. In addition the following colors of parisites have been observed: yellow (inclusion in emerald of Muzo) [16], honey-yellow [10, 17], brownish amber (greisen) [9], orange-red [18], brown-red (accumulations) [17], brownish pink to raspberry-pink (in fluorite) [19], honey-brown to dark brown (dolomite, Binnatal) [20], pale rose (replacing synchysite with the same color) [21], pink or slightly greenish [17], and light blue-gray (covered with white bastnaesite) [7]. The streak is yellowish white [3, 5].

In transmitted light parisite is colorless, see table p. 151 and [4], almost colorless to very pale yellow [8], but also yellowish brown [5] to reddish, due to inclusions of Fe hydroxide [17].

Pleochroism is weakly developed with absorption E>O [1, 3], with O light yellow, E golden-yellow [1, 8], or O pale yellowish brown, E yellow-brown [5].

The optical character of parisite is uniaxial positive [1, 3, 22]; most of the parisites listed in table p. 151, including a hydroxyl-containing variety (No. 18) and a parisite-(Y) (No. 19), have this character.

In the following table, color and **refractive indices** of parisites are compiled together with the values for the measured density D(meas.). The samples are arranged according to their genesis. Mother rocks of the parisites are: No. 1 alkaligranite pegmatite, Nos. 2 to 6 alkali rocks (syenites, nepheline syenites, and their pegmatites), No. 7 a Sn deposit (mesothermal), No. 8 a hydrothermal carbonate vein, Nos. 9 and 10 a secondary hydrothermal vein, Nos. 11 to 16 carbonatites; No. 17 is parisite pseudomorphous after allanite; No. 18 is a hydroxyl-containing parisite from a carbonatite, No. 19 parisite-(Y) from granite pegmatites; No. 20 is the synthetic compound $2CeFCO_3 \cdot CaCO_3$.

No.	Locality	Color	Refractive indices n_o	n_e	D(meas.) g/cm³	Ref.
1	Quincy, Massachusetts	colorless to pale yellow	1.676 ±0.002	1.757	4.320	[8], cf. [1]
2	Narsarsuk, South Greenland	—	1.6742[1)]	1.7701[1)]	—	[23]
3		—	1.6730[1)]	1.7690[1)]	—	[23]
4		—	1.670 ±0.002[2)]	1.765 ±0.005[2)]	—	[23]
5	Marathon area, Ontario	pink to red	1.676	1.770	—	[24]
6	Khodzha Achkan river, Alai range, Kirgizia	white, grayish green, rose, brown	1.674	1.755	—	[4]
7	Potgietersrus Tin Fields, Transvaal	reddish	1.682	1.760	—	[11]
8	Tannu Ola, Tuva, Siberia	lemon-yellow	1.674	—	—	[25]
9	Muzo district, Colombia	—	1.675	1.755	4.302[3)]	[26]
10		—	1.6717	1.7712	—	[1]
11	Mountain Pass, California	—	1.667	1.760	4.32	[27]
12	Vuoriyarvi, Kola, USSR	rose	1.678	1.780	4.47	[6]
13	East Sayan, Siberia	wax-yellow to rose, grayish	1.672 ±0.002	1.760 ±0.005	4.37 ±0.01	[12]
14	North Baikal area, USSR	wax-yellow to reddish brown	1.670 ±0.002	1.754 ±0.002	4.07 ±0.02	[5]
15	Siberia	—	1.678	1.780	4.16[4)]	[22]
16	Tundulu, southern Malawi	colorless[5)]	1.670	~1.770	—	[28]
17	Akzhailyau massif, Tarbagatai, Kazakhstan	colorless[5)]	1.670 ±0.005	1.767	—	[29]
18	Eastern Siberia	white, greenish yellow, greenish	1.655	1.736	3.90	[30]
19	Adun-Cholon, Transbaikalia	—	1.643 ±0.002	1.755 ±0.002	—	[31]
20	—	—	1.667[6)]	1.768[6)]	4.29	[23]

[1)] Values listed under synchysite by [1], but ascribed to parisite by [23]. – [2)] Values measured by Jaffe [unpublished]. – [3)] D from [8]. – [4)] D from [17]. – [5)] Color in transmitted light. – [6)] Calculated from ionic refractivities.

The measured mean index of refraction of parisite (with 10.10 wt% CaO) $n_m = (2n_o + n_e)/3 = 1.698$ deviates by only 0.003 units from the calculated value $dK + 1 = 1.695$, obtained by the rule of Gladstone and Dale [27]; the agreement between both values is remarkable [32].

The **IR spectrum** of parisite is characterized by the vibrations of the CO_3^{2-} groups and shows bands of ν_1 to ν_4; the doubling of ν_1 and ν_4 indicates two nonequivalent carbonate groups [33], which probably cause a lower site symmetry for parisite [34]. The following table indicates wave numbers in cm^{-1}:

Number	ν_3	ν_1	ν_2	ν_4	Ref.
1	1449	1088, 1079	870	746, 734	[33, 34]
2	1470	1100	885, 860	750	[35], cf. [36]

No. 1 is a parisite from Colombia; the values were recalculated from the wavelength in μm. No. 2, probably from a Russian locality, furthermore shows weak bands of molecular water at 1660 cm^{-1}.

Upon heating of parisite to 400 to 500°C the absorption band of the CO_3 groups between 1400 and 1500 cm^{-1} is split, probably caused by the Ca ion [36].

The **absorption spectrum in visible light** of parisite is characterized by two weak bands at 13350 and 20200 cm^{-1}, an intermediate band at 19000 cm^{-1}, and a strong band with splitting at 17250 cm^{-1}; the splitting is caused by different symmetry around the Nd^{3+} cations [37].

Parisites from pegmatites (three samples) show olive to deep dark green **fluorescence** [38].

References for 10.5.2.2.4:

[1] C. Palache, H. Berman, C. Frondel (The System of Mineralogy, 7th Ed., Vol. 2, Wiley, New York 1951, pp. 1/1124, 283, 287). – [2] J. Murdoch, R. W. Webb (Bull. Calif. Div. Mines Geol. No. 173 [1956] 1/452, 245). – [3] E. I. Semenov (in: K. A. Vlasov, Geochemistry and Mineralogy of Rare Elements and Genetic Types of Their Deposits, Vol. 2: Mineralogy of Rare Elements, Israel Program Sci. Trans., Jerusalem 1966, pp. 249/328, 271/2 [Russian Original, Moscow 1964]). – [4] B. I. Omel'yanenko, N. A. Sirotinina (Mater. Geol. Rudn. Mestorozhd. Petrogr. Mineral. Geokhim. Akad. Nauk SSSR **1959** 414/22, 421). – [5] V. V. Gerasimovskii (Tr. Mineral. Muzeya Akad. Nauk SSSR No. 15 [1964] 194/202, 200).

[6] Yu. L. Kapustin (in: K. A. Vlasov, Mineralogiya i Geneticheskie Osobennosti Shchelochnykh Massivov, Moscow 1964, pp. 135/94, 170). – [7] L. Marble, A. Regis (Rocks Minerals **54** [1979] 4/25, 13). – [8] C. Palache, H. Warren (Am. J. Sci. [4] **31** [1911] 533/57, 536, 538/42). – [9] J. J. Glass, A. H. Koschmann, J. S. Vhay (Econ. Geol. **53** [1958] 65/84, 66, 79). – [10] Y. Moëlo, J.-P. Choutier, E. Melon (from R. Pierrot, Bull. Soc. Franc. Mineral. Crist. **97** [1974] 521/4, 523).

[11] C. A. Strauss (S. Africa Geol. Surv. Mem. **46** [1954] 1/241, 157). – [12] M. Ya. Somina, A. G. Bulakh (Mineral. Geokhim. **1972** No. 4, pp. 64/73, 72). – [13] R. Lasmanis (Mineral. Rec. **8** [1977] 83/6). – [14] J. R. Houston, R. G. Bates, R. S. Velikanje, H. Wedow (U.S. Geol. Surv. Bull. No. 1058-A [1958] 1/31, 12). – [15] M. Gramaccioli (Mineral. Rec. **8** [1977] 287/93, 288).

[16] J. Walton (Gemmologist **19** [1950] 123/5 from Mineral. Abstr. **11** [1951] 289). – [17] V. S. Gaidukova, T. B. Zdorik (Geol. Mestorozhd. Redk. Elem. No. 17 [1962] 86/117, 112; Ref. Zh. Geol. **1963** No. 10 V 121). – [18] V. Anderson (Mineral. Rec. **13** [1982] 107/12, 107). – [19] V. I. Kuzmenko (Dopov. Akad. Nauk Ukr.RSR No. 3 [1940] 35/40, 38/9). – [20] S. Graeser (Schweiz. Mineral. Petrogr. Mitt. **45** [1965] 597/795, 774).

[21] O. B. Dudkin, L. I. Polezhaeva, F. V. Minakov (Veshchestvennyi Sostav Shchelochnykh Intrusivnykh Kompleksov Kol'skogo Poluostrova, Apatity 1981, pp. 73/81, 80; Ref. Zh. Geol. **1981** No. 12 V 179). – [22] Yu. L. Kapustin (Mineralogiya Karbonatitov, Moscow 1971, pp. 1/288, 184/5). – [23] G. Donnay, J. D. H. Donnay (Am. Mineralogist **38** [1953] 932/63, 960). – [24] J. Mandarino, W. M. Tovell (Can. Mineralogist **7** [1963] 819). – [25] A. P. Khomyakov (in: K. A. Vlasov, Mineralogiya Geneticheskie Osobennosti Shchelochnykh Massivov, Moscow 1964, pp. 56/82, 67).

[26] C. W. Beck (Am. Mineralogist **35** [1950] 985/1013, 1000). – [27] H. W. Jaffe (Am. Mineralogist **41** [1956] 757/77, 764, 773). – [28] M. S. Garson (Nyasaland Geol. Surv. Dept. Mem. No. 2 [1962] 1/248, 109). – [29] V. I. Chernov, O. F. Krol', A. V. Stepanov (Geol. Geokhim. Mineral. Mestorozhd. Redk. Elem. Kaz. **1966** 138/44, 139). – [30] V. A. Basina, E. K. Vasil'ev, A. A. Kashaev, A. V. Sharova (Vopr. Mineral. Gorn. Porod Rud Vost. Sibiri **1972** 148/52, 148; Ref. Zh. Geol. **1973** No. 9 V 324).

[31] E. I. Nefedov (Dokl. Akad. Nauk SSSR **32** [1941] 361/4). – [32] G. Donnay, J. D. H. Donnay (Bull. Geol. Soc. Am. **66** [1955] 1551). – [33] H. H. Adler, P. F. Kerr (Am. Mineralogist **48** [1963] 839/53, 849/51). – [34] W. B. White (in: V. C. Farmer, Infrared Spectra of Minerals, Vol. 4, The Carbonate Minerals, Mineral. Soc. Monogr., London 1974, pp. 1/539, 227/84, 272). – [35] M. V. Akhmanova, L. P. Orlova (Geokhimiya **1966** 571/8; Geochem. Intern. **3** [1966] 444/51, 447).

[36] A. S. Pavlenko, L. P. Orlova, M. V. Akhmanova, K. I. Tobelko (Zap. Vses. Mineral. Obshchestva **94** [1965] 105/13, 107/9). – [37] A. N. Platonov, O. A. Krasil'shchikova, A. N. Tarashchan (Konst. Svoistva Mineral. **9** [1975] 58/64, 59, 61). – [38] A. Kipfer (Schweiz. Strahler **3** No. 9 [1975] 370/84, 377, 383).

10.5.2.2.5 Other Physical Properties

Hardness H_M (mean value) = 4.5 [1, 2]; $H_M \approx 4$ to 5 of parisite from Binnatal, Swiss Alps [3]. Microhardness of parisite 360 to 470 kg/mm^2 corresponds to H_M = 4.4 to 4.9 [4, 5]. Parisite from the northern Baikal area has microhardness 313 kg/mm^2, given as H_M = 4.6 [6].

The measured **density** of parisites from various localities is given in the table p. 151. Further values D(meas.) in g/cm^3 are: 4.36 [1, 2]; parisite from Greenland 4.364 [7], from Switzerland 4.03 [3], and from Siberia 4.16 [8] or 4.1 to 4.5 [9]. The values D(meas.) = 4.26, 4.39, and 4.41 ($\pm$0.1 each) g/cm^3 are probably of polycrystals [10].

The calculated densities of parisites D(calc.) = 4.37 to 4.38 g/cm^3 agree very well with data from the literature and are only somewhat larger than most of the measured values [11]; D(calc.) = 4.38 [1] and 4.40 g/cm^3 [4, 5, 12]. Ordered parisite has D(calc.) = 4.29, disordered 4.35 g/cm^3 [10].

The **fracture** of parisite is semiconchoidal to splintery [4, 5], uneven [3], or irregular conchoidal [6]. The mineral is brittle [1 to 6].

For parisite from the Ukraine the **magnetic susceptibility** $\chi = 4.1 \times 10^{-6}$ cm^3/g is due to the strong paramagnetic RE ions [13]. **Piezoelectricity** of parisite is weak or completely absent [11]. **Radioactivity** is weak and probably due to U content [14] or U and Th contents [15], perhaps caused by inclusions of thorite [6].

10.5.2.2.6 Chemical Behavior and Thermal Analyses

Parisite dissolves in strong acids [4, 5], especially in hot strong acids [1]; it is also soluble in dilute hydrochloric acid [16], without any heating [17], but only very slowly [8].

The DTA curve of parisite from Muzo, Colombia, shows an endothermal effect due to decomposition with loss of CO_2, which begins at 470°C and reaches its maximum at 660°C; this endothermal curve merges with an exothermal curve, that reaches a peak at 720°C; it is due to oxidation of $Ce^{3+} \rightarrow Ce^{4+}$. At 975°C an endothermal break begins slowly [18]. The DTA curve of parisite from the Ukraine shows the endothermal effect from 440°C up to a maximum at 580°C, followed by two exothermal effects between 850 and 880°C and between 1060 and 1090°C [4, 5].

The thermogram of parisite from the East Sayan, Siberia, shows the standard behavior of this mineral with an endothermal peak at 600°C. The TG curve of this parisite indicates a weight loss of 20.6% at 700°C and of 21.1% at 1100°C [19].

Calcination of parisite reduces the mineral to a cubic phase of the same type as that which forms upon calcination of bastnaesite (CeO_2 and Ln_2O_3) [4, 5], cf. p. 136.

References for 10.5.2.2.5 and 10.5.2.2.6:

[1] C. Palache, H. Berman, C. Frondel (The System of Mineralogy, 7th Ed., Vol. 2, Wiley, New York 1951, pp. 1/1124, 283). – [2] J. Murdoch, R. W. Webb (Bull. Calif. Div. Mines Geol. No. 173 [1956] 1/452, 245). – [3] S. Graeser (Schweiz. Mineral. Petrogr. Mitt. **45** [1965] 597/795, 774). – [4] E. I. Semenov (Mineralogiya Redkikh Zemel', Moscow 1963, pp. 1/412, 103). – [5] E. I. Semenov (in: K. A. Vlasov, Geochemistry and Mineralogy of Rare Elements and Genetic Types of Their Deposits, Vol. 2: Mineralogy of Rare Elements, Israel Program Sci. Trans., Jerusalem 1966, pp. 249/328, 271 [Russian Original, Moscow 1964]).

[6] V. V. Gerasimovskii (Tr. Mineral. Muzeya Akad. Nauk SSSR No. 15 [1964] 194/202, 200/1). – [7] O. B. Bøggild (Medd. Groenland **149** No. 3 [1953] 1/442, 158). – [8] V. S. Gaidukova, T. B. Zdorik (Geol. Mestorozhd. Redk. Elem. No. 17 [1962] 86/117, 112; Ref. Zh. Geol. **1963** No. 10 V 121). – [9] Yu. L. Kapustin (in: L. N. Ovchinnikov, N. A. Solodov, Mestorozhdeniya Litofil'nykh Redkikh Metallov, Nedra, Moscow 1980, pp. 243/85, 271). – [10] G. Donnay, J. D. H. Donnay (Am. Mineralogist **38** [1953] 932/63, 936).

[11] I. Oftedal (Z. Krist. **79** [1931] 437/64, 440/1, 464). – [12] E. I. Semenov (Tr. Inst. Mineral. Geokhim. Redk. Elem. Akad. Nauk SSSR No. 2 [1959] 181/6, 184). – [13] E. I. Semenov, A. P. Khomyakov (Diagnost. Diagnost. Svoistva Mineral. **1981** 88/93, 88, 92/3; Ref. Zh. Geol. **1982** No. 3 V 187). – [14] M. G. White, J. M. Stevens (U.S. Geol. Surv. Circ. No. 279 [1953] 1/19, 12, 19). – [15] E. W. Heinrich (Mineralogy and Geology of Radioactive Raw Materials, Mc Graw-Hill, New York 1958, pp. 1/654, 126/7).

[16] R. Lasmanis (Mineral. Rec. **8** [1977] 83/6). – [17] B. I. Omel'yanenko, N. A. Sirotinina (Mater. Geol. Rudn. Mestorozhd. Petrogr. Mineral. Geokhim. Akad. Nauk SSSR **1959** 414/22, 421). – [18] C. W. Beck (Am. Mineralogist **35** [1950] 985/1013, 999/1000, 1009). – [19] M. Ya. Somina, A. G. Bulakh (Mineral. Geokhim. **1972** No. 4, pp. 64/73, 72).

10.5.2.3 Roentgenite $Ca_2Ce_3[F_3|(CO_3)_5]$

Roentgenite is a member of the bastnaesite series, which mostly occurs as intimate syntaxic intergrowths with the other members of this series, often as polycrystals, see pp. 90/1. The mineral corresponds to the morphological species "parisite β" previously described as one of three modifications of parisite [1], cf. [2].

Occurrence. Roentgenite has first been found at Narsarsuk, southern Greenland, where it is intimately associated with synchysite, parisite, and bastnaesite; it commonly occurs in syntaxic intergrowths with each of these species, simulating single crystals [3], and has been formed in alkali syenite pegmatites during the hydrothermal stage of the pegmatite process [4, 5]. Roentgenite, intergrown with parisite, also occurs in one specimen from Pyrites, Montana [6, pp. 933, 940]. Further, a crystal from Quincy, Massachusetts, formerly described as parisite [7], is essentially roentgenite [6, pp. 951, 956].

A RE-F carbonate which has been found together with fluorite, zircon, apatite, and rarely with monazite, in schlichs of rocks of liparitic composition (spherulitic tuffaceous lavas and porphyritic quartz liparites) of the southwestern spurs of the Chatkal Mountains, USSR, is assumed to be roentgenite according to some of its properties (amount of RE, correlation between RE and density, extrapolated refractive indices); some of its properties are, however, compatible with parisite-(Y) [8].

Chemistry. Because of the almost universal intergrowth of roentgenite with the other members of the bastnaesite series not enough pure material was available for a chemical or even a spectrographic analysis. The composition therefore was derived from a plot of the cell dimension c‴ (in Å) against mol% $CaCO_3$; a straight line drawn through the known c‴ values*) of bastnaesite and parisite indicates for the observed c‴ of roentgenite almost exactly 40.0 mol% $CaCO_3$, so that the formula is $3\ CeFCO_3 \cdot 2\ CaCO_3$ [3], [6, pp. 943/6], or $(Ce,La)_3Ca_2(CO_3)_5F_3$ [9]. This formula corresponds to a theoretical composition (in wt%) of 57.24 Ce_2O_3, 13.12 CaO, 25.77 CO_2, and 6.63 F [10, 11]. An analysis with 27.03% Ce_2O_3, 34.52% ΣLa_2O_3, 2.41% ΣY_2O_3, and 10.98% CaO, besides 2.40% Fe_2O_3, has been given for the mineral of the Chatkal Mountains [8].

In roentgenite, Ce is replaced by other RE, as it is the case in the associated RE-F carbonates [3]. The RE composition of the mineral from the USSR ($\Sigma RE_2O_3 = 100$) is: La 18.4, Ce 42.4, Pr 3.8, Nd 16.8, Sm 6.1, Eu 0.2, Gd 4.6, Tb 0.4, Dy 3.2, Er 0.2, Yb 0.6, and Y 3.3 [8].

Crystal forms. Roentgenite is trigonal-pyramidal with a morphological axial ratio c/a = 9.74 ± 0.03 [3], which is the same as the structural c/a [6, p. 956]. The axial ratio of "parisite β" (= roentgenite), c/a = 10.094 [1], corresponds to 9c‴/a′ [6, p. 956].

Forms of the available crystals are $\{10\bar{1}0\}$ (dominant), $\{01\bar{1}2\}$, {0001} (or {0003} in accordance with the R criterion), $\{11\bar{2}0\}$, and $\{10\bar{1}4\}$ [6, pp. 951/2, 956], cf. [3, 4].

The habit of roentgenite is similar to that of parisite and synchysite [3, 4]; most of the crystals of southern Greenland are singly terminated, with $\{10\bar{1}1\}$ and $\{01\bar{1}2\}$, truncated by a small basis. The two pyramids are striated horizontally [3]. Roentgenite is twinned by reticular merohedry in consequence of its rhombohedral symmetry; the twin operation is defined by reflection in the twin plane and composition plane (0001). The twins are doubly terminated and show twin symmetry $\bar{6}m2$ [3], [6, p. 956]. A cleavage has not been observed [3].

Syntaxic intergrowths between roentgenite, parisite, and synchysite are common on crystals from Greenland. The boundary surface between two species is planar, irregular, or both [3]. Microsyntaxic intergrowths between the three minerals in narrow regions of ~100 Å become evident from electron microscope images [12].

Structure. The unit cell constants of roentgenite are a = 7.131 ± 0.02, c = 69.41 ± 0.2 Å, c/a = 9.733, space group R3-C_3^4 (No. 146), Z = 9 [3], [6, pp. 936, 938], cf. [4, 11]. The unit cell constants a and c are computed from the pseudo-periods a′ = 4.117 Å (by $a = a'\sqrt{3}$) and c‴ = 4.627 (by c = 15c‴), or from c′ = 23.14 Å (by c = 3c′ with c′ = 5c‴). The cell volume of the true cell ac V = 203.8 × 15 Å³ [6, pp. 936/7].

The structure of roentgenite consists of horizontal Ca layers alternating with Ce-F layers, the Ca ions are placed above the Ce ions; the ac′ cell contains five layers of metals (three of Ce, two of Ca), substitution of Ca and Ce for each other is ruled out. The pronounced pseudo-period c′ indicates, that the heavy atoms repeat along the c axis at intervals close in length to c′. The disorder observed in parisite can probably exist also in roentgenite [6, pp. 939, 943/6]. In accordance with the X-ray diffraction data the electron microscope line image indicates a spacing of ~23 Å. The sequence of lines is represented by an alternation of lines with a larger and with a smaller spacing, the repeat distance in each case is 23 Å. With four weak spots

*) In [3] given as c″.

between strong spots in c* direction on the electron diffraction pattern the pseudo-period $d_{0005} = c/5$ [12].

Optical properties. Roentgenite crystals from Greenland are transparent to translucent, wax-yellow to brown, in transmitted light colorless to light yellow; they are optical uniaxial positive, with refractive indices $n_o = 1.662 \pm 0.002$, $n_e = 1.756 \pm 0.005$ [3], cf. [6, p. 960], [11]. The mineral from the Chatkal Mountains is brown, rarely orange, with $n_o = 1.658$, $n_e = 1.756$ [8]; its refractive indices are similar to those calculated from the ionic refractivities of $3CeFCO_3 \cdot 2CaCO_3$, $n_o = 1.654$, $n_e = 1.755$ [6, p. 960]. The mean refractive index $(2n_o + n_e)/3 = 1.693$ of the Greenland mineral deviates by only 0.001 units from the calculated value $dK + 1 = 1.694$, obtained by the rule of Gladstone and Dale [13]. In a diagram of refractive indices against composition ($CeFCO_3$-$CaCO_3$), the points for n_o and n_e of the mineral from the Chatkal Mountains lie intermediate between those of parisite and synchysite on a straight line, near the point of roentgenite [8].

Other physical properties. The fracture of roentgenite from Greenland is conchoidal. Hardness and density could not be determined owing to too small crystals; the calculated density D(calc.) = 4.19 g/cm^3 [3], cf. [6, p. 960]. For the mineral of the Chatkal Mountains D(meas.) = 4.21 to 4.23 g/cm^3 [8].

Chemical behavior and DTA. Roentgenite from Greenland dissolves in strong acids, e.g. in 6n nitric acid; the solubility is intermediate between those of synchysite and parisite [3]. The mineral from the Chatkal Mountains is soluble in dilute hydrochloric acid (1:10). The DTA curve of this mineral shows an endothermal peak at 560°C which is typical for RE-F carbonates; the peak is due to a dissociation of roentgenite with loss of CO_2, which mainly occurs between 450 and 600°C and comprises 13.1 to 19% of the total loss [8].

References for 10.5.2.3:

[1] H. Ungemach (Z. Krist. **91** [1935] 1/22, 10, 12, 15). – [2] N. W. Thibault (Am. Mineralogist **29** [1944] 249/78, 327/62, 359). – [3] G. Donnay (Am. Mineralogist **38** [1953] 868/70). – [4] E. I. Semenov (in: K. A. Vlasov, Geochemistry and Mineralogy of Rare Elements and Genetic Types of Their Deposits, Vol. 2: Mineralogy of Rare Elements, Israel Program Sci. Trans., Jerusalem 1966, pp. 249/328, 272/3 [Russian Original, Moscow 1964]). – [5] E. I. Semenov (Mineralogiya Redkikh Zemel', Moscow 1963, pp. 1/412, 105).

[6] G. Donnay, J. D. H. Donnay (Am. Mineralogist **38** [1953] 932/63). – [7] C. Palache, C. H. Warren (Am. J. Sci. [4] **31** [1911] 533/57, 537). – [8] K. U. Urunbaev, A. K. Kasymov, L. A. Skachkova, A. N. Khan (Uzb. Geol. Zh. **14** No. 3 [1970] 21/4). – [9] M. Fleischer (Can. Mineralogist **16** [1978] 361/3). – [10] E. I. Semenov (Tr. Inst. Mineral. Geokhim. Kristallokhim. Redk. Elem. Akad. Nauk SSSR No. 2 [1959] 181/6, 184).

[11] P'ei-Shan Chang [Pei-Shan Chzhan] (Sci. Sinica [Peking] **11** [1962] 251/8, 252, 256; C.A. **57** [1962] 10799). – [12] J. van Landuyt, S. Amelinckx (Am. Mineralogist **60** [1975] 351/8, 353, 355/7).– [13] H. W. Jaffe (Am. Mineralogist **41** [1956] 757/77, 764, 773).

10.5.2.4 **Synchysite** $Ca(RE,Y)[F|(CO_3)_2]$

10.5.2.4.1 Name, Synonyms, and Varieties

The name synchysite (synonym synchisite) has been given to a mineral from Narsarsuk, Greenland, originally erroneously determined to be parisite, but discerned to be an uniquely new species by Flink in 1900, see [1 to 3]. The polytypic species parisite α is also synchysite [4], cf. [5, 6] and parisite p. 137.

The common synchysite is generally characterized by a predominance of Ce-earth, mostly with Ce selectivity, see Table 5, p. 165; it is denoted as synchysite in the following text. Synchysites characterized by significant contents of RE of the yttrium-group ± yttrium have been named yttrosynchysite in the Russian literature [7 to 9], but doverite in the American literature [10, 11]. A close similarity of doverite with synchysite has been shown by a detailed study including X-ray data [11, 12]. Consequently the name doverite has been questioned even by the authors themselves [11], cf. [13]. Both these minerals are identical except of some differences in unit cell parameters apparently due to RE composition [15].

For a mineral from Predazzo, Italy, with remarkable quantities of La and Y besides Ca and Ce, "yttrium-bearing synchysite" has been thought to be a better name than "yttrosynchysite" [13]. The mineral X found by Semenov in northern Kirgizia may also be yttrosynchysite [7], cf. [14].

According to the system of nomenclature of RE minerals approved by the IMA, the correct name of the Y variety of synchysite is synchysite-(Y); the term doverite should be dropped [16].

In accordance with this nomenclature a recently found Nd dominated synchysite from northern Bohemia, Czechoslovakia (ČSSR), which has been extensively investigated by electron microprobe analysis and X-rays, has been designated synchysite-(Nd) [17]. The occurrence of a synchysite-(Nd) had already been established by Vorma et al. who found this mineral in the Pyörönmaa pegmatite in SW Finland [18].

References for 10.5.2.4.1:

[1] C. Palache, H. Berman, C. Frondel (The System of Mineralogy, 7th Ed., Vol. 2, Wiley, New York 1951, pp. 1/1124, 287/8). – [2] E. I. Semenov (in: K. A. Vlasov, Geochemistry und Mineralogy of Rare Elements and Genetic Types of Their Deposits, Vol. 2: Mineralogy of Rare Elements, Israel Program Sci. Trans., Jerusalem 1966, pp. 249/328, 273/5 [Russian Original, Moscow 1964]). – [3] O. B. Bøggild (Medd. Groenland **149** No. 3 [1953] 1/442, 157). – [4] H. Ungemach (Z. Krist. **91** [1935] 1/22, 10/2). – [5] N. W. Thibault (Am. Mineralogist **29** [1944] 249/78, 327/62, 359).

[6] G. Donnay, J. D. H. Donnay (Am. Mineralogist **38** [1953] 932/63, 939). – [7] E. I. Semenov (Tr. Inst. Mineral. Geokhim. Kristallokhim. Redk. Elem. No. 2 [1959] 181/6, 185). – [8] E. I. Semenov (in: R. P. Tikhonenkova, E. I. Semenov, Mineralogiya Pegmatitov i Gidrotermalitov Shchelochnykh Massivov, Nauka, Moscow 1967, pp. 1/168, 36/40). – [9] E. P. Gurov, E. P. Gurova (Tr. Mineral. Muzeya Akad. Nauk SSSR No. 24 [1975] 171/4). – [10] W. L. Smith, J. Stone, D. D. Riska, H. Levine (Science **122** [1955] 31).

[11] W. L. Smith, J. Stone, D. R. Ross, H. Levine (Am. Mineralogist **45** [1960] 92/8, 94). – [12] A. A. Levinson, R. A. Borup (Am. Mineralogist **47** [1962] 337/43, 338). – [13] F. Emiliani, G. Gandolfi (Mineralog. Petrogr. Acta **11** [1965] 123/31, 129). – [14] J. D. H. Donnay, G. Donnay (Bull. Soc. Franc. Mineral. Crist. **84** [1961] 25/9). – [15] N. I. Aleksandrova, I. V. Maksimova, N. M. Svirskaya (Izv. Vysshikh Uchebn. Zavedenii Geol. Razvedka **20** No. 10 [1977] 80/7, 82, 85).

[16] A. A. Levinson (Am. Mineralogist **51** [1966] 152/8, 154, 156). – [17] B. Scharm, P. Kühn (Neues Jahrb. Mineral. Monatsh. **1983** 201/10, 202/3, 207). – [18] A. Vorma, P. Ojanperä, V. Hoffrén, J. Siivola, A. Löfgren (Bull. Comm. Geol. Finlande No. 222 [1966] 241/74, 245, 257).

10.5.2.4.2 Occurrence, Paragenesis, and Alteration

Occurrence and Paragenesis

Common Synchysite

Common synchysite mainly occurs in granites and their pegmatites, in nepheline syenites and their pegmatites, and in Alpine fissure veins [1, 2]. Synchysite is a rare mineral in typical carbonatites, but it occurs, associated with sulfides, parisite, and quartz, in late ankeritic carbonatites of Africa and Siberia [3]. For these and further occurrences see the following compilation of genetically ordered localities of common synchysite.

Magmatic and metasomatic rocks. Synchysite has been found in a granite of the Urals, USSR [4], and an yttrium-bearing synchysite occurs (together with ancylite) in the granite of Predazzo, northern Italy [5]. For synchysite in cavities of rhomb porphyries see p. 159. Together with parisite synchysite occurs in the pegmatite of Quincy, Massachusetts, and with parisite and pyrite in the altered effusives of Pyrites, Ravalli County, Montana [6], cf. [7, pp. 933, 951, 957]. In its type locality, Narsarsuk, Julianehaab district, southern Greenland, synchysite occurs in pegmatitic veins of alkali syenites, associated with neptunite, epididymite, aegirine, albite, elpidite, fluorite, cordylite, and polylithionite [8], and intimately associated and in syntaxic intergrowths with other members of the bastnaesite series, especially roentgenite [7, pp. 939/40], [9]. In the nepheline syenite pegmatite of Mont St. Hilaire, Quebec, Canada, synchysite is associated with siderite, zircon, and pyrite in vugs of albite [10].

Hydrothermal occurrences. Synchysite occurs at the Zaaiplaats tin mine in the Potgietersrus district, Transvaal, South Africa, where the hydrothermally formed RE minerals are closely associated with cassiterite and molybdenite [11]; for formation and paragenesis of this occurrence see pp. 138/9.

Synchysite is a rare mineral in thorium veins associated with alkali rocks at the Lemhi Pass, Lemhi County, Idaho, and at Powder Horn, Gunnison County, Colorado [12], where it has been identified by X-ray methods in several samples together with thorite [13]; cf. "Rare Earth Elements" A 8, 1984, Chapters 11.7.2.3.3.1 and 11.7.2.3.3.2, respectively, for these two occurrences of thorium veins.

In Alpine fissure veins transecting cavernous, mostly granitic country rocks, synchysite, frequently together with other RE minerals, occurs in distinct localities in masses of small crystals [14]; it seems to be even more common than monazite and to be the prevalent cerium mineral in the fissures in granite-like rocks, whereas monazite prevails in mica schists [15]. The first find of synchysite at the Piz Blas and Piz Rondadura, Val Nalps, Graubünden, Switzerland [16], was followed by other finds in this valley; together with gadolinite and monazite [17, pp. 134, 136, 141, 151/3] synchysite was found in the Val Tavetsch, at the top of La Bianca [18], and at the Oberaar power station in the Grimsel area, Switzerland [19]. Synchysite further occurs at the Cina Zianca, Graubünden, at the Furka base tunnel, Uri/Wallis, at Tête Noir and at the Fiescher glacier, Wallis [20]. Together with anatase it has been found in the Aeginental, Goms, Wallis [21], and at the Dent de Salantin, Lower Wallis [22]. In Ticino, Switzerland, synchysite has been found at Fibbia [20], in 14 different parageneses in the Rotondo granite at Ronco, Val Bedretto [23], and, together with anatase and brookite, at Lodrino near Biasca [15]. Several finds in the Aare granite are not localized [17, p. 151]. In Italy synchysite together with anatase and brookite has been observed at Beura, Ossola, Piemonte [24]. In the region of Salzburg, Austria, synchysite occurs together with monazite at Bodenhaus, Rauris [25], as a nearly common mineral, and at Böckstein, Gasteiner Tal, in fissures and vugs in pegmatitic granite gneiss [26]; recently it has been found in cavernous gneiss on the Hopffeldboden, Obersulzbachtal [27].

In southern Norway synchysite has been found at several localities, always associated with anatase: a more or less close relation to the similar parageneses in Alpine fissure veins is indicated for synchysite of the Kongsberg Ag deposit [28]; in the Drammen biotite granite west of the Oslofjord, besides anatase brookite also accompanies synchysite, which is embedded in calcite in miarolitic cavities and in pegmatitic druses; both, a formation of synchysite as a hydrothermal mineral directly from solutions (in Kongsberg) or by decomposition of titanite by CO_2- and F-bearing siliceous solutions (in the Drammen granite) seems to be possible [29, pp. 232, 235/6, 238], cf. [30]. Polycrystals of synchysite-parisite, together with anatase, occur in cavities of rhomb porphyries near Sandungen, Asker, and a similar paragenesis has been found in the rhomb porphyry flow of the Holmestrand area, southern Norway [29, p. 237].

Carbonatites. In the Tomtor carbonatite massif, northern Siberian Platform, accessory synchysite is closely intergrown with carbonates and Fe hydroxides and associated with pyrochlore und ilmenorutile; it has been deposited after calcite and before ankerite [31]. In the complex carbonatite core of the Khibina massif, Kola, synchysite closely associated with parisite occurs together with calcite, strontianite, analcime, pyrite, ancylite, and huanghoite in veinlets of natrolite (hydronepheline) and of calcite, which transect biotitized basic rocks and altered foyaites [32]; the association represents the fifth stage of carbonatite formation [33]; cf. p. 139. In carbonatites on Chilwa Island, Lake Chilwa, southern Malawi, synchysite occurs together with quartz in sövite, in altered dikes of nephelinitic composition, in quartz veinlets transecting quartz fluorite veins, and in the quartz of these veins [34], cf. synchysite included in quartz and calcite or in druses and cavities of white carbonatites, in younger, lighter bands in carbonatite, and in dark carbonatite dikes [35]. In the Tundulu carbonatite ring complex, southern Malawi, synchysite occurs at the Tundulu Hill in several carbonatites, in a dark brown fluorite carbonate assemblage [35], in silicate sövite and in apatite sövite dikes; at the Kamilala Hill it has been found in siderite sövite, closely associated with a mineral supposed to be cordylite; at the Nathace Hill (to the southwest of Tundulu Hill) synchysite mainly occurs in apatite-poor bastnaesite carbonatite dikes of the intermediate and outer ring structure, associated with bastnaesite and parisite, and together with quartz, apatite, florencite, pyrochlore, pyrite, and anatase. Locally synchysite is very abundant and reaches about 12 wt% in one calcite-siderite-synchysite rock [36].

At Glenover, Thabazimbi district, Transvaal, South Africa, synchysite is a prominent constituent of mineralized quartz-synchysite veinlets of a late stage occurring as stringers in the carbonatite; synchysite is enclosed by fluorite, quartz, and dolomite. In discontinuous carbonatite veinlets accessory synchysite is accompanied by apatite, hematite, and deposited after magnetite, phlogopite, pyrite, quartz, barite, fluorite, Na-hornblende, and late calcite. Synchysite is very rare in beforsite, and only in one type of calcite-beforsite do scarlet blebs consist of synchysite, calcite, barite, and ferric hydrates [37], cf. [38].

In a metasomatically formed, hematite-bearing carbonate rock of the Fen area, Telemark, southern Norway, synchysite is associated with anatase [29, p. 237].

Sedimentary and metamorphic rocks. Minute amounts of synchysite have been identified by X-rays in calcite veinlets and joints cutting dark gray shale and siltstone beds within the pyrite zone of the Nonesuch shale in northern Michigan [39]. A single crystal of synchysite has been found in a talc quarry of Trimouns at Luzenac, Département Ariège, France [40].

Synchysite-(Nd) and Synchysite-(Y)

The occurrence of synchysite-(Nd) is characterized by a secondary authigenic formation or by the replacement of calciogadolinite in pegmatitic environment. Synchysite-(Y) predominantly occurs in pegmatites and metasomatites, often also as a replacement product of older RE- and/or Y-bearing minerals. See "alteration" p. 161.

In the Upper Cretaceous uranium sandstone-type deposits of Holičky near Česka Lipa, ČSSR, the Nd-dominant member of the synchysite group, synchysite-(Nd), has recently been found as a very rare authigenic mineral in the cement of the coarse to medium-grained sandstone, associated with florencite-(La), sphalerite, manganoan siderite, pyrite, kaolinite, chalcedony, and an insufficiently identified Al-RE-arsenate [41]. In the karstic bauxite deposit of Grebnik, Yugoslavia, synchysite-(Nd) has been found by back-scattered electron images; the mineral occurs just above the footwall limestone as secondary space fillings or micropore fillings in the cement; it has been formed during the bauxitization process by removal of RE and Y from clay particles "per descensum" and has been sedimentarily deposited as synchysite-(Nd) [42, 43]. In the Pyörönmaa pegmatite at Kangasala, southwestern Finland, synchysite-(Nd) occurs among several Y-minerals [44].

Synchysite-(Y) (doverite) was first found in the Scrub Oaks iron mine at Dover, Morris County, New Jersey, where it occurs in brownish red aggregates, mixed with xenotime, hematite, and quartz, sometimes rimmed by bastnaesite, and further associated with magnetite, apatite, albite, zircon, and rutile in a pegmatitic phase [45], cf. [46]. A red mineral in late veins of this occurrence is probably a late generation of synchysite-(Y), deposited from carbonate solutions [47]. For a description of the Scrub Oaks mine see "Rare Earth Elements" A 8, 1984, Chapter 11.7.2.3.2.

In the White Cloud pegmatite near South Platte, Jefferson County, Colorado, synchysite-(Y) is the most abundant product of a late stage fluoride and fluorocarbonate mineralization, replacing primary pegmatitic minerals; the mineral associated with xenotime und cyrtolite has also been observed as one of the pegmatite minerals in the Big Bertha feldspar mine west of Long Scraggy Peak, Jefferson County [13], cf. for these two occurrences [48]. In the Henry pegmatite at Cotopaxi, Fremont County, Colorado, synchysite-(Y) occurs intimately associated and to some extent intergrown with kainosite (cenosite) in a replacement unit of the pegmatite [49]; besides quartz, microcline, and other feldspars this pegmatite contains other RE minerals; synchysite-(Y) especially occurs in purple fluorite chiefly replacing microcline [50]. A similar occurrence of synchysite-(Y) has been found as inclusions in zonal fluorite and associated with rose microcline in an Ukrainian pegmatite; it has been formed in a late stage of the pegmatitic process at 460 to 490°C [51] and amounts to 8 to 25% in the core of the fluorite in vugs of the pegmatite [52]. Synchysite-(Y) has been further found together with bastnaesite, thorite, and zircon in a radioactive pegmatitic zone in mixed pegmatite-aplite rocks on the northern side of the alkali granite of the Jabal Sayid complex, Saudi Arabia [53], and rarely in the Evans-Lou Pegmatite, Quebec [54].

Synchysite-(Y), besides synchysite, is a typical accessory mineral of the alkali-free metasomatic or pegmatitic quartz-feldspar veins in the exocontact of the apogranitic Bol'shoi Vykhod massif in the northwestern Tarbagatai, Kazakhstan [55], cf. synchysite-(Y) in alkali granites, their pegmatites and metasomatites of Kazakhstan und of Transbaikalia, especially in metasomatically altered granophyres, which are genetically bound with alaskitic granite massifs [56]. The mineral X from northern Kirgizia supposed to be synchysite-(Y) occurs in late stage hydrothermal and greisen deposits [57], cf. [51].

Alteration and Inclusions

Synchysite is one of the **alteration products** of allanite, its deposition depends on the original composition of the allanite and of the solutions. During the alteration of allanite (cf. "bastnaesitization" p. 103) in radioactive granites of the Canadian Shield, synchysite under alkaline conditions uses the Ca^{2+} ions available at the breakdown of the allanite structure and forms larger domains than bastnaesite within the altered allanite, or it tends to migrate towards the grain boundary and into fractures around the altered allanite [58]. Synchysite replaces monazite in pegmatites on the southeastern shore of Shatford Lake, southeastern Manitoba [59], quartz in veinlets in carbonatite of Glenover, Thabazimbi district, Transvaal, South Africa [37], and quartz at the edge of altered allanite at Sharbot Lake, Ontario [58]. Further, synchysite is given as the end product of the alteration series gagarinite → synchysite-(Y) → synchysite in metasomatic alkali granitic rocks of Kazakhstan. The pseudomorphous replacement of gagarinite by synchysite-(Y) and by bastnaesite-(Y) is common, but a synchysitization of fluorosilicates only rarely occurs [55].

In another series synchysite-(Y) is the final product, occurring together with tengerite and thorogummite in the series hellandite → kainosite I → synchysite-(Y) in the Evans-Lou pegmatite, Quebec [54]. Fine-grained synchysite-(Y) together with hematite and sericite replaces kainosite at Cotopaxi, Fremont County, Colorado [50]. The mineral X (thought to be synchysite-(Y)) of a hydrothermal greisen deposit in northern Kirgizia pseudomorphously replaces monazite [57], cf. [49]. In the Scrub Oaks mine, Morris County, New Jersey, synchysite-(Y) locally replaces albite and oligoclase [47].

Synchysite-(Nd) is a replacement product of calcio-gadolinite in the Pyörönmaa pegmatite, SW Finland [44].

Synchysite itself is replaced by parisite, see p. 140, synchysite-(Y) is replaced by synchysite, see the series above.

Synchysite-(Nd) of northern Bohemia frequently contains unidentified **inclusions** and intergrowths [41]; in synchysite-(Y) of the Scrub Oaks mine, New Jersey, inclusions of hematite have been observed [47].

References for 10.5.2.4.2:

[1] C. Palache, H. Berman, C. Frondel (The System of Mineralogy, 7th Ed., Vol. 2, Wiley, New York 1951, pp. 1/1124, 287/8). – [2] E. I. Semenov (in: K. A. Vlasov, Geochemistry and Mineralogy of Rare Elements and Genetic Types of Their Deposits, Vol. 2: Mineralogy of Rare Elements, Israel Program Sci. Trans., Jerusalem 1966, pp. 249/328, 274/5 [Russian Original, Moscow 1964]). – [3] Yu. L. Kapustin (Mineralogiya Karbonatitov, Moscow 1971, pp. 1/288, 186). – [4] V. A. Chashchukhina (in: P. Ya. Yarosh, Mineraly Gornykh Porod i Rud Urala, Sverdlovsk 1980, pp. 1/149, 58/60 from Ref. Zh. Geol. **1981** No. 12V185). – [5] F. Emiliani, G. Gandolfi (Mineral. Petrogr. Acta **11** [1965] 123/31, 127, 129).

[6] I. Oftedal (Z. Krist. **79** [1931] 437/64, 443). – [7] G. Donnay, J. D. H. Donnay (Am. Mineralogist **39** [1953] 932/63). – [8] O. B. Bøggild (Medd. Groenland **149** No. 3 [1953] 1/442, 155/7). – [9] G. Donnay (Am. Mineralogist **38** [1953] 868/70). – [10] L. Marble, A. Regis (Rocks Minerals **54** [1979] 4/25, 13).

[11] C. A. Strauss (S. Africa Geol. Surv. Mem. No. 46 [1954] 1/241, 117). – [12] M. H. Staatz (Econ. Geol. **69** [1974] 494/507, 498). – [13] E. B. Eckel (U.S. Geol. Surv. Bull. No. 1114 [1961] 1/399, 127, 318). – [14] Anonymous (Mineralienfreund **14** [1976] 89/120, 116). – [15] M. Gramaccioli (Mineral. Rec. **8** [1977] 287/93, 288).

[16] R. L. Parker, F. de Quervain, F. Weber (Schweiz. Mineral. Petrogr. Mitt. **19** [1939] 293/306, 295). – [17] A. Kipfer (Schweiz. Strahler **3** [1973] 133/69). – [18] M. Weibel (Schweiz. Mineral. Petrogr. Mitt. **44** [1964] 133/5). – [19] Y. Iitaka, H. A. Stalder (Schweiz. Mineral. Petrogr. Mitt. **41** [1961] 485/8). – [20] Anonymous (Mineralienfreund **15** No. 2 [1977] 29/31).

[21] A. Kipfer (Schweiz. Strahler **5** [1981] 397/417, 400, 405). – [22] A. Frey (Schweiz. Strahler **4** [1977] 205/7). – [23] A. Kipfer (Schweiz. Strahler **5** [1979] 45/81, 67). – [24] E. Huen, C. M. Gramaccioli (Natura [Milan] **60** [1969] 193/4). – [25] K. Kontrus, G. Niedermayr (Tschermaks Mineral. Petrogr. Mitt. [3] **13** [1969] 355/9).

[26] H. Weninger (Lapis **4** No. 10 [1979] 26/8). – [27] K. Schebesta (Lapis **7** No. 1 [1982] 9/20, 9/12). – [28] P. C. Sæbø, H. Neumann (Norsk Geol. Tidsskr. **41** [1961] 247/54, 247, 252). – [29] G. Raade (Norsk Geol. Tidsskr. **49** [1969] 227/39). – [30] G. Raade (Mineral. Rec. **3** No. 1 [1972] 7/11, 8/9).

[31] I. N. Kuznetsova, E. L. Rozinova, K. S. Mishchenko (in: L. S. Egorov, Shchelochnoi Magmatizm i Apatitonosnost Severa Sibiri, Leningrad 1980, pp. 101/11; Ref. Zh. Geol. **1981** No. 7V260). – [32] O. B. Dudkin, F. V. Minakov (in: I. V. Bel'kov, Materialy po Mineralogii i Geokhimii Shchelochnykh Kompleksov Porod Kol'skogo Poluostrova, Apatity 1975, pp. 159/74, 162/3, 169). – [33] O. B. Dudkin, L. I. Polezhaeva, F. V. Minakov (in: O. B. Dudkin et al., Veshchestvennyi Sostav Shchelochnykh Intrusivnykh Kompleksov Kol'skogo Poluostrova, Apatity 1981, pp. 73/81, 78; Ref. Zh. Geol. **1981** No. 12V179). – [34] M. S. Garson, W. C. Smith (Nyasaland Geol. Surv. Dept. Mem. No. 1 [1957] 1/127, 40/1, 53, 73, 83). – [35] W. C. Smith (Bull. Brit. Mus. [Nat. Hist.] Mineral. **1** [1950/64] 97/119, 105/7, 110/6).

[36] M. S. Garson (Nyasaland Geol. Surv. Dept. Mem. No. 2 [1962] 1/248, 84/5, 88/9, 100/3, 108/9). – [37] W. J. Verwoerd (S. Africa Geol. Surv. Handbook **6** [1967] 1/452, 49, 59/61). – [38] W. J. Verwoerd (S. Africa Geol. Surv. Annual Rept. **2** [1963] 119/35, 122). – [39] W. S. White, J. C. Wright (Econ. Geol. **61** [1966] 1171/90, 1178). – [40] Y. Moëlo, J.-P. Choutier, E. Melon (from R. Pierrot, Bull. Soc. Franc. Mineral. Crist. **97** [1974] 521/4, 523).

[41] B. Scharm, P. Kühn (Neues Jahrb. Mineral. Monatsh. **1983** 201/10, 202/4). – [42] Z. Maksimović, G. Pantó (4th Intern. Congr. Study Bauxites Alumina Alum. Preprints, Athens 1978, Vol. 2, pp. 540/52, 540; Mineral. Abstr. **31** [1980] No. 80-2783). – [43] Z. Maksimović, G. Pantó (Acta Geol. Acad. Sci. Hung. **24** [1981] 217/22, 219, 221). – [44] A. Vorma, P. Ojanperä, V. Hoffrén, J. Siivola, A. Löfgren (Bull. Comm. Geol. Finlande No. 222 [1966] 241/74, 241, 245, 257). – [45] W. L. Smith, J. Stone, D. D. Riska, H. Levine (Science **122** [1955] 31).

[46] W. L. Smith, J. Stone, D. R. Ross, H. Levine (Am. Mineralogist **45** [1960] 92/8, 92/3). – [47] H. Klemic, A. V. Heyl, A. R. Taylor, J. Stone (U.S. Geol. Surv. Bull. No. 1082-B [1959] 29/60, 47, 50/2). – [48] C. V. Haynes (in: S. M. Del Rio, Mineral Resources of Colorado, First Sequel, Mineral Resources Board, Denver, Colo., 1960, pp. 370/85, 376/7). – [49] A. A. Levinson, R. A. Borup (Am. Mineralogist **47** [1962] 337/43, 338/40). – [50] E. W. Heinrich, R. A. Borup, C. A. Salotti (Am. Mineralogist **47** [1962] 328/36, 329, 331/2).

[51] E. P. Gurov, E. P. Gurova (Tr. Mineral. Muzeya Akad. Nauk SSSR **24** [1975] 171/4). – [52] E. P. Gurov, E. P. Gurova, L. G. Loginova, Yu. A. Lavitskaya (in: P. M. Tatarinov et al., Mineraly i Paragenezisy Mineralov Metasomaticheskikh i Metamorficheskikh Gornykh Porod, Leningrad 1975, pp. 106/14, 107; C. A. **84** [1976] No. 108517). – [53] A. R. A. Turkistany, K. Koyama (IAG Bull. No. 3 [1980] 99/104, 99/100). – [54] D. D. Hogarth (Mineral. Rec. **3** [1972] 69/77, 73, 75/7). – [55] D. A. Mineev (Geokhimiya Apogranitov i Redkometal'nykh Metasomatitov Severo-Zapadnogo Tarbagataya, Moscow 1968, pp. 1/183, 34, 85, 96, 130).

[56] N. I. Aleksandrova, I. V. Maksimova, N. M. Svirskaya (Izv. Vysshikh Uchebn. Zavedenii Geol. Razvedka **20** No. 10 [1977] 80/7, 81/2). – [57] E. I. Semenov (Tr. Inst. Mineral. Geokhim. Kristallokhim. Redk. Elem. Akad. Nauk SSSR No. 2 [1959] 181/6, 181/2, 185). – [58] A. L. Littlejohn (Geol. Surv. Paper Geol. Surv. Can. No. 81-1 B [1981] 95/104, 95, 100). – [59] P. Černý, N. A. Bristol (Can. Mineralogist **11** [1971/73] 560/3).

10.5.2.4.3 Chemistry

For a compilation of **analyses** and $(RE,Y)_2O_3$ contents, mean and variance of 7 synchysites (including 4 yttrosynchysites) and 3 "doverites", see "Seltenerdelemente" A 4, 1979, pp. 30, 5 and pp. 18, 6. A new analysis of a synchysite from the Khibina massif, Kola, with 52.7% RE_2O_3 [1]

is in the range of these synchysites. The $(RE,Y)_2O_3$ content of a synchysite-(Nd) from the ČSSR is 50.0 wt% (41.0% RE_2O_3, 9.0% Y_2O_3) [2].

Formulas given for synchysite are $CeCa[F|(CO_3)_2]$ [3 to 5], $(Ce,La)Ca[F|(CO_3)_2]$ [6 to 8], $(Ce,La,Dy)Ca[F|(CO_3)_2]$ [9], or in the oxide form $CeFCO_3 \cdot CaCO_3$ [10] and (Ce, La, Dy)$FCO_3 \cdot CaCO_3$ [11]. The formula calculated from the new analysis of synchysite from the Khibina massif, Kola, with 15.6% CaO and 1.5% SrO results in $(Ca_{0.96}Sr_{0.04})(La_{0.42}Ce_{0.51}(Pr\ to\ Sm)_{0.12})[CO_3]_{2.02}F_{1.02}$; this is close to the theoretical composition [1].

For synchysite-(Nd) from Finland with Nd dominating and Ca, Ce, Y, and La as other major elements the formula $NdCa[F|(CO_3)_2]$ is given [12]. Synchysite-(Nd) from the Grebnik bauxite, Yugoslavia, has $Ca_{1.05}(Nd_{0.36}La_{0.29}Y_{0.10}Pr_{0.09}Sm_{0.06}Ce_{0.04}Gd_{0.04}Dy_{0.02})[CO_3]_{1.90}F_{1.04}$ [13], cf. a similar formula in [14]. Synchysite-(Nd) from northern Bohemia, ČSSR, including 0.9% U replacing Ca, yields the formula $(Ca_{1.03}U_{0.01})(Nd_{0.33}Y_{0.25}Gd_{0.14}Sm_{0.13}Dy_{0.04}Eu_{0.04}Pr_{0.03}La_{0.02}Ce_{0.01}Ho_{0.01})[F_{1.13}|(CO_3)_{1.94}]$, which may be idealized to Ca(Nd, Y, Gd...)$[F|(CO_3)_2]$ [2].

The general formula of synchysite-(Y) is assumed to be $YCa[F|(CO_3)_2]$ [5, 15, 16], (Y, Ce) $Ca[F|(CO_3)_2]$ [7, 17], or in the oxide form $YFCO_3 \cdot CaCO_3$ [18, 19], cf. [10, 17, 20].

The theoretical content of CaO is 17.60% in synchysites with Ce : La = 1 : 1 [6] and 17.62% for the pure Ce member [17, 21]; the CaO contents of two minerals from Greenland are 17.13 and 16.63% [9, 6], those of synchysite-(Nd) from Yugoslavia, 18.2% [13], and from the ČSSR, 18.3% [2]. The theoretical CaO value of synchysite-(Y) with Y : Dy = 2 : 1 is 19.2% [15]. Four samples of synchysite-(Y) from the USSR with 13.6 to 20.05% CaO show a relative broad range [22]. The mineral from Kirgizia thought to be synchysite-(Y) contains 17.12% CaO besides 0.28% SrO [17], cf. the low Ca-value and a marked SrO-content for the mineral from Kola, above, and spectrographically recorded values of Sr und Ba besides Mn, Fe, and Zn in synchysite from a Siberian carbonatite massif [4]. The mineral from Kirgizia contains 0.94% ThO_2 [17].

In synchysites the content of F, theoretically 5.96% (with Ce : La = 1 : 1) [6] or 5.98% (for Ce only) [17], is often somewhat low, e.g., in two samples from Greenland 5.82 and 5.04% F and 2.10% H_2O for the second sample [6, 9]. For synchysite-(Y) from Cotopaxi, Colorado, the recalculated analysis also shows a relatively low value of 5.7% F compared with the theoretical value of 6.5% F for $YCaF(CO_3)_2$ with Y : Dy = 2 : 1; the ratio Ca : F = 1 : 0.9 [15].

RE composition. Common synchysites are strongly Ce selective minerals [7], cf. [5]. Synchysite from the Hopffeldboden, Salzburg, Austria, contains besides Ce relatively high amounts of Sm and Nd [23], synchysite from Trimouns, France, besides Ce, La, and Nd has subordinate amounts of Gd, Pr, Sm, and Dy [24]; an elevated ratio La/Ce in synchysite is, in most cases, connected with a higher H_2O content in the mineral [25].

Synchysite-(Nd) of Finland is characterized by Nd prevailing over Ce, Y, and La [12], cf. the predominance of Nd over the other Ce-earth in two samples of synchysite-(Nd) in Table 5, p. 165. Synchysite-(Y) from Colorado, after elimination of 10.8% kainosite and 1.0% Fe_2O_3, contains 24% Y_2O_3 and (by difference) 22.1% remaining RE_2O_3; this corresponds to an atomic ratio Y : RE $\approx$ 2 : 1 [15].

For an "yttrium-bearing" synchysite from Predazzo, Italy, with remarkable quantities of La and Y besides Ce, see p. 157.

Table 4, p. 164, shows the RE distribution in at.% together with Σ (La + Ce + Pr) and the ratios between Y and RE for synchysite and synchysite-(Y). The means of the few analyses show a less marked increase in light lanthanides from granites and granite pegmatites to hydrothermal rocks for synchysite-(Y) compared with those of bastnaesites and parisites (cf. pp. 111 and 144, respectively) [7].

Table 4

At.% of La and Lanthanides in Synchysites and Synchysites-(Y) from Various Rocks from [7].

Genetic group	La	Ce	Pr	Nd	Sm	Eu	Gd	Tb	Dy	Ho	Er	Tm	Yb	Lu	$\frac{100Y}{Y+RE}$	Σ(La + Ce + Pr)	La/Nd
Synchysites																	
Alkaline rocks (2)[1]	20.5	41.7	6.0	22.8	3.7	0.5	2.5	0.3	0.9	0.2	0.5	—	0.4	—	7.2(1)[2]	68.2	0.90
Synchysites-(Y)																	
Granites and granite pegmatites (6)	3.7	9.5	2.0	6.6	4.4	0.8	11.3	1.7	18.9	3.8	18.2	1.4	16.0	1.7	71.5(1)	15.2	0.56
Hydrothermal occurrences (2)	16.2	38.3	5.3	18.3	5.4	0.1	4.6	0.6	4.7	0.9	2.5	0.3	2.3	0.5	39.3(2)	59.8	0.89

[1] Numbers in parentheses indicate the number of analyses. – [2] Number of analyses, in which Y is reported.

Table 5
Distribution of Rare Earth Elements in Synchysites.

No.	Locality	La	Ce	Pr	Nd	Sm	Eu	Gd	Tb	Dy	Ho	Er	Tm	Yb	Lu	Y	Ref.
	Common synchysites																
*1	Narsarsuk, Greenland	20.0	41.0	5.2	20.0	3.5	0.4	2.6	0.4	0.7	0.2	0.4	—	0.6	—	5.0[1)]	[26]
2	Khibina massif, Kola	23.0	49.8[2)]	5.8	14.4	2.0	—	1.8[3)]	—	3.2[4)]	—	—	—	—	—	—	[1]
3	Tomtor, north Siberia[5)]	31.1	47.8	3.6	15.5	1.1	—	0.2	—	—	—	—	—	—	—	0.7	[27]
4	Siberia	19	39	6.5	25	4.0	0.6	2.7	0.2	1.5	0.1	0.7	—	—	—	—	[28]
	Synchysites-(Nd)																
5	Grebnik, Yugoslavia	29.5	4.6	8.4	36.1	6.8	—	4.8	—	2.7	—	—	—	—	—	7.2	[13]
6	Česká Lipa, ČSSR	1.8	1.4	3.6	35.2	13.8	4.6	15.6	—	5.0	0.6	—	—	—	0.4	18	[2]
	Synchysites-(Y)																
7	Ukraine[6)]	14.4	25.8	3.0	15.1	3.5	—	3.0	—	2.6	0.7	2.2	0.4	3.0	0.5	25.8	[29]
*8	Bol'shoi Vykhod, Kazakhstan	2.6	8.4	2.6	8.2	7.4	0.3	16.4	2.1	18.7	4.0	13.2	1.6	12.4	2.1	—	[30]
*9		0.4	1.6	0.2	0.8	4.1	3.1	11.2	2.6	22.4	6.3	25.1	2.1	17.1	3.6	—	[30]
*10		9.2	22.1	4.2	11.7	6.0	—	6.4	2.4	12.3	3.5	10.0	1.6	9.1	1.5	—	[30]
*11		4.8	19.1	2.9	14.3	3.8	—	9.5	1.0	11.9	2.9	11.8	1.3	16.7	—	—	[30]
*12	Kazakhstan	2.6	0.4	1.1	1.9	3.8	0.8	8.6	1.9	21.0	7.1	24.8	2.6	19.2	4.1	—	[31]
*13	Kirgizia, USSR	8.0	27.0	4.2	11.0	4.5	0.2	4.0	1.0	4.8	0.8	2.0	0.1	1.0	0.3	30[1)]	[17]
*14		11.4	22.8	1.1	19.0	5.7	0.4	1.9	1.1	9.5	1.5	2.7	0.8	2.7	0.4	18	[32]
15	Middle Asia, USSR	9.5	19.1	1.0	15.9	3.5	<0.3	3.2	1.0	4.5	1.3	3.2	0.3	1.9	0.3	35	[33]

[1)] Y calculated by difference. – [2)] CeO_2. – [3)] Gd + Eu. – [4)] Dy + Tb + Y. – [5)] Synchysite concentrate. – [6)] Intergrowths of synchysite-(Y) with fluorite with 90% of ΣRE in synchysite-(Y).

In Table 5, p. 165, the composition of RE ± Y of common synchysite, synchysite-(Nd) and synchysite-(Y) from various rocks and of different genesis is given as $(RE \pm Y)_2O_3 = 100$ or as RE ± Y = 100 (numbers with an asterisk). The numbers in the table refer to: Nos. 1 to 4 synchysites of alkali rocks; Nos. 5 and 6 authigenic synchysites-(Nd) from bauxite (5) and uranium sandstone (6); Nos. 7 to 15 synchysites-(Y) from pegmatite (7), from metasomatic alkali rocks, most of them replacing gagarinite (8 to 12), from a hydrothermal greisen deposit in granite (13), of a metasomatically altered granophyr (14), and of a nonspecified deposit (15).

References for 10.5.2.4.3:

[1] O. B. Dudkin, L. I. Polezhaeva, F. V. Minakov (in: O. B. Dudkin et al., Veshchestvennyi Sostav Shchelochnykh Intrusivnykh Kompleksov Kol'skogo Poluostrova, Apatity 1981, pp. 73/81, 76, 80; Ref. Zh. Geol. **1981** No. 12V179). – [2] B. Scharm, P. Kühn (Neues Jahrb. Mineral. Monatsh. **1983** 201/10, 206). – [3] H. Strunz, C. Tennyson (Mineralogische Tabellen, 6th Ed., Geest and Portig, Leipzig 1977, pp. 1/621, 242). – [4] Yu. L. Kapustin (Mineralogiya Karbonatitov, Moscow 1971, pp. 1/288, 186). – [5] E. I. Semenov (in: K. A. Vlasov, Geochemistry and Mineralogy of Rare Elements and Genetic Types of Their Deposits, Vol. 2: Mineralogy of Rare Elements, Israel Program Sci. Trans., Jerusalem 1966, pp. 249/328, 273, 275 [Russian Original, Moscow 1964]).

[6] C. Palache, H. Berman, C. Frondel (The System of Mineralogy, 7th Ed., Vol. 2, Wiley, New York 1951, pp. 1/1124, 287/8). – [7] M. Fleischer (Can. Mineralogist **16** [1978] 361/3). – [8] L. Marble, A. Regis (Rocks Minerals **54** [1979] 4/25, 13). – [9] O. B. Bøggild (Medd. Groenland **149** No. 3 [1953] 1/442, 155, 157). – [10] J. D. H. Donnay, G. Donnay (Bull. Soc. Franc. Mineral. Crist. **84** [1961] 25/9).

[11] M. S. Garson (Nyasaland Geol. Surv. Dept. Mem. No. 2 [1962] 1/248, 109). – [12] A. Vorma, P. Ojanperä, V. Hoffrén, J. Siivola, A. Löfgren (Bull. Comm. Geol. Finlande No. 222 [1966] 241/74, 257). – [13] Z. Maksimović, G. Pantó (Acta Geol. Acad. Sci. Hung. **24** [1981] 217/22, 220). – [14] Z. Maksimović, G. Pantó (4th Intern. Congr. Study Bauxites Alumina Alum. Preprints, Athens 1978, Vol. 2, pp. 540/52, 544; Mineral. Abstr. **31** [1980] No. 80-2783). – [15] A. A. Levinson, R. A. Borup (Am. Mineralogist **47** [1962] 337/43, 341).

[16] A. R. A. Turkistany, K. Koyama (IAG Bull. No. **3** [1980] 99/104, 100). – [17] E. I. Semenov (Tr. Inst. Mineral. Geokhim. Kristallokhim. Redk. Elem. No. 2 [1959] 181/6, 181, 184/5). – [18] W. L. Smith, J. Stone, D. D. Riska, H. Levine (Science **122** [1955] 31). – [19] W. L. Smith, J. Stone, D. R. Ross, H. Levine (Am. Mineralogist **45** [1960] 92/8, 94). – [20] E. B. Eckel (U.S. Geol. Surv. Bull. No. 1114 [1961] 1/399, 127).

[21] P'ei-Shan Chang [Pei-Shan' Chzhan] (Sci. Sinica [Peking] **11** [1962] 251/8, 252; C.A. **57** [1962] 10799). – [22] V. V. Delyageid, S. V. Yakovleva, R. N. Morzheedova, T. V. Shvedova, G. D. Evlanova, R. V. Kotlyarov, Z. M. Grigor'eva (Nauchn. Tr. Gos. Nauchn. Issled. Proektn. Inst. Redkometal. Prom. **45** [1972] 75/8; C.A. **85** [1976] No. 163454). – [23] K. Schebesta (Lapis **7** No. 1 [1982] 9/20, 16). – [24] Y. Moëlo, J.-P. Choutier, E. Melon (from R. Pierrot, Bull. Soc. Franc. Mineral. Crist. **97** [1974] 521/4, 523). – [25] N. V. Svyazhin (Tr. Inst. Geol. Akad. Nauk SSSR Ural'sk. Filial No. **70** [1965] 249/52).

[26] E. I. Semenov (Mineralogiya Redkikh Zemel', Moscow 1963, pp. 1/412, 106). – [27] I. N. Kuznetsova, E. L. Rozinova, K. S. Mishchenko (in: L. S. Egorov, Shchelochnoi Magmatizm i Apatitonosnost Severa Sibiri, Leningrad 1980, pp. 101/11, 104; Ref. Zh. Geol. **1981** No. 7V260). – [28] Yu. L. Kapustin (in: L. N. Ovchinnikov, N. A. Solodov, Mestorozhdeniya Litofil'nykh Redkikh Metallov, Nedra, Moscow 1980, pp. 243/85, 268). – [29] E. P. Gurov, E. P. Gurova (Tr. Mineral. Muzeya Akad. Nauk SSSR No. 24 [1975] 171/4). – [30] D. A. Mineev (Geokhimiya Apogranitov i Redkometal'nykh Metasomatitov Severo-Zapadnogo Tarbagataya, Moscow 1968, pp. 1/183, 126).

[31] D. A. Mineev [Mineyev] (Geokhimiya **1963** 1082/100; Geochemistry [USSR] **1963** 1129/49, 1138, 1140). – [32] N. I. Aleksandrova, I. V. Maksimova, N. M. Svirskaya (Izv. Vysshikh Uchebn. Zavedenii Geol. Razvedka **20** No. 10 [1977] 80/7, 84). – [33] I. T. Aleksandrova, G. A. Sidorenko (Rentgenogr. Mineral'n. Syr'ya No. 11 [1977] 19/26, 20; Ref. Zh. Geol. **1977** No. 7V484).

10.5.2.4.4 Crystal Forms and Crystal Structure

Morphology

Symmetry and forms. Although synchysite is markedly pseudohexagonal, X-ray data indicate a lower symmetry than hexagonal, either orthorhombic C-centered or monoclinic with an angle of 120°. The morphological observations in the literature ([1])*), can be interpreted as due to orthorhombic symmetry [2, pp. 938, 951, 957]. The rare occurrence of pyramidal crystals indicates the absence of a center of symmetry and the presence of a polar axis [3].

The morphology of synchysite is controlled by the pseudocell a′c″ (cf. p. 90) with an axial ratio $c''/a' = 2.223$; the orthorhombic axial ratio $a:b:c = a':a:c'' = 1/\sqrt{3}:1:2.223/\sqrt{3} = 0.577:1:1.283$ [2, p. 957].

Crystal forms of synchysite are generally given with hexagonal indices; compilations of forms and figures of crystals see [2, pp. 953/5, 957], [4, 5]. Besides the dominant base c $\{0001\}$ and the prisms m $\{10\bar{1}0\}$ and/or $\{11\bar{2}0\}$, synchysite crystals show a great number of rhombohedrons, which, according to the apparent orthorhombic symmetry of the mineral, are more probably dipyramids [2, p. 957]. Investigations of crystals from Alpine fissure veins have confirmed the existence of $\{10\bar{1}1\}$, $\{11\bar{2}1\}$, $\{60\bar{6}1\}$, and $\{22\bar{4}1\}$ besides c and m [6] (cf. [7]), and of $\{33\bar{6}1\}$ [8]; for figures of ten crystals see [9]. Sometimes the base is developed only on one side of the crystal [7].

The **habit** of synchysite varies between acute rhombohedral with a small base and thick tabular $\{0001\}$ [4], cf. [10, 11]; this is typical also for synchysites from Alpine fissure veins, where the crystals are flat prismatic to acute rhombohedric and often spindle-shaped, sometimes acute pyramidal without a base [7], cf. spindle-shaped hexagonal-bipyramidal habit [12], sometimes slender to short prismatic crystals conically tapering [13], spatulate forms [9], and spindles without any development of a base [8]. These are subordinate compared with short prismatic crystals with a base [14]. Synchysite in Alpine fissure veins also occurs in hexagonal prismatic crystals [15] and more rarely in pyramidal crystals [3]. In other occurrences prismatic crystals are terminated by very acute dipyramids [16]; or small prisms are only slightly modified by pyramidal faces [17]; hexagonal prismatic crystals have rounded faces [18]; flat plates show well developed prisms and pyramids [19], or consist of a packing of fine lamellas [20]. Flat, thin tabular or platy hexagonal crystals are common or even typical in carbonatites; in thin section they occur as laths or fibers, especially in carbonatites of Africa [21 to 25] and of Kola, USSR [26]. Distorted crystals [8], skeletal development (on large crystals) [5], and grains (in granites) [16] have been observed; a step-like edifice of synchysite crystals is due to alternating "rhombohedrons" of different steepness [7, 8, 27].

Synchysite often occurs aggregated to subparallel groups, radiating bundles, rosettes, and fibrous bunches of crystals [24, 25], to semiradiate sheaves, bundles, and clusters [22], to tufts of acicular crystals or of radiating needles and to bunches of radiating prisms in druses [21]; cf. sheaves and bunches of thin plates and blades and other small crystals mostly in druses and drusy cavities [23]. There also occur spidery aggregates [7, 8], cellular aggregates of plates [28],

*) Fig. 3 in [1] is identified to be synchysite by [2, pp. 951, 957].

felty aggregates of needle-shaped or fine lamellar crystals [29], densely matted fibers [25], crystal lawns [14], and disordered groups [9].

Synchysite-(Nd) occurs in "lepispheres" or groups of intergrown "lepispheres", consisting of blade-like crystals or very thin platelets [30]. Synchysite-(Y) forms prismatic, barrel-shaped or spherical crystals with hexagonal cross section or rounded form, some crystals show asymmetric dendritic or branching form [31]; at Dover fine crystalline mixtures in the form of platy aggregates or in irregular, rounded to nodular masses contain synchysite-(Y) together with other minerals [32].

Synchysites show a horizontal **striation** on lateral faces [4, 10]; it occurs on prism faces of synchysites of Alpine fissure veins [6, 33], due to a step-like formation parallel to the base [3], cf. [9].

Twinning of synchysite on {0001} is common and may be lamellar [4]. An internal twinning of 4-layer synchysite blocks is strongly suggested by the observation of different contrasts exhibited by successive pairs of synchysite slabs in dark-field electron microscope images [34].

Cleavage and a separation parallel to {0001} have been observed [10]; both are due to alteration [4]. Fresh synchysite shows no cleavage parallel to the base [5], cf. no preferred cleavage [17], or an indistinct one on $\{10\bar{1}0\}$ [35]. Synchysite-(Y) has a cleavage on {0001} [31], for synchysite-(Nd) a cleavage has not been observed [30].

Syntaxic **intergrowths** between the end members of the bastnaesite series, bastnaesite and synchysite, which have not been oberserved by the thorough investigation of [2, pp. 939/42, 957/8], have been found by X-ray powder patterns of synchysites from Alpine fissure veins [12]; they also occur between synchysite and another member of the bastnaesite series in these veins [7] and between synchysite and parisite in the Khibina massif, Kola [26]. Electron microscope images show a perfect syntaxy of synchysite with parisite and roentgenite in layers with a thickness of only ~100 Å for each mineral [34]. In a synchysite crystal included between roentgenite tips, the a axes of the minerals are parallel [2, pp. 951, 958].

Intergrowths of synchysite-(Y) with fluorite show coincidence of (001) of synchysite-(Y) with (111) of the surrounding fluorite; the synchysite-(Y) crystals are arranged perpendicularly to the growth zone of the fluorite, with deviations of 3° to 8° for single crystals and a disordered behavior of an insignificant number of crystals [31].

Crystal Structure

The **unit cell constants** of synchysites from several localities are given in the following table, Nos. 1 to 8 are common synchysites from: granite (1), alkali rocks (2 to 4), Alpine fissure veins (5 to 7), and a paragenesis similar to these veins (8); No. 9 is a synchysite-(Nd) from a sedimentary sandstone-type uranium deposit; Nos. 10 to 15 are synchysites-(Y) from: pegmatites and pegmatitic rocks (10 to 13), a nonspecified deposit (14) and hydrothermal greisen (15). Most values are recalculated from the subcells a′c‴ and a′c′ to the real unit cell ac ($a = a' \cdot \sqrt{3}$, $c = 3c' = 6c'' = 12c'''$); some from orthorhombic setting a_{or}, b_{or}, c_{or} ($a_{or} = a'$, $b_{or} = a$, $c_{or} = c''$). The original data of the authors are given in footnotes.

No.	Locality	a in Å	c in Å	Ref.
1	Predazzo, Italy	7.087 ±0.005	54.54[1)]	[16]
2	Narsarsuk, South Greenland	7.10[2)]	54.72[2)]	[36]
3		7.107	54.72	[2, p. 936]

No.	Locality	a in Å	c in Å	Ref.
4	Quincy, Massachusetts	7.084[3)]	54.48[3)]	[36]
5	Grimsel area, Switzerland	7.067[4)]	54.72	[12]
6	Val Nalps, Switzerland	7.103[5)]	54.6[5)]	[6]
7	Beura, Piedmont, Italy	7.119[6)]	54.48[6)]	[37]
8	Kongsberg	7.049[7)]	54.48[7)]	[17]
9	Northern Bohemia, ČSSR	6.984(5)	54.27(4)	[30]
10	Cotopaxi, Colorado	6.928[8)]	54.00[8)]	[38]
11	Dover, New Jersey	7.049[9)]	54.72[9)]	[32]
12	Saudi Arabia	6.946[10)]	54.01[10)]	[39]
13	Ukraine, USSR	6.94	54.00[11)]	[31]
14	Middle Asia, USSR	6.96[12)]	54.36[12)]	[40]
15	Kirgizia, USSR	6.946[13)]	53.88[13)]	[41]

[1)] $c' = 18.18 \pm 0.03$ Å. – [2)] $a' = 4.098 \pm 0.003$, $c''' = 4.56 \pm 0.01$ Å. – [3)] $a' = 4.090 \pm 0.005$, $c''' = 4.54 \pm 0.01$ Å. – [4)] $a' = 4.08$ Å. – [5)] $a' = 4.101$, $c' = 18.2$ Å. – [6)] $a' = 4.11$, $c' = 18.16$ Å. – [7)] $a' = 4.07 \pm 0.01$, $c''' = 4.54 \pm 0.01$ Å. – [8)] $a_{or} = 4.00$, $b_{or} = 6.92$, $c_{or} = 9.00$ Å. – [9)] $a_{or} = 4.07 \pm 0.01$, $b_{or} = 7.06 \pm 0.02$, $c_{or} = 9.12 \pm 0.02$ Å. – [10)] $a_{or} = 4.01 \pm 0.01$, $b_{or} = 6.97 \pm 0.01$, $c_{or} = 9.002 \pm 0.003$ Å. – [11)] $c'' = 9.00$ Å. – [12)] $a_{or} = 4.02 \pm 0.02$, $b_{or} = 7.02 \pm 0.02$, $c_{or} = 9.06 \pm 0.02$ Å. – [13)] $a' = 4.01 \pm 0.01$, $c''' = 4.49 \pm 0.01$ Å.

For the pseudoperiods in c direction ($c = 3c' = 6c'' = 12c'''$) of synchysites compared with the other members of the bastnaesite series, see p. 90 and Fig. 5, p. 91.

For synchysite-(Y) ("doverite" and the mineral from Kirgizia [41]) the unit cell constants are significantly lower than those of synchysite [38], cf. table pp. 168/9. A substantial difference in the b_{or} dimension between two synchysites-(Y) may be interpreted as being due to the existence of two lattices in the pseudohexagonal system, a true C-centered orthorhombic lattice and a pseudo-orthorhombic one [39]. Differences in unit cell parameters of other synchysites-(Y) have been ascribed to deviations in RE composition [11].

Synchysite-(Nd) in orthorhombic (or monoclinic) setting with $a' = 4.039(2)$, $a = b = 6.984(5)$, $c'' = 9.045(6)$ Å has the axial ratio a : b : c = 0.578 : 1 : 1.295 [30]. The space group of synchysite [2, p. 936], and of synchysite-(Nd) [30] could not be determined. Synchysite from the Grimsel area has the symmetry class D_{6h}-6/mmm [12]. Synchysite has (real unit cell) Z = 18, V = 12 × 199.5 Å [2, p. 936]; synchysite-(Nd) Z = 12 (true cell), Z = 2 (pseudocell), and V = 1531(2) and 255.1(2) Å^3, respectively [30]; cf. synchysite-(Y) (= doverite) with V = 262.1 Å^3 [32].

Structure description. The structure of synchysite is characterized by clearly separated layers of RE-F and of Ca, arranged vertically to the c axis; the distance between a RE-F layer and a Ca layer is 4.55 Å. The CO_3 groups (triangles) lie parallel to the c axis between the RE-F and Ca layers [36]. The Ce and F ions alternate at the vertices of regular hexagons and the Ca ions are placed above Ce ions in the alternating layers; the content of the ac′ cell of synchysite leads to four layers (two of Ce, two of Ca); the existence of a disordered form of synchysite is improbable, because the vertical positions of carbonate groups between adjacent Ca layers are very unlikely [2, pp. 945/6, 950]. In the plot c‴ against mol% $CaCO_3$ the observed c‴ of synchysite = 4.560 Å corresponds exactly to 50 mol% $CaCO_3$ [2, pp. 943/5], [42]. The thickness of a synchysite layer (repeat distance) is 9.12 Å [34].

References for 10.5.2.4.4:

[1] C. Palache, C. H. Warren (Am. J. Sci. [4] **31** [1911] 533/57, 537). – [2] G. Donnay, J. D. H. Donnay (Am. Mineralogist **38** [1953] 932/63). – [3] M. Gramaccioli (Mineral. Rec. **8** [1977] 287/93, 288). – [4] C. Palache, H. Berman, C. Frondel (The System of Mineralogy, 7th Ed., Vol. 2, Wiley, New York 1951, pp. 1/1124, 287). – [5] O. B. Bøggild (Medd. Groenland **149** No. 3 [1953] 1/442, 156/7).

[6] R. L. Parker, E. Brandenberger (Schweiz. Mineral. Petrogr. Mitt. **26** [1946] 12/8, 13/4). – [7] A. Kipfer (Schweiz. Strahler **3** [1973] 133/69, 141, 151/3). – [8] A. Kipfer (Schweiz. Strahler **5** [1979] 45/81, 68, 72/3). – [9] K. Schebesta (Lapis **7** No. 1 [1982] 9/20, 16, 18/9). – [10] E. I. Semenov (in: K. A. Vlasov, Geochemistry and Mineralogy of Rare Elements and Genetic Types of Their Deposits, Vol. 2: Mineralogy of Rare Elements, Israel Program Sci. Trans., Jerusalem 1966, pp. 249/328, 274 [Russian Original, Moscow 1964]).

[11] N. I. Aleksandrova, I. V. Maksimova, N. M. Svirskaya (Izv. Vysshikh Uchebn. Zavedenii Geol. Razvedka **20** No. 10 [1977] 80/7, 82, 85). – [12] Y. Iitaka, H. A. Stalder (Schweiz. Mineral. Petrogr. Mitt. **41** [1961] 485/8). – [13] Anonymous (Mineralienfreund **14** [1976] 89/120, 116). – [14] A. Kipfer (Schweiz. Strahler **5** [1981] 397/417, 405). – [15] A. Frey (Schweiz. Strahler **4** [1977] 205/7).

[16] F. Emiliani, G. Gandolfi (Mineral. Petrogr. Acta **11** [1965] 123/31, 127/9). – [17] P. C. Sæbø, H. Neumann (Norsk Geol. Tidsskr. **41** [1961] 247/54, 248/50, 253/4). – [18] Yu. L. Kapustin (Mineralogiya Karbonatitov, Moscow 1971, pp. 1/288, 186). – [19] L. Marble, A. Regis (Rocks Minerals **54** [1979] 4/25, 13). – [20] Y. Moëlo, J.-P. Choutier, E. Melon (from R. Pierrot, Bull. Soc. Franc. Mineral. Crist. **97** [1974] 521/4, 523).

[21] M. S. Garson, W. C. Smith (Nyasaland Geol. Surv. Dept. Mem. No. 1 [1957] 1/127, 37, 40/1, 53, 83). – [22] M. S. Garson (Nyasaland Geol. Surv. Dept. Mem. No. 2 [1962] 1/248, 88, 109). – [23] W. C. Smith (Bull. Brit. Mus. [Nat. Hist.] Mineral. **1** [1950/64] 97/119, 105/7, 110/6). – [24] W. J. Verwoerd (S. Africa Geol. Surv. Ann. Rept. **2** [1963] 119/35, 122). – [25] W. J. Verwoerd (S. Africa Geol. Surv. Handbook **6** [1967] 1/452, 50, 61).

[26] O. B. Dudkin, F. V. Minakov (in: I. V. Bel'kov, Materialy po Mineralogii i Geokhimii Shchelochnykh Kompleksov Porod Kol'skogo Poluostrova, Apatity 1975, pp. 159/74, 169). – [27] R. L. Parker, F. de Quervain, F. Weber (Schweiz. Mineral. Petrogr. Mitt. **19** [1939] 293/306, 295). – [28] O. B. Dudkin, L. I. Polezhaeva, F. V. Minakov (in: O. B. Dudkin et al., Veshchestvennyi Sostav Shchelochnykh Intrusivnykh Kompleksov Kol'skogo Poluostrova, Apatity 1981, pp. 73/81; Ref. Zh. Geol. **1981** No. 12V179). – [29] I. N. Kuznetsova, E. L. Rozinova, K. S. Mishchenko (in: L. S. Egorov, Shchelochnoi Magmatizm i Apatitonosnost Severa Sibiri, Leningrad 1980, pp. 101/11, 103; Ref. Zh. Geol. **1981** No. 7V260). – [30] B. Scharm, P. Kühn (Neues Jahrb. Mineral. Monatsh. **1983** 201/10, 203/4, 207).

[31] E. P. Gurov, E. P. Gurova (Tr. Mineral. Muzeya Akad. Nauk SSSR No. 24 [1975] 171/4). – [32] W. L. Smith, J. Stone, D. R. Ross, H. Levine (Am. Mineralogist **45** [1960] 92/8, 93/4). – [33] M. Weibel (Schweiz. Mineral. Petrogr. Mitt. **44** [1964] 133/5). – [34] J. van Landuyt, S. Amelinckx (Am. Mineralogist **60** [1975] 351/8, 353, 355/8). – [35] Anonymous (Mineralienfreund **15** [1977] 29/31).

[36] I. Oftedal (Z. Krist. **79** [1931] 437/64, 438, 443, 445, 450/3). – [37] E. Huen, C. M. Gramaccioli (Natura [Milan] **60** [1969] 193/4). – [38] A. A. Levinson, R. A. Borup (Am. Mineralogist **47** [1962] 337/43, 340/1). – [39] A. R. A. Turkistany, K. Koyama (IAG Bull. **3** [1980] 99/104,

100, 103). – [40] I. T. Aleksandrova, G. A. Sidorenko (Rentgenogr. Mineral'n. Syr'ya No. 11 [1977] 19/26, 20; Ref. Zh. Geol. **1977** No. 7V484).

[41] E. I. Semenov (Tr. Inst. Mineral. Geokhim. Kristallokhim. Redk. Elem. No. 2 [1959] 181/6, 184). – [42] G. Donnay (Am. Mineralogist **38** [1953] 868/70).

10.5.2.4.5 Physical Properties

Optical Properties

The **luster** of synchysite is vitreous to greasy or subadamantine [1]; in contrast to the other faces only the base is always glossy and shining [2], cf. a brilliant base and the other faces being dull and slightly altered [3, 4]. The luster of synchysite-(Nd) is dull [5].

Synchysite is translucent [1, 4] or transparent and clear [4, 6 to 8], central parts of larger crystals are opaque and ash-gray, the other parts semitransparent [2]. Synchysite in granite of Predazzo, Italy, more often occurs as turbid grains, rarely as transparent crystals [9]; synchysite in carbonatites forms turbid fibers [10]. Synchysite-(Nd) is translucent, only rarely (in very thin platelets) is it transparent [5].

An investigation of more than 1000 synchysites from ten different parageneses in Ticino, Switzerland, shows a lamellar **zonal coloring** on many crystals [7], cf. zonal coloring on crystals from other Alpine fissure veins [8], with colors passing gradually from one to the other [6]. Under the microscope synchysite crystals from Greenland in their central parts show a concentric arrangement of differently colored layers [2]. Crystals of synchysite-(Y) from the Ukraine in cross section sometimes show a zonality expressed in a slight variation of the refractive indices [11].

Color. Synchysites are characterized by yellowish, reddish, and brownish colors and may be also colorless to white; pale colors predominate in synchysites of carbonatites, see table p. 172. Synchysites from Alpine fissure veins have a similar broad spectrum of colors ranging from orange-yellow, orange-pink, yellowish brown, and pale pink to colorless [8], cf. yellowish, reddish, and orange colors [4, 12] besides colorless and rose crystals [7], red-brown [6, 13] or pale brown-red [14], brownish yellow [15], and rarely rose und greenish gray [16], or milky to milky red colors [6]. Also pale rose (carbonatite) [17], wax-yellow to brown, grayish yellow [1], and white to tan (alkali rocks) [18], and yellowish crystals (in talc) [19] have been observed. A red color may be due to finely dispersed iron oxide [20], cf. dark brown or dark red synchysite due to contamination with hydrous compounds of Fe [21, 22].

Synchysite-(Nd) under the stereoscopic microscope shows a light grayish blue color [5], synchysite-(Y) is characterized by darker colors than synchysite, predominantly by brownish colors, see table p. 172; three synchysites-(Y) from a deposit in the USSR are milk-white, blue-lilac and red-brown while a sample from a deposit in the European USSR is yellow-gray [23].

The streak of synchysite-(Nd) is white [5]. In transmitted light synchysite is colorless to pale yellow [1], cf. colorless, white, and pale yellow in thin plates and in thin section, table p. 172. Synchysite-(Nd) (under the polarization microscope) in very thin platelets is light violet to colorless, in thicker parallel intergrown plates light brown to grayish brown [5]. Synchysite-(Y) in thin section is colorless to light yellowish or brown (Ukraine) [11].

Pleochroism of synchysite is weak in yellow tints with absorption E>O [24], cf. [1], with O yellowish, E nearly colorless [25], or O light brownish yellow, E brownish yellow with a greenish tint [26]. Synchysite-(Nd) is not pleochroic [5].

The **optical character** of most synchysites is uniaxial positive [1, 3, 9 to 11, 20, 25], but perceptibly biaxial positive for synchysite from South Africa [21, 22], and very rarely for synchysite from Predazzo, Italy, which shows 2V values less than 10° [9]. Synchysite-(Nd) is also biaxial positive [5].

In the following table color and **refractive indices** of synchysites are compiled together with the values for the measured density D(meas). The samples are arranged according to their genesis. Mother rocks of the synchysites are: No. 1 granite, Nos. 2 and 3 alkali rocks, No. 4 an Sn deposit (mesothermal), Nos. 5 to 7 Alpine fissure veins, No. 8 a paragenesis similar to Alpine fissure veins, Nos. 9 to 15 carbonatites, No. 16 is an alteration product of titanite in granite; Nos. 17 to 19 are synchysites-(Y) from pegmatites (17 and 18) and from a hydrothermal greisen rock (19); No. 20 is the synthetic compound $CeCa[F|(CO_3)_2]$.

No.	Locality	Color	Refractive indices n_o	n_e	Ref.
1	Predazzo, Italy	brownish red, rarely pale rose	1.648	1.745	[9]
2	Narsarsuk, South Greenland	gray to brown	1.6742	1.7701	[2]
3		—	1.650	1.745	[27]
4	Potgietersrus district, Transvaal, South Africa	—	1.644	1.744	[1]
5	Val Nalps, Switzerland	wax-yellow	1.649	>1.74	[26]
6	Grimsel area, Switzerland	light brownish yellow	1.649 ±0.001	1.750 ±0.002	[28]
7	Beura, Piedmont, Italy	pale yellow	1.64	1.74	[25]
8	Kongsberg, southern Norway	light yellow	>1.653	1.745	[3]
9	Tundulu Hill, southern Malawi	colorless, white, pale yellow[1)]	1.6455 ±0.002	—	[10, 29]
10		pale pinkish	1.648	1.735	[10]
11		pearly white	1.645	1.733	[10]
12	Kamilala Hill, southern Malawi	colorless, pearly white	1.645	1.748	[10]
13	Glenover, South Africa	buff to dark red[2)]	1.647 ±0.003	1.750 ±0.003	[21, 22]
14	Northern Siberia	brown-rose	≈1.698	≈1.767	[30]
15	Siberia	rose	1.673	1.769	[31]
16	Oslo region, Norway	white or red	≈1.656	1.742 ±0.002	[20]
17	Cotopaxi, Colorado	reddish brown[3)]	1.643 ±0.002	1.73 ±0.01	[32]
18	Ukraine	colorless, light yellowish[1)]	1.650	1.748	[11, 33]
19	Northern Kirgizia	brownish rose	≈1.63	≈1.72	[35]
20	—	—	1.641[4)]	1.744[4)]	[27]

[1)] Color in transmitted light. – [2)] Due to inclusions, see p. 171. – [3)] Color of aggregates. – [4)] Calculated from ionic refractivities.

The refractive indices of the biaxial synchysite-(Nd) are given: $1.61 < n_x$, $n_y < 1.66 < n_z < 1.74$ [5].

The **IR spectrum** of synchysite-(Nd) from the ČSSR is characterized by the vibrations of the CO_3^{2-} group and of admixtures. The wave numbers (cm^{-1}) of the maxima of the bands correspond to the following CO_3^{2-} bands: 1472(4) and 1420(4) = ν_3, 1083(3) and 1034(2) = ν_1, 871(1) = ν_2, and 747(2) = ν_4; further bands of florencite-(La) and of an unknown admixture are present [5]. The IR spectrum of synchysite-(Y) corresponds to those of other fluorocarbonates; the investigations are discussed for samples heated to 530 to 800°C; they show no bands characteristic for CO_3^{2-}, but only such of RE oxyfluorides in the range between 400 and 600 cm^{-1} (530°C) and that of RE_2O_3 at 570 cm^{-1} (600°C) [36].

Luminescence. In unfiltered UV light synchysites show fluorescence in olive to leek-green or deep green [13], brilliant green (short wave UV) [16], or intense grass-green [14]; the deep green colors are due to the Nd-content of the mineral [7]; a formerly observed phosphorescence could not be confirmed [13].

Synchysite-(Nd) shows no luminescence in UV light and no cathodic luminescence [5].

Other Physical Properties

The **Hardness** of synchysite $H_M = 4.5$ [1, 2], 4 [24, 37], 4.5 to 5 [4], 5 [38]. For synchysite-(Y) aggregates $H_M \approx 4$ [34], for synchysite-(Nd) H_M is only about 1 due to the very weak cohesion of the aggregates; the hardness of the very small individual grains could not be determined [5].

The measured **density** of synchysite D(meas.) in $g/cm^3 = 3.90$ (Potgietersrus, Transvaal) [1], 3.902 (Narsarsuk, southern Greenland) [2], >3.94 (Val Nalps, Switzerland) [26], 3.95 (unknown locality) [35], and 4.05 ± 0.1 and 4.15 ± 0.1 (unknown locality); the last values are probably obtained on polycrystals [27]. D(calc.) = 3.88 [1], 3.99 [27], 3.96 (Greenland), and ≈4.0 g/cm^3 (Quincy) [39]. D(calc.) of the synthetic compound $CeCa[F|(CO_3)_2] = 3.99$ g/cm^3 [27].

Synchysite-(Y) from Kazakhstan has D(meas.) = 3.52 g/cm^3 [34]. D(calc.) = 3.92 g/cm^3 of synchysite-(Y) of Kirgizia [35]. D(calc.) = 4.14 g/cm^3 of synchysite-(Nd) from northern Bohemia [5].

The **fracture** of synchysite is subconchoidal to splintery [1, 24] or conchoidal [3, 4]; the mineral is brittle [1, 3, 4, 24]. Synchysite crystals (transparent and turbid) are **magnetic** and concentrate in the magnetic fraction at 0.6 to 1.0 A [9]. The **radioactivity** of synchysite from Transvaal is slight [21, 22], that of synchysite-(Y) from Dover, New Jersey, moderate [40].

References for 10.5.2.4.5:

[1] C. Palache, H. Berman, C. Frondel (The System of Mineralogy, 7th Ed., Vol. 2, Wiley, New York 1951, pp. 1/1124, 287/8). – [2] O. B. Bøggild (Medd. Groenland **149** No. 3 [1953] 1/442, 156/8). – [3] P. C. Sæbø, H. Neumann (Norsk Geol. Tidsskr. **41** [1961] 247/54, 248, 250). – [4] Anonymous (Mineralienfreund **15** [1977] 29/31). – [5] B. Scharm, P. Kühn (Neues Jahrb. Mineral. Monatsh. **1983** 201/10, 203/4, 208).

[6] A. Kipfer (Schweiz. Strahler **3** [1973] 133/69, 145, 151, 153). – [7] A. Kipfer (Schweiz. Strahler **5** [1979] 45/81, 67/8). – [8] K. Schebesta (Lapis **7** No. 1 [1982] 9/20, 16). – [9] F. Emiliani, G. Gandolfi (Mineral. Petrogr. Acta **11** [1965] 123/31, 127). – [10] M. S. Garson (Nyasaland Geol. Surv. Dept. Mem. No. 2 [1962] 1/248, 84/5, 88, 109).

[11] E. P. Gurov, E. P. Gurova (Tr. Mineral. Muzeya Akad. Nauk SSSR No. 24 [1975] 171/4). – [11] A. Frey (Schweiz. Strahler **4** [1977] 205/7). – [13] A. Kipfer (Schweiz. Strahler **3** [1975] 370/84, 376/8). – [14] A. Kipfer (Schweiz. Strahler **5** [1981] 397/417, 405). – [15] M. Weibel (Schweiz. Mineral. Petrogr. Mitt. **44** [1964] 133/5).

[16] H. Weninger (Lapis **4** No. 10 [1979] 26/8). – [17] O. B. Dudkin, L. I. Polezhaeva, F. V. Minakov (in: O. B. Dudkin et al., Veshchestvennyi Sostav Shchelochnykh Intrusivnykh Kompleksov Kol'skogo Poluostrova, Apatity 1981, pp. 73/81, 80; Ref. Zh. Geol. **1981** No. 12V179). – [18] L. Marble, A. Regis (Rocks Minerals **54** [1979] 4/25, 13). – [19] Y. Moëlo, J.-P. Choutier, E. Melon (from R. Pierrot, Bull. Soc. Franc. Mineral. Crist. **97** [1974] 521/4, 523). – [20] G. Raade (Norsk Geol. Tidsskr. **49** [1969] 227/39, 232).

[21] W. J. Verwoerd (S. Africa Geol. Surv. Ann. Rept. **2** [1963] 119/35, 122). – [22] W. J. Verwoerd (S. Africa Geol. Surv. Handbook **6** [1967] 1/452, 49, 61, 67). – [23] V. V. Delyageid, S. V. Yakovleva, R. N. Morzheedova, T. F. Shvedova, G. D. Evlanova, R. V. Kotlyarov, Z. M. Grigor'eva (Nauchn. Tr. Gos. Nauchn. Issled. Proektn. Inst. Redkometal. Prom. **45** [1972] 75/8; C.A. **85** [1976] No. 163454). – [24] E. I. Semenov (in: K. A. Vlasov, Geochemistry and Mineralogy of Rare Elements and Genetic Types of Their Deposits, Vol. 2: Mineralogy of Rare Elements, Israel Program Sci. Trans., Jerusalem 1966, pp. 249/328, 274 [Russian Original, Moscow 1964]). – [25] E. Huen, C. M. Gramaccioli (Natura [Milan] **60** [1969] 193/4).

[26] R. L. Parker, F. de Quervain, F. Weber (Schweiz. Mineral. Petrogr. Mitt. **19** [1939] 293/306, 295/6). – [27] G. Donnay, J. D. H. Donnay (Am. Mineralogist **38** [1953] 932/63, 936, 960). – [28] Y. Iitaka, H. A. Stalder (Schweiz. Mineral. Petrogr. Mitt. **41** [1961] 485/8). – [29] W. C. Smith (Bull. Brit. Mus. [Nat. Hist.] Mineral. **1** [1950/64] 97/119, 105). – [30] I. N. Kuznetsova, E. L. Rozinova, K. S. Mishchenko (in: L. S. Egorov, Shchelochnoi Magmatizm i Apatitonosnost Severa Sibiri, Leningrad 1980, pp. 101/11, 103).

[31] Yu. L. Kapustin (Mineralogiya Karbonatitov, Moscow 1971, pp. 1/288, 186). – [32] A. A. Levinson, R. A. Borup (Am. Mineralogist **47** [1962] 337/43, 340). – [33] E. P. Gurov, E. P. Gurova, L. G. Loginova, Yu. A. Lavitskaya (in: Mineraly i Paragenezisy Mineralov Metasomaticheskikh i Metamorficheskikh Gornykh Porod, Leningrad 1975, pp. 106/14, 107; C.A. **84** [1976] No. 108517). – [34] D. A. Mineev (Geokhimiya Apogranitov i Redkometal'nykh Metasomatitov Severo-Zapadnogo Tarbagataya, Moscow 1968, pp. 1/183, 96). – [35] E. I. Semenov (Tr. Inst. Mineral. Geokhim. Kristallokhim. Redk. Elem. No. 2 [1959] 181/6, 182, 184).

[36] I. T. Aleksandrova, G. A. Sidorenko (Rentgenogr. Mineral'n. Syr'ya No. 11 [1977] 19/26, 20, 22; Ref. Zh. Geol. **1977** No. 7V484). – [37] N. I. Aleksandrova, I. V. Maksimova, N. M. Svirskaya (Izv. Vysshikh Uchebn. Zavedenii Geol. Razvedka **20** No. 10 [1977] 80/7, 83). – [38] A. S. Povarennykh (Crystal Chemical Classification of Minerals, Vol. 2, translated from Russian by J. E. S. Bradley, Plenum, New York 1972, pp. 459/766, 616). – [39] I. Oftedal (Z. Krist. **79** [1931] 437/64, 441, 443). – [40] W. L. Smith, J. Stone, D. R. Ross, H. Levine (Am. Mineralogist **45** [1960] 92/8, 94).

10.5.2.4.6 Chemical Behavior and Thermal Analyses

Synchysite is relatively easily soluble in acids [1], cf. [2]; it dissolves slowly in hydrochloric acid with effervescence [3, 4, 5] and also with weak effervescence in strong hydrochloric acid [6]. Synchysite-(Y) [7] and synchysite-(Nd) with vigorous effervescence [8] are also soluble in dilute hydrochloric acid.

Results of a thermal analysis of synchysite from granite of the Urals are announced in an abstract [9].

The heating curve of synchysite-(Y) from Kirgizia shows an endothermal effect at 500°C, caused by thermal dissociation of the mineral [7], that is at a lower temperature than for other RE-F carbonates [2]; the TGA shows loss of CO_2 between 450 and 550°C [7]. Synchysite-(Y) from a deposit in Middle Asia shows maxima of endothermal effects at 510, 580, and 660°C, caused by decomposition of the mineral; the originating phases are: at 500°C cubic phases (oxyfluorides of the type LnOF) which are stable up to 600°C; upon heating to 800°C a cubic phase (type

RE_2O_3), at 900°C two other phases (YOF and HoOF) together with CeO_2 [10]. Synchysite-(Y) from an unknown locality in the USSR shows an endothermal effect in the range between 440 and 580°C; the occurrence of the oxyfluorides CeOF and YOF has been observed upon heating to 600°C [11]. The DTA curve of three differently colored samples of synchysite-(Y) from a deposit of the USSR shows a clear endothermal maximum at 480°C for the monomineralic milky-white variety; the other two samples, both contaminated with thermo-inert fluorite, show two endothermal effects superimposing that of the milky-white variety. The endothermal effect with maximum at 450 to 600°C of all three samples is caused by the dissociation of the mineral with loss of CO_2. According to TGA the loss of CO_2 in this region amounts to 8%, an additional loss of 2.2% between 750 and 950°C is due to oxidation of Ce^{3+} to Ce^{4+}. At 500°C oxyfluorides of La and Y are present besides fluorite [12].

References for 10.5.2.4.6:

[1] C. Palache, H. Berman, C. Frondel (The System of Mineralogy, 7th Ed., Vol. 2, Wiley, New York 1951, pp. 1/1124, 288). – [2] E. I. Semenov (in: K. A. Vlasov, Geochemistry and Mineralogy of Rare Elements and Genetic Types of Their Deposits, Vol. 2: Mineralogy of Rare Elements, Israel Program Sci. Trans., Jerusalem 1966, pp. 249/328, 274/5 [Russian Original, Moscow 1964]). – [3] O. B. Bøggild (Medd. Groenland **149** No. 3 [1953] 1/442, 157). – [4] F. Emiliani, G.Gandolfi (Mineral. Petrogr. Acta **11** [1965] 123/31,127).–[5] O.B. Dudkin, L.I. Polezhaeva, F. V. Minakov (in: O. B. Dudkin et al., Veshchestvennyi Sostav Shchelochnykh Intrusivnykh Kompleksov Kol'skogo Poluostrova, Apatity 1981, pp. 73/81, 80; Ref. Zh. Geol. **1981** No. 12 V 179).

[6] P. C. Sæbø, H. Neumann (Norsk Geol. Tidsskr. **41** [1961] 247/54, 249). – [7] E. I. Semenov (Tr. Inst. Mineral. Geokhim. Kristallokhim. Redk. Elem. No. 2 [1959] 181/6, 182). – [8] B. Scharm, P. Kühn (Neues Jahrb. Mineral. Monatsh. **1983** 201/10, 206). – [9] V. A. Chashchukhina (in: P. Ya. Yarosh, Mineraly Gornykh Porod i Rud Urala, Sverdlovsk 1980, pp. 1/149, 58/60 from Ref. Zh. Geol. **1981** No. 12V185). – [10] I. T. Aleksandrova, G. A. Sidorenko (Rentgenogr. Mineral'n. Syr'ya No. 11 [1977] 19/26, 20/2; Ref. Zh. Geol. **1977** No. 7V484).

[11] N. I. Aleksandrova, I. V. Maskimova, N. M. Svirskaya (Izv. Vysshikh Uchebn. Zavedenii Geol. Razvedka **20** No. 10 [1977] 80/7, 85/7). – [12] V. V. Delyageid, S. V. Yakovleva, R. N. Morzheedova, T. F. Shvedova, G. D. Evlanova, R. V. Kotlyarov, Z. M. Grigor'eva (Nauchn. Tr. Gos. Nauchn. Issled. Proektn. Inst. Redkometal. Prom. **45** [1972] 75/8; C.A. **85** [1976] No. 163454).

10.5.2.5 Cordylite $(Ce,La)_2Ba[F_2|(CO_3)_3]$

10.5.2.5.1 Synonyms, Occurrence, and Paragenesis

Besides the English name cordylite the German name kordylite is used in the literature; the old designation barium parisite [1] has been discredited [2] and was transfered to a new species found in China with a deviating RE/Ba cation ratio, see p. 185.

Cordylite was first found at Narsarsuk in the Julianehaab district, Greenland, in pegmatitic veins in nepheline syenite; it is associated with aegirine, ancylite, synchysite, neptunite [1], and astrophyllite [3]. In this locality it seems to be the youngest mineral [4], deposited in the hydrothermal stage of the pegmatitic process [3]. Cordylite also occurs in a pegmatitic dike in nepheline syenite of the Mont St. Hilaire, St. Lawrence lowlands, Quebec, associated with aegirine, analcime, albite and with minor amounts of biotite, rinkite, calcite, apatite, catapleiite, pyrochlore, and pyrophanite [5]; the cordylite is here extremely rare and further associated with ancylite and synchysite [6]. Single grains of cordylite have been observed as an accessory mineral in the late K-alkaline basaltic rocks of the Verkhnedunkel'dykskii massif, South Pamir,

USSR [7], and cordylite closely intergrown with bastnaesite and parisite in microscopic grains occurs in alkaline rocks of the East Pamir [8]. Cordylite was further found in a hydrothermal metasomatic iron-ore deposit of North China, associated with "barium-parisite", Na-rich pyroxene, fluorite, barite, and apatite [9].

Cordylite as an accessory mineral has been found in carbonatites of the Kola Peninsula, USSR [10], where it occurs in the inner zone of the carbonatitic core of the alkaline Sallanlatva massif, together with pyrochlore, carbonates, barite, vermiculite, and sulfides [11]; it occurs in the silicified dolomite carbonatites of Lesnaya Varaka [12] and in a nonspecified post-mesozoic deposit, where it occurs in lenses or small veins of calcite-ankerite composition, associated with biotite, amphibole, barite, quartz, Mn-bearing calcite, several RE carbonates, vaterite, pyrochlore, and sulfides [13]. A mineral, indicated as cordylite by X-ray powder pattern, has been found in the Tundulu carbonatite ring complex in Southern Malawi; it is closely associated and interleaved with synchysite in siderite sövites, and associated with bastnaesite in apatite sövites and carbonatites of the second ring structure centered around Nathace Hill, southwest of the Tundulu complex [14].

Upon weathering, cordylite from Greenland alters to bastnaesite [10].

References for 10.5.2.5.1:

[1] C. Palache, H. Berman, C. Frondel (The System of Mineralogy, 7th Ed., Vol. 2, Wiley, New York 1951, pp. 1/1124, 285/6). – [2] H. Strunz, C. Tennyson (Mineralogische Tabellen, 6th Ed., Geest and Portig, Leipzig 1977, pp. 1/621, 507). – [3] E. I. Semenov (in: K. A. Vlasov, Geochemistry and Mineralogy of Rare Elements and Genetic Types of Their Deposits, Vol. 2: Mineralogy of Rare Elements, Israel Program Sci. Trans., Jerusalem 1966, pp. 249/328, 275/6 [Russian Original, Moscow 1964]). – [4] O. B. Bøggild (Medd. Groenland **149** No. 3 [1953] 1/442, 158). – [5] T. T. Chen, G. Y. Chao (Can. Mineralogist **13** [1975] 93/4).

[6] L. Marble, A. Regis (Rocks Minerals **54** [1979] 4/25, 11). – [7] A. S. Lomteva, E. A. Dmitriev (Dokl. Akad. Nauk Tadzh.SSR **19** No. 7 [1976] 53/6). – [8] E. A. Dmitriev (Dokl. Akad. Nauk Tadzh.SSR **4** No. 3 [1961] 29/31). – [9] H.-C. Wang, H.-T. Su, C.-T. Kuo, S.-P. Li, C.-K. Wang (Ti Ch'iu Hua Hsueh **1973** 31/8; Geochim. [Peking] **1973** 30/7, 30). – [10] E. I. Semenov (Mineralogiya Redkikh Zemel', Moscow 1963, pp. 1/412, 107).

[11] B. I. Serba (Magmaticheskie Obrazovaniya Kol'sk. Poluostrova, Moscow 1962, pp. 58/64, 63). – [12] Yu. L. Kapustin (Mineralogiya Karbonatitov, Moscow 1971, pp. 1/288, 186). – [13] Yu. L. Kapustin (in: L. N. Ovchinnikov, N. A. Solodov, Mestorozhdeniya Litofil'nykh Redkikh Metallov, Nedra, Moscow 1980, pp. 243/85, 269). – [14] M. S. Garson (Nyasaland Geol. Surv. Dept. Mem. No. 2 [1962] 1/248, 89, 100, 108/9).

10.5.2.5.2 Chemistry

The RE_2O_3 contents in cordylites of various localities are: 49.39, Greenland [1], 47.36, North China [2, p. 33], and 51.32%, Quebec (microprobe) [3].

Recalculation of the analysis of Greenland cordylite gives the theoretical formula $(Ce,La)_2Ba[CO_3]_3F_2$ [4], simplified to $Ce_2Ba[CO_3]_3F_2$ by [5]. From the analysis of cordylite from Quebec the formula $(Ba_{0.92}Ca_{0.07}Sr_{0.04})(Ce_{1.37}La_{0.59}Th_{0.01})[CO_3]_3F_{1.58}$ has been calculated [3], and from that of cordylite of North China the formula $[(Ce,La,Pr,Nd\ldots)_{2.01}Ca_{0.25}](Ba_{1.10}Sr_{0.01})[CO_3]_{2.82}F_{1.33}(OH)_{0.45}$ [2, p. 32].

The ratio $\Sigma Ce/Ba$ of cordylite has the theoretical value of 2 : 1; it deviates for the mineral from Greenland to a higher value and for the mineral from North China to a lower value, which is mainly due to the variance in the Ba content [2, pp. 34/5].

Besides the main cations RE and Ba, and some Ca and Sr replacing these cations, see formulas p. 176, all three analyses show tenths of a percent of ThO_2 [4, 3], [2, p. 32], which is supposed to replace in part RE [5, 6], cf. formula for cordylite from Quebec.

The deficiency of F in the empirical formula of the mineral from Quebec is most likely due to the omission of H_2O in the analysis, which is probably present in the mineral as a hydroxyl group (replacing F) [5], cf. deficiency of F in the mineral from North China with 0.58% H_2O^+ [2, p. 32].

The structures of parisite and synchysite deviate so much from that of cordylite, that any isomorphism is impossible, see p. 179.

The relative amounts of the RE in cordylites from various localities, calculated as $\Sigma RE_2O_3 = 100$, are given in the following table:

Locality	La	Ce	Pr	Nd	Sm	Eu	Gd	Yb	Ref.
Greenland	29	45	5.5	15	3.2	0.1	2.1	—	[7]
North China	19.4	51.9	6.4	19.4	1.94	0.35	0.6	—	[2, p. 33]
Quebec	30.0	69.8	—	—	—	—	—	0.2	[3]

The ratio (La to Eu): Gd = 166 for the mineral from North China, but is only 47 for that from Greenland [2, p. 33]. Cordylite, as other RE carbonates, shows an increasing ratio La/Ce with increasing content of water [8].

References for 10.5.2.5.2:

[1] O. B. Bøggild (Medd. Groenland **149** No. 3 [1953] 1/442, 159/60). – [2] H.-C. Wang, H.-T. Su, C.-T. Kuo, S.-P. Li, C.-K. Wang (Ti Ch'iu Hua Hsueh **1973** 31/8; Geochim. [Peking] **1973** 30/7). – [3] T. T. Chen, G. Y. Chao (Can. Mineralogist **13** [1975] 93/4). – [4] C. Palache, H. Berman, C. Frondel (The System of Mineralogy, 7th Ed., Vol. 2, Wiley, New York 1951, pp. 1/1124, 285/6). – [5] E. I. Semenov (in: K. A. Vlasov, Geochemistry and Mineralogy of Rare Elements and Genetic Types of Their Deposits, Vol. 2: Mineralogy of Rare Elements, Israel Program Sci. Trans., Jerusalem 1966, pp. 249/328, 275/6 [Russian Original, Moscow 1964]).

[6] E. I. Semenov (Mineralogiya Redkikh Zemel', Moscow 1963, pp. 1/412, 107). – [7] E. I. Semenov, R. L. Barinskii (Geokhimiya **1958** 314/33; Geochemistry [USSR] **1958** 398/419, 401). – [8] N. V. Svyazhin (Tr. Inst. Geol. Akad. Nauk SSSR Ural'sk. Filial No. 70 [1965] 249/52, 252).

10.5.2.5.3 Crystal Forms and Crystal Structure

Cordylite is hexagonal [1], with goniometric axial ratio a : c = 1 : 1.1288 [2]. Crystals with dominant m $\{10\bar{1}0\}$ and subordinate r $\{20\bar{2}3\}$, p $\{40\bar{4}5\}$, q $\{10\bar{1}1\}$, and c $\{0001\}$ are common, see **Fig. 8**, p. 178, from [2]. According to the structural axial ratio a : c = 1 : 4.525 (checking the morphological axial ratio 4 c/a = 4.5152, see above), the observed crystal forms are reindexed to $\{0001\}$, $\{10\bar{1}0\}$, $\{10\bar{1}1\}$, $\{10\bar{1}2\}$, $\{10\bar{1}4\}$, $\{10\bar{1}5\}$ [3].

The crystals from Greenland are short prismatic, often with enlarged terminations on a slender prism, and sometimes scepter-like [2]; the crystals from Quebec have hexagonal tabular habit; they display a rough pinacoid, a prism and at least one pyramid (under electron microscope) [4]; the crystals from Kola are hexagonal tables [5]. At Tundulu, Malawi, cordylite forms sheaves and elongated prisms [6].

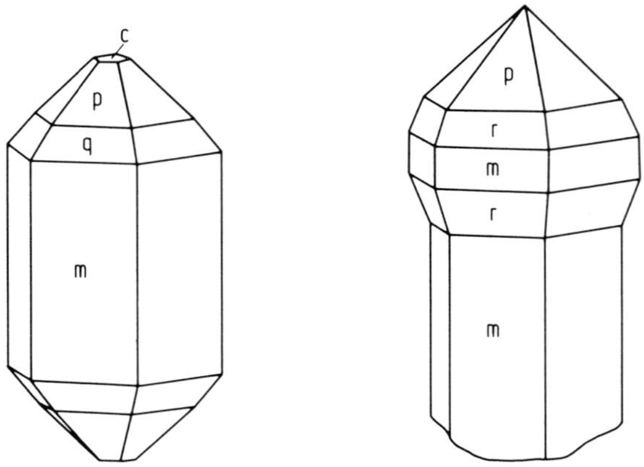

Fig. 8. Habit and forms of cordylite from [2].

The cleavage parallel to {0001} [1] is good [4] or distinct; however, it might be a parting due to alteration [2].

The unit cell constants, the axial ratio c/a, the number of formula units in the cell Z, and the calculated density values of cordylites from the three occurrences Narsarsuk, Julianehaab district, southern Greenland (1), Mont St. Hilaire, Quebec (2), and an iron ore deposit in North China (3) are given in the following table:

No.	a in Å	c in Å	c/a	Z	D(calc.) (g/cm^3)	Ref.
1	4.36	22.8	5.23	2	5.6	[7, p. 458]
	7.55*)	22.8	3.02	6*)	—	[8]
	5.100(15)	23.08(7)	4.525	2	4.06	[3]
2	5.098(1)	23.050(5)	4.5214	2	3.97	[4]
3	5.08	23.35	4.60	2	—	[9]

*) According to the structure discussion, which requires a turning of 30°, $a = 4.35 \cdot \sqrt{3}$ and Z = 6 [8].

The cell volume of cordylite of Greenland V = 523 $Å^3$ [10], that of cordylite of northern China, 522 $Å^3$ [9].

The space group of cordylite is $P6_3/mmc\text{-}D^4_{6h}$ (No. 194) or $P\overline{6}2c\text{-}D^4_{3h}$ (No. 190) [3, 4]; the first group is also given by [10 to 12]; it is considered to be the pseudo space group in the smaller cell with Z = 2 [8].

The structure of cordylite consists of horizontal layers of RE atoms and F atoms (as in bastnaesite, see p. 123), alternating with layers of Ba atoms, lying above the gaps of the RE layers and vice versa. The CO_3 groups have positions between two RE-F layers parallel to the c axis and between RE-F layers and Ba layers perpendicular to the c axis [8]; cf. alternating hexagonal Ba layers and puckered Ce-F layers interleaved by horizontal CO_3 groups [3], and a part of the groups of CO_3 perpendicular, another part parallel to the c axis [7, p. 461]. In cordylite a bastnaesite cell alternates with a cell of smaller size along the c axis [12]. The change of a large fraction of the CO_3 groups into an orientation perpendicular to the c axis causes a marked change in the c parameter compared with that of minerals of the bastnaesite series [13]. This special orientation of the CO_3 groups is further the cause of a strong negative birefringence [7, p. 461], cf. [3, 8, 12]. The distance of 4.86 Å between two RE atoms in c direction, observed for parisite and bastnaesite (not synchysite), which corresponds to the pseudo c axis (½ c axis) of bastnaesite, is characteristic also for cordylite [7, pp. 450, 452, 456, 463]. The structure is related to that of parisite, however quite different from this structure; the different values of the a axis of the two minerals cordylite and parisite exclude isomorphism between the two minerals; parallel intergrowths between cordylite and the minerals of the bastnaesite group seem to be impossible [7, pp. 462/4].

References for 10.5.2.5.3:

[1] O. B. Bøggild (Medd. Groenland **149** No. 3 [1953] 1/442, 158/9). – [2] C. Palache, H. Berman, C. Frondel (The System of Mineralogy, 7th Ed., Vol. 2, Wiley, New York 1951, pp. 1/1124, 285/6). – [3] G. Donnay, J. D. H. Donnay (Bull. Geol. Soc. Am. **66** [1955] 1551). – [4] T. T. Chen, G. Y. Chao (Can. Mineralogist **13** [1975] 93/4). – [5] Yu. L. Kapustin (Mineralogiya Karbonatitov, Moscow 1971, pp. 1/288, 186).

[6] M. S. Garson (Nyasaland Geol. Surv. Dept. Mem. No. 2 [1962] 1/248, 100, 109). – [7] I. Oftedal (Z. Krist. **79** [1931] 437/64). – [8] Strukturbericht, Vol. 2, 1928/32 [1937], pp. 75/8, 402. – [9] H.-C. Wang, H.-T. Su, C.-T. Kuo, S.-P. Li, C.-K. Wang (Ti Ch'iu Hua Hsueh **1973** 31/8; Geochim. [Peking] **1973** 30/7, 31). – [10] E. I. Semenov (in: K. A. Vlasov, Geochemistry and Mineralogy of Rare Elements and Genetic Types of Their Deposits, Vol. 2: Mineralogy of Rare Elements, Israel Program Sci. Trans., Jerusalem 1966, pp. 249/328, 276 [Russian Original, Moscow 1964]).

[11] H. Strunz, C. Tennyson (Mineralogische Tabellen, 6th Ed., Geest and Portig, Leipzig 1977, pp. 1/621, 243). – [12] A. S. Povarennykh (Crystal Chemical Classification of Minerals, Vol. 2, translated from Russian by J. E. S. Bradley, Plenum, New York 1972, pp. 459/766, 616). – [13] A. S. Povarennykh (Crystal Chemical Classification of Minerals, Vol. 1, translated from Russian by J. E. S. Bradley, Plenum, New York 1972, pp. 1/458, 181/2).

10.5.2.5.4 Physical Properties

The luster of cordylite is greasy to adamantine, pearly on {0001} [1 to 3], waxy on crystal faces, vitreous on freshly exposed surfaces [4]. The base is always clear and shining, the faces immediatly adjacent to it have a somewhat lesser quality [5]. The mineral is transparent, when fresh, but often dull due to surface alterations [1], later formed crystals from Greenland are quite clear with beautifully developed faces. In cross section the crystals show a regular zonal structure [5]. Basal sections sometimes show concentric hexagonal zones of colorless and cloudy or opaque material together with radial sets of striae [1]. Cordylite from the Tundulu carbonatite complex, southern Malawi, shows prisms with pale brown color with dark brown cores or with yellow and yellowish brown bands in vertical section [6].

The cordylite from Greenland is colorless to wax-yellow [1 to 3], generally pale yellow, but varying from nearly white to brownish yellow [5], often ocher-yellow due to surface alteration [1]. The cordylite crystals from Mont St. Hilaire are pale yellow [4] to yellow-green [7], the mineral from Lesnaya Varaka, Kola, is yellowish [8], the mineral from southern Malawi is colorless to yellowish brown [6]. In transmitted light cordylite from Greenland is colorless to pale yellow, weakly dichroic with O greenish yellow, E brownish yellow; it is uniaxial negative [1], cf. pleochroism and negative optical character [2, 3, 5].

The refractive indices of cordylites from Greenland (No. 1) and from Quebec (No. 2) are given in the following table:

No.	n_o	n_e	Ref.
1	1.764	1.577	[1 to 3, 9]
	1.7640	1.5772	[5]
2*)	1.773(2)	1.576(1)	[4]

*) Determined at 25°C.

A high birefringence was also found for cordylite from southern Malawi [6]. The birefringence of cordylite from Greenland observed by [5] is inexplicably larger than the birefringence calculated from the molar refractivities of the mineral [10].

The hardness of cordylite from Greenland $H_M = 4$ [2], 4.5 [1, 3, 5], for that from Quebec ≈4.5 [4], cf. a range of 4.5 to 5 for cordylites given by [11]. Cordylite has a conchoidal fracture and is rather brittle [1 to 3], cf. [4].

The density of cordylite from Greenland D(meas.) = 4.31 g/cm^3 [1 to 3, 5] is probably too low [1] due to measuring on impure material [12] or on altered material [13]. A measurement with the Berman balance gives a value of only 4.10 g/cm^3 [10]. The calculated density of cordylite from Greenland D(calc.) = 5.61 g/cm^3 [1] differs appreciably from more recent values for cordylites from Greenland and Quebec, see table on p. 178. According to the rule of Gladstone and Dale D≈4.76 g/cm^3 has been calculated [14].

10.5.2.5.5 Chemical Behavior

Cordylite from Greenland is easily soluble in acids [1 to 3]; cordylite from Quebec is soluble with effervescence in hydrochloric acid at room temperature [4].

Upon heating the cordylite from Greenland apparently dissociates to BaF_2, CeO_2, and Ln_2O_3 [3].

References for 10.5.2.5.4 and 10.5.2.5.5:

[1] C. Palache, H. Berman, C. Frondel (The System of Mineralogy, 7th Ed., Vol. 2, Wiley, New York 1951, pp. 1/1124, 286). – [2] E. I. Semenov (Mineralogiya Redkikh Zemel', Moscow 1963, pp. 1/412, 407). – [3] E. I. Semenov (in: K. A. Vlasov, Geochemistry and Mineralogy of Rare Elements and Genetic Types of Their Deposits, Vol. 2: Mineralogy of Rare Elements, Israel Program Sci. Trans., Jerusalem 1966, pp. 249/328, 276 [Russian Original, Moscow 1964]). – [4] T. T. Chen, G. Y. Chao (Can. Mineralogist **13** [1975] 93/4). – [5] O. B. Bøggild (Medd. Groenland **149** No. 3 [1953] 1/442, 158/9).

[6] M. S. Garson (Nyasaland Geol. Surv. Dept. Mem. No. 2 [1962] 1/248, 103, 109). – [7] L. Marble, A. Regis (Rocks Minerals **54** [1979] 4/25, 11). – [8] Yu. L. Kapustin (Mineralogiya Karbonatitov, Moscow 1971, pp. 1/288, 186). – [9] P'ei-Shan Chang [Chzhan Pei-Shan] (Sci. Sinica [Peking] **11** [1962] 251/8, 256 [Russian]; C.A. **57** [1962] 10799). – [10] G. Donnay, J. D. H. Donnay (Bull. Geol. Soc. Am. **66** [1955] 1551).

[11] A. S. Povarennykh (Crystal Chemical Classification of Minerals, Vol. 2, translated from Russian by J. E. S. Bradley, Plenum, New York 1972, pp. 459/766, 616). – [12] I. Oftedal (Z. Krist. **79** [1931] 437/64, 458). – [13] C. Doelter (Handbuch der Mineralchemie, Vol. 1, Steinkopff, Dresden 1912, pp. 1/1008, 534). – [14] H. W. Jaffe (Am. Mineralogist **41** [1956] 757/77, 761).

10.5.2.6 Huanghoite $CeBa[F|(CO_3)_2]$

10.5.2.6.1 Name, Occurrence, and Paragenesis

The name of this new mineral is derived from the river Hwang Ho [Yellow River], in the area in which it has been first found [1 to 4]. According to different transcriptions from Russian the following writings are also known: chuanchite [5], khuankhite [6], huanchite [4], or huanhite [2].

At the type locality in Inner Mongolia, northern China, huanghoite is relatively widely distributed in hydrothermal veins of magnetite-hematite sheet deposits [4], which are genetically associated with alkali granosyenites; it also occurs in the surrounding reworked dolomites. Associated minerals are aegirine, fluorite, magnetite, hematite, monazite, bastnaesite, aeschynite, parisite, et al. Sometimes observed symplectic intergrowths between bastnaesite and barite are probably pseudomorphs after huanghoite; in this case the two minerals may be products of endogenic dissociation of huanghoite [1 to 4]. A second occurrence of huanghoite lies in Siberia, where it is sporadically distributed in vuggy parts of a late veinlike body of carbonatite, which consists predominantly of ankerite and dolomite. It is intimately associated with quartz and fluorite; further minerals are RE fluorocarbonates, sulfides, siderite, cleiophane, and monazite. Huanghoite is confined to areas devoid of parisite and bastnaesite, possibly it was replaced by these minerals. In those parts of the rocks which contain fluorite, huanghoite is peripherically replaced by a dense powdery aggregate of bastnaesite, barite, and quartz. In the zone of supergene alteration huanghoite is only relatively stable, locally it is covered by powdery bastnaesite [6], cf. [7], cf. below. The occurrence of huanghoite in paragenesis with witherite, strontianite, burbankite, bastnaesite, sulfides, and magnetite in late dolomite-ankerite-carbonatite veins of the carbonatite complex Arbarastakh, Yakutia [8], is probably identical with this Siberian locality, described as the first find in the USSR.

Huanghoite has been further found in foyaites of the Khibina massif, Kola, where it occurs as fine segregations together with synchysite and strontianite or with Mn calcite in carbonate veins cutting biotitized basic rocks. It also occurs associated with synchysite, hydronepheline, and natrolithe ("spreustein"), formed after the natrolithe in the altered foyaites. In some carbonate veins huanghoite amounts to 5% of the mass; these veins contain besides calcite also dawsonite, nahcolite, cryolite, and marcasite in cavities [9]. In a later paper this locality is more precisely designated as the depression of the river Tuliok; the huanghoite occurs together with synchysite, parisite, ancylite, fluorite, strontianite, sulfides, and other minerals in the fifth stage of carbonatite formation of the complex core of this massif [10].

For a compilation of the occurrences of huanghoite in China and in Siberia, and its paragenesis in these localities, see also [11].

As a phase, not as a mineral, huanghoite has been found in rounded or tabular form among other phases, such as ewaldite, in mineral precipitates obtained by evaporation of aqueous extracts of sphalerite and quartz from the Madjarovo ore deposit, eastern Rhodope Mountains, Bulgaria [12].

Huanghoite is unstable under surface conditions; it is readily replaced by yellow and brown bastnaesite ochers [1 to 4].

10.5.2.6.2 Chemistry

For a compilation of analyses and the content of RE_2O_3 (mean and variance) in three samples of huanghoite from China and one sample from Siberia, see "Seltenerdelemente" A 4, 1979, pp. 22, 6. Three additional analyses of huanghoite from China show values from 38.52 to 39.71, with a mean value of 39.22% RE_2O_3, very close to the theoretical value of 39.39% RE_2O_3 [13, pp. 32/3].

The analyses of huanghoite from China closely fit the theoretical composition (ideal formulas) $CeBa[CO_3]_2F$ or $CeFCO_3 \cdot BaCO_3$*) [1 to 4]. An analysis of huanghoite from China (after exclusion of impurities) leds to the empirical formula

$$[(Ce, La, Pr, Nd \ldots)_{1.036}Th_{0.008}Ca_{0.026}Mg_{0.060}](Ba_{0.98}Sr_{0.02})[CO_3]_{1.95}F_{0.95}$$

[13, pp. 31/2]; the analysis of huanghoite from Siberia yields the formula

$$Ba_{1.05}Ca_{0.04}RE_{1.02}[CO_3]_2F_{1.04} \text{ [6, 7].}$$

The mineral is a double salt of $CeCO_3F$*) and $BaCO_3$ with $\Sigma Ce/Ba = 1:1$ [13, pp. 31, 34/5]; for the variability of the ratio Ce/Ba in the "Ba series" between bastnaesite and witherite, see p. 92.

The RE are possibly replaced by Th, and Ba by Sr [1 to 4]. Very small amounts of Fe_2O_3 and H_2O are probably due to finely disseminated hematite or goethite [6], cf. Fe^{3+} probably present in form of hematite in the Siberian huanghoite [7]. Spectral analyses further show the presence of thousandth of a percent of Ti and Zr in the Chinese mineral [1, 2], and traces of Mg, Mn, Cu, and Pb in huanghoite of Siberia [6].

The relative amounts of the RE ± Y in huanghoites from two localities are given in the following table, calculated from wt% RE ± Y = 100 (Nos. 1 to 5) or $RE_2O_3 = 100$ (No. 6):

No.	Locality	La	Ce	Pr	Nd	Sm	Eu	Gd	Ref.
1	China	22	49.8	6.4	20	1	0.1	0.5	[1, 2]
1a		22	50	6.4	20	1	0.1	0.5	[3]
2		24	46.9	6	19	1.4	0.2	1.2	[1, 2]
3		27.2	50.0	5.1	15.5	1.42	0.26	0.46	[13, p. 33]
4		34.5	48.8	3.9	11.5	0.3	0.2	0.5	[13, p. 33]
5		19.5	52.1	6.7	17.9	1.4	—	0.3	[13, p. 33]
6	Siberia	20	53	7	17	2	—	1	[6, 7]

No.	Locality	Tb	Dy	Ho	Er	Yb	Y	Ref.
1	China	—	0.2	—	—	—	—	[1, 2]
1a		—	—	—	0.2	—	—	[3]
2		0.2	0.2	—	0.4	0.5	—	[1, 2]
3		—	—	—	—	—	0.04	[13, p. 33]
4		—	0.1	—	—	—	0.2	[13, p. 33]
5		0.1	0.5	0.1	0.6	0.5	0.3	[13, p. 33]
6	Siberia	—	—	—	—	—	—	[6, 7]

Nos. 1 and 1a are two analyses from the same sample.

As can be seen from the table, huanghoite is characterized by an essentially ceric composition [1 to 4, 7], with very small amounts of heavy lanthanides [2].

*) The authors use Ce for RE ± Y, see table on this page.

10.5.2.6.3 Crystal Forms and Crystal Structure

Huanghoite is hexagonal [1 to 4, 7]. Besides narrow faces of the hexagonal prism on some crystals, the basic pinacoid {0001} is best developed [6]. Huanghoite from China is frequently characterized by tabular to lamellar aggregations or by irregular forms [1 to 4], huanghoite from Siberia shows flaky crystals, fan-shaped intergrowths, lamellar aggregations, radial-fibrous concretions grown on quartz and rarely granular masses [7]; huanghoite from Kola shows plates up to 2 cm in diameter [9], or is lamellar in nonterminated grains [10]. On poorly developed crystals the faces are uneven, step-like, and covered with etching figures [6], cf. [7].

A cleavage lamellar to {0001} is present on the mineral from China [1 to 4], and a distinct one on the mineral from Siberia [6, 7].

The unit cell constants a and c and the number of molecules in the unit cell Z for huanghoite are given in the following table:

No.	Locality	a (in Å)	c (in Å)	Z	Ref.
1	China	5.1	19.6	3[1)]	[1 to 4]
		5.11	19.6	3	[15]
2	China	5.07[2)]	38.6[2)]	—	[16]
3	China	5.070(3)	38.408(5)	6	[17]
4	Siberia	5.1	19.6	6[3)]	[6, 7]

[1)] Cell volume V = 440 $Å^3$. – [2)] Determined by electron diffraction method. – [3)] Z = 3 in [6].

The space group has been recently determined to be $R3m{-}C_{3v}^5$ (No. 160) [17]. The X-ray powder pattern of huanghoite from Siberia is analogous to that of huanghoite from China [6, 7], and also the interplanar spacings of huanghoite from Kola are similar to those of huanghoite from China [9].

The structure of huanghoite (as that of cordylite, see p. 178) is characterized by a bastnaesite cell alternating with a smaller cell along the c axis. The Ba present in huanghoite causes the change in the orientation of a large fraction of the CO_3 groups to a position perpendicular to the c axis in the smaller cell, which in turn leads to a marked change in the c parameters and to the negative optical sign of the mineral [15]. For a discussion of the relation between superstructure and pseudostructure of a huanghoite crystal, see [17]. Stacking faults in huanghoite can be directly observed by the electron diffraction method [16].

Huanghoite is isostructural with benstonite [3, 4].

10.5.2.6.4 Physical Properties

Huanghoite has a greasy luster and is subtransparent [1 to 4], the mineral from Siberia is macroscopically opaque with a few transparent areas; under the microscope it is characterized by numerous dust-like inclusions of hematite or goethite [6], cf. [7]. Huanghoite is uniaxial negative [1 to 4, 6, 7]. The color, the refractive indices, and the measured densities of huanghoites from China (No. 1), Siberia (No. 2), and Kola (No. 3) are given in the following table:

No.	Color	Refractive indices n_o	n_e	D(meas.) g/cm³	Ref.
1	honey-yellow, yellowish green	1.765	1.603	4.51 to 4.67	[1 to 4]
2	yellow, orange	1.766	1.588	4.58	[6, 7]
3	cinnamon	1.770	1.610	—	[10]

The mineral is weakly pleochroic in greenish yellow tints [1 to 4].

The hardness H = 4.5 to 5 [15]; the mineral from Siberia has H = 4.5 [6] or 5.5 [7]; the mineral from China has a microhardness of 411 kg/mm^2 corresponding to Mohs' scale hardness 4.7 [1 to 3]. For a microhardness of 382 to 448 kg/mm^2 see [4]. The fracture is uneven [1 to 4]. The density is calculated as 4.72 g/cm^3 by [3, 4], however only as 4.49 g/cm^3 by [18].

10.5.2.6.5 Chemical and Thermal Behavior

Huanghoite is easily soluble in hydrochloric acid [1 to 4].

The heating curve of huanghoite shows three endothermal effects in the intervals between 400 and 470, 620 and 680, and 760 and 800°C. The thermogravimetric curve shows the strongest loss at the second effect. The main products of decomposition (thermal dissociation) of huanghoite according to X-ray determination are two cubic phases with fluorite structure, namely the oxides CeO_2 and ΣRE_2O_3 with a ≈ 5.54 Å (possibly in solid solution with BaO with a = 5.54 Å), and BaF_2 with a = 6.21 Å together with CO_2 [1 to 4]. A further DTA curve which deviates from the one previously discussed shows deep endothermal valleys at 705, 778, and 794°C. The first effect is irreversible and mainly due to the release of CO_2 during the dissociation of the $CeCO_3F$ component, the second effect is reversible and mainly results from the change of $BaCO_3$ (witherite) [13, pp. 34/6].

References for 10.5.2.6:

[1] E. I. Semenov, Chzhan Pei-Shan (Kexue Tongbao **1961** No. 6, pp. 46/8). – [2] E. I. Semenov, Chzhan Pei-Shan (Sci. Sinica [Peking] **10** [1961] 1007/11 [Russian]). – [3] E. I. Semenov (Mineralogiya Redkikh Zemel', Moscow 1963, pp. 1/412, 107/9). – [4] E. I. Semenov (in: K. A. Vlasov, Geochemistry and Mineralogy of Rare Elements and Genetic Types of Their Deposits, Vol. 2: Mineralogy of Rare Elements, Israel Program Sci. Trans., Jerusalem 1966, pp. 249/328, 276/8 [Russian Original, Moscow 1964]). – [5] C. Hintze, K. F. Chudoba (Handbuch der Mineralogie, Suppl. Vol. 3, de Gruyter, Berlin 1968, pp. 73, 140).

[6] Yu. L. Kapustin (Dokl. Akad. Nauk SSSR **202** [1972] 422/5; Dokl. Earth Sci. Sect. **202** [1972] 116/9). – [7] Yu. L. Kapustin (Mineralogiya Karbonatitov, Moscow 1971, pp. 1/288, 170, 186/8). – [8] A. G. Zhabin, N. G. Shumyatskaya, N. S. Samsonova (Tr. Mineral. Muzeya Akad. Nauk SSSR No. 20 [1971] 202/4). – [9] O. B. Dudkin, F. V. Minakov (in: I. V. Bel'kov, Materialy po Mineralogii i Geokhimiya Shchelochnykh Kompleksov Porod Kol'skogo Poluostrova, Apatity 1975, pp. 159/74, 163/4, 166, 168/9). – [10] O. B. Dudkin, L. I. Polezhaeva, F. V. Minakov (Veshchestvennyi Sostav Shchelochnykh Intruzivnykh Kompleksov Kol'skogo Poluostrova, Apatity 1981, pp. 73/81, 76, 78, 80; Ref. Zh. Geol. **1981** No. 12 V 179).

[11] Yu. L. Kapustin (in: L. N. Ovchinnikov, N. A. Solodov (Mestorozhdeniya Litofil'nykh Redkikh Metallov, Nedra, Moscow 1980, pp. 243/85, 245, 271, 274). – [12] V. Dimov, V. Breskovska, M. Maleev (Geokhim. Mineral. Petrol. **13** [1980] 37/44). – [13] H. C. Wang, H.-T. Su, C.-T.

Kuo, S.-P. Li, C. K. Wang (Ti Ch'iu Hua Hsueh **1973** 31/8; Geochim. [Peking] **1973** 30/7; C.A. **79** [1973] No. 116895). – [14] M. Fleischer (Can. Mineralogist **16** [1978] 361/3). – [15] A. S. Povarennykh (Crystal Chemical Classification of Minerals, Vol. 1, translated from Russian by J. E. S. Bradley, Plenum, New York 1972, pp. 1/458, 181, Vol. 2, pp. 459/766, 616).

[16] F. Li, H. Fan, D. Yang, P. Fu, Y. Kong (Wuli Xuebao **31** [1982] 571/6, 576; C.A. **97** [1982] No. 112625). – [17] J. Qian, P. Fu, Y. Kong, G. Gong (Wuli Xuebao **31** [1982] 577/84 from C.A. **97** [1982] No. 112626). – [18] C. P'ei-Shan [Chzhan Pei-Shan] (Sci. Sinica [Peking] **11** [1962] 251/8, 256; C.A. **57** [1962] 10799; Ref. Zh. Geol. **1963** No. 2 V 145).

10.5.2.7 "Barium-Parisite" ("Cebaite") $Ce_2Ba_3[F_2|(CO_3)_5]$

Name and Occurrence

The name "barium-parisite", formerly a synonym for cordylite [1, 2], was given anew to a mineral species of the Ba-RE fluorocarbonate series with $\Sigma Ce:Ba = 2:3$. The original Chinese paper describes a mineral $Ce_2Ba_3(CO_3)_5F_2$ without giving a name, see [4], but the English translation of the original uses the name barium-parisite for this component [3, p. 30]. In the English abstract of a later Chinese paper on zhonghuacerite (see p. 186), the name "cebaite" was used for the mineral formerly named barium-parisite without any reference [6], cf. [5]. Though being inappropriate, see below, the name "barium-parisite" will be used in the following.

"Barium-parisite" was found together with cordylite and huanghoite in a hydrothermal metasomatic rare earth-iron ore deposit in North China, where it is mainly associated with Na pyroxene, fluorite, barite, and apatite [3, p. 30].

Chemistry

"Barium-parisites" (5 samples) contain 29.86 to 32.90%, mean 31.37% RE_2O_3 (theoretical value 31.86%). Two chemical analyses with 30.81 and 31.89% RE_2O_3, after deduction of impurities, lead to the following formula:

$$[(Ce,La,Pr,Nd\ldots)_{1.98}Ca_{0.05}Mg_{0.08}Th_{0.01}](Ba_{2.81}Sr_{0.19})[CO_3]_{4.73}F_{2.01}(OH)_{0.19}.$$

This approaches the theoretical formula given in the heading. "Barium-parisite" is a double salt, composed of the components $CeCO_3F$*) and $BaCO_3$, with a simple integer proportion rather than a series of similar materials with interchangeable Ba and RE. The ratio $Ba:\Sigma Ce = 3:2$ is relatively constant [3, p. 32]. This ratio and the ratio of the anions $F:CO_3 = 2:5$ distinguishes the mineral from parisite with $Ba:Ce = 2:1$ and $F:CO_3 = 3:2$.

The distribution of the RE and Y is characterized by a Ce maximum; the $\Sigma Ce/\Sigma Y$ ratio is given as ranging from 90 to 332. The complete distribution of the RE ($\Sigma RE \pm Y = 100$) in five samples of "barium-parisite" from North China is as follows [3, pp. 32/3]:

No.	La	Ce	Pr	Nd	Sm	Eu	Gd	Tb	Dy	Y
1	32.2	45.4	4.8	15.7	0.7	0.1	0.7	—	0.1	0.3
2	32.4	51.7	4.0	10.5	0.7	0.13	0.3	0.07	—	0.11
3	27.7	51.5	4.9	13.9	1.1	0.2	0.5	—	—	0.2
4	32.0	49.4	3.9	12.5	0.9	0.3	0.4	—	0.2	0.4
5	29.9	54.3	4.6	10.2	0.7	—	0.3	—	—	—

*) The authors use Ce for RE ± Y, see table on this page.

Unit Cell and Physical Properties

The mineral is hexagonal and has a platy habit. The unit cell constants are a = 4.96, c = 20.20 Å, c/a = 4.07, cell volume V = 434 $Å^3$, Z = 1. The X-ray powder pattern is basically the same as that of hexagonal huanghoite (see p. 183) [3, pp. 30/1].

The mineral has a vitreous to oily luster, is translucent to transparent, and has a wax-yellow color; it is uniaxial negative, sometimes biaxial negative with an axial angle of 5° to 10°. The refractive indices are n_o = 1.740 to 1.748, n_e = 1.598 to 1.604. The hardness H_M = 4.5, the fracture is conchoidal, density D(meas.) = 4.46 g/cm^3 [3, pp. 30/1].

Chemical Behavior and Thermoanalysis

The mineral is readily soluble in various inorganic acids [3, p. 30].

The thermoanalyses (DTA and TGA) indicate for "barium-parisite" a strong endothermal effect at 683 and another endothermal effect at 778°C, both similar to those of huanghoite. The first effect is irreversible and mainly due to a weight loss of 13.9% by the release of CO_2 between 670 and 710°C, the second effect is reversible and results mainly from a phase change of witherite ($BaCO_3$). Part of the CO_2 (about ⅖ of the total amount) is very stable and not released until 1100°C. The X-ray powder pattern shows the dissociation products of barium-parisite at 700 to 800°C to be mainly the phases CeO_2, BaF_2, and $BaCO_3$ [3, pp. 34/6].

References for 10.5.2.7:

[1] C. Palache, H. Berman, C. Frondel (The System of Mineralogy, 7th Ed., Vol. 2, Wiley, New York 1951, pp. 1/1124, 285). – [2] H. Strunz, C. Tennyson (Mineralogische Tabellen, 6th Ed., Geest and Portig, Leipzig 1977, pp. 1/621, 507). – [3] H.-C. Wang, H.-T. Su, C.-T. Kuo, S.-P. Li, C.-K. Wang (Ti Ch'iu Hua Hsueh **1973** 31/8; Geochim. [Peking] **1973** 30/7). – [4] G. Y. Chao (Am. Mineralogist **60** [1975] 738). – [5] P.-S. Zhang, K.-J. Tao (Sci. Geol. Sinica [Ti Chih K'o Hsueh] **1981** 195/6; Mineral. Abstr. **33** [1982] No. 82M/0707).

[6] G. Y. Chao (Am. Mineralogist **67** [1982] 1078).

10.5.2.8 Zhonghuacerite $CeBa_2[F|(CO_3)_3]$

Zhonghuacerite has been found in the Proterozoic metamorphic dolomite from Bayan Obo in northern China, associated with "cebaite" ("barium-parisite", cf. p. 185), ferro-dolomite, phlogopite, riebeckite, magnetite, quartz, and others. This new mineral has its name from the locality and from the chemical composition. It is a member of the RE-Ba fluorocarbonate mineral series, see p. 92, with the theoretical content of 26.71% RE_2O_3. According to three analyses it contains 22.02 to 24.85, mean = 23.26% RE_2O_3. The ideal formula given above is derived from the three following empirical formulas*⁾:

1) $(Ba_{1.67}Ca_{0.24}Sr_{0.08})(Ce_{0.39}Ln_{0.49}Fe_{0.18}Mg_{0.12}Al_{0.01}Th_{0.01})C_{2.78}O_{9.76}F_{0.95}$
2) $(Ba_{1.74}Ca_{0.29}Sr_{0.10})(Ce_{0.38}Ln_{0.43}Fe_{0.17}Mg_{0.01}Al_{0.02}Th_{0.01})C_{2.76}O_{9.79}F_{1.22}$
3) $(Ba_{1.73}Ca_{0.32}Sr_{0.11})(Ce_{0.36}Ln_{0.41}Fe_{0.11}Mg_{0.32}Al_{0.01}Th_{0.01})C_{2.73}O_{9.88}F_{1.55}$

An electron probe analysis shows Ce as the dominating element among the RE [1].

*⁾ Ln = La, Nd, and other lanthanides [1]. The formulas are from the Chinese text, they somewhat differ from those given in the English text by higher values for Ba and other cations; oxygen atoms per formula unit are too high in all formulas [2].

Zhonghuacerite forms extremely fine grains, which are composed to granular aggregates. The grains show no cleavage [1].

Zhonghuacerite is trigonal, the parameters of the hexagonal subcell are a′ = 5.07, c′ = 9.82 Å, V′ = 218.6 $Å^3$, Z = 1 [1]. The strongest X-ray diffraction lines are very close to those of huanghoite [2]. The electron diffraction pattern and the X-ray energy spectrum are, however, quite different from those of similar minerals, such as cordylite, huanghoite, and "cebaite" [1].

Zhonghuacerite has vitreous or greasy luster and is translucent. The color of the aggregates is pale yellowish, the streak (powder) is light yellowish; the mineral is uniaxial negative with refractive indices $n_o = 1.745$, $n_e = 1.565$*) [1].

The hardness $H_M \approx 4.6$, microhardness is 289 to 309 kg/mm^2. Density D(meas.) = 4.20 to 4.40 g/cm^3, D(calc.) = 4.66 g/cm^3 [1].

The mineral is soluble in hydrochloric, sulfuric, nitric, and perchloric acid [1].

The DTA curve shows the mineral to be quite different from cordylite, huanghoite, and "cebaite"; the first pronounced endothermal effect, which indicates decomposition with release of CO_2 appears at 645°C and not at 680 to 690°C as it does for the last two minerals; the second pronounced endothermal effect, indicating decomposition of $BaCO_3$, occurs at 800°C for all three minerals [1].

References for 10.5.2.8:

[1] P.-S. Zhang, K.-J. Tao (Sci. Geol. Sinica [Ti Chih K'o Hsueh] **1981** 195/6; Mineral. Abstr. **33** [1982] No. 82M/0707). – [2] G. Y. Chao (Am. Mineralogist **67** [1982] 1078).

10.5.2.9 Ambatoarinite $Sr_5(Ce,La,Dy)_{10}[CO_3]_{17}O_3$

Ambatoarinite occurs enclosed in plates of celestine in veins of manganoan calcite and quartz in metamorphosed limestone, associated with aegirine, biotite, microcline, galena, monazite, and apatite at Ambatoarina, near Ambositra, Madagascar [1, 2]. It has also been found as accessory mineral together with cordylite, pyrochlore, carbonates, barite, vermiculite, and sulfides in the inner zone of the carbonatite core of the alkaline Sallanlatva massif, Kola, USSR [3]. A mineral, occurring rarely together with ancylite in the latest dolomite-calcite carbonatites of Vuoriyarvi, Kola, originally supposed to be also ambatoarinite [4], according to mineralogical, chemical and X-ray investigations has been shown to be carbocernaite [5], cf. [6, 7], see also p. 71.

The analysis of material from Madagascar, separated from associated minerals, shows the mineral to be a Ce-Sr carbonate [1], "Seltenerdelemente" A 1, 1938, pp. 59/60, with the formula given in the heading [2]. The modified formula $Sr(Ce,La,Nd)[O|(CO_3)_3]$ indicates a "strontium-oxyparisite" [8].

Deviating from the otherwise similar ancylite, the mineral contains no water and has the opposite optical sign (negative) [6, 7].

The mineral is probably orthorhombic, reddish to blackish due to ferruginous impurities; it is optically negative with $n_x > 1.658$ and a birefringence of about 0.08; the axial plane is parallel to (010) with X = c, in thin section the prism angle is near 95° [1, 2], cf. [7].

The mineral needs further investigation, particularly in relation to ancylite [2].

*) In English text $n_e = 1.568$.

References for 10.5.2.9:

[1] A. Lacroix (Bull. Soc. Franc. Mineral. **38** [1915] 265/71, 268/70). – [2] C. Palache, H. Berman, C. Frondel (The System of Mineralogy, 7th Ed., Vol. 2, Wiley, New York 1951, pp. 1/1124, 293). – [3] B. I. Serba (Magmaticheskie Obrazovaniya Kol'skogo Poluostrova, Moscow 1962, pp. 58/64, 63; Ref. Zh. Geol. **1963** No. 4 V 216). – [4] L. V. Kozyreva, G. A. Il'inski (Mater. Mineral. Kol'sk. Poluostrova **1** [1959] 69/76, 70, 72). – [5] A. G. Bulakh, V. V. Kondrat'eva, E. N. Baranova (Zap. Vses. Mineral. Obshchestva **90** [1961] 42/9, 42).

[6] E. I. Semenov (Mineralogiya Redkikh Zemel', Moscow 1963, pp. 1/412, 93/4). – [7] E. I. Semenov (in: K. A. Vlasov, Geochemistry and Mineralogy of Rare Elements and Genetic Types of Their Deposits, Vol. 2: Mineralogy of Rare Elements, Israel Program Sci. Trans., Jerusalem 1966, pp. 249/328, 262/4 [Russian Original, Moscow 1964]). – [8] H. Strunz, C. Tennyson (Mineralogische Tabellen, 6th Ed., Geest and Portig, Leipzig 1977, pp. 1/621, 243).

10.5.2.10 Unnamed Orthorhombic Y-Ca-RE Carbonate

A mineral, occurring in the form of spherulitic crusts, outgrowths, and veinlets together with tengerite, fluorite, synchysite, and opal, was found in a granite pegmatite of Kazakhstan. It is not identical with any of the known minerals.

Due to insufficient material, a chemical analysis was not possible; X-ray spectral analysis gives: CaO 26.32, Y_2O_3 35.55, Dy_2O_3 3.21, Ho_2O_3 1.12, Er_2O_3 3.66, and Yb_2O_3 2.51; the composition of the RE is yttric. The presence of carbonate and silica is shown by chemical tests.

The prismatic crystals are somewhat elongated parallel to c and lamellar; they show twins and a complete cleavage. The X-ray examination of monocrystals and by powder patterns show the mineral to be orthorhombic with unit cell constants a = 13.32, b = 14.02, c = 9.73 Å, a : b : c = 0.950 : 1 : 0.695, V = 1817.5 Å^3, Z = 4. The space group is $C222_1$–D_2^5 (No. 20). The molecular weight of 793.7 is calculated from unit cell data, Z, and density.

The mineral is transparent, colorless or yellowish to brownish due to films of Fe hydroxide. It is biaxial negative, with small axial angle 2V and refractive indices n_x = 1.577, n_z = 1.611; it shows direct extinction and no pleochroism. The hardness is 4.5, the mineral is brittle. Density D(meas.) = 2.9 g/cm^3.

The mineral is soluble in concentrated hydrochloric acid at heating with development of CO_2 and formation of a silicic residue.

Reference for 10.5.2.10:

Yu. V. Shipovalov, A. V. Stepanov (Issledovaniya v Oblasti Khimicheskikh i Fizicheskikh Metodov Analiza Mineral'nogo Syr'ya, Alma-Ata 1971, pp. 186/8; Zap. Vses. Mineral. Obshchestva **102** [1973] 448 [Abstract]).

10.5.3 Hydrous Carbonates

The group of hydrous RE carbonates consists of minerals formed during supergene weathering of primary RE minerals and occurring as ochers. These minerals are readily dissolved in acids.

A discrimination between only 2 types of hydrates, namely those with 8 and those with 4 molecules of water, is not in accordance with reports on crystal hydrates with 2, 6 etc. molecules of water in the chemical literature, E. I. Semenov (in: K. A. Vlasov, Geochemistry and

Mineralogy of Rare Elements and Genetic Types of Their Deposits, Vol. 2: Mineralogy of Rare Elements, Israel Program Sci. Trans., Jerusalem 1966, pp. 249/328, 278 [Russian Original, Moscow 1964]).

Several new minerals of this type found in recent time confirm, with contents of about 1.5 to 3 (in 3 cases) molecules of water per formula unit, the supposed broader range of water of crystallization in this mineral group.

10.5.3.1 **Calkinsite** $(RE)_2[CO_3]_3 \cdot 4H_2O$

Occurrence and Paragenesis

Calkinsite has been found originally in Vermiculite Prospects at the head of Big Sandy Creek, Bearpaw Mountains, Montana. It occurs in silicate-carbonate veins in shonkinite, mafic monzonite, and syenite, being most abundant in partly altered areas of a shonkinite. Here it occurs together with burbankite-ancylite intergrowths, with lanthanite, barite, goethite, sulfides and a number of unidentified alteration products. The mineral forms by decomposition of burbankite, most probably by weathering. Some calkinsites, occurring in isolated vugs without signs of weathering, may be formed under late hydrothermal conditions [1, pp. 1171/2, 1183], cf. [2, 3].

A second occurrence of calkinsite has been found in burbankite-calcite veins in the massif Vuoriyarvi, Kola, where it is grown on the surface of altered burbankite, associated and intergrown with barite, strontianite, and pyrite. It only occurs to a depth of 70 m in a sphere without weathering [4]. In a sequence of carbonatite minerals calkinsite is indicated as one of the minerals of the calcitic phase of the second genetic group of carbonatites [5].

A synthesis by the reaction of La and Ce hydroxides with $NaHCO_3$ solution saturated with CO_2 shows calkinsite to form much more readily than lanthanite [6], which is a more common mineral in nature, see p. 191.

Chemistry

According to one analysis with 62.73% $(RE)_2O_3$ (after deduction of impurities and recalculation to 100) the formula of calkinsite can be expressed as $(RE)_2[CO_3]_3 \cdot 4H_2O$; recalculating the chemical analysis from theoretically perfect ratios the probable composition of pure calkinsite corresponds to the formula $(La_{0.64}Ce_{1.02}Pr_{0.09}Nd_{0.25})[CO_3]_3 \cdot 4H_2O$. This formula shows the close relationship between calkinsite and lanthanite, that differs only by its content of $8H_2O$ [1, pp. 1178, 1181], cf. p. 193.

The RE distribution of calkinsites in at.% is shown in the following table:

Locality	La	Ce	Pr	Nd	Sm	Gd	Tb	Dy	Er	Yb	Ref.
Montana[1)]	31.6	51.1	4.6	12.7	—	—	—	—	—	—	[1, p. 1178]
Vuoriyarvi	27	56	6	9	1	1	—	—	—	—	[5]
Vuoriyarvi[2)]	32	50.2	4	12	0.5	0.2	0.1	0.2	0.2	0.6	[4]

[1)] Recalculated from the RE oxides. – [2)] $\Sigma RE_2O_3 = 55.21\%$.

Crystal Forms, Crystal Structure, and Physical Properties

Calkinsite is orthorhombic; the axial ratio (from structure data) a:b:c = 0.757:1:0.707. The crystal forms are c{001}, b{010}, a{100}, d{102}, q{122}, r{132}. The habit is typically platy parallel to {010}, the plates are commonly twinned, compound, and elongated parallel to the a axis; the twinning plane is {101}; see **Fig. 9**. The cleavage is {010} perfect, {101} distinct and {001} possibly a parting [1, pp. 1176/7, 1182]. The platy crystals from Vuoriyarvi have a distinct cleavage parallel to their flattening [4].

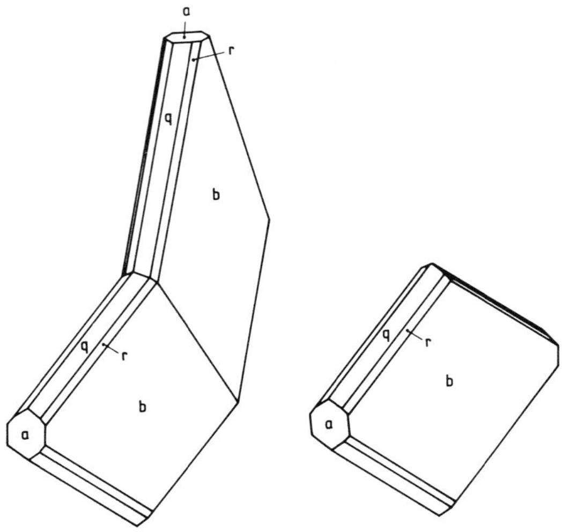

Fig. 9. Clinographic projection of twinned (left) and untwinned (right) crystals of calkinsite from [1, p. 1177].

The unit cell parameters are a = 9.57 ± 0.02, b = 12.65 ± 0.08, and c = 8.94 ± 0.02 Å, the space group is $P2_122_1$-D_2^3 (No. 18, standard setting $P2_12_12$); Z = 4 and cell volume V = 1083 $Å^3$ [1, p. 1180]. The X-ray powder pattern indexed by use of the cell dimensions and obtained from the single crystal study shows good agreement of the calculated and observed interplanar spacings for {030}, {040}, and {060}, but poor agreement for {020} perhaps by absorption [1, pp. 1179/81]. The X-ray powder pattern of the mineral from Vuoriyarvi is similar to that of calkinsite from Montana [4], which is given by [7].

Calkinsite from Vuoriyarvi is transparent with a bright yellow color [4], the crystals from Montana are pale yellow with refractive indices $n_x = 1.569 \pm 0.002$, $n_y = 1.657 \pm 0.002$, $n_z = 1.686 \pm 0.002$; 2V negative 54° (calc.) and 57.2° (est.), dispersion $r < v$ is perceptible. Pleochroism X = b (yellow), Y = c (colorless), Z = a (colorless). The mean index of refraction calculated according to the rule of Gladstone and Dale is $n_m = 1.619$ compared with the mean index calculated from indices of refraction n = 1.636 [1, pp. 1176/7, 1181].

The hardness on Mohs' scale is about 2.5, the measured density D = 3.28 ± 0.01 g/cm^3 [1, p. 1176]; the mineral is brittle [4].

Calkinsite is soluble in cold dilute hydrochloric acid with strong effervescence [1, p. 1176].

References for 10.5.3.1:

[1] W. T. Pecora, J. H. Kerr (Am. Mineralogist **38** [1953] 1169/83). – [2] E. I. Semenov (Mineralogiya Redkikh Zemel', Nauka, Moscow 1963, pp. 1/412, 112). – [3] C. Hintze (Handbuch der Mineralogie, Suppl. Vol. 2, de Gruyter, Berlin 1960, pp. 1/958, 677/8). – [4] Yu. L. Kapustin (Mineralogiya Karbonatitov, Nauka, Moscow 1971, pp. 1/288, 170, 188/91). – [5] Yu. L. Kapustin (Geokhimiya **1966** 1311/21; Geochem. Intern. **3** [1966] 1054/64, 1055, 1060).

[6] P. C. Sæbø (Norsk Geol. Tidsskr. **41** [1961] 311/7, 315/6). – [7] H. Neumann, K. O. Bryn (Avhandl. Norske Videnskaps Akad. Oslo I Mat. Natur. Klasse **1958** No. 1, pp. 1/15, 5, 8).

10.5.3.2 Lanthanite and Lanthanite-(Nd) $(RE)_2[CO_3]_3 \cdot 8H_2O$

10.5.3.2.1 Name and Synonyms

The name lanthanite has been given by Haidinger in 1845 to a mineral from Bastnäs, Sweden, because of its predominance of light RE; this, however, is erroneous, as its La content is not higher as in other RE minerals, see "Seltenerdelemente" A 1, 1938, p. 59, and [1, 2]. A later study found an Nd-predominant member of the lanthanite group from Curitiba, Paraná, Brazil, with Nd > La, and rich in Sm and Pr, but nearly free from Ce; this mineral has been named "coutinite" by Fujimori [3], but lanthanite-(Nd) by Roberts et al. [4] in accordance with the nomenclature of RE minerals proposed by [5]. The last name is approved by IMA and has priority [6].

A phase $Ln_2[CO_3]_3 \cdot 8H_2O$ is stable for the elements La to Nd in the ternary heterogeneous system Ln_2O_3-H_2O-CO_2 under hydrothermal conditions; it is isostructural with lanthanite [7].

References for 10.5.3.2.1:

[1] C. Hintze (Handbuch der Mineralogie, Vol. 1, Div. 3, Part 1, de Gruyter, Berlin 1930, pp. 2675/3656, 3498/9). – [2] E. I. Semenov (in: K. A. Vlasov, Geochemistry and Mineralogy of Rare Elements and Genetic Types of Their Deposits, Vol. 2: Mineralogy of Rare Elements, Israel Program Sci. Trans., Jerusalem 1966, pp. 249/328, 282 [Russian Original, Moscow 1964]). – [3] K. Fujimori (Anais Acad. Brasil. Cienc. **53** [1981] 147/52, 150). – [4] A. C. Roberts, G. Y. Chao, F. Cesbron (Geol. Surv. Paper Geol. Surv. Can. No. 80-1C [1980] 141/2). – [5] A. A. Levinson (Am. Mineralogist **51** [1966] 152/8, 153/4).

[6] M. Fleischer (Am. Mineralogist **67** [1982] 414). – [7] J. A. K. Tareen, T. R. N. Kutty, I. Mohamed (Indian Mineralogist **21** [1980] 43/8, 43).

10.5.3.2.2 Occurrence and Paragenesis

Lanthanite is a common supergene RE mineral [1], that often is formed as an alteration product, especially by oxidation and weathering after several RE minerals such as cerite and allanite [2], mostly allanite [3], but also burbankite [4] or burbankite and ancylite [5]. As a late secondary mineral it forms at very low temperatures probably from solutions which contain RE hydroxides besides alkalies and CO_2 [5]. As a product of weathering of RE minerals in weathered zones of carbonatites, lanthanite may be more widely distributed than presently described. It is, however, relatively rarely observed, as such incrustations are constantly contaminated with supergene powdery monazite; thus pure material is not easily obtained [4].

The first discovery of lanthanite at Riddarhyttan, Bastnäs, Västmanland, Sweden, shows it to be an alteration product of cerite [5], occurring as crystalline film on cracks and fissures in the cerium ore and sometimes in druses formed later than the other Ce minerals [6]. In this Mg metasomatic deposit, which is the type locality for five cerium minerals, lanthanite is commonly associated with bastnaesite, sulfides, and magnetite; as surficial, secondary product it replaces the dolomite [7], and probably bastnaesite [6]. Lanthanite may be also the H_2O-rich, white alteration product of large bastnaesite crystals from Madagascar [8] and of bastnaesite and tysonite from an occurrence near Manitou, El Paso Co., Colorado [9].

As an alteration product of allanite, lanthanite has been observed in a pegmatite at Baringer Hill, Llano County, Texas [10], in RE-bearing pegmatite in the Adirondack Mountains, coating cleavage planes of allanite as microscopic film [3], and in the magnetite deposit of the Sandford Mine, Moriah [10], both Essex County, New York. The same origin is supposed for lanthanite contained in an ochery soil from the zinc mine at Friedensville, Saucon Valley [11], known as occurrence in oxidized zinc ore from Bethlehem, Lehigh County, Pennsylvania [10]. Lanthanite found in small cavities in the pegmatite from Seteråsen on Mjaerum farm, Hobøl, Østfold district, Norway [12], has been formed as a late secondary mineral in the surroundings of altered allanite [5].

In silicate-carbonate-sulfide deposits in the Bearpaw Mountains, Montana, lanthanite occurs, closely associated with calkinsite (see p. 189) and other minerals, as a weathering product of burbankite and ancylite; together with barite it is most abundant in completely altered material. It is still richer in RE than the altered complex RE carbonates [13]. In Vuoriyarvi, Kola, RE minerals often are replaced under supergene conditions by lanthanite and rhabdophane [14]; as preferred weathering product lanthanite is here observed at the surface of altered burbankite, associated with aragonite and celestite [4].

Lanthanite-(Nd) from the Curitiba basin, Paraná, Brazil, occurs as crusts on calcite nodules or on the surface of outcropping sediments with coatings on bedding planes, or as irregular films in fissures of the sediment; the lanthanite is epigenetically formed by evaporation of salt-containing solutions [15]. In the sediments of the same locality also minute crystals are observed on fissure surfaces of clay-carbonate concretions [16]. Another sedimentary lanthanite occurs in the slate of the Campbell Medford quarry, 30 miles to the west of Baltimore, Maryland, together with limonite [17]. For an association of lanthanite with rhabdophane in tertiary sandstones of the European USSR, see [1]. For a single lanthanite crystal found at the Grimsel Pass, Switzerland, the paragenesis is not given [18].

References for 10.5.3.2.2:

[1] E. I. Semenov (in: K. A. Vlasov, Geochemistry and Mineralogy of Rare Elements and Genetic Types of Their Deposits, Vol. 2: Mineralogy of Rare Elements, Israel Program Sci. Trans., Jerusalem 1966, pp. 249/328, 283 [Russian Original, Moscow 1964]). – [2] K. Fujimori (Anais Acad. Brasil. Cienc. **53** [1981] 147/52, 147). – [3] E. B. Rowley (Rocks and Minerals **37** [1962] 341/7, 341, 347, 453/60, 458). – [4] Yu. L. Kapustin (Mineralogiya Karbonatitov, Nauka, Moscow 1971, pp. 1/288, 191/3). – [5] P. C. Sæbø (Norsk Geol. Tidsskr. **41** [1961] 311/7, 311/3, 315/6).

[6] P. Geijer (Sveriges Geol. Unders. Årsbok Ser. C **14** No. 6 [1920/21] 1/24, 12, 20). – [7] P. Geijer (Arkiv Mineral. Geol. **3** [1962/65] 99/105, 100). – [8] J. Behier (Rep. Malgache Ann. Geol. Madagascar No. 29 [1960] 1/78, 18). – [9] E. B. Eckel (U.S. Geol. Surv. Bull. No. 1114 [1961] 1/399, 202). – [10] C. Palache, H. Berman, C. Frondel (The System of Mineralogy, 7th Ed., Vol. 2, Wiley, New York 1951, pp. 1/1124, 242).

[11] R. G. Wells (Rocks Minerals **53** [1978] No. 3, pp. 108/11). – [12] H. Neumann (Norsk Geol. Tidsskr. **39** [1959] 231/6, 233). – [13] W. T. Pecora, J. H. Kerr (Am. Mineralogist **38** [1953] 1169/83, 1171/2, 1183). – [14] Yu. L. Kapustin (Geokhimiya **1966** 1311/21; Geochem. Intern. **3** [1966] 1054/64, 1056). – [15] F. Cesbron, M.-C. Sichère, H. Vachey, J.-P. Cassedanne, J. O. Cassedanne (Bull. Mineral. **102** [1979] 342/7, 345/6).

[16] J. M. V. Coutinho (Univ. São Paulo Fac. Fil. Cienc. Letras Bol. Mineral. No. 13 [1955] 119/26 from C.A. **1956** 4728). – [17] A. Roe (Mineral. Rec. **6** [1975] 299; C.A. **84** [1976] No. 124463). – [18] M. Gramaccioli (Mineral. Rec. **8** [1977] 287/93, 288).

10.5.3.2.3 Chemistry

Analyses and formulas. Compilation of analyses, mean and variance of $(RE,Y)_2O_3$ (4 samples), see "Seltenerdelemente" A 4, 1979, pp. 23, 5. The content in Y_2O_3 of lanthanite from Bastnäs, Sweden, is 0.79% [1], of lanthanite-(Nd) from Curitiba, Paraná, Brazil, 0.22% [2] and (mean of 3 samples) 0.20% [3]. New analyses of lanthanite-(Nd) from Curitiba lead to the following formulas $(Nd_{0.850}La_{0.740}Sm_{0.280}Pr_{0.117}Y_{0.004})[CO_3]_3 \cdot nH_2O$ with $n \approx 8$ or abbreviated $(Nd, La, Pr, Sm, Y)_2[CO_3]_3 \cdot nH_2O$ [3] and $(Nd_{0.397}La_{0.365}Pr_{0.096}Sm_{0.072}Gd_{0.029}Eu_{0.028}Dy_{0.007}Y_{0.006})_2[CO_3]_3 \cdot 8H_2O$ or theoretical $(Nd, La)_2[CO_3]_3 \cdot 8H_2O$ (not taking account of 0.03% Ce_2O_3 and ThO_2 each) [5, 2]. The formula of lanthanite from Campinhas, São Paulo, Brazil, is given as $(La, Y, Nd)_2[CO_3]_3 \cdot nH_2O$ and that of the samples from Bastnäs, Sweden, and from Pennsylvania as $(La, Ce)_2[CO_3]_3 \cdot 8H_2O$ [3], cf. [1, 4]. For a deviating calculation of the water content of lanthanite assuming a stoichiometric amount of CO_2, see [6].

Distribution of RE. Lanthanite from Bastnäs [1] and from Kola [7] are characterized by a strong dominance of Ce-earths over Y-earths. Lanthanite from the Medford quarry west of Baltimore has major La and Ce, minor Sm and Nd [8]. Lanthanite-(Nd) from Curitiba contains besides Nd > La more Eu than most RE minerals, but practically no Ce [5]. For the distribution of RE in % in lanthanites from various localities, see the following table:

Locality	$\Sigma(RE \pm Y)_2O_3 = 100$									Ref.
	La	Ce	Pr	Nd	Sm	Eu	Gd	Dy	Y	
Curitiba	36.38	—	6.44	40.91	9.16	1.39	5.72	—	—	[6]
Curitiba[1)]	38.47	—	5.89	42.01	13.29	—	—	—	0.35	[3]
Curitiba	35.60	0.05	9.50	40.00	7.50	3.00	3.10	0.80	0.40	[5, 2]
Curitiba	41.97	0.04	9.28	39.70	7.12		~1.8[2)]		—	[10]
Vuoriyarvi	27	55	5	11	1	—	1	—	—	[9]
Kola	33	50	6	10	1	—	—	—	—	[7]

[1)] Mean of 3 samples. – [2)] Eu, Gd, Dy, Ho, and Yb.

References for 10.5.3.2.3:

[1] C. Palache, H. Berman, C. Frondel (The System of Mineralogy, 7th Ed., Vol. 2, Wiley, New York 1951, pp. 1/1124, 242). – [2] A. C. Roberts, G. Y. Chao, F. Cesbron (Geol. Surv. Paper Geol. Surv. Can. No. 80-1C [1980] 141/2). – [3] K. Fujimori (Anais Acad. Brasil. Cienc. **53** [1981] 147/52, 148/9, 152). – [4] P. C. Sæbø (Norsk Geol. Tidsskr. **41** [1961] 311/7, 311). – [5] F. Cesbron, M.-C. Sichère, H. Vachey, J.-P. Cassedanne, J. O. Cassedanne (Bull. Mineral. **102** [1979] 342/7, 344).

[6] H. G. Ansell, G. J. Pringle, A. C. Roberts (Geol. Surv. Paper Geol. Surv. Can. No. 76-1B [1976] 353/5). – [7] Yu. L. Kapustin (Mineralogiya Karbonatitov, Nauka, Moscow 1971, pp. 1/288, 170, 193). – [8] A. Roe (Mineral. Rec. **6** [1975] 299). – [9] Yu. L. Kapustin (Geokhimiya **1966** 1311/21; Geochem. Intern. **3** [1966] 1054/64, 1060). – [10] D. P. Svisero, Y. Mascarenhas (3rd Atas Simp. Reg. Geol., Curitiba, Braz., 1981, Vol. 1, pp. 295/304 from C. A. **97** [1982] No. 130656).

10.5.3.2.4 Crystal Forms and Crystal Structure

Lanthanite is orthorhombic [1, 2, 3]. The axial ratio a : b : c = 0.5568 : 1 : 0.5629 [1] according to unit cell parameters a : b : c = 0.5594 : 1 : 0.5279 [2], 0.557 : 1 : 0.527 [3].

The crystal forms known from Bastnäs and Bethlehem, a{100}, b{010}, c{001}, d{101}, and o{121} [1, 4], are observed on crystals from Curitiba together with the form {111} [5] or with {122} [6]; see **Fig. 10**, p. 194, and a restricted number of these forms for Hobøl, Norway [4].

The habit of well developed crystals is platy to thick tabular {010} [1, 4], cf. [7, 8]. The crystals of sedimentary lanthanite-(Nd) are platy [5, 2], or platy to thick tabular [9], they are rarely well developed, mostly they are micaceous, flattened to the cleavage plane {010} [6], or to the basic plane {001} [10]. Further there occur lath-like crystals by extension [001] [1], thin lath-like radiating spears, greatly resembling strengite [11], irregular fan-shaped aggregates of tabular crystals [4], and radiating tufted groups or crusts of matted microscopic acicular crystals [12], cf. fine nodules of radiated fibrous structure under the microscope [13]. Also fine granular to earthy forms exist [1].

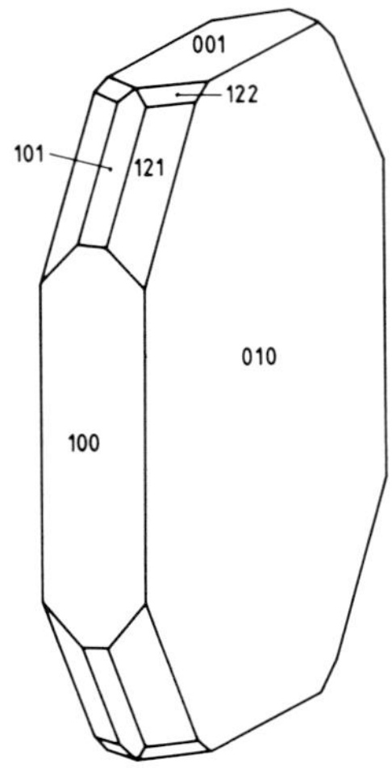

Fig. 10. Habit and forms of lanthanite-(Nd) from Curitiba basin, Paraná, Brazil, from [6].

Striation on {010} is caused by twinning after {101} [6], cf. [9, 2]. A faint lineation on {010} parallel to the edge of {101} of crystals from Hobøl may represent submicroscopic twinning [4]. The twin and composition plane is {101} [1], cf. twin plane {101} (Montana) [14], often found also in Hobøl, where sometimes cross-shaped twins occur [4].

The cleavage is {010} perfect and {101} very good [2] or distinct [14], {010} is well developed only for Hobøl [4].

The unit cell parameters for lanthanites and lanthanite-(Nd) from various occurrences and calculated density values D(calc.) are given in the following table:

Locality	a in Å	b in Å	c in Å	cell volume in $Å^3$	D (calc.) in g/cm^3	Ref.
Bastnäs, Sweden	9.504(4)	16.943(6)	8.937(5)	1439.1	2.78[1)]	[15]
Bethlehem, USA	9.52	17.13	9.02	—	—	[1]
Curitiba, Brazil[2)]	10.01	9.50[2)]	17.04[2)]	—	—	[10]
	9.470(7)	16.902(7)	8.929(6)	1429.1	—	[9]
	9.476(4)	16.940(8)	8.942(4)	1435.4	2.816[3)]	[6, 2]
	9.460(5)	16.884(9)	8.905(5)	—	2.83	[18]
Synthetic	9.580(4)	17.00(1)	8.984(4)	1463.1	2.732	[16, 9]
$La_2[CO_3]_3 \cdot 8H_2O$	9.58	17.22	9.00	—	—	[9]

[1)] For $(La,Ce)[CO_3]_3 \cdot 8H_2O$. – [2)] b and c values are exchanged, see space group below. – [3)] For the analytical formula of [2] (see p. 193).

The space group of lanthanite is Pbnb-D_{2h}^{10} (No. 56, standard setting Pccn) [2, 15, 6, 9]; or (with exchanged b and c) Pccn [10]; Z = 4 [15, 2, 3]. Lanthanite-(Nd) from Curitiba with Nd > La has a cell volume ~2% smaller than that of the synthetic compound; lanthanite from Bastnäs with Ce_2O_3 = 25.52%, but unknown relative amounts of La : Nd, has an intermediate cell volume. The a value is obviously most sensitive for replacement of different RE by each other [6].

X-ray powder patterns of lanthanite from Bastnäs, Sweden, check that from Hobøl, Norway [17], respectively those from Medford west of Baltimore and from Saucon Valley, Pennsylvania [11], and that from Curitiba, Paraná [9]. A third pattern of lanthanite from Bastnäs given by Sabina and Trail does not match the patterns of lanthanite from Curitiba [9] and from Medford; the deviations are evidently caused by an error in the published data [11]. The parameters of lanthanite from Curitiba are comparable with those of lanthanite from Bethlehem (cf. [1]) and, after transformation, with those of 2 synthetic lanthanites [9], cf. [16] and table p. 194.

The principal structural feature of lanthanite consists of infinite layers of RE-O coordination polyhedra and CO_3^{2-} groups lying parallel to the a-c plane and being connected to one another by hydrogen bonds only. This feature accounts for the micaceous cleavage {010}. The 2 independent RE atoms have a tenfold coordination: RE(1) is surrounded by 6 oxygen atoms belonging to the carbonate groups and by 4 water molecules; RE(2) is bonded to 8 carbonate oxygens and 2 water molecules. The corresponding coordination polyhedra are similar and cannot be described in terms of regular geometric solids. For the setting of the coordination polyhedra, see original paper Fig. 3 [15]. For stacking faults along the cleavage plane {001}, observed in precession diagrams of lanthanite-(Nd), see [10].

References for 10.5.3.2.4:

[1] C. Palache, H. Berman, C. Frondel (The System of Mineralogy, 7th Ed., Vol. 2, Wiley, New York 1951, pp. 1/1124, 241/2). – [2] A. C. Roberts, G. Y. Chao, F. Cesbron (Geol. Surv. Paper Geol. Surv. Can. No. 80-1C [1980] 141/2). – [3] E. I. Semenov (Mineralogiya Redkikh Zemel', Nauka, Moscow 1963, pp. 1/412, 112/3). – [4] P. C. Sæbø (Norsk Geol. Tidsskr. **41** [1961] 311/7, 313/4). – [5] J. M. V. Coutinho (Univ. São Paulo Fac. Fil. Cienc. Letras Bol. Mineral. No. 13 [1955] 119/26 from C.A. **1956** 4728).

[6] F. Cesbron, M.-C. Sichère, H. Vachey, J.-P. Cassedanne, J. O. Cassedanne (Bull. Mineral. **102** [1979] 342/7, 342/3). – [7] M. Gramaccioli (Mineral. Rec. **8** [1977] 287/93, 288). – [8] R. G. Wells (Rocks Minerals **53** [1978] No. 3, pp. 108/11). – [9] H. G. Ansell, G. J. Pringle, A. C. Roberts (Geol. Surv. Paper Geol. Surv. Can. No. 76-1B [1976] 353/5). – [10] K. Fujimori (Anais Acad. Brasil. Cienc. **53** [1981] 147/52, 149, 152).

[11] A. Roe (Mineral. Rec. **6** [1975] 299; C.A. **84** [1976] No. 124463). – [12] E. B. Rowley (Rocks Minerals **37** [1962] 341/7, 347). – [13] Yu. L. Kapustin (Mineralogiya Karbonatitov, Nauka, Moscow 1971, pp. 1/288, 191/3). – [14] W. T. Pecora, J. H. Kerr (Am. Mineralogist **38** [1953] 1169/83, 1177). – [15] A. Dal Negro, G. Rossi, V. Tazzoli (Am. Mineralogist **62** [1977] 142/6, 142/4, 146).

[16] D. B. Shinn, H. A. Eick (Inorg. Chem. **7** [1968] 1340/5). – [17] H. Neumann, K. O. Bryn (Avhandl. Norske Videnskaps. Akad. Oslo I Mat. Natur. Klasse **1958** No. 1, pp. 1/15, 5, 7). – [18] D. P. Svisero, Y. Mascarenhas (3rd Atas Simp. Reg. Geol., Curitiba, Braz., 1981, Vol. 1, pp. 295/304 from C.A. **97** [1982] No. 130 656).

10.5.3.2.5 Physical Properties

Optical Properties

Lanthanite has a pearly [1, 2], sometimes pearly to vitreous luster [3, 4]; it is transparent [1, 2], but turns to a dull milky white on exposure to the air [4], cf. [5].

The mineral is colorless, white, rose, or yellowish [1, 2], often white in Colorado [6], Adirondack Mountains [7], and Kola [8], yellowish at the Grimsel Pass, Alps [9], pink in Maryland [10], Pennsylvania [11], Hobøl, Norway [12], and bright pink in Curitiba [4], cf. [3, 5]. In transmitted light it is colorless [1, 4], the streak is white [4]. The optical character of lanthanite is biaxial negative [1, 2, 4]. Values for the axial angles 2V, the dispersion r to v, and the refractive indices of lanthanites and lanthanite-(Nd) from several occurrences, see in the following table:

Occurrence	n_x	n_y	n_z	2V	Dispersion r to v	Ref.
Bastnäs, Sweden	1.52	1.587	1.613	63°	—	[2, 8, 13]
Bethlehem, USA	1.52 ± 0.01	1.587 ± 0.003	1.613 ± 0.003	~63°[1)]	r<v (weak)	[1, 13]
Montana	—	1.587	—	60°	r<v (perceptible)	[14]
Hobøl, Norway	1.520 ± 0.01[2)]	1.586 ± 0.004	1.615 ± 0.005	63° ± 1°	—[3)]	[12]
Kola, USSR	1.515	—	1.613	—	—	[8]
Campinas, São Paulo	1.518	1.588	1.613	62°	—	[15]
Curitiba, Paraná	1.514[2)]	1.589 ± 0.003	1.612 ± 0.003	60°	—	[16, 13]
	1.532(1)	1.590(1)	1.614(1)	61°[4)]	—	[4]

[1)] 62° are given by [17]. – [2)] Calculated value. – [3)] The usual dispersion r<v was not found in this lanthanite. – [4)] 2V(calc.) = 63.5°.

The calculated mean index of refraction (after the law of Gladstone and Dale) n = 1.587, that is a deviation of +0.014 from the mean index $(n_x + n_y + n_z)/3 = 1.573$ of lanthanite from Bethlehem [18].

The optical orientation is X = b, Y = c, Z = a [1, 4, 14], see, however, X = c, Y = a [15].

The physical refractive energy (Kp) for D(meas.) is 0.2059 [4]. Lanthanite is nonfluorescent in both long- and short-wave ultraviolet light [4].

Other Physical Properties

The hardness $H_M = 2.5$ to 3 [1, 3], 3 [2].

The measured densities for lanthanites in g/cm³ are 2.69 to 2.74 (Bastnäs), 2.61 (Bethlehem) [1], 2.81 ± 0.03 (lanthanite-(Nd), Curitiba, mean of five determinations) [3], cf. [4], 2.72 (synthetic material) [3].

The calculated density is 2.73 g/cm³ for a mean atomic weight of RE = 140 [1], and 2.82 g/cm³ for $(Nd,La)_2[CO_3]_3 \cdot 8H_2O$ [3]. For further values see table p. 194.

References for 10.5.3.2.5:

[1] C. Palache, H. Berman, C. Frondel (The System of Mineralogy, 7th Ed., Vol. 2, Wiley, New York 1951, pp. 1/1124, 242). – [2] E. I. Semenov (in: K. A. Vlasov, Geochemistry and Mineralogy of Rare Elements and Genetic Types of Their Deposits, Vol. 2: Mineralogy of Rare Elements, Israel Program Sci. Trans., Jerusalem 1966, pp. 249/328, 283 [Russian Original, Moscow 1964]). – [3] H. G. Ansell, G. J. Pringle, A. C. Roberts (Geol. Surv. Paper Geol. Surv. Can. No. 76-1B [1976] 353/5). – [4] A. C. Roberts, G. Y. Chao, F. Cesbron (Geol. Surv. Paper Geol. Surv. Can. No. 80-1C [1980] 141/2). – [5] F. Cesbron, M.-C. Sichère, H. Vachey, J.-P. Cassedanne, J. O. Cassedanne (Bull. Mineral. **102** [1979] 342/7, 342).

[6] E. B. Eckel (U.S. Geol. Surv. Bull. No. 1114 [1961] 1/399, 202). – [7] E. B. Rowley (Rocks Minerals **37** [1962] 341/7, 347). – [8] Yu. L. Kapustin (Mineralogiya Karbonatitov, Nauka, Moscow 1971, pp. 1/288, 191/2). – [9] M. Gramaccioli (Mineral. Rec. **8** [1977] 287/93, 288). – [10] A. Roe (Mineral. Rec. **6** [1975] 299).

[11] R. G. Wells (Rocks Minerals **53** [1978] No. 3, pp. 108/11). – [12] P. C. Sæbø (Norsk Geol. Tidsskr. **41** [1961] 311/7, 312, 315). – [13] K. Fujimori (Anais Acad. Brasil. Cienc. **53** [1981] 147/52, 149, 152). – [14] W. T. Pecora, J. H. Kerr (Am. Mineralogist **38** [1953] 1169/83, 1177). – [15] A. J. Melfi [1963] from [13].

[16] J. M. V. Coutinho (Univ. São Paulo Fac. Fil. Cienc. Letras Bol. Mineral. No. 13 [1955] 119/26 from C.A. **1956** No. 4728). – [17] E. S. Larsen, H. Berman (U.S. Geol. Surv. Bull. No. 848 [1934] 1/266, 164). – [18] H. W. Jaffe (Am. Mineralogist **41** [1956] 757/77, 764).

10.5.3.2.6 Chemical Behavior and DTA

Lanthanite dissolves in acids with effervescence and decomposes to a white powder in boiling water. It is easy to prepare by reacting hydroxides with a CO_2-saturated solution of $NaHCO_3$ [1], cf. [2]. It dissolves readily and completely in dilute hydrochloric acid [3, 4], e.g., 0.2n HCl [5].

At heating lanthanite (Kola) loses a part of its water and turns to calkinsite at 80°C; the remaining water is liberated at 100 to 200°C, thereafter the mineral dissociates [6]. The thermal behavior of lanthanite-(Nd) from Curitiba is characterized by 3 endothermal maxima, which occur at higher temperatures in the DTA curve than in the DTG curve and are interpreted as follows [7], cf. [4]:

DTG maxima heating rate 300°C/h	DTA maxima heating rate 600°C/h	Interpreted reaction
90°C	119°C	continuous loss of 8 H_2O molecules
483°C	500°C	loss of 2 CO_2 molecules
677°C	766°C	loss of 1 CO_2 molecule

Calcination for 2 h at 1100°C results in a weight loss of 44%, evidently caused by loss of CO_2 and H_2O; this loss conforms with DTA peaks at 185, 520, and 860°C at a heating rate of 960°C/h [5]. For synthetic $RE_2[CO_3]_3 \cdot 8H_2O$ the most characteristic endothermal peaks are at 150 and 500°C [8].

References for 10.5.3.2.6:

[1] E. I. Semenov (in: K. A. Vlasov, Geochemistry and Mineralogy of Rare Elements and Genetic Types of Their Deposits, Vol. 2: Mineralogy of Rare Elements, Israel Program Sci.

Trans., Jerusalem 1966, pp. 249/328, 283 [Russian Original, Moscow 1964]). – [2] C. Palache, H. Berman, C. Frondel (The System of Mineralogy, 7th Ed., Vol. 2, Wiley, New York 1951, pp. 1/1124, 242). – [3] H. G. Ansell, G. J. Pringle, A. C. Roberts (Geol. Surv. Paper Geol. Surv. Can. No. 76-1B [1976] 353/5). – [4] A. C. Roberts, G. Y. Chao, F. Cesbron (Geol. Surv. Paper Geol. Surv. Can. No. 80-1C [1980] 141/2). – [5] K. Fujimori (Anais Acad. Brasil. Cienc. **53** [1981] 147/52, 147/9).

[6] Yu. L. Kapustin (Mineralogiya Karbonatitov, Nauka, Moscow 1971, pp. 1/288, 193). – [7] F. Cesbron, M.-C. Sichère, H. Vachey, J.-P. Cassedanne, J. O. Cassedanne (Bull. Mineral. **102** [1979] 342/7, 344). – [8] E. I. Semenov (Mineralogiya Redkikh Zemel', Nauka, Moscow 1963, pp. 1/412, 113).

10.5.3.3 Donnayite $NaCaSr_3Y[CO_3]_6 \cdot 3H_2O$

Donnayite occurs in small quantities of very small crystals from 0.05 to 1 (or 2) mm in the pegmatite dikes, miarolitic cavities and interstices in the nepheline syenite at Mont St. Hilaire, Quebec. It is usually associated with numerous silicates, carbonates, among them ancylite, further oxides, and sulfides [1], cf. [2].

The electron microprobe analysis of one crystal only, with the CO_2 and H_2O values calculated on the basis of $6CO_3^{2-}$ and $3H_2O$ per formula by analogy with the isomorphous weloganite, corresponds to the formula $Na_{0.93}(Ca_{0.87}Nd_{0.09}La_{0.03})(Sr_{2.95}Ba_{0.05})Y_{0.99}[CO_3]_6 \cdot 3H_2O$ or ideally $NaCaSr_3Y[CO_3]_6 \cdot 3H_2O$. The mineral is apparently isomorphous with weloganite and mackelveyite (see p. 200), the latter being the Ba analogue of donnayite, with which it forms a solid solution of unknown extent. The ratio Na:Ca is very close to 1:1. The content of 13.1% Y_2O_3, 1.83% Nd_2O_3, and 0.45% La_2O_3 in donnayite, recalculated to $\Sigma(Y+RE)_2O_3 = 100\%$ corresponds to 85.2% Y_2O_3, 11.9% Nd_2O_3, and 2.9% La_2O_3 [1].

The apparent symmetry of donnayite is trigonal or hexagonal, but X-ray determinations (precession photographs) show that it is triclinic, pseudorhombohedral, and isomorphous with weloganite [1], cf. trimorphous with weloganite and mackelveyite [2]; with $\beta = 59°59'$ instead of 60° it is distinctly pseudotrigonal [3].

Various habits from platy, tabular, saucer-shaped, columnar, barrel-shaped to irregular granular are observed [1], cf. [2]. Some crystals are clearly hemimorphic, with curved trigonal pyramidal faces converging toward a point on one end and terminated with a flat pedion on the other. Twinning in donnayite is the rule rather than the exception. Two twin laws are operative either individually or simultaneously: rotation about $[103]_{120°}$ or $[001]^*_{120°}$ and a reflection across (010), $(30\bar{1})$, or $(\bar{3}\bar{3}1)$. The twin axis corresponds to the pseudo threefold axis and the twin planes correspond to the pseudo mirror planes of the pseudorhombohedral lattice [1].

Donnayite commonly forms syntactic intergrowths with other minerals of similar structure and chemical composition such as ewaldite, an Sr analogue of ewaldite, mackelveyite, and rarely synchysite; it may occur also as a "cap" on ewaldite [1], cf. [2]. The cleavage is indistinct to fair basal [1].

The unit cell parameters of donnayite are a = 9.000(1), b = 8.999(1), c = 6.793(1) Å, α = 102.77(1)°, β = 116.28(1)°, γ = 59.99(1)°, space group P1-C_1^1 (No. 1), Z = 1, cell volume V = 427.4 Å^3. The space group is P1-C_1^1 (No. 1) by virtue of the hemimorphic crystal morphology; a nonconventional cell is used here in order to bring out the strong pseudosymmetry (a = b, γ = 60°); due to this there are 2 equally acceptable choices of the reduced cell, which may be derived from the adopted cell by the matrices $\bar{1}0\bar{1}/1\bar{1}0/001$ and $101/0\bar{1}0/00\bar{1}$; the cell constants of these cells are given [1].

Donnayite is sometimes transparent, especially near the rim, but almost opaque toward the core; transparent crystals have vitreous luster, but the white and gray varieties are opaque and earthy. It is colorless, white, pale yellow to yellow, gray, and rarely reddish brown due to inclusions of hematite. The streak is white [1]. Optically donnayite is biaxial negative with $n_x = 1.551$ to 1.561(2), $n_y = 1.646(2)$, $n_z = 1.652(2)$ [1]. The axial angle 2V varies from crystal to crystal and from point to point on the same crystal from 20° to almost 0°, the most commonly observed values being 5° to 10°. X is nearly parallel to c*, with a maximum deviation from parallelism of 5°. Prismatic sections of well-formed crystals always show sharp and nearly parallel extinctions between crossed polars, whereas the basal sections either show patchy extinctions or no extinctions at all. The orientation of the optic plane in basal sections varies from point to point corresponding to rotations of 60° or 120° about c*, strongly suggesting twinning, see above [1].

The hardness on Mohs' scale of crystals with vitreous luster is about 3, earthy material is apparently softer. The density, determined by the flotation method, is 3.30(1) g/cm^3, and calculated from the empirical formula, 3.266 g/cm^3 [1].

The mineral dissolves rapidly in hydrochloric acid 1:1 with strong effervescence [1].

References for 10.5.3.3:

[1] G. Y. Chao, P. R. Mainwaring, J. Baker (Can. Mineralogist **16** [1978] 335/40). – [2] L. Marble, A. Regis (Rocks Minerals **54** [1979] 4/25, 11/2). – [3] H. Strunz (Aufschluß **31** [1980] 317/72, 338).

10.5.3.4 Mackelveyite $NaCaBa_3Y[CO_3]_6 \cdot 3H_2O$

10.5.3.4.1 Name, Occurrence, and Paragenesis

Mackelveyite is the approved name following the recommendation of the Commission on New Minerals and Mineral Names of the IMA [1], cf. [2], instead of the original spelling „mckelveyite" [3, p. 597].

Mackelveyite is a rare subsurface mineral from the Green River Formation, found at 4 different localities in the Sweetwater County, Wyoming, USA [3, pp. 593/7], in the Wilkins Peak Member of the Green River Formation [4]. It occurs in paragenesis with burbankite and is associated with labuntsovite, leucosphenite, and searlesite; further it contains as intimately associated impurities biotite, acmite, quartz and 3 up to 8.7 wt% organic matter, which is apparently syngenetically formed with mackelveyite [3, pp. 593/5, 598/602]; the organic matter (carbonaceous material) is preferably included in the black crystals [2]. Sometimes mackelveyite together with pyrite is embedded in nahcolite [4]. Very often the mineral is intergrown with ewaldite in polycrystals; at a new locality in Sweetwater County, the Texas Gulf Sulfur Mine Shaft, however, clear euhedral crystals have been found, which are apparently free from carbonaceous inclusions [2] and from intergrown ewaldite [5].

A second occurrence of mackelveyite is in calc-silicate inclusions of the Mont St. Hilaire, Quebec, Canada, where it is also intergrown with ewaldite [6] and with donnayite [7], see p. 198.

10.5.3.4.2 Chemistry

According to 2 analyses (a massive material with 9.0% $(RE,Y)_2O_3$ and a crystalline one with 11.0% $(RE,Y)_2O_3$) and assuming 32 oxygen atoms in one formula unit the originally calculated formula was $Na_{1.9}Ba_{4.0}Ca_{1.1}Sr_{0.2}RE_{1.5}U^{4+}_{0.3}[CO_3]_9 \cdot 5H_2O$ [3, p. 606]. A recalculation of the original analyses of mackelveyite on the basis of $6CO_2$ per formula unit gives $(Na_{1.29}K_{0.02})(Ca_{0.73}U_{0.17})(Ba_{2.72}Sr_{0.17})(Y_{0.70}RE_{0.32})[CO_3]_6 \cdot 3.48H_2O$ and is, despite the analytical difficulties due to large amounts of inclusions, very close to the expected ideal formula $NaCaBa_3Y[CO_3]_6 \cdot 3H_2O$. This formula confirms the isomorphous relationship of mackelveyite with donnayite and weloganite, see p. 198, which is supposed from syntactic intergrowth between the first 2 minerals and from the strong similarity of the powder diffraction data of all 3 minerals [7]. The stoichiometric relationships in mackelveyite are appreciably disturbed by diadochic replacement of Na by Ca and Y by RE [8], the ratio Na:Ca, however, is very close to 1:1 and it is reasonable to assume that Na and Ca atoms are ordered in the 2 distinct positions occupied by Na in the weloganite structure. Mackelveyite is clearly the Ba analogue of donnayite and the 2 minerals probably form a solid solution, but the extent of the solid solution is not yet known [7]. Electron microprobe studies of 4 green polycrystals mackelveyite-ewaldite show homogeneous distribution of the elements Ba, Ca, Y, and Na in the polycrystals except for the regions occupied by inclusions. These data indicate dimorphism for both minerals, except for possibly different water contents [2].

The excess of total CO_2 over acid-evolved CO_2 in black colored mackelveyite crystals is attributed to the presence of organic material [3, p. 598], cf. p. 199.

Mackelveyite is characterized by a virtually complete segregation of Y-earths due to the crystal chemical character, not to the geochemical environment [3, p. 610], cf. [4].

In mackelveyite (free from acmite, biotite, and quartz) the spectrographically determined values for groups of RE in crystals (I) and massive material (II) from the Green River Formation are: $\Sigma(RE+Y)_2O_3 = 13.38$ (I) and 11.37% (II), with Ce-earths 1.09 (I) and 0.92% (II), Y-earths 4.60 (I) and 3.94% (II), and Y = 7.69 (I) and 6.51% (II). The relative distribution of the RE plus Y, given in wt% of the oxides, can be seen in the following table [3, pp. 604/5]:

	La	Ce	Pr	Nd	Sm	Eu	Gd	Tb	Dy	Ho	Er	Tm	Yb	Lu	Y
I	0.67	1.20	0.37	1.94	2.54	1.42	8.82	2.84	7.47	2.09	7.10	0.89	4.56	0.60	57.55
II	0.62	1.23	0.44	1.85	2.55	1.41	8.88	2.81	7.56	2.11	7.12	0.88	4.57	0.70	57.25

References for 10.5.3.4.1 and 10.5.3.4.2:

[1] M. Fleischer (Am. Mineralogist **51** [1966] 1248/357, 1290). – [2] G. Donnay, J. D. H. Donnay, M. H. Hey (Tschermaks Mineral. Petrogr. Mitt. [3] **15** [1971] 185/200, 186/7, 192/3). – [3] C. Milton, B. Ingram, J. R. Clark, E. J. Dwornik (Am. Mineralogist **50** [1965] 593/612). – [4] J. Fitzpatrick, A. Pabst (Am. Mineralogist **62** [1977] 158/63, 159, 162). – [5] C. Milton (Univ. Wyoming Contrib. Geol. **10** [1971] 57/63, 57, 60).

[6] L. Marble, A. Regis (Rocks Minerals **54** [1979] 4/25, 12). – [7] G. Y. Chao, P. R. Mainwaring, J. Baker (Can. Mineralogist **16** [1978] 335/40, 337/8, 340). – [8] A. S. Povarennykh (Crystal Chemical Classification of Minerals, Vol. 2, Plenum, New York 1972, pp. 459/766, 619).

10.5.3.4.3 Crystal Forms and Crystal Structure

Based on morphological evidence for crystals of mackelveyite the trigonal system has been assumed [1, p. 606], cf. [2]; a lower (monoclinic) symmetry, but with a markedly pseudo-hexagonal lattice simulated by intimate twinning, has been discussed, as the trigonal morphology can easily be described in monoclinic terms, when the habit of each individual in the trilling is referred to a postulated monoclinic C-centered lattice [3, pp. 194/6]. New data from a single crystal of mackelveyite obtained by repeatedly splitting a twinned crystal show triclinic symmetry and confirm isomorphism with donnayite (see p. 198) and weloganite [4, pp. 335, 340]. Therefore, mackelveyite, so far conceived to be trigonal, is according to the isotypy with these 2 minerals triclinic, but markedly pseudotrigonal ($\beta = 59°59'$) [5].

In the trigonal system the axial ratio is c/a (goniometric) = 2.077 [2] or = 2.071 and c/a (X-rays) = 2.087 [3, pp. 196, 198]. The forms are c$\{0001\}$, r$\{10\bar{1}1\}$, e$\{01\bar{1}2\}$, m$\{10\bar{1}0\}$, and $\overline{m}\{01\bar{1}0\}$ [1, p. 608], sometimes also z$\{01\bar{1}1\}$ is observed and with a maximum of these 6 forms all crystals are very simple in development [2]; in monoclinic setting the combination of forms is c$\{001\}$, c'$\{00\bar{1}\}$, b$\{020\}$, p$\{111\}$, and e$\{02\bar{1}\}$ [3, p. 196].

Although several different habits of mackelveyite crystals from Wyoming are observed, they all tend to be platy on $\{0001\}$, some with a hexagonal outline formed by rhombohedra, others bound by prisms with a rhombohedron as terminal form [1, pp. 594/6, 608]. The crystals from the Texas Gulf Sulfur Mine Shaft have pyramidal habit, the black crystals from the other localities in Sweetwater County have the habit of a twin (trilling), even though the X-ray diffraction shows the edifice to be a single crystal. The crystals must therefore be the products of skeletal growth, if they belong to point group 3m; if they have a lower symmetry (e.g., monoclinic) with pseudohexagonal lattice, intimate twinning by pseudo-merohedry could be expected to simulate the morphological point group 31m and the X-ray space group P31m [3, pp. 194/6].

Mackelveyite forms syntactic intergrowths with ewaldite in Wyoming [6], [3, pp. 186/7], and with donnayite at the Mont St. Hilaire, Quebec, where in one case mackelveyite and donnayite form caps on different sides of an ewaldite crystal [4, pp. 337/8]. The habit of the polycrystals mackelveyite-ewaldite, regardless of composition, is tabular to pyramidal with forms characteristic of mackelveyite [3, pp. 197/8]. The polar distribution of faces is that described by Desautels, who observed all crystals, obviously polycrystals, cf. [3, p. 197], to be decidedly hemimorphic with sixfold symmetry and observed 2 types of malformation suggesting trigonal symmetry or pseudo-orthorhombic appearance [2], the latter being observed also for mackelveyite from Mont St. Hilaire, Quebec [7]. All forms of one polycrystal can be indexed with reference to a single crystal lattice; they can, however, be referred to mackelveyite or ewaldite, see transformation matrix mackelveyite → ewaldite p. 88 and [3, pp. 197/8].

A strong striation is observed in monoclinic setting on the pinacoid b$\{020\}$ parallel with the elongation a; one of the two monohedra (c) is smooth, whereas the other (c') is rough with shagreen surface; p and e are also smooth [3, p. 196].

Twinning of mackelveyite in trigonal setting is characterized by merohedry, giving twin symmetry 6'm'm; the simultaneous twin operations are the reflections in the three $(11\bar{2}0)$ planes and rotations about the c axis through 60°, 180°, and 300°. The twin planes in monoclinic notation are $(1\bar{1}1)$ and $(11\bar{1})$ [3, pp. 192, 196].

No perfect cleavage could be noted, but imperfect fractures can be produced approximately parallel to (0001) and to $(11\bar{2}1)$ [1, p. 608].

The more abundant, shiny black crystals consist of mackelveyite only, while the dull bluish green specimens contain mackelveyite with dominant ewaldite in syntactic intergrowth [3, pp. 186/7], see p. 88 and above.

The unit cell constants, angles, space group, number of formula units in unit cell Z, cell volume V (in $Å^3$) and density values D(meas.) and D(calc.) for mackelveyite from Wyoming in different settings (symmetries) are given in the following table:

Symmetry	Triclinic	Trigonal	Monoclinic	Trigonal
a in Å	9.170(3)	9.174 ± 0.002	9.164	9.164(10)
b in Å	9.169(3)	—	15.818	—
c in Å	7.075(2)	19.154 ± 0.007	21.314	19.126(18)
α	102.50°(3)	—	—	—
β	115.63°(3)	—	115.6°	—
γ	59.99°(3)	—	—	—
V ($Å^3$)	464.5	1396.1 ± 0.7	—	1391
Z	1	3[1)]	—	9
Space group	P1-C_1^1 (No. 1)	P$\bar{3}$-C_{3i}^1 (No. 147)	—	P31m-C_{3v}^2 (No. 157)
D(calc.) in g/cm^3	—	3.62	—	3.37
D(meas.) in g/cm^3	—	3.25[2)]	—	3.25 ± 5
Ref.	[4, p. 338]	[1, pp. 606/7], cf. [8]	[3, p. 196]	[3, pp. 192, 198]

[1)] Z = 2 in [7]. – [2)] For crystals. For corrections see p. 203.

The space groups originally given to mackelveyite by [1, p. 606], [2] have been determined both on material which has later been shown to consist of polycrystals [3, pp. 186/7], cf. [6]. The studies on black shiny single crystals of mackelveyite confirm the space group given by [3, pp. 191/2]. A recalculation of the powder data given by [1, p. 607] by least-squares refinement leads to a triclinic cell and to the space group P1-C_1^1 (No. 1), in accordance with the confirmed isomorphism of mackelveyite with the triclinic minerals donnayite (see p. 198) and weloganite [4, p. 340].

The polycrystals of mackelveyite-ewaldite have coinciding c axes; the 2 sets of a axes are turned through 30° with respect to each other about the common c axis. The nets $(0001)_n$ common to both minerals show the coincidence of the P net of mackelveyite with the H supernet of ewaldite (at n = 0, 3, 6 ...); further the "H-centering effect" accounts for the observed syntactic intergrowth with the irregular contact surface and the simulated single crystal habit of the polycrystals [3, p. 193], see syntactic relations at ewaldite p. 88.

Mackelveyite, as compared with ewaldite, is the original mineral; it is the more disordered mineral with the higher free energy and has a larger cell and a lower symmetry than ewaldite. The transformation of mackelveyite to ewaldite is an ordering transformation, which accounts for the fact that the composition surface between the 2 phases in a polycrystal cannot be seen (cf. p. 88). The ordering transformation is not observed in the "pure" black mackelveyite crystals; this is apparently a consequence of the amorphous carbonaceous inclusions in it [3, p. 199].

References for 10.5.3.4.3:

[1] C. Milton, B. Ingram, J. R. Clark, E. J. Dwornik (Am. Mineralogist **50** [1965] 593/612). – [2] P. E. Desautels (Am. Mineralogist **52** [1967] 860/4). – [3] G. Donnay, J. D. H. Donnay, M. H. Hey (Tschermaks Mineral. Petrogr. Mitt. [3] **15** [1971] 185/200). – [4] G. Y. Chao, P. R. Mainwaring, J. Baker (Can. Mineralogist **16** [1978] 335/40). – [5] H. Strunz (Aufschluß **31** [1980] 317/72, 338).

[6] C. Milton (Univ. Wyoming Contrib. Geol. **10** [1971] 57/63, 60). – [7] L. Marble, A. Regis (Rocks Minerals **54** [1979] 4/25, 12). – [8] A. S. Povarennykh (Crystal Chemical Classification of Minerals, Vol. 2, Plenum, New York 1972, pp. 459/766, 618).

10.5.3.4.4 Optical and other Physical Properties

The luster is submetallic on (00$\bar{1}$), the dihedra p and e (monoclinic setting) show a fairly good reflecting power. The black crystals of mackelveyite are opaque even at very small grains; only the crystals from the Texas Gulf Sulfur Mine Shaft are clear and transparent in thin section, they appear to be quite free from carbonaceous material. The faces of the bluish green to greenish blue polycrystals reflect poorly if at all [1, pp. 193/4, 196/7).

Pure crystals from Wyoming are lime-yellow, other crystals are dark in color to black owing to carbonaceous inclusions; the polycrystals mackelveyite-ewaldite are dull bluish green to greenish blue, the color changing with the content of ewaldite [1, pp. 186, 194, 196] and with the size of the crystals: large crystals are pale green, smaller, more platy crystals very dark green [2]. The crystals from Mont St. Hilaire, Quebec, are also lime-yellow to black [3].

In thin section the lime-yellow crystals are colorless [1, p. 194], the black crystals with much organic matter appear brown rather than colorless in basal sections [4].

For lime-yellow mackelveyite the refractive index $n_o = 1.644 \pm 2$ (measured at 22°C) [1, p. 194]. Refractive indices $n_o = 1.66$ and $n_e = 1.57$ were measured on optical uniaxial negative crystals. Basal sections are pale green and nonpleochroic, prismatic sections are strongly pleochroic in green with O > E [4]. These data are determined on polycrystals, see pp. 201/2. A scaly fragment of lime-yellow mackelveyite parallel (0001) exhibits a distinctly biaxial interference figure with small axial angle and shows the pseudo-uniaxial character of the mineral. It suggests the presence of sectors in the sample, but does not prove that the mackelveyite pyramid is a trilling of a biaxial mineral [1, p. 194].

The values for the measured density in g/cm^3 are 3.14 (for crystals) and 2.87 (for massive material); corrected for the observed impurities (acmite, biotite, quartz, organic matter) they become 3.46 and 3.58 g/cm^3. Further correction, considering the calculated D value = 3.62 give D = 3.25 g/cm^3 (for crystals) and 2.89 g/cm^3 (for massive material) [4]. The measured density for several pure mackelveyite crystals of 3.25 ± 0.05 g/cm^3 (at 24°C) is also considerably less than a calculated D value of 3.54 g/cm^3 for a cell content of $2(Na_2Ca)(Ba_4RE_{1.7}U_{0.3})[CO_3]_9$ [5]. Lower values of the measured as compared with the calculated density are ascribed preferably to the carbonaceous material [1, pp. 192/3], cf. for ewaldite p. 89.

Piezoelectric tests for mackelveyite were negative [4].

Mackelveyite from the Green River Formation is (due to its remarkable U content) the only mineral with a strong radioactivity; all the remaining material of the cores shows quite low radioactivity [4].

References for 10.5.3.4.4:

[1] G. Donnay, J. D. H Donnay, M. H. Hey (Tschermaks Mineral. Petrogr. Mitt. [3] **15** [1971] 185/200). – [2] P. E. Desautels (Am. Mineralogist **52** [1967] 860/4). – [3] L. Marble, A. Regis (Rocks Minerals **54** [1979] 4/25, 12). – [4] C. Milton, B. Ingram, J. R. Clark, E. J. Dwornik (Am. Mineralogist **50** [1965] 593/612, 608, 610). – [5] G. Donnay, J. D. H. Donnay (Carnegie Inst. Washington Yearbook **67** [1969] 218/9).

10.5.3.5 Tengerite $Y[CO_3]_3 \cdot 3H_2O$

10.5.3.5.1 Definition and Related Minerals

It appears that at least 2 different minerals have been included under the name tengerite [1], or even 3 different mineral species, resembling each other in occurrence, chemistry, and refractive indices. The minerals can only be distinguished with certainty by X-rays [2]. For X-ray identity of the Skandinavian tengerites with the sample from Japan and with 2 synthetic yttrium carbonates, see p. 207. Deviating structural data have been observed for the mineral lokkaite, which is intimately associated with tengerite [2, 3], see p. 210. For 2 minerals, which are similar with each other, and have been found in Kazakhstan and South Africa, see p. 210, for a monoclinic mineral described as "tengerite" from Kazakhstan, see p. 212, and for a new orthorhombic SiO_2-containing RE mineral from Kazakhstan, see p. 188. The mineral, supposed to be tengerite, from Ampasipoana, Itasy, Madagascar [4], cf. [5], proved to be churchite on new examinations [1]. Be-rich tengerites, regarded as a species apart [6], cf. [7], are very probably caused by mechanical admixtures of a Be-containing mineral, see p. 206.

A tengerite-like phase $Ln_2[CO_3]_3 \cdot 3H_2O$, isostructural with tengerite and confirmed by comparison with the IR spectrum of tengerite, is stable at low temperatures (<200°C) in the ternary heterogeneous systems Ln_2O_3-H_2O-CO_2 under hydrothermal conditions [8].

References for 10.5.3.5.1:

[1] W. J. Verwoerd (S. Africa Geol. Surv. Ann. Rept. **2** [1963] 119/35, 128). – [2] V. Perttunen (Bull. Geol. Soc. Finland **43** [1971] 67/72, 69/70). – [3] D. D. Hogarth (Mineral. Rec. **3** [1972] 69/77, 73). – [4] J. Behier (Rep. Malgache Ann. Geol. Madagascar No. 29 [1960] 1/78, 64). – [5] J. Behier (Haut Comm. Madagascar Dépendances Trav. Bur. Geol. No. 61 [1954] from C. A. **1955** 14588).

[6] A. Vorma, P. Ojanperä, V. Hoffrén, J. Siivola, A. Löfgren (Bull. Comm. Geol. Finlande No. 222 [1966] 241/74, 269). – [7] M. H. Hey (An Index of Mineral Species and Varieties Arranged Chemically, 2nd Ed., Brit. Museum, London 1955, pp. I/XXIV, 1/728, 80). – [8] J. A. K. Tareen, T. R. N. Kutty, I. Mohamed (Indian Mineralogist **21** [1980] 43/8, 44/6).

10.5.3.5.2 Occurrence and Paragenesis

Tengerite is mostly found as a secondary mineral in pegmatites, associated with other RE- and Y minerals, feldspars, and other minerals; it replaces especially numerous Y-rich primary minerals. It occurs as an alteration product in the form of rims, coatings, radiating spherulitic aggregates or powdery fillings in fractures and on cleavage planes of the Y minerals and feldspars.

Tengerite is a product of reprecipitation of yttrium and calcium (beryllium) from solutions produced by weathering of Y minerals; the Y-earths are in general readily extractable from any radioactive Y minerals by alkali carbonate solutions [1]. As a mineral, characterized by the Yb subgroup, and by the less common Er subgroup, tengerite belongs to the minerals deposited in the acid stage of an alkali metasomatism ("early albitization") in alkali granites of Kazakhstan [2], especially as an accessory mineral in vein-like quartz-feldspar exocontact metasomatites of the northwestern Tarbagatai, East Kazakhstan [14]. In RE metasomatites tengerite is a rare mineral, which occurs besides other Y-rich RE minerals, monazite and cerianite, and besides Be and W minerals [3].

Tengerite occurs intimately associated with lokkaite and many other primary and secondary RE minerals (such as allanite, fergusonite, britholite, thalenite, bastnaesite, and synchysite) in the pegmatite at Pyörönmaa in Kangasala, SW-Finland [4], as powdery fillings in the fractures of other minerals and on their surfaces, especially on albite [5].

Tengerite further occurs together with minor amounts of lokkaite as a white powder coating exposed surfaces, fracture openings and cavities of primary Y minerals such as hellandite, allanite, fergusonite, euxenite, and kainosite, associated with primary microcline and a great number of rare minerals in the Evans Lou Pegmatite, Quebec, Canada [6]. In another small pegmatite dike (consisting dominantly of perthitic feldspar) from the West Portland Township, Quebec, tengerite is found as a thin radiated dendritic crust on fractures or cleavage planes in albite enclosing monazite [7]. Tengerite, associated with other RE minerals, occurs further in the Roscoe Dike, Clear Creek Canyon, Jefferson County, Colorado [8], coating gadolinite [9].

For several occurrences, mostly pegmatites, the replacing character of tengerite as an alteration or weathering product is quite clear and the replaced primary Y mineral is known. Thus tengerite occurs on gadolinite in the original locality at Ytterby, northeast of Stockholm, Sweden [10, 11], in Lövböle, Kemio, SW-Finland [5], in Barringer Hill [10, 11], and at the Clear Creek, both Burnet County, Texas [12]. Tengerite pseudomorphous after thalenite occurs at Åskagen, Vermland, Sweden [10, 11], tengerite originating from hellandite at Kragerö, South Norway [10], and from hellandite, fergusonite, and perhaps from yttrian sphene, but also from kainosite as an intermediate product, in alteration paths in the Evans Lou Pegmatite, Quebec [6]. Further tengerite occurs as an alteration product of yttrialite at Iisaka, Japan [1, p. 832], [10, 11], and of allanite in the mineralized zone inside the pegmatite dike of Rømteland, South Norway [13].

References for 10.5.3.5.2:

[1] T. Iimori (Sci. Papers Inst. Phys. Chem. Res. [Tokyo] **34** [1937/38] 832/41, 832, 840/1). – [2] D. A. Mineev [Mineyev] (Geokhimiya **1963** 1082/100, 1093; Geochemistry [USSR] **1963** 1129/49, 1140). – [3] D. A. Mineev (in: L. N. Ovchinnikov, N. A. Solodov, Mestorozhdeniya Litofil'nykh Redkikh Metallov, Nedra, Moscow 1980, pp. 178/84, 179, 183). – [4] V. Perttunen (Bull. Geol. Soc. Finland **43** [1971] 67/72, 67/8). – [5] A. Vorma, P. Ojanperä, V. Hoffrén , J. Siivola, A. Löfgren (Bull. Comm. Geol. Finlande No. 222 [1966] 241/74, 268, 270).

[6] D. D. Hogarth (Mineral. Rec. **3** [1972] 69/77, 72/3, 75/7). – [7] H. S. Spence, O. B. Muench (Am. Mineralogist **20** [1935] 724/32, 725/6, 728). – [8] C. V. Haynes (in: S. M. Del Rio, Mineral Resources of Colorado First Sequel, Mineral Resources Board, Bolder, Colo., 1960, pp. 370/85, 376). – [9] E. B. Eckel (U.S. Geol. Surv. Bull. No. 1114 [1961] 1/399, 323). – [10] C. Palache, H. Berman, C. Frondel (The System of Mineralogy, 7th Ed., Vol. 2, Wiley, New York 1951, pp. 1/1124, 275/6).

[11] E. I. Semenov (Mineralogiya Redkikh Zemel', Nauka, Moscow 1963, pp. 1/412, 114). – [12] W. W. Crook (Mineral. Rec. **8** [1977] 88/90). – [13] T. L. Sverdrup (Norg. Geol. Unders. No. 211 [1959] 124/96, 170, 175). – [14] D. A. Mineev (Geokhimiya Apogranitov i Redkometal'nykh Metasomatitov Severo-Zapadnogo Tarbagataya, Moscow 1968, pp. 1/183, 34/5).

10.5.3.5.3 Chemistry

Tengerite has been insufficiently examined, its identity is doubtful. The chemical analyses show considerable discrepancies, the number of H_2O molecules and the range of RE composition have not been established [1, 2]; the mineral needs reinvestigation [3].

Because the majority of the samples named tengerite, especially those from Skandinavia and Japan show X-ray identity with the synthetic hydrous yttrium carbonate, see p. 207, the composition of tengerite is supposed to be close to this compound with the average molecular ratio $Y_2O_3:CO_2:H_2O = 1.00:2.90:2.56$ and with the formula $Y_2[CO_3]_3 \cdot nH_2O$ ($n \sim 2$ to 3) [4], cf. [1]. The generally accepted formula $Y_2[CO_3]_3 \cdot 3H_2O$ [5, 6] is an idealized one; the amount of water may be lower than $3H_2O$ [7]. According to a diagram of [8] that shows a weight loss of 7.5% at 110°C, the actual formula of this dehydrated, but structurally unchanged compound is calculated to be $Y_2[CO_3]_3 \cdot 1.28\,H_2O$ [7]. Formulas including other cations have been given for tengerites from Japan and Texas. For the Japanese occurrence the general formula $Y_3M[CO_3]_{5-x}(OH)_{1+2x} \cdot 3\,H_2O$ has been simplified to $Y_3Ca[CO_3]_4(OH)_3 \cdot 3\,H_2O$ [9, pp. 839, 841], cf. [10, 11]; it is, however, questioned by [4].

Essential amounts of Be (9.7 wt% BeO) and Fe (4.0 wt% Fe_2O_3) found in the sample from Barringer Hill, Burnet County, Texas [10], have been regarded to indicate a species apart with a formula $BeYCO_3(OH)_3 \cdot H_2O$ [12], cf. [13], or about $(Y,Ce)BeCO_3(OH)_3$ [11]. This Be-rich sample, however, may be a mixture of 2 hypergene minerals: a red colloidal mineral with the formula $Be_2Fe^{III}_2Si_2O_9 \cdot 8\,H_2O$ and Be- and Fe-free bastnaesite-(Y), which have been observed in pseudomorphs after gadolinite [1]. The tengerites of Pyörönmaa, Finland, and of Iisaka, Japan, both are devoid of Be [13].

The Ca found in tengerites is supposed to replace Y in the mineral [5]; it seems, however, to be an unessential constituent in the tengerite [7]. A tengerite from Kazakhstan is properly a "calcito-tengerite", because it is composed of tengerite and calcite in defined proportions [14]. There exist, however, 2 tengerite-like minerals with Ca contents, see pp. 210/1.

The relative amounts of the RE ($\Sigma RE = 100\%$) in tengerites have been determined in 4 samples of the northwestern Tarbagatai, East Kazakhstan: Nos. 1 and 2 are tengerites with calcite ("calcito-tengerite") from a metasomatite, No. 3 is a tengerite pseudomorphous after gagarinite in a metasomatite, No. 4 is a spherulite of tengerite in voids in altered gagarinite from a pegmatitic vein [14]:

No.	La	Ce	Pr	Nd	Sm	Eu	Gd	Tb	Dy	Ho	Er	Tm	Yb	Lu
1	2.6	0.4	1.1	1.9	3.8	0.8	8.6	1.9	21.0	7.1	24.8	2.6	19.2	4.1
2	—	—	—	—	2.1	—	6.9	3.2	20.7	6.7	28.8	4.3	21.3	—
3	9.8	17.9	4.9	13.5	6.9	0.1	7.4	2.9	13.8	4.2	10.1	1.5	5.3	1.5
4	3.9	9.3	1.7	8.4	5.6	—	5.1	2.3	21.2	4.4	13.4	1.7	23.0	—

The chemical analysis of tengerite from Japan shows 49.0 wt% Y-earths and 1.3 wt% Ce-earths, and 0.3 wt% ThO_2. The RE distribution, measured qualitatively from the intensities of X-ray spectral lines, shows Y to be decisive predominant, Gd, Dy, and Er to be abundant and the remaining elements of the Yb- and Tb groups (Yb, Lu, Tb, and Ho) and of the Ce-group (Ce, Pr, Nd, Sm) to be quite deficient [9, pp. 837/9].

References for 10.5.3.5.3:

[1] E. I. Semenov (Mineralogiya Redkikh Zemel', Nauka, Moscow 1963, pp. 1/412, 114). – [2] E. I. Semenov (in: K. A. Vlasov, Geochemistry and Mineralogy of Rare Elements and Genetic Types of Their Deposits, Vol. 2: Mineralogy of Rare Elements, Israel Program Sci. Trans., Jerusalem 1966, pp. 249/328, 228/9 [Russian Original, Moscow 1964]). – [3] W. J. Verwoerd (S. Africa Geol. Surv. Ann. Rept. **2** [1963] 119/35, 128). – [4] K. Nagashima, H. Wakita (Nippon Kagaku Zasshi **89** [1968] 856/9, 858, A50 [Engl. Abstr.]). – [5] A. S. Povarennykh (Crystal Chemical Classification of Minerals, Vol. 2, Plenum, New York 1972, pp. 459/766, 617).

[6] D. D. Hogarth (Mineral. Rec. **3** [1972] 69/77, 73). – [7] V. Perttunen (Bull. Geol. Soc. Finland **43** [1971] 67/72, 71). – [8] L. P. Domingues, R. L. Wilfong, L. R. Furlong (U.S. Bur. Mines Rept. Invest. No. 6029 [1962] 1/19, 9). – [9] T. Iimori (Sci. Papers Inst. Phys. Chem. Res. [Tokyo] **34** [1937/38] 832/41). – [10] C. Palache, H. Berman, C. Frondel (The System of Mineralogy, 7th Ed., Vol. 2, Wiley, New York 1951, pp. 1/1124, 276).

[11] M. Ross (U.S. Geol. Surv. Profess. Papers No. 468 [1962] 1/34, 22). – [12] M. H. Hey (An Index of Mineral Species and Varieties Arranged Chemically, 2nd Ed., Brit. Museum, London

1955, pp. I/XXIV, 1/728, 80). – [13] A. Vorma, P. Ojanperä, V. Hoffrén, J. Siivola, A. Löfgren (Bull. Comm. Geol. Finlande No. 222 [1966] 241/74, 269). – [14] D. A. Mineev (Geokhimiya Apogranitov i Redkometal'nykh Metasomatitov Severo-Zapadnogo Tarbagataya, Moscow 1968, pp. 1/183, 96, 127).

10.5.3.5.4 Crystal Forms and Crystal Structure

Tengerite is orthorhombic [1] as is the synthetic phase $Y_2[CO_3]_3 \cdot nH_2O$ [2].

The single crystals of the synthetic phase show the forms {100}, {010}, and {011}; the faces are smooth. Penetration twins of 2 to 6 individuals, penetrating each other with their c axes, are frequent [2].

The natural tengerite mostly occurs in form of radial fibrous aggregates, sometimes mamillary or as thin coatings, or in earthy aggregates [3], cf. [1, 4]. The tengerite from Iisaka, Japan, forms globular concretions, the single crystals (discernible under the microscope) are lamellar and show tabular habit, rarely tabular prisms [5]. The tengerite from West Portland Township, Quebec, occurs in thin radiated, dendritic crusts [6].

Radiating fibers of tengerite are intergrown with some fibers of lokkaite in rosettes or minute spheres in the Evans Lou pegmatite, Quebec [7]. Tengerite is also intimately associated with lokkaite in white powdery incrustations at Pyörönmaa, Finland; the mutual relationship of these 2 minerals, however, is not known [8, pp. 68, 71].

Unit cell constants are determined for synthetic materials only: The parameters of $Y_2[CO_3]_3 \cdot nH_2O$, a = 6.17, b = 9.23, and c = 15.31 Å are similar to those for $Nd_2[CO_3]_3 \cdot 2$ to $3H_2O$, a = 6.27, b = 9.471, c = 15.628 Å [2]. These values, however, deviate from the values a = 9.18, b = 11.34, and c = 7.60 Å calculated by [9] from the d values of a synthetic compound with $Y_2O_3 : CO_2 : H_2O = 1.00 : 2.90 : 2.56$ (average of 4 analyses) [10, p. 858].

The structure of tengerite apparently is the same as that of synthetic hydrous yttrium carbonate; a loss of water does not change the structure. The X-ray powder patterns of tengerites from Rosas, Iveland, Norway, from Pyörönmaa and Lövböle, Finland, from Iisaka, Japan [8, pp. 70/1], [10, pp. 858, A50], from Ytterby, Sweden, and Hobøl, Norway [11], are identical with the X-ray powder patterns of 2 synthetic hydrous yttrium carbonates. Deviations, however, exist between these patterns and those of the mineral named "tengerite" from Kazakhstan [8, pp. 70/1] and the unnamed RE carbonate from Goudini, Transvaal [12]. These 2 minerals have similar X-ray powder patterns, but quite different ones from those of tengerite and lokkaite [8, p. 71], see p. 211.

References for 10.5.3.5.4:

[1] E. I. Semenov (in: K. A. Vlasov, Geochemistry and Mineralogy of Rare Elements and Genetic Types of Their Deposits, Vol. 2: Mineralogy of Rare Elements, Israel Program Sci. Trans., Jerusalem 1966, pp. 249/328, 229 [Russian Original, Moscow 1964]). – [2] J. A. K. Tareen, T. R. Narayanan Kutty, K. V. Krishnamurty (J. Crystal Growth **49** [1980] 761/5, 762/4). – [3] C. Palache, H. Berman, C. Frondel (The System of Mineralogy, 7th Ed., Vol. 2, Wiley, New York 1951, pp. 1/1124, 275). – [4] E. I. Semenov (Mineralogiya Redkikh Zemel', Nauka, Moscow 1963, pp. 1/412, 114). – [5] T. Iimori (Sci. Papers Inst. Phys. Chem. Res. [Tokyo] **34** [1937/38] 832/41, 834/7).

[6] H. S. Spence, O. B. Muench (Am. Mineralogist **20** [1935] 724/32, 728). – [7] D. D. Hogarth (Mineral. Rec. **3** [1972] 69/77, 73). – [8] V. Perttunen (Bull. Geol. Soc. Finland **43** [1971] 67/72). – [9] A. S. Povarennykh (Crystal Chemical Classification of Minerals, Vol. 2, Plenum, New York

1972, pp. 459/766, 617). – [10] K. Nagashima, H. Wakita (Nippon Kagaku Zasshi **89** [1968] 856/9, A50 [Engl. Abstr.]).

[11] H. Neumann, K. O. Bryn (Avhandl. Norsk Videnskaps Akad. Oslo I Mat. Natur. Klasse **1958** No. 1, pp. 1/15, 6, 9). – [12] W. J. Verwoerd (S. Africa Geol. Surv. Ann. Rept. **2** [1963] 119/35, 128).

10.5.3.5.5 Physical Properties

The luster of tengerite is pearly [1 to 3], sometimes dull, chalky [4]. The synthetic crystals of $Y_2[CO_3]_3 \cdot nH_2O$ are transparent to translucent [5].

The color of tengerite is white [2, 4, 6, 7], as are the synthetic crystals [5].

The refractive indices of 3 tengerites from different localities are given in the following table:

Occurrence	n_x	n_z	Ref.
Iisaka, Japan	1.622	1.642	[1, 8]
Ytterby, Sweden[1)]	1.555	1.585	[4, 8]
Pyörönmaa, Finland[2)]	1.569	1.627	[3]

[1)] For this occurrence $n_x=1.555$, $n_y=1.57$, $n_z=1.585$ are given by [9]. – [2)] This tengerite is intimately associated with lokkaite [10].

The large differences in refractive indices of tengerites, e.g., from Japan and Sweden are probably caused by unequal contents of crystal water or by admixtures [8].

The tengerite from Sweden is biaxial positive with large axial angle 2V; X is parallel to the elongation [4], cf. [9].

The IR absorption spectra of the tengerite from Iisaka, Japan, and of the synthetic hydrated yttrium carbonate ($Y_2O_3:CO_2:H_2O=1.00:2.90:2.56$) are very similar with bands at 1400 to 1500 cm^{-1}, at 1650 cm^{-1}, and at 3500 cm^{-1} [11]. The IR spectrum of tengerite-like crystals of $Y_2[CO_3]_3 \cdot (2\ to\ 3)H_2O$ shows no sharp OH vibrational and stretching modes around 3600 and 630 cm^{-1}. This indicates, that the synthetic tengerite crystals have only the water of crystallization, unlike the natural tengerite, which is shown by [12] to have hydroxyl groups [5].

The hardness of tengerite $H_M \approx 3.5$ [13]. The measured density of tengerite from Ytterby, Sweden, ranges from 3.1 to 3.2 g/cm^3 [9], that of tengerite from Iisaka, Japan, is 3.12 g/cm^3 [1, 4, 8].

References for 10.5.3.5.5:

[1] T. Iimori (Sci. Papers Inst. Phys. Chem. Res. [Tokyo] **34** [1939] 832/41, 836/7). – [2] H. S. Spence, O. B. Muench (Am. Mineralogist **20** [1935] 724/32, 728). – [3] A. Vorma, P. Ojanperä, V. Hoffrén, J. Siivola, A. Löfgren (Bull. Comm. Geol. Finlande No. 222 [1966] 241/74, 268/70). – [4] C. Palache, H. Berman, C. Frondel (The System of Mineralogy, 7th Ed., Vol. 2, Wiley, New York 1951, pp. 1/1124, 275/6). – [5] J. A. K. Tareen, T. R. Narayanan Kutty, K. V. Krishnamurty (J. Crystal Growth **49** [1980] 761/5, 762, 764).

[6] T. L. Sverdrup (Norg. Geol. Unders. No. 211 [1959] 124/96, 170). – [7] W. W. Crook (Mineral. Rec. **8** [1977] 88/90). – [8] E. I. Semenov (Mineralogiya Redkikh Zemel', Nauka,

Moscow 1963, pp. 1/412, 114). – [9] E. S. Larsen, H. Berman (U.S. Geol. Surv. Bull. No. 848 [1934] 1/266, 106). – [10] V. Perttunen (Bull. Geol. Soc. Finland **43** [1971] 67/72, 68).

[11] K. Nagashima, H. Wakita (Nippon Kagaku Zasshi **89** [1968] 856/9, 859, A50 [Engl. Abstract]). – [12] Neumann, Bergstol from [5]. – [13] A. S. Povarennykh (Crystal Chemical Classification of Minerals, Vol. 2, Plenum Press, New York 1972, pp. 459/766, 617).

10.5.3.5.6 Chemical Behavior and Thermoanalyses

The tengerite from Iisaka, Japan, is easily soluble with rigorous effervescence in dilute hydrochloric or nitric acid [1].

The thermal decomposition of the tengerite from Iisaka, Japan, and of the synthetic hydrated yttrium carbonate, investigated by DTA and TGA, shows the similarity between the 2 materials, which was already found by X-ray diffraction patterns [2]. The synthetic $Y_2[CO_3]_3 \cdot 3H_2O$ shows upon heating to 110°C a weight loss of 7.5% without change in its structure [3]; the complete decomposition takes place below 600°C at a heating rate of 2.5°C/min; above 620°C the thermogram shows a slight deviation from the horizontal in the direction of increasing weight loss, which is possibly caused by the slow release of residual adsorbed CO_2 by the Y_2O_3, which is present above 620°C [4].

References for 10.5.3.5.6:

[1] T. Iimori (Sci. Papers Inst. Phys. Chem. Res. [Tokyo] **34** [1937/38] 832/41, 832). – [2] K. Nagashima, H. Wakita (Nippon Kagaku Zasshi **89** [1968] 856/9, 858/9, A50 [Engl. Abstract]). – [3] V. Perttunen (Bull. Geol. Soc. Finland **43** [1971] 67/72, 68, 71). – [4] L. P. Domingues, R. L. Wilfong, L. R. Furlong (U.S. Bur. Mines Rept. Invest. No. 6029 [1962] 1/19, 9).

10.5.3.6 Lokkaite, $\approx(Y,Ca)_2[CO_3]_3 \cdot 1.58\,H_2O$

The supergene hydrous RE carbonate lokkaite was first found by Vorma et al. in the Pyörönmaa pegmatite in Kangasala, SW-Finland, and taken for tengerite [1]. Later on Perttunen showed it to be a new mineral and named it lokkaite. The mineral occurs as isolated white radial disk-like aggregates of 0.5 to 1.0 mm in diameter on the surface and in fissures of albite or as white powdery incrustations intimately associated with tengerite [2, pp. 67/8], cf. [3]. A second occurrence of lokkaite as a white fibrous mineral is the Evans Lou Pegmatite, Quebec. Here it is common at 5 localities, normally also together with tengerite and with the primary minerals microcline, hellandite, allanite, and less frequently with fergusonite, euxenite, kainosite (cenosite) and a great number of rare minerals, thereby 3 U- and 10 Y minerals [4].

The analysis of lokkaite from Finland (cations by electron microprobe analyzer, CO_2, H_2O, and F by conventional procedures) suggests the formula $Ca_{0.23}RE_{1.58}[CO_{2.87}]_3 \cdot 1.58\,H_2O$ [2, p. 68], simplified to the formula given above by [3]. According to microprobe analyses (Y in 3, La to Lu in 6 determinations) the mineral contains 29.0 wt% Y_2O_3, 19.7 wt% heavy RE oxides (Gd_2O_3 to Lu_2O_3), and 4.3 wt% light RE oxides (La_2O_3 to Sm_2O_3) [1], [2, p. 68]. The composition of the RE (Y_2O_3 plus RE oxides recalculated to 100%) is the following (tr = traces) [1], [2, p. 68]:

La	Ce	Pr	Nd	Sm	Eu	Gd	Tb	Dy	Ho	Er	Tm	Yb	Lu	Y
0.38	1.32	0.57	2.45	3.40	tr	8.68	2.26	12.83	tr	7.55	1.70	4.15	tr	54.72

Lokkaite is orthorhombic. The crystals in the radial aggregates are straight fibers, but even the smallest fibers are actually crystal bundles; the c axis of each individual crystal is parallel to the fiber. The grain size in powdery incrustations of lokkaite plus tengerite is less than 0.01 mm [2, p. 69].

X-ray powder pattern and single crystal methods give the following unit cell constants of lokkaite: a = 39.07, b = 6.079, c = 9.19 Å. The Laue-class of lokkaite is mmm, the possible space groups are B222-D_2^6 (No. 21, standard setting C222), Bm2m-C_{2v}^{11} (No. 35, standard setting Cmm2), Bmmm-D_{2h}^{19} (No. 65, standard setting $C\frac{2}{m}\frac{2}{m}\frac{2}{m}$), or B2mm-$C_{2v}^{14}$ (No. 38, standard setting Amm2). The mineral has a distinct superstructure with the dimensions of the subcell: a′ = ½ a, b′ = b, and c′ = ½ c [2, p. 69]. Tengerite as impurity enhances some lines of lokkaite; all of the stronger reflections of tengerite, however, except in 7.6 Å, have 1 or 2 counterparts among the reflections of lokkaite. The distinct superstructure of lokkaite suggests the presence of an order-disorder phenomenon, either substitutional (RE, Ca, vacancy) or rotational (CO_3^{2-}); it is improbable that the essential difference between lokkaite and tengerite is in the degree of hydration only [2, pp. 69, 71], cf. pp. 205/6.

The color of lokkaite from Finland is white, sometimes stained to yellow; the mineral has the refractive indices $n_x = 1.569$, $n_y = 1.592$, $n_z = 1.620$; 2V is not measurable, the elongation is positive and Z||c [2, p. 69]. The mineral from Quebec is white with $n_z \sim 1.626$, it is optically similar to tengerite, from which it is distinguishable by X-rays only [4].

The crystals of lokkaite (fibers) are brittle; the density is not measurable [2, pp. 69/70].

References for 10.5.3.6:

[1] A. Vorma, P. Ojanperä, V. Hoffrén, J. Siivola, A. Löfgren (Bull. Comm. Geol. Finlande No. 222 [1966] 241/74, 268/71). – [2] V. Perttunen (Bull. Geol. Soc. Finland **43** [1971] 67/72). – [3] C. Hintze, K. F. Chudoba (Handbuch der Mineralogie, Suppl. Vol. 4, Pt. 1, de Gruyter, Berlin 1974, pp. 1/373, 370/1). – [4] D. D. Hogarth (Mineral. Rec. **3** [1972] 69/77, 70, 72/3).

10.5.3.7 Unnamed Y-Ca Carbonates with H_2O

In the following two Y-Ca carbonates are described. One of them, named tengerite, has been found in Kazakhstan, USSR, the other is an unidentified Y-Ca carbonate from Transvaal, South Africa. According to X-ray data, see p. 211, both deviate from tengerite.

The "tengerite" from Kazakhstan occurs associated with gagarinite and elpidite in alkaline granites and is apparently a secondary product of alteration of gagarinite, forming spherulitic incrustations or reniform outgrowths on fissures in this mineral, or dense to cryptocrystalline aggregates. The gagarinite crystals are replaced by tengerite together with yttrofluorite, opal, and other minerals [1, pp. 147, 154], or by aggregates of tengerite, synchysite, and yttrofluorite [2].

The mineral from the Goudini volcano, Zeerust District, Transvaal, South Africa, occurs in a quartz-barite vein within carbonatites. It forms isolated white spheroids (<1 mm in diameter) on ferruginous opal and is usually encrusted by colorless quartz [3]; it is found especially in drusy cavities in the barite veins [4].

For the mineral from Kazakhstan analyses of pure material are not available; a sample (examined by X-rays) contains according to spectral analyses >1% Y, Yb, Ca, Mg, and Al; 0.3 to 1.0% Si and Fe; 0.1 to 0.3% Ce, La, Na, and Sr; and 0.003 to 0.01% Pb [1, p. 155].

The IR spectrographic examination shows nearly identical amounts of OH^- and H_2O besides CO_2. The amount of Y and Y-earths predominates over the amount of Ce-earths, the RE composition of the purest sample (crusts and nodules examined by X-rays), recalculated to 100%, is La, Ce, Pr, and Nd = 0.00, Sm = 1.58, Gd = 5.18, Tb = 2.40, Dy = 19.88, Ho = 4.97, Er = 21.59, Tm = 3.19, Yb = 15.90, Lu = 0.00, and Y = 25.36 [1, pp. 155/6]. For the mineral from South Africa, the spectrographic analysis shows Ca and Y as main and Yb, Er, Dy, Gd, Nd(?), and Pr(?) as minor constituents. The presence or absence of H_2O and F in this mineral is not confirmed [3].

The mineral from Kazakhstan is dense to cryptocrystalline in spherulitic aggregates with concentric layers in coarse spherulites; single grains (under the microscope) are elongated prisms [1, p. 155]. The spheroids from South Africa have a radial structure and also concentric zoning, consisting of tapering crystals with parallel extinction and negative elongation [3].

The unit cell parameters of the mineral from Kazakhstan are $a = 8.375 \pm 0.005$, $c = 8.72$ Å, $a/c = 0.96$, apparently belonging to the tetragonal system, while tengerites from several localities are thought to belong to the orthorhombic system [1, p. 156], cf. p. 207. The X-ray powder pattern of the mineral from South Africa is not comparable with available data of any known mineral and especially not with those of tengerites from Japan and Scandinavia [3]. The X-ray powder patterns of both minerals (from Kazakhstan and South Africa) are, however, similar with each other and different not only from those of tengerites but also from that of lokkaite [5, p. 71], confirming a new species as already assumed by [3].

The mineral from Kazakhstan has a vitreous to silky luster; the color of spherulites is white, sometimes turning to yellowish; central parts of coarse spherulites are dark colored with brownish tints, caused by a fine-pulverized brown pigment (apparently Fe hydroxide); a yellow-greenish color is due to an argillaceous mineral [1, p. 155]. The spheroids from South Africa are white [3]. The following refractive indices are observed for the minerals from Kazakhstan [1, p. 155] and South Africa [3]:

Occurrence	n_x	n_z	$n_z - n_x$
Kazakhstan, white[1)]	1.620 ± 0.02	1.642	0.022
Kazakhstan, yellowish[2)]	1.611	1.636	—
South Africa, white	—	1.64[3)]	≈0.02

[1)] From fissures. – [2)] From nodular outgrowths. – [3)] Refractive index of the ray vibrating parallel to the elongation.

The refractive indices of the white minerals from Kazakhstan and South Africa are very similar to those of tengerite from Iisaka, Japan, see p. 208. The prisms of the mineral from Kazakhstan show direct extinction, sometimes also undulating extinction. The mineral shows no luminescence in UV light nor with cathode rays [1, p. 155].

The IR spectroscopic investigation shows deep and wide zones of absorption at 3.0 and 3.5 μm, corresponding to OH^- and H_2O, and at 7.0 μm, corresponding to CO_2 (with admixed opal) [1, p. 155].

The mineral is brittle. The hardness on Mohs' scale of aggregates is about 2. The measured density of aggregates (intergrowths) is 2.85 g/cm³. The mineral is electromagnetic and weakly radioactive [1, p. 155].

The mineral grains from Kazakhstan decompose totally, with violent development of CO_2, in dilute hydrochloric acid; the grains burst in the flame and give a yellowish color with phosphorus salt (borax pearl), quickly decoloring at cooling [1, p. 155]. The mineral from South Africa is insoluble in water, but dissolves with effervescence in dilute acids. The thermal decomposition product is Y_2O_3 (determined by X-rays) [3].

References for 10.5.3.7:

[1] A. V. Stepanov (Tr. Kaz. Nauchn. Issled. Inst. Mineral. Syr'ya **5** [1961] 147/61). – [2] A. V. Stepanov, E. A. Severov (Dokl. Akad. Nauk SSSR **141** [1961] 954/7; Dokl. Earth Sci. Sect. **139/141** [1961] 1290/3). – [3] W. J. Verwoerd (S. Africa Geol. Surv. Ann. Rept. **2** [1963] 119/35, 127/8). – [4] W. J. Verwoerd (S. Africa Geol. Surv. Handbook **6** [1967] 1/452, 147). – [5] V. Perttunen (Bull. Geol. Soc. Finland **43** [1971] 67/72, 71).

10.5.3.8 Monoclinic RE Carbonate with H_2O

A mineral, forming fine radiating spherulitic aggregates, was found in amphibolized pegmatites with rare metal mineralization in Kazakhstan. On account of its similar physico-chemical properties to those of tengerites from Scandinavia it was named tengerite. As there are, however, appreciable deviations in symmetry, structure, optical and other physical properties between this mineral and tengerite (pp. 207/8), the mineral is considered as an unnamed new mineral.

The chemistry of this mineral is characterized (in wt%) by $RE_2O_3 + Y_2O_3$ 55.61, CaO 2.23, Al_2O_3 2.24, FeO 0.40, CO_2 24.17, H_2O 2.30, loss on ignition 11.59, insoluble residue 2.14; the rate of Ce-earths to Y-earths is about 1:1, the composition of the RE ($\Sigma RE + Y = 100\%$) is: La 0.95, Ce 13.53, Nd 20.96, Pr 1.14, Sm 12.95, Gd 5.71, Tb 1.52, Dy 3.81, Ho 1.71, Er 2.67, Tm 1.33, Y 33.72.

The mineral is monoclinic (in contrast to the orthorhombic tengerite, see p. 207), the fine radiating aggregates consist of well developed monocrystals, which are suitable for determining the unit cell parameters: a = 10.80, b = 8.78, c = 7.78, $\beta = 103°06'$, a:b:c = 1.227:1:0.886, V = 720 Å^3, Z = 2. The space group is $P2_1/m\text{-}C_{2h}^2$ (No. 11) or $P2_1\text{-}C_2^2$ (No. 4). The X-ray pattern is similar to those of tengerites from Norway and Sweden.

The mineral has a vitreous luster, single plates are transparent. The color is white; the mineral is optically biaxial negative and has the refractive indices $n_x = 1.582$, $n_z = 1.623$, which are similar to those of tengerite from Sweden. The measured density is 2.8, the hardness on Mohs' scale is approximately 3.

Reference for 10.5.3.8:

Yu. V. Shipovalov, A. V. Stepanov (Issled. Obl. Khim. Fiz. Metodov Anal. Mineral. Syr'ya No. 1 [1971] 184/5).

10.5.4 Hydrous Carbonates with Additional Anions

10.5.4.1 Ancylite and Calcio-ancylite $(RE)_x(Sr, Ca)_{2-x}(CO_3)_2(OH)_x \cdot (2-x) H_2O$

10.5.4.1.1 Name and Varieties

The species name ancylite is applied to those members of the isomorphic series with Sr > Ca, while the Ca-dominated species with Ca > Sr is named calcio-ancylite [1], cf. [2]. These two members are named strontian ancylite and calcian ancylite, respectively, by [3]. The subvarieties ferroan calcio-ancylite and manganoan calcio-ancylite with Fe and Mn contents of 5.36% and 5.55%, respectively [1], are mentioned as "synonyms" by [4] and [5]. These names may, however, be caused by analyses of impure material and are therefore questioned by [6] and [7]. The Ba-strontio-ancylite variety mentioned by [7] is, however, confirmed by several analyses in recent times, see pp. 218/9.

Two types of the hydroxycarbonate $RE(OH)CO_3$ have been synthesized, one of them, the A type, is isostructural with ancylite [8, 9]. For the B type, see p. 94.

References for 10.5.4.1.1:

[1] C. Palache, H. Berman, C. Frondel (The System of Mineralogy, 7th Ed., Vol. 2, Wiley, New York 1951, pp. 1/1124, 292). – [2] H. Strunz, C. Tennyson (Mineralogische Tabellen, 6th Ed., Geest and Portig, Leipzig 1977, pp. 1/621, 248). – [3] A. Dal Negro, G. Rossi, V. Tazzoli (Am. Mineralogist **60** [1975] 280/4, 280). – [4] E. I. Semenov (Mineralogiya Redkikh Zemel', Moscow 1963, pp. 1/412, 109). – [5] E. I. Semenov (in: K. A. Vlasov, Geochemistry and Mineralogy of Rare Elements and Genetic Types of Their Deposits, Vol. 2: Mineralogy of Rare Elements, Israel Program Sci. Trans., Jerusalem 1966, pp. 249/328, 279 [Russian Original, Moscow 1964]).

[6] E. I. Semenov, M. E. Kazakova (Tr. Mineral. Muzeya Akad. Nauk SSSR No. 11 [1961] 196/8). – [7] A. S. Povarennykh (Crystal Chemical Classification of Minerals, Vol. 2, translated from Russian by J. E. S. Bradley, Plenum, New York 1972, pp. 459/766, 620). – [8] J. Sawyer, P. Caro, L. Eyring (Rev. Chim. Minerale **10** [1973] 93/104, 93/4). – [9] J. A. K. Tareen, T. R. N. Kutty, I. Mohamed (Indian Mineralogist **21** [1980] 43/8, 43).

10.5.4.1.2 Occurrence, Paragenesis, and Alteration

Occurrence and Paragenesis

Ancylite is widespread in the two following environments: 1) In alkaline rocks and pegmatites, where it is a late mineral predominantly in nepheline-syenite pegmatites; 2) in carbonatites, where it occurs preferably in the late dolomite-ankerite- or calcite-zeolite stages. In these rocks it possibly is hydrothermally deposited, often replacing older RE minerals, especially burbankite. Ancylite is only rarely observed in acid magmatic and metasomatic rocks and in one case in an amphibolite.

The mineral was first found in druses in pegmatitic veinlets in nepheline syenite at Narsarsuk, Julianehaab district, Greenland, associated with aegirine, albite, microcline, zircon, synchysite, cordylite, and eudidymite [1]. Most of the hitherto known occurrences of ancylite, nearly all associated with carbonates and with alkaline igneous rocks, chiefly syenitic pegmatites, are situated in Russia, some in USA and Canada [2]. Ancylites in **acid magmatic rocks** are known from the granite of Predazzo, North Italy, where it was found in the magnetic fraction [3], and from cataclastic micropegmatitic granites from the Northern Urals, USSR, where it is supposed to be a product of hydrothermal reworking and carbonatization of acid rocks; the mineral here occurs in paragenesis with Ca zeolites, not with Na zeolites [4]. Ancylite in very small quantities and extremely rare in unaltered form, is one of the latest minerals in druses of the nordmarkite ("Grefsen syenite") from the Grorud district, Oslo region, Norway; it is associated with calcite and quartz and nearly completely altered to bastnaesite [5], cf. [6]. In an alkaline eckermannite-microcline pegmatite of the Inagli massif, Aldan province, South Yakutia, Siberia, ancylite is a mineral of the zeolitization stage; it occurs as inclusion in natrolite in interstices between feldspars and is associated with thomsonite and neptunite [7].

A characteristic accessory hydrothermal mineral is ancylite in **nepheline syenites** and related **alkali pegmatites** or **nepheline syenite pegmatites** [8]. To this type belong the original locality at Narsarsuk, Julianehaab district, Greenland, the rocks from the Langesundfjord, Norway, and from Khibina and Lovozero, Kola, USSR, where ancylite is found in drusy cavities in pegmatites, together with natrolite, albite, aegirine, epididymite, synchysite, and other minerals [9]. The ancylite from Greenland occurs in druses in pegmatitic veinlets in nepheline-syenite [1], it is associated with aegirine, albite, and zircon in crusts or filling cavities between other minerals together with the younger cordylite [10]. In South Norway ancylite is a very rare mineral in syenite pegmatites of the Langesundfjord at Saga (Tvedal), and at Bratthagen, Lågendal [11]; it is one of the latest minerals, but very rarely present in unaltered condition [12]. On the Kola Peninsula the variety calcio-ancylite occurs in a natrolite pegmatite of the Vukisvumchorr Mountains, Khibina, together with ilmenite, fluorite, and sphalerite forming garlands in vugs of an unknown outleached mineral [8], cf. [1]; at Lovozero ancylite in pegmatites of poikilitic syenites is closely associated with katapleite and zeolite [13]. At the Mont St. Hilaire, Quebec, Canada, a new species mentioned by [51] was identified as ancylite, which occurs together with burbankite as an accessory mineral in veins and silicate vugs in the nepheline syenite [14], commonly associated with microcline, aegirine, and minor amounts of calcite and siderite, which are all embedded in analcime and natrolite in miaroles and pegmatitic dikes in the nepheline syenite [15]. Further associated minerals are donnayite [16] and aegirine, albite, and rhodochrosite in altered pegmatites and silicate vugs [17].

Ancylite occurs sparingly in massive white analcime and in granular masses of calcite and K-feldspar in igneous dikes and nonspecified sills, which intrude Ordovician limestone. It is associated with natrolite and, less commonly, with burbankite at the Miron quarry, Montreal, and is intimately intergrown with stillwellite in coarsely crystalline calcite at the property of Desmont Mining Corporation Ltd. near Wilberforce, Ontario [18].

Ancylite together with burbankite and bastnaesite are the accessory RE carbonates of miaskitic aegirine-albite **metasomatites** [19]. In an alkaline metasomatite of the Southern Urals ancylite, associated with calcite, sulfides, and ramsaite, is observed in the third, most alkaline zone of this metasomatite [20].

Ancylite as a rare accessory mineral is apparently typomorphic for late stage ankerite-dolomite **carbonatites**; it is characteristic particularly for carbonatites poor in fluorine and develops after burbankite and carbocernaite directly on these minerals [21]. According to the conditions of its formation, ancylite is one of the latest minerals in the Fe-richest dolomitic stage of each individual series of carbonatitic rocks [22], where it occurs as a hydrothermal deposition [23], or as an alteration product [9]. Ancylite is present in all carbonatite massifs on the Kola Peninsula, USSR, and in some massifs of Siberia [21]. It is the most widespread RE mineral on the Kola Peninsula, found in ankerite-dolomite veins of the massifs Vuoriyarvi, Kovdor, Lesnaya Varaka, and Sebl'-Yavr, in the alkali pegmatitic massifs of Khibina and Lovozero, and in Sallanlatva, North Karelia [23]. Ancylite was formerly attached to the zeolite-calcite stage as the latest stage of carbonatite formation by Kapustin [24]. The occurrence of ancylite shows differences for several carbonatite provinces of the USSR: It is widespread and associated with other carbonates and fluorocarbonates of the RE in the carbonatites of the Kola Peninsula; it forms a characteristic mineralization in carbonatites of the Maimecha Kotui province, North Siberia; it is only subordinate relative to the dominating parisite and smaller amounts of bastnaesite in carbonatites of Eastern Sayan, South Siberia [25].

Barite, strontianite, chlorite, galena, and thorite, forming impregnations in cavities are commonly associated with ancylite in carbonatites [21]; yellow barite, galena, and hisingerite occur together with ancylite in carbonatites of Kola [23]. For minerals associated with ancylite in numerous carbonatites, predominantly in the late ankerite-dolomite or Fe-rich stage, and in carbonatitic veins in alkaline rocks, see in the following.

In the Khibina massif, Kola, a Ca-rich ancylite occurs in carbonatites with Mn-calcite and Mn-siderite ("oligonite"), representing a fifth stage of formation. It is associated with synchysite, parisite, and huanghoite, fluorite, strontianite, and sulfides [26]. Ancylite, observed in veinlets of natrolite which transect porphyric rocks of this massif, is associated with synchysite, pyrite, and calcite; it is supposed to be a regenerated relic of a carbonatite [27]. At Vuoriyarvi,

Kola, ancylite is rare and one of the latest specific accessory minerals, associated with carbocernaite, barite, quartz, microcline, and zeolites in small vugs of the dolomite-calcite carbonatite veins [28]. In these veins it is a mineral of the second stage of succession, deposited after quartz, zeolites, and feldspar and before podolite, anhydrite, and marcasite [29]. In the Kovdor massif, Kola, ancylite is associated with labuntsovite, chlorite, and quartz [23]; in the Sallanlatva massif, North Karelia, ancylite is a specific hydrothermal phase of low temperature, deposited in all three types of carbonatites (dolomite-siderite I, calcite with phlogopite and magnetite II, and calcite with natrolite III). The largest amounts of ancylite are situated in the peripheral zone of the central part of the carbonatites in type II, and in parts rich in chlorite and phlogopite in type I; ancylite occurs associated with zircon, pyrochlore, and lueshite and is, together with the latter, sometimes the most widespread accessory mineral [30], cf. [23, 31]. In the massif Sebl'-Yavr, Kola, accessory ancylite, celestine-barite, strontianite, and pyrochlore, occur in ankerite carbonatites together with chlorite, sulfides, barite, and biotite [32]. In the Odikhincha intrusion, Lower Kotui, northern margin of the Siberian Platform, a mineral containing inclusions of aegirine that probably is calcio-ancylite, was observed [33]. In the Bor Uryakh massif, Maimecha-Kotui province, Krasnoyarskii Krai, Northern Siberia, Ca-rich ancylite is regularly connected with white dolomite veins, occurring in vugs of lens-shaped gray dolomite [22]. In an East Siberian carbonatite massif ancylite is formed in the fourth hydrothermal stage (aegirine-dolomite to ankerite) together with further carbonates, aegirine, strontianite, barite, monazite, sulfides, and RE-Nb-Ta minerals which are pseudomorphs after burbankite [34], see p. 216. In a less eroded massif of ultrabasic alkaline rocks and carbonatites ancylite occurs together with parisite, ankerite, strontianite, barite, and pyrite in only one of four varieties observed in the fourth stage of formation of carbonatites, the dolomite-ankerite carbonatites [35]. In a late veinlike ankerite-carbonatite body in a multiple pluton of Siberia, ancylite is associated with quartz, which formed together with siderite, sulfides, cleiophane, and the accessory minerals monazite, parisite, and bastnaesite [36], cf. [19].

Ancylite was found in **calcitic veins** in the massif Afrikanda and in dolomitic veinlets in older carbonatites at Sallanlatva, North Karelia [23]. In southern Africa ancylite is a late accessory mineral, occurring in irregular patches and intimately associated with calcite veins in carbonatites. At Spitzkop, East Transvaal, it occurs in an amphibole-rich stripe in the central beforsite; at Ondurakorume, Namibia, it is a common constituent of soevite and beforsite, associated with apatite and replacing other carbonates (see p. 216) [37, 38]. At Ondurakorume it also occurs in a lenticular mass near the summit of dike-like iron ore bodies together with strontianite, quartz, and calcite [38]. Very fine-grained ancylite, intimately intergrown with burbankite, occurs associated with subordinate sanidine, biotite, aegirine, pyrrhotite, and several other RE carbonates in veins with dominant calcite intersecting shonkinite in the Bearpaw Mountains, Montana [39]. In the Sheep Creek vein, belonging to the tabular or lens-shaped carbonatic Nb-RE deposits in the Ravalli County, Montana, which are similar to those of the Bearpaw Mountains, ancylite is intimately intergrown with quartz, Sr-Ba calcite, and some barite. The amount of ancylite in the fine-grained ancylite-quartz(-carbonatite) aggregates is indicated by increasing coloring of the specimens. The ancylite has been formed in the sixth low temperature metasomatic replacement stage ("ancylite stage") together with chlorite, muscovite, quartz II, fersmite, hematite, fluorite, and sometimes monazite; it has not incorporated, however, the RE present in monazite or in older allanite [40]. In an association resembling that of carbonatites in the northern Lemhi County, Idaho, ancylite, found together with abundant other carbonates, chiefly calcite, but also bastnaesite, is formed later than the silicates (including allanite), apatite, and monazite and before barite, oxides, and sulfides [41]; for a description of this deposit, see "Rare Earth Elements" A 8, 1984, Chapter 11.7.2.3.3.1. In the McClure Mountain Complex, Colorado, carbonate dikes, principally composed of calcite and dolomite, contain ancylite together with two Ce minerals, one of which may be bastnaesite [42]. Minor amounts of ancylite associated with carbocernaite, calcite, fluorite, quartz, barite, and sulfides were observed in an albite-biotite-calcite-carbonatite vein in an alkaline syenite at Sturgeon Narrows, Sturgeon Lake, district of Thunder Bay, Ontario [43]. In the mafic-ultramafic complex of the Gem Park, Custer and Fremont Counties, Colorado, ancylite is a common accessory mineral in the **carbonate veins**, such as dolomite-barite-monazite veins, associated with calcite and predominating in strontianite-rich zones, but occurring also in porous, cellular parts with limonite. It is only a rare accessory mineral in dikes with dolomite, blue amphibole and pyrochlore, and in late-stage carbonate veinlets in fenites together with monazite [45]. Ancylite together with parisite, is one of the minerals occurring in nonspecified rocks of the Oka Carbonatite and Alkaline Complex, Quebec [44].

At Cornog, Chester County, Pennsylvania, a calcio-ancylite occurs together and sometimes intergrown with calcite, pyrite, and actinolite and closely associated with apophyllite and barite in an **amphibolite**; the ancylite is limited to one narrow rock zone paralleling the layering of the biotite-gneiss [2].

Inclusions

A great amount of inclusions of aegirine needles is observed in the ancylites from Greenland [10] and from Odikhincha, Siberia [33]; such as dustlike Fe hydroxides contaminate ancylite from Eastern Siberia [34], and an unknown dark material occurs in great amounts in the coarse ancylite crystals from Vuoriyarvi, Kola [29].

Pseudomorphs

Ancylite is often a product of alteration of earlier formed RE minerals, and sometimes it forms pseudomorphs predominantly after burbankite.

In ankerite-dolomite veins, the final stages of the carbonatite formation, ancylite replaces older RE minerals, especially burbankite and carbocernaite [23]; it develops after and directly on burbankite and carbocernaite [21], and replaces burbankite, e.g., in ankerite-dolomite carbonatites of Vuoriyarvi, Kola [52, 53].

In the alkali pegmatite of Khibina, Kola, the mineral, replaced by ancylite, was formerly supposed to be a RE-containing apatite or mosandrite ("rinkolite") [8]; later, the hexagonal-prismatic mineral pseudomorphically replaced by ancylite was shown to be burbankite or belovite in several pegmatites of Kola [9], especially burbankite in those of Khibina [46] and Lovozero [31]. At Vuoriyarvi, besides pseudomorphs of ancylite after burbankite, fine crystalline accumulations of ancylite in leaching cavities of burbankite are observed [23], the latter being actively replaced by ancylite [31] or by ancylite, barite, and calcite at later stages [47]. In an East Siberian carbonatite ancylite replaces burbankite II in the process of "ancylitization" to produce a fine-grained product of hydration. Sometimes complete pseudomorphs are formed [34], cf. [48]. Further ancylite replaces burbankite partly at the Mont St. Hilaire, Quebec [15], and is intimately intergrown with burbankite or forms veins in burbankite as a locally formed hydration product in the Bearpaw Mountains, Montana [39]. In North Vietnam ancylite develops as secondary mineral on aggregates of carbocernaite [49], at the Sheep Creek, Ravalli County, Montana, it replaces together with quartz Sr-Ba calcite and barite [40], at Ondurakorume, Namibia, ancylite replaces magnesio-dolomite together with 5 other minerals, among them carbocernaite [38], ferromagnesite (pistomesite), quartz, and calcite [37].

Ancylite, together with bastnaesite in pseudomorphs, rarely replaces zircon in nepheline syenite from the Langesundfjord, South Norway [12]; it is further the main component of vudyavrite, an alteration product of rinkolite from Khibina [9]. The formation of ancylite as a weathering product of nordite in Lovozero, Kola [9, 31] is improbable, as it is unstable under such conditions [8].

Alteration

Ancylite itself is replaced by bastnaesite in the form of pseudomorphs named "weibyeite" [12], cf. [5] and p. 105. It further decomposes to the RE-richer minerals calkinsite and lanthanite [50], cf. p. 192.

References for 10.5.4.1.2:

[1] C. Palache, H. Berman, C. Frondel (The System of Mineralogy, 7th Ed., Vol. 2, Wiley, New York 1951, pp. 1/1124, 292). – [2] F. A Keidel, A. Montgomery, C. W. Wolfe, R. P. Christian (Mineral. Rec. **2** No. 1 [1971] 18/25, 36, 18, 20/3). – [3] F. Emiliani, G. Gandolfi (Mineral. Petrogr. Acta **11** [1965] 123/31, 126). – [4] I. P. Nikitenko, E. P. Kalinin, N. P. Yushkin (in: Geologiya i Poleznye Iskopaemye Severo-Vostoka Evropeiskoi Chasti SSSR Severa Urala, Vol. 2, Syktyvkar 1973, pp. 430/3 from Ref. Zh. Geol. **1974** No. 2 V 192). – [5] I. Oftedal, P. C. Sæbø (Norsk Geol. Tidsskr. **45** [1965] 171/5, 173).

[6] G. Raade (Mineral. Rec. **3** No. 1 [1972] 7/11, 8/9). – [7] A. E. Efimov, S. M. Kravchenko, E. V. Vlasova (Tr. Inst. Mineral. Kristallokhim. Redk. Elem. Akad. Nauk SSSR No. 16 [1963] 141/75, 145, 171). – [8] E. I. Semenov, M. E. Kazakova (Tr. Mineral. Muzeya Akad. Nauk SSSR No. 11 [1961] 196/8). – [9] E. I. Semenov (Mineralogiya Redkikh Zemel', Moscow 1963, pp. 1/412, 111). – [10] O. B. Bøggild (Medd. Groenland **149** No. 3 [1953] 1/442, 161).

[11] G. Raade, J. Haug, R. Kristiansen, A. O. Larsen (Lapis **5** No. 10 [1980] 22/8, 24, 26). – [12] P. C. Sæbø (Norsk Geol. Tidsskr. **43** [1963] 441/3). – [13] K. A. Vlasov, M. V. Kuz'menko, E. M. Es'kova (Lovozerskii Shchelochnoi Massiv, Moscow 1959, pp. 1/624, 498, 566). – [14] G. Y. Chao, D. C. Harris, A. W. Hounslow, J. A. Mandarino, G. Perrault (Can. Mineralogist **9** [1967] 109/23, 112/3, 116). – [15] T. T. Chen, G. Y. Chao (Can. Mineralogist **12** [1973/74] 342/5, 342).

[16] G. Y. Chao, P. R. Mainwaring, J. Baker (Can. Mineralogist **16** [1978] 335/40, 337). – [17] L. Marble, A. Regis (Rocks Minerals **54** [1979] 4/25, 11). – [18] A. P. Sabina (Geol. Surv. Paper Geol. Surv. Can. No. 78-1A [1978] 253/8, 253/4). – [19] Yu. L. Kapustin (in: L. N. Ovchinnikov, N. A. Solodov, Mestorozhdeniya Litofil'nykh Redkikh Metallov, Nedra, Moscow 1980, pp. 243/96, 245, 292). – [20] A. F. Efimov, E. M. Es'kova, Z. T. Kataeva (Tr. Mineral. Muzeya Akad. Nauk SSSR No. 19 [1969] 165/9).

[21] Yu. L. Kapustin (Mineralogiya Karbonatitov, Moscow 1971, pp. 1/288, 187/8). – [22] L. M. Dakhiya (Zap. Vses. Mineral. Obshchestva **98** [1969] 737/9). – [23] Yu. L. Kapustin (in: K. A. Vlasov, Mineralogiya i Geneticheskie Osobennosti Shchelochnykh Massivov, Moscow 1964, pp. 135/94, 171, 173, 191/2). – [24] Yu. L. Kapustin (Geokhimiya **1966** 1311/21; Geochem. Intern. **3** [1966] 1054/64, 1055). – [25] M. Ya. Somina, A. G. Bulakh (Mineral. Geokhim. **1972** No. 4, pp. 64/73, 64).

[26] O. B. Dudkin, L. I. Polezhaeva, F. V. Minakov (Veshchestvennyi Sostav Shchelochnykh Intrusivnykh Kompleksov Kol'skogo Poluostrova, Apatity 1981, pp. 73/81, 78). – [27] O. B. Dudkin, F. V. Minakov (in: I. V. Bel'kov, Materialy po Mineralogii Geokhimii Shchelochnykh Kompleksov Porod Kol'skogo Poluostrova, Apatity 1975, pp. 159/74, 162, 173). – [28] A. G. Bulakh, V. V. Kondrat'eva, E. N. Baranova (Zap. Vses. Mineral. Obshchestva **90** [1961] 42/9, 43). – [29] L. V. Kozyreva, G. A. Il'inskii (in: Materialy po Mineralogii Kol'skogo Poluostrova, Kirovsk 1959, pp. 69/76, 70, 72, 75). – [30] M. P. Orlova, Yu. P. Rozhdestvenskii, E. N. Baranova (Tr. Vses. Nauchn. Issled. Geol. Inst. **96** [1963] 3/20, 6, 10, 12/3).

[31] E. I. Semenov (in: K. A. Vlasov, Geochemistry and Mineralogy of Rare Elements and Genetic Types of Their Deposits, Vol. 2: Mineralogy of Rare Elements, Israel Program Sci. Trans., Jerusalem 1966, pp. 249/328, 281 [Russian Original, Moscow 1964]). – [32] A. G. Bulakh (Mater. Vses. Nauchn. Issled. Geol. Inst. **1961** No. 45, pp. 3/14 from Ref. Zh. Geol. **1962**

No. 1 D 196). – [33] V. F. Motychko (Tr. Nauchn. Issled. Inst. Geol. Arktiki No. 107 [1959] 23/41, 23, 28). – [34] T. B. Zdorik (Tr. Mineral. Muzeya Akad. Nauk SSSR No. 17 [1966] 60/75, 61, 68, 73). – [35] A. A. Frolov, A. Ya. Volzhenkova, E. A. Nechaeva (Geol. Rudn. Mestorozhdenii **1969** No. 1, pp. 27/39, 37/8).

[36] Yu. L. Kapustin (Dokl. Akad. Nauk SSSR **202** [1972] 422/5; Dokl. Earth Sci. Sect. **202** [1972] 116/9). – [37] W. J. Verwoerd (S. Africa Geol. Surv. Ann. Rept. **2** [1963] 119/35, 123/4). – [38] W. J. Verwoerd (S. Africa Geol. Surv. Handbook **6** [1967] 1/452, 32, 35, 224/5). – [39] W. T. Pecora, J. H. Kerr (Am. Mineralogist **38** [1953] 1169/83, 1171/2, 1183). – [40] E. W. Heinrich, A. A. Levinson (Am. Mineralogist **46** [1961] 1424/47, 1427/8, 1432, 1439/40, 1443).

[41] A. L. Anderson (Econ. Geol. **55** [1960] 1179/201, 1180, 1184, 1189, 1196). – [42] R. L. Parker, F. A. Hildebrand (U.S. Geol. Surv. Profess. Papers No. 450-E [1962] 8/10). – [43] D. C. Harris (Can. Mineralogist **11** [1971/73] 812/8, 812/3). – [44] D. P. Gold (Papers Proc. Gen. Meeting Intern. Mineral. Assoc., New Delhi 1964, Vol. 4, pp. 109/25, 116). – [45] R. L. Parker, W. N. Sharp (U.S. Geol. Surv. Profess. Papers No. 649 [1970] 1/24, 8/9, 12/4, 17).

[46] R. P. Tikhonenkova (in: O. V. Vershkovskaya, E. I. Semenov, Novye Dannye po Mineralogii i Geokhimii Redkikh Elementov, Moscow 1974, pp. 89/92). – [47] L. S. Borodin, Yu. L. Kapustin (Dokl. Akad. Nauk SSSR **147** [1962] 462/5; Dokl. Earth Sci. Sect. **147** [1962] 144/7). – [48] T. B. Zdorik (Geol. Mestorozhd. Redkikh Elementov Vses. Nauchn. Issled. Inst. Mineral. Syr'ya No. 30 [1966] 121/31, 123/6). – [49] A. G. Bulakh, E. P. Izokh (Dokl. Akad. Nauk SSSR **175** [1967] 175/7; Dokl. Earth Sci. Sect. **175** [1967] 118/20). – [50] P. C. Sæbø (Norsk Geol. Tidsskr. **41** [1961] 311/7, 312).

[51] J. A. Mandarino, D. C. Harris, J. Bradley (Can. Mineralogist **8** [1964/66] 398). – [52] A. S. Kirillov (Dokl. Akad. Nauk SSSR **159** [1964] 1048/50; Dokl. Earth Sci. Sect. **159** [1964] 93/5). – [53] A. S. Kirillov (Zap. Vses. Mineral. Obshchestva **95** [1966] 51/9, 58).

10.5.4.1.3 Chemistry

For a compilation of analyses and $(RE,Y)_2O_3$ contents, mean and variance of 17 analyses, see "Seltenerdelemente" A 4, 1979, pp. 13, 6.

The general **formula** $(RE)_x(Sr,Ca)_{2-x}[CO_3]_2(OH)_x \cdot (2-x)H_2O$ is derived from the data of an electron microprobe analysis of an ancylite crystal from Mont St. Hilaire, Quebec; the data are recalculated, taking into account the total weight loss of 28.36%, and are in good agreement with the results of the structure analysis and with the experimental density measured by flotation. The detailed chemical formula is $La_{0.56}Ce_{0.82}Sr_{0.40}Ca_{0.22}[CO_3]_2(OH)_{1.38} \cdot 0.62H_2O$ [1]. The older formulas, recalculated from a chemical analysis of the original, Ca-poor ancylite from Narsarsuk, Greenland, are given as $Sr_3(Ce,La,Dy)_4[CO_3]_7(OH)_4 \cdot 3H_2O$ [2], or $Sr_3Ce_4[CO_3]_7(OH)_4 \cdot 3H_2O$, simplified to $SrCe[CO_3]_2(OH) \cdot H_2O$ [3], cf. [4 to 7]. The formula $(Ce,La)_4(Sr,Ca)_3[CO_3]_7(OH)_4 \cdot 3H_2O$ considers the constant occurrence of considerable amounts of Ca, e.g., up to 13.01% CaO [8], cf. up to 6.51% CaO in a compilation of analyses from Kola [6], to 8.35% CaO in calcio-ancylite from Cornog, Pennsylvania [9], and >10% Ca (spectrographically determined) in calcio-ancylite from Odikhincha, Siberia [10].

Besides Ca, elevated contents of Ba have been observed in five of six analyses with 3.33 to 14.36% BaO of ancylites from Kola; the highest value corresponds to an atomic ratio Ca : Sr : Ba of about 1:1:1 in a sample from Vuoriyarvi, Kola. Three analyses of ancylites from this locality correspond to the following formulas:

1) $(Ca_{0.03}Sr_{0.85}Ba_{0.12})Ce_{1.26}[CO_3]_2(OH)_{1.40} \cdot 1.20\,H_2O$
2) $(Sr_{0.40}Ba_{0.35}Ca_{0.34})Ce_{1.00}[CO_3]_2(OH)_{1.00} \cdot 1.05\,H_2O$
3) $(Sr_{0.73}Ba_{0.12}Ca_{0.12}Ce_{0.03})Ce_{1.06}[CO_3]_2(OH)_{1.00} \cdot 1.00\,H_2O$ [6].

For an ancylite from Sallanlatva, North Karelia, with 2.57% P_2O_5 besides 6.51% CaO, 3.33% BaO and smaller amounts of Na_2O and K_2O the formula

$$RE_{3.95}(Ca_{1.65}Sr_{1.62}Ba_{0.30}Na_{0.22}K_{0.02})([CO_3]_{6.49}[PO_4]_{0.51}) \cdot (OH_{3.28}O_{0.72}) \cdot 3\,H_2O$$

is given; the existence of P in the mineral was confirmed by analyses of thoroughly selected mineral grains; nevertheless the existence of PO_4 in a carbonate is still hypothetical [11].

In accordance with the analyses compiled by [8] the **diadochic replacement** between Sr and Ca seems to be perfect and leads to the varieties strontio-ancylite and calcio-ancylite [12]. Mutual substitution is, however, supposed to exist between the four elements Sr, Ca, Fe^{2+}, and Mn^{2+} [8]; in calcio-ancylite Ca is replaced by Mn up to 5.6% (and by Y up to 1%), in strontio-ancylite Sr is replaced by Ba up to 1.6% [12], cf. Ba replacing Sr [4, 5]. Essential amounts of Ba were first observed in ancylite from Khibina, Kola [3], they are still higher in ancylites from Vuoriyarvi, Kola, reaching 4.13% BaO [7] and 14.36% BaO. These values indicate a constant diadochic replacement of Sr by Ba in ancylites [6] and led to the inclusion of Ba in the formulas, see above.

The surplus of RE compared with the simplified formula indicates a diadochy

$$3\,Sr^{2+} \rightarrow 2\,Ce^{3+} \text{ [4] or } Ce^{3+} + O^{2-} \rightarrow Sr^{2+} + (OH)^- \text{ [5].}$$

An analysis of a high-grade ore specimen (with about 44 wt% ancylite besides Sr-Ba calcite and quartz) from the Sheep Creek vein, Ravalli County, Montana, in which all of the RE are assigned to ancylite, gives a ratio wt% Sr/wt% RE = 0.25; this is considerably lower than the ratio wt% Sr/wt% RE = 0.47 for pure strontium ancylite and indicates sufficient substitution of Ca for Sr in this ancylite to regard it as an calcian ancylite [13].

Th partially replaces the RE [4, 5], especially La and Ce up to 0.2% [12].

The Ca, Ba, and P containing ancylite from Sallanlatva, North Karelia, shows in selected mineral grains besides the main components (in wt%) 2.21 Fe_2O_3, 0.14 Al_2O_3, 0.50 Na_2O, 0.07 K_2O and, determined by spectral analyses, about 0.01 Mn and Mg; only Fe and Al are supposed to belong to admixtures [11]. Further spectral analyses of ancylites constantly show Mn, Al, Pb, Be, Zn, and Yb [7]; ancylite from Khibina has weak lines of Mn, Ti, and Be [3], and ancylite from Odikhincha, Siberia, besides the main components, contains (in wt%) 0.5 Fe, 0.4 Si, 0.2 Y, 0.1 Al, <0.1 Mg, 0.01 Zr, <0.01 Mn and Ti, 0.001 Cu, and <0.001 Be; additionally 0.15 (?) Th and 2.5 (?) U [10].

The **RE distribution** in ancylites shows strong predominance of the cerium group [6]; the mineral is selectively ceric [4]. Besides Ce often La is also enriched [5] and both elements are the main representatives of the RE [1]. This strong relative enrichment of La (up to 39 wt%) and Ce (up to 54 wt%) is characteristic for ancylites from nepheline syenite pegmatites, hydrothermalites and their weathering zones [4], but also for ancylites from carbonatites, see Table 6, p. 220. In ancylite from Yum'echorr, Khibina massif, Kola, the ratio Ce/La is 1.8, the sum of the cerium group $\Sigma(La+Ce+Pr+Nd)$ is 98.9 wt% of all RE [14]. Ancylite from the Aldan, Yakutia, also shows a strong predominance of the cerium group [15], and besides Nd also Sm and Gd are enriched [4], see also enrichment of Nd in two Ca-rich ancylites, Table 6, p. 220.

Table 6

Relative Amounts of RE in Ancylites and Calcio-ancylites from Several Localities, $\Sigma(RE)_2O_3 = 100\%$.

Occurrence	Genetic type	La	Ce	Pr	Nd	Sm	Eu	Gd	Tb	Dy	Er	Yb	Ref.
Ancylites													
Khibina, Kola	Nepheline syenite pegmatite	32	54	4.1	9.5	0.3	—	—	—	—	—	—	[4, 5]
Yum'echorr Mountains, Khibina	Aegirine-feldspar pegmatite	28.4	51.9	4.7	13.9	0.8	—	0.3[1]	—	—	—	—	[14]
Lovozero, Kola	Nepheline syenite pegmatite	30	48	4.7	11	5.5	0.3	0.8	—	—	—	—	[4, 5]
Inagli, Aldan, Yakutia	Nepheline syenite pegmatite	13	46.3	8	25	4.7	0.5	2	0.15	0.3	—	—	[15]
Vuoriyarvi, Kola	Carbonatite	36	51	3	9	0.4	0.1	0.2	—	0.2	0.1	—	[16]
Kovdor, Kola	Carbonatite	35	50	4	9	1	0.1	0.3	0.1	0.2	0.2	0.2	[16]
Sallanlatva, North Karelia	Carbonatite	34	51	3	11	0.5	0.1	0.2	0.1	—	—	—	[16]
Ravalli County, Montana	Carbonatic rock	38	50	3	9[1]	—	—	—	—	—	—	—	[13]
Kola	Hydrothermalite[2]	36	51	3.4	9	0.3	—	—	—	—	—	—	[4]
Lovozero, Kola	Supergene	39	49	4.3	8.4	—	—	—	—	—	—	—	[4, 5]
Calcio-ancylites													
Sallanlatva, North Karelia	Carbonatite	25.6	50.5	—	17.2	0.8	5.9[3] (Eu + Gd)		—	—	—	—	[11]
Cornog, Chester County, Pennsylvania	Amphibolite	28.8	48.5	5.0	17.1	0.7	—	—	—	—	—	—	[9]

[1] RE given in at.%. – [2] In an ultrabasic alkali rock. – [3] $\Sigma RE = 100\%$.

References for 10.5.4.1.3:

[1] A. Dal Negro, G. Rossi, V. Tazzoli (Am. Mineralogist **60** [1975] 280/4, 280/1). – [2] O. B. Bøggild (Medd. Groenland **149** No. 3 [1953] 1/442, 160). – [3] E. I. Semenov, M. E. Kazakova (Tr. Mineral. Muzeya Akad. Nauk SSSR No. 11 [1961] 196/8). – [4] E. I. Semenov (Mineralogiya Redkikh Zemel', Moscow 1963, pp. 1/412, 109/11). – [5] E. I. Semenov (in: K. A. Vlasov, Geochemistry and Mineralogy of Rare Elements and Genetic Types of Their Deposits, Vol. 2: Mineralogy of Rare Elements, Israel Program Sci. Trans., Jerusalem 1966, pp. 249/328, 279/80 [Russian Original, Moscow 1964]).

[6] Yu. L. Kapustin (Mineralogiya Karbonatitov, Moscow 1971, pp. 1/288, 170, 187/9). – [7] Yu. L. Kapustin (in: K. A. Vlasov, Mineralogiya i Geneticheskie Osobennosti Shchelochnykh Massivov, Moscow 1964, pp. 135/94, 171, 173). – [8] C. Palache, H. Berman, C. Frondel (The System of Mineralogy, 7th Ed., Vol. 2, Wiley, New York 1951, pp. 1/1124, 292). – [9] F. A. Keidel, A. Montgomery, C. W. Wolfe, R. P. Christian (Mineral. Rec. **2** No. 1 [1971] 18/25, 36, 25). – [10] V. F. Motychko (Tr. Nauchn. Issled. Inst. Geol. Arktiki No. 107 [1959] 23/41, 28).

[11] M. P. Orlova, Yu. P. Rozhdestvenskii, E. N. Baranova (Tr. Vses. Nauchn. Issled. Geol. Inst. **96** [1963] 3/20, 14, 16/7). – [12] A. S. Povarennykh (Crystal Chemical Classification of Minerals, Vol. 2, translated from Russian by J. E. S. Bradley, Plenum, New York 1972, pp. 459/766, 620). – [13] E. W. Heinrich, A. A. Levinson (Am. Mineralogist **46** [1961] 1424/47, 1441). – [14] B. E. Borutskii, G. M. Varshal, V. I. Pavlutskaya, M. N. Sokolova, Z. V. Shlyukova (in: F. V. Chukhrov, B. E. Borutskii, M. N. Mozgova, Izomorfism v Mineralakh, Moscow 1975, pp. 221/46, 238/9). – [15] A. F. Efimov, S. M. Kravchenko, E. V. Vlasova (Tr. Inst. Mineral. Kristallokhim. Redk. Elem. Akad. Nauk SSSR No. 16 [1963] 141/75, 173).

[16] Yu. L. Kapustin (Geokhimiya **1966** 1311/21; Geochem. Intern. **3** [1966] 1054/64, 1060).

10.5.4.1.4 Crystal Forms and Crystal Structure

Ancylite is orthorhombic-dipyramidal with a goniometric axial ratio a:b:c = 0.571:1:0.534 [1] or a:b:c = 0.666:1:1.108 corresponding with the ratio from X-ray data a:b:c = 0.686:1:1.175 [2]. The **forms** of ancylite, originally determined on crystals from Narsarsuk, Greenland, and formerly thought by Flink to be a combination of two domes {101} and {011} [3], are a combination of a prism d{120} with a rhombic dipyramid e{111} [1, 4, 5]. The prism d{120} is observed at ancylites from several localities, e.g., together with e{111} on those from North Siberia [6], from Cornog, Pennsylvania, where both forms are equally developed or d > e, together with the subordinate forms a {100}, b{010}, and h{101} [2], and from Yakutia, together with the rhombic dipyramid {477}, a rhombic prism {h0l} and the pinacoid {100} [7]. The ancylites from the Sebl'-Yavr massif, Kola, are characterized by d{120} and {477} [5, 7], those from Sallanlatva, North Karelia, by {120} and {101} [8], sometimes together with {021} and {010} [5]; see **Fig. 11**, p. 222.

Ancylite crystals usually have pseudohexagonal-dipyramidal **habit** [9] and are short prismatic parallel to [001] [1], cf. [4], sometimes with equally developed prism faces [11]. Other crystals are long prismatic [6], elongated four-sided or short stubby (Mont St. Hilaire) [12], prismatic or bladed (Gem Park Complex) [13], or flattened rhombic pyramidal (Ural) [14]. Some crystals appear pseudo-octahedral [1], e.g., crystals in crusts on the walls of voids in dolomitic rock at Vuoriyarvi, Kola [15], very tiny prismatic crystals included in natrolite from Inagli, Yakutia [7], and crystals with slender prismatic or stout to elongated prismatic habit at Cornog, Pennsylvania [2]. Minute crystals resembling octahedra occur in druses in nordmarkite from South Norway [16]. Furthermore tabular crystals are observed [10]. Druses of prismatic crystals in vugs are generally characteristic for carbonatites [17], cf. the well developed crystals in druses, but also granular aggregates of ancylites in a carbonatite from Siberia [6]. From two generations of ancylites in the carbonatite of Sallanlatva, North Karelia, the first consists of prismatic or short prismatic crystals in calcitic carbonatites, the second consists of long prismatic or lance-shaped crystals, often forming crosses or irregular growths in vugs of the dolomite-siderite carbonatites [8]. Ancylite is also observed in form of aggregates of radial fibers in granite of Predazzo [18] and (in few cases) in the Sheep Creek vein [19]; it forms rosettes of tabular crystals at Kovdor, Kola [5], radiating groups at Mont St. Hilaire [20], and aggregates of plates in Ontario [21], further concretions or aggregates of rounded crystals [9, 4, 22], and disseminated clusters in dolomite-barite-monazite dikes of the Gem Park Complex, Colorado [13].

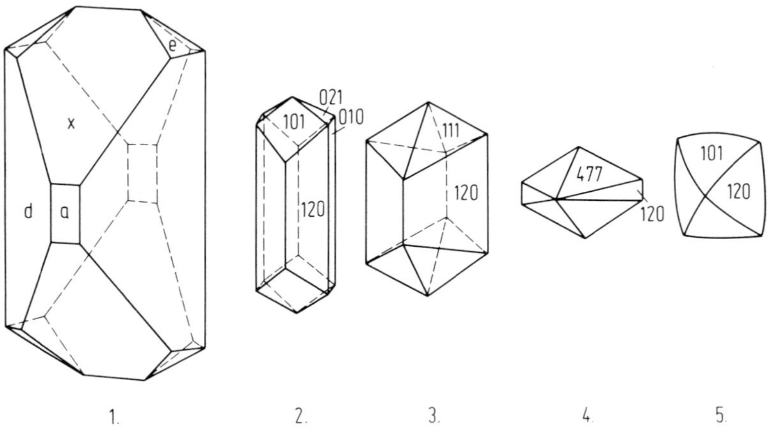

Fig. 11. Forms and habit of ancylite crystals.
1. Inagli, Yakutia, from [7]. 2. and 5. Sallanlatva, North Karelia;
3. Narsarsuk, Greenland; 4. Sebl'-Yavr, Kola, from [5].

Ancylite crystals often show **rounded** or curved **faces** and edges [4], cf. rounded faces and distorted character of the crystals [1]. In allusion to this property, first observed at the original locality in Greenland [3], the mineral has received its name [1, 4]. In recent times rounded features have been observed for most of the measured faces of the crystals from Cornog, Pennsylvania, especially on e{111}, so that pairs of these faces appear to be single faces [2]. Such features further occur in several other localities of the USA [23], Canada [12], Siberia [1, 6, 11], and at Khibina, Kola [7].

Ancylites of nearly all occurrences show no **cleavage** [1, 3, 4, 6, 7, 11, 15]. On ancylite from the granite of Predazzo, North Italy, however, two sets of cleavage seem recognizable: one perpendicular to the optical direction X and the other inclined on X and parallel to Y [18]. For the ancylite of the Mont St. Hilaire, Quebec, a perfect prismatic cleavage is mentioned only in a preliminary report [20].

The phenomenon of syntaxic intergrowths, well known from the bastnaesite series (see pp. 90/1), is supposed to exist also for ancylite [27].

The **unit cell constants** in Å and axial ratios of ancylite from Odikhincha, Siberia (I), from Mont St. Hilaire, Quebec (II), of calcio-ancylite from Cornog, Pennsylvania (III), and of synthetic $Nd(OH)CO_3$ (A type) (IV) are given in the following table:

	a in Å	b in Å	c in Å	a:b:c	Ref.
I	5.00 ± 0.02	8.48 ± 0.04	7.01 ± 0.06	0.590:1:0.827	[4, 9, 11]
II[1)]	7.24	8.49	5.02	0.8527:1:0.5912	[20]
II	5.03(1)	8.53(1)	7.29(1)	0.5897:1:0.8546	[24]
III[2)]	4.97 ± 0.05	7.24 ± 0.05	8.50 ± 0.05	0.686:1:1.175	[2]
IV	4.953	8.477	7.210	0.584:1:0.851	[28, p. 139]

[1)] Setting (choice of axes) in conventional orientation with $c<a<b$ for orthorhombic crystals [2]. – [2)] b and c exchanged.

The space group of ancylite is Pmcn-D_{2h}^{16} (No. 62, standard setting Pnma), Z = 2 [24], cf. Z = 2, cell volume V = 297.5 Å^3, calculated for ancylite from Odikhincha, Siberia, by Semenov [22, 9]. Z = 1 is given by Povarennykh [25]. The d values of ancylite from Narsarsuk, Greenland, determined by [26] correspond exactly to the d values of ancylite from Norway [16], while the set of X-ray powder data of ancylite from Predazzo agrees very well with another determination of ancylite from Greenland, given in the Catalogue of X-ray Diffraction Patterns; the d spacings are, however, mostly lower [18]. Identical d values have been observed for ancylites from several localities on Kola and from Odikhincha, Siberia [5, 11], for the ancylites from Khibina [15] and from Inagli, South Yakutia [7], with that of Vuoriyarvi [15], and for that from Bor Uryakh, North Siberia, with those from other massifs [6]. Very characteristic for many ancylites, among them crystals from Yakutia, is the failure of X-ray photographs; it is probably caused by displacement of single lattice blocks relative to each other [7].

Ancylite shares the space group Pmcn with the orthorhombic carbonates aragonite, strontianite, and witherite. The dimensions of its a and c cell edges are very close to the corresponding ones in strontianite. As in the orthorhombic carbonates, the triangular carbonate groups of ancylite are normal to the c cell edge and are grouped in pairs, in which one triangle is superposed on the other one after a rotation of 60°. The carbonate ions are connected to one another by the tenfold coordinated cations. The **crystal structure** of ancylite can ideally be derived from that of the orthorhombic carbonates by introducing in the latter a $(OH)^-$ group or a water molecule lying on the mirror plane and bounded to the heavy cations M = (La, Ce, Sr, Ca). These cations are bounded to eight oxygen atoms belonging to carbonate ions and to two water molecules; the mean M-O distance is 2.61 Å. For a detailed description of the ancylite structure, further for M-O bond distances, and discussion of the impossibility of coexistence of ancylite-type layers with aragonite-type layers, see the original paper [24].

The synthetic compound $RE(OH)CO_3$, isostructural with ancylite, represents the limiting composition where all M^{2+} ions have been replaced by RE^{3+} ions; the isostructural relation between the two phases of different composition is explained by the random distribution of ions of approximately the same size over the four sites (Ca^{2+}, Sr^{2+}), RE^{3+}, OH^-, and H_2O [27]. The structures of ancylite and this synthetic compound (A type) are exactly the same, the coordination number for Nd in $Ln(OH)CO_3$ is, however, 9. As ancylite has a very complex chemical composition, it is possible that the structures of ancylite and of the synthesized hydroxycarbonate are different but both have a similar cationic frame [28, pp. 135, 138/40].

References for 10.5.4.1.4:

[1] C. Palache, H. Berman, C. Frondel (The System of Mineralogy, 7th Ed., Vol. 2, Wiley, New York 1951, pp. 1/1124, 291/2). – [2] F. A. Keidel, A. Montgomery, C. W. Wolfe, R. P. Christian (Mineral. Rec. **2** No. 1 [1971] 18/25, 36, 21/4). – [3] O. B. Bøggild (Medd. Groenland **149** No. 3 [1953] 1/442, 161). – [4] E. I. Semenov (in: K. A. Vlasov, Geochemistry and Mineralogy of Rare

Elements and Genetic Types of Their Deposits, Vol. 2: Mineralogy of Rare Elements, Israel Program Sci. Trans., Jerusalem 1966, pp. 249/328, 279 [Russian Original, Moscow 1964]). – [5] Yu. L. Kapustin (in: K. A. Vlasov, Mineralogiya i Geneticheskie Osobennosti Shchelochnykh Massivov, Moscow 1964, pp. 135/94, 171/2).

[6] L. M. Dakhiya (Zap. Vses. Mineral. Obshchestva **98** [1968] 737/9). – [7] A. F. Efimov, S. M. Kravchenko, E. V. Vlasova (Tr. Inst. Mineral. Kristallokhim. Redk. Elem. Akad. Nauk SSSR No. 16 [1963] 141/75, 171/2). – [8] M. P. Orlova, Yu. P. Rozhdestvenskii, E. N. Baranova (Tr. Vses. Nauchn. Issled. Geol. Inst. **96** [1963] 3/20, 13). – [9] E. I. Semenov (Mineralogiya Redkikh Zemel', Moscow 1963, pp. 1/412, 110). – [10] O. B. Dudkin, L. I. Polezhaeva, F. V. Minakov (Veshchestvennyi Sostav Shchelochnykh Intrusivnykh Kompleksov Kol'skogo Poluostrova, Apatity 1981, pp. 73/81, 80; Ref. Zh. Geol. **1981** No. 12 V 179).

[11] V. F. Motychko (Tr. Nauchn. Issled. Inst. Geol. Arktiki No. 107 [1959] 23/41, 28). – [12] L. Marble, A. Regis (Rocks Minerals **54** [1979] 4/25, 11). – [13] R. L. Parker, W. N. Sharp (U.S. Geol. Surv. Profess. Paper No. 649 [1970] 1/24, 13). – [14] I. P. Nikitenko, E. P. Kalinin, N. P. Yushkin (in: Geologiya i Poleznye Iskopaemye Severo-Vostoka Evropeiskoi Chasti SSSR Severa Urala, Vol. 2, Syktyvkar 1973, pp. 430/3; Ref. Zh. Geol. **1974** No. 2 V 192). – [15] L. V. Kozyreva, G. A. Il'inskii (Mater. Mineral. Kolsk. Poluostrova **1** [1959] 69/76, 71/2).

[16] I. Oftedal, P. C. Sæbø (Norsk Geol. Tidsskr. **45** [1965] 171/5). – [17] Yu. L. Kapustin (Mineralogiya Karbonatitov, Moscow 1971, pp. 1/288, 188). – [18] F. Emiliani, G. Gandolfi (Mineral. Petrogr. Acta **11** [1965] 123/31, 126/7). – [19] E. W. Heinrich, A. A. Levinson (Am. Mineralogist **46** [1961] 1424/47, 1432/3). – [20] J. A. Mandarino, D. C. Harris, J. Bradley (Can. Mineralogist **8** [1964/66] 398).

[21] A. P. Sabina (Geol. Surv. Paper Geol. Surv. Can. No. 78-1A [1978] 253/8, 254). – [22] E. I. Semenov, M. E. Kazakova (Tr. Mineral. Muzeya Akad. Nauk SSSR No. 11 [1961] 196/8). – [23] A. L. Anderson (Econ. Geol. **55** [1960] 1179/201, 1189). – [24] A. Dal Negro, G. Rossi, V. Tazzoli (Am. Mineralogist **60** [1975] 280/4). – [25] A. S. Povarennykh (Crystal Chemical Classification of Minerals, Vol. 2, translated from Russian by J. E. S. Bradley, Plenum, New York 1972, pp. 459/766, 620).

[26] H. Neumann, K. O. Bryan (Avhandl. Norske Videnskaps. Akad. Oslo I Mat. Naturw. Klasse No. 1 [1958] 1/15, 4, 12). – [27] J. Sawyer, P. Caro, L. Eyring (Rev. Chim. Minerale **10** [1973] 93/104, 93, 100/1, 103). – [28] H. Dexpert, E. Antic-Fidancev, J. P. Coutures, P. Caro (J. Cryst. Spectrosc. Res. **12** No. 2 [1982] 129/42).

10.5.4.1.5 Optical and Other Physical Properties

The luster of ancylite is vitreous on crystal faces, greasy on fracture surfaces [1], cf. [2], weakly vitreous on the faces [3], vitreous to distinctly greasy on fracture surfaces [4], vitreous, sometimes greasy or dull [5]. The mineral is translucent [4 to 7] to subtranslucent [1], translucent in thin splinters [5]; sometimes it is opaque [8, p. 13] or almost opaque in thin section [9, 10]. Fine crystals are transparent [11, 12], semitransparent to nontransparent [3]. Semitransparency of crystals from Greenland is due to innumerable enclosed aegirine needles [13]. The crystals from Sallanlatva, northern Karelia, are sometimes transparent, but commonly nontransparent; transparent colorless crystals with a brownish colored cap are observed [14]. Ancylite crystals from Yakutia sometimes have a zonal coloring with outer parts darker than the inner parts [5].

The color of ancylite is commonly pale yellow with a tinge of orange, also yellowish brown to brown or gray [1], cf. [7], light yellow or brown [6]. The ancylite from Greenland is yellowish or brownish in various shades [13], that from Lovozero, Kola, light yellow to greenish yellow [15], that from Mont St. Hilaire, Quebec, is colorless to dirty white, flesh-colored to rose [16], that from Montreal is dark green to greenish brown [2]. At the Gem Park Complex, Colorado, the opaque ancylites are cream to pale pink, others are pink or dark reddish brown [8, pp. 8, 13, 17], cf. rose tints for ancylite from East Siberia [17], and different colors up to black for ancylites from several localities in the table on pp. 225/6. In the fine-grained ancylite-quartz aggregate of the Sheep Creek vein, Montana, with increasing amounts of ancylite, the color deepens in reddish tints from faintly pink to reddish lilac at a content slightly less than 50% ancylite [18], cf. p. 215. The pink color observed for ancylite from Cornog, Pennsylvania, obviously is connected with a larger absorption in the green and yellow parts compared with the orange and red parts of the spectrum for incandescent light or daylight [4]. Black or dark brown colors may be caused by structural defects resulting from U- and Th contents in the mineral [19]. The streak of ancylite is white [1, 7]. In transmitted light ancylite is colorless, but sometimes clouded by inclusions [1], buff-colored [9, 10], also brown-rose with a pseudo-absorption $Z \gg X$ [14].

For ancylites and calcio-ancylites from several localities the color, the refractive indices, the values for the axial angle 2V, and the density D(meas.) are given in the following table:

Locality	Color	Refractive indices n_x	n_y	n_z	Axial angle 2V(−)	D (meas.) in g/cm³	Ref.
Narsarsuk, Greenland	brown, yellow-green, orange	1.625	1.700	1.735	66°	3.95	[1, 13]
Ravalli County, Montana	indian red to white	1.656	1.719	1.759	71° to 75°	—	[21]
Cornog, Pennsylvania 1)	pale pink; rarely whitish, grayish, brownish	1.654 ±0.003 2)	1.733 ±0.003 2)	1.722 ±0.003 2)	60° to 70°	3.95	[4]
	lavender-pink	1.656 ±0.003	1.734 ±0.003	1.776 ±0.003	70°	—	[21]
Bearpaw Mountains, Montana	—	1.615 ±0.003	1.695 ±0.003	1.730 ±0.003	70° ±5°	—	[22]
Mont St. Hilaire, Quebec	lavender-pink to brownish pink	1.676 ±0.003	1.727 ±0.003	1.766 ±0.003	75° to 80°	—	[21]
	pink to lilac	1.680 2)	1.725 2)	1.768 2)	90° 86°(calc.)	4.15	[23]
Grorud, South Norway	dull yellow	—	1.720	—	≈70°	—	[24]
Predazzo, North Italy	pale pink	—	1.705	—	≈70°	—	[25]

Locality		Color	Refractive indices n_x	n_y	n_z	Axial angle 2V(−)	D (meas.) in g/cm³	Ref.
Khibina, Kola[3]								
	I	honey yellow	1.628	1.705	1.740	68°	—	[3]
		to brownish	1.638	1.705	1.740	68°	3.987	[6, 7, 26]
	II	honey yellow to brownish	1.636	—	1.748	—	3.82	[26]
	III	white	1.620	—	1.730	—	—	[27]
Vuoriyarvi, Kola								
	I	brown	1.629	1.699	1.737	70°	—	[11]
	II[4]	colorless	1.635	1.732	1.764	—	4.10	[28]
	III[4]	yellowish	1.630	—	1.760	72°	4.15	[28]
	IV	yellowish	1.630	1.732	1.762	72°	3.72 to 4.1	[29]
Kovdor, Kola								
	I	rose	1.621	1.718	1.745	70°	4.00	[28]
	II	rose	1.620	1.695	1.720	71°	3.82	[29]
Sebl'-Yavr, Kola		brown	1.658	1.724	1.747	—	3.61	[29]
Afrikanda, Kola		yellowish	1.640	1.708	1.743	70°	3.99	[29]
Sallanlatva, North Karelia[5]								
	I	yellowish, red-brown, brown, colorless	1.658	1.718	1.743	60° to 70°	3.90 to 4.04[6]	[14]
	II	red	1.625	1.730	1.754	71°	3.78	[28, 29]
	III	rose	1.631	—	1.760	—	3.97	[28]
Torgovaya, North Urals		—	—	1.690	1.746	—	3.91	[12]
Odikhincha, North Siberia[7]		colorless, pale yellow, rose	1.625	1.693	1.737	large	—	[30]
Bor Uryakh, North Siberia		light yellowish tints	1.656	1.727	1.758	60° to 70°	3.94	[3]
Inagli, Yakutia		yellow, rose, brownish to black	1.636	—	1.740	68°	3.74	[5]

[1] Calcio-ancylite with 8.35% CaO. – [2] Measured in white light according to [21]. – [3] I with 3.78, II with 4.36, and III with 4.10% CaO. – [4] Ba-rich ancylites with 4.13 (II) and 14.36 (III)% BaO. –

5) I and II are calcio-ancylites with 6.51% CaO, III contains besides 2.08% CaO also 4.31% BaO. – 6) D = 3.90 g/cm^3, yellow crystals; D = 4.04 g/cm^3, red-brown crystals. – 7) Contains >10% Ca according to a spectral analysis.

The optical character of all the samples enumerated in the table is negative; the brown crystals from Narsarsuk, Greenland, determined by Larsen [20] to have an optical positive character, $2V \approx 0°$ and $n_x = 1.865$, apparently are not ancylite [1]. The calculated mean index of refraction (after the law of Gladstone and Dale) $dK + 1 = 1.692$, that is a deviation of +0.005 from the mean index $n = 1.687$ $(n_x + n_y + n_z)/3$ of ancylite from Greenland [31].

A direct dependence between the refractive index n_z and the content of CaO and a reverse between n_z and the content of SrO in ancylites is shown by corresponding diagrams n_z to CaO and n_z to SrO, respectively; the refraction is, however, apparently not influenced by the amounts of Ce_2O_3 and BaO [3]. In contrast the difference in refractive indices (values lower by three or more units in the second decimal) are not explainable by replacement of Ca by Sr or by the ratios of the individual RE; only a change in the water content (water of composition or of hydration), which is not a simple dehydration, but connected with a solid solution or a series of closely related hydrates, would probably meet this requirement [21].

The dispersion of the optic axes is $r < v$ for the crystals from the Bearpaw Mountains, Montana [22], but barely perceptible for that from Cornog, Pennsylvania [4]; the dispersion is slight for ancylites from Ravalli County, Montana, and from Mont St. Hilaire, Quebec [21], but high for the latter in white light [23], cf. [21]. The orientation determined for crystals from Greenland and Quebec is X = a, Y = b, Z = c [1, 21], cf. [34], but X = a, Z = b for ancylite from Quebec [23].

Absorption spectra of ancylite show weak bands at 13500 and 19000 cm^{-1} corresponding to Nd^{3+} [19]; the spectrum of calcio-ancylite from Cornog, Pennsylvania, in ordinary light shows narrow absorption bands near the center of the visible range at 5700 to 5800 Å and at about 7400 Å; it is typical for materials containing Nd and subordinate Pr [4].

Ancylites and calcio-ancylites show no luminescence at 3600, 3200 to 2800, and 2500 Å, and also no cathodic luminescence [32]. The pale pink color (in daylight) of calcio-ancylite from Cornog dissappears under the fluorescent lamp and the mineral displays a pale bluish gray tinge; a few crystals remain whitish, grayish, or brownish under any illumination [4].

The hardness of ancylites $H_M = 4$ to 4.5 [1, 33], cf. 4 [6, 7], about 4 [15], 4.5 [13, 34]. The microhardness VHN of ancylite was determined to be 259 to 303 kg/mm^2, mean value 286 kg/mm^2 (load 50 g), corresponding with a Mohs value of 4.6; this is sufficiently close to the measured values on the Mohs' scale and some deviation may be caused by the high brittleness of the mineral [35].

The fracture of ancylite is splintery [7], splintery, but rather tough [1], conchoidal [6], uneven [3, 11], or irregular [14]. The mineral is somewhat brittle [1], highly brittle [35], see above.

The ranges of density values given for ancylites from 3.8 to 4.3 g/cm^3 [33], 3.82 to 3.95 g/cm^3 [15], or 3.95 to 3.99 g/cm^3 (theoretical value 4.26 g/cm^3) [6, 7], are in the range of the density values (3.61 to 4.15 g/cm^3) measured for ancylites from several occurrences, see the table on pp. 225/6; the highest value is measured for a sample with 14.36% BaO [28]. For the structurally determined ancylite from the Mont St. Hilaire, Quebec, the measured density is 4.1, the calculated one 4.15 g/cm^3 [36].

References for 10.5.4.1.5:

[1] C. Palache, H. Berman, C. Frondel (The System of Mineralogy, 7th Ed., Vol. 2, Wiley, New York 1951, pp. 1/1124, 291/3). – [2] A. P. Sabina (Geol. Surv. Paper Geol. Surv. Can. No. 78-1A

[1978] 253/8, 253/4). – [3] L. M. Dakhiya (Zap. Vses. Mineral. Obshchestva **98** [1968] 737/9). – [4] F. A. Keidel, A. Montgomery, C. W. Wolfe, R. P. Christian (Mineral. Rec. **2** No. 1 [1971] 18/25, 36, 21/2, 24/5). – [5] A. F. Efimov, S. M. Kravchenko, E. V. Vlasova (Tr. Inst. Mineral. Kristallokhim. Redk. Elem. Akad. Nauk SSSR No. 16 [1963] 141/75, 173).

[6] E. I. Semenov (Mineralogiya Redkikh Zemel', Moscow 1963, pp. 1/412, 110). – [7] E. I. Semenov (in: K. A. Vlasov, Geochemistry and Mineralogy of Rare Elements and Genetic Types of Their Deposits, Vol. 2: Mineralogy of Rare Elements, Israel Program Sci. Trans., Jerusalem 1966, pp. 249/328, 279 [Russian Original, Moscow 1964]). – [8] R. L. Parker, W. N. Sharp (U.S. Geol. Surv. Profess. Papers No. 649 [1970] 1/24). – [9] W. J. Verwoerd (S. Africa Geol. Surv. Ann. Rept. **2** [1963] 119/35, 124). – [10] W. J. Verwoerd (S. Africa Geol. Surv. Handbook **6** [1967] 1/452, 35).

[11] L. V. Kozyreva, G. A. Il'inskii (Mater. Mineral. Kol'sk. Poluostrova **1** [1959] 69/76, 71/2). – [12] I. P. Nikitenko, E. P. Kalinin, N. P. Yushkin (in: Geologiya i Poleznye Iskopaemye Severo-Vostoka Evropeiskoi Chasti SSSR Severa Urala, Vol. 2, Syktyvkar 1973, pp. 430/3 from Ref. Zh. Geol. **1974** No. 2 V 192). – [13] O. B. Bøggild (Medd. Groenland **149** No. 3 [1953] 1/442, 160/1). – [14] M. P. Orlova, Yu. P. Rozhdestvenskii, E. N. Baranova (Tr. Vses. Nauchn. Issled. Geol. Inst. **96** [1963] 3/20, 13/4, 16). – [15] K. A. Vlasov, M. V. Kuz'menko, E. M. Es'kova (Lovozerskii Shchelochnoi Massiv, Moscow 1959, pp. 1/624, 497/8).

[16] L. Marble, A. Regis (Rocks Minerals **54** [1979] 4/25, 11). – [17] T. B. Zdorik (Tr. Mineral. Muzeya Akad. Nauk SSSR No. 17 [1966] 60/75, 68). – [18] E. W. Heinrich, A. A. Levinson (Am. Mineralogist **46** [1961] 1424/47, 1432). – [19] A. N. Platonov, O. A. Krasil'shchikova, A. N. Tarashchan (Konst. Svoistva Mineral. No. 9 [1975] 58/64, 60/1, 63). – [20] E. S. Larsen (U.S. Geol. Surv. Bull. No. 679 [1921] 1/294, 39).

[21] W. M. D. Bryant (Microscope **21** [1973] 39/47, 41, 44/5). – [22] W. T. Pecora, J. H. Kerr (Am. Mineralogist **38** [1953] 1169/83, 1172). – [23] J. A. Mandarino, D. C. Harris, J. Bradley (Can. Mineralogist **8** [1964/66] 398). – [24] I. Oftedal, P. C. Sæbø (Norsk Geol. Tidsskr. **45** [1965] 171/5, 173). – [25] F. Emiliani, G. Gandolfi (Mineral. Petrogr. Acta **11** [1965] 123/31, 126).

[26] E. I. Semenov, M. E. Kazakova (Tr. Mineral. Muzeya Akad. Nauk SSSR No. 11 [1961] 196/8). – [27] O. B. Dudkin, L. I. Polezhaeva, F. V. Minakov (Veshchestvennyi Sostav Shchelochnykh Intrusivnykh Kompleksov Kol'skogo Poluostrova, Apatity 1981, pp. 73/81, 76; Ref. Zh. Geol. **1981** No. 12 V 179). – [28] Yu. L. Kapustin (Mineralogiya Karbonatitov, Moscow 1971, pp. 1/288, 189). – [29] Yu. L. Kapustin (in: K. A. Vlasov, Mineralogiya i Geneticheskie Osobennosti Shchelochnykh Massivov, Moscow 1964, pp. 135/94, 173). – [30] V. F. Motychko (Tr. Nauchn. Issled. Inst. Geol. Arktiki No. 107 [1959] 23/41, 28).

[31] H. W. Jaffe (Am. Mineralogist **41** [1956] 757/77, 764). – [32] G. P. Barsanov, V. A. Seveleva (Tr. Mineral. Muzeya Akad. Nauk SSSR No. 5 [1953] 56/89, 62, 70). – [33] A. S. Povarennykh (Crystal Chemical Classification of Minerals, Vol. 2, translated from Russian by J. E. S. Bradley, Plenum, New York 1972, pp. 459/766, 620). – [34] E. S. Larsen, H. Berman (U.S. Geol. Surv. Bull. No. 848 [1934] 1/266, 188). – [35] A. S. Povarennykh, A. D. Lebedeva (Konst. Svoistva Mineral. No. 5 [1971] 39/46, 44/5).

[36] A. Dal Negro, G. Rossi, V. Tazzoli (Am. Mineralogist **60** [1975] 280/4, 281).

10.5.4.1.6 Chemical Behavior and DTA

Ancylite is easily soluble in acids [1, 2], it readily decomposes and evolves CO_2 [3, 4, 5]. Hydrochloric acid [6] dissolves the mineral in the cold [7], in the diluted state [8], and converts it to a grayish powdery material [9].

The thermograms of ancylites from three different localities show two endothermal effects, that of a fourth ancylite analysed up to 1200°C (Sallanlatva) shows a third endothermal effect at 1080°C:

Locality	Main endothermal effects in °C	weight loss in %	range in °C	Ref.
Khibina, Kola	300 to 460, 920	27.16	300 to 1000	[9]
Inagli, Yakutia	420 to 450, 925	—	—	[5]
Bor-Uryakh, Siberia	510, 810	—	—	[10]
Sallanlatva, North Karelia	≈500, ≈820, and ≈1080	17	20 to 900	[11]
Mont St. Hilaire, Quebec	—	28.4	25 to 1100	[12]

The range of the "main endothermal effect" between 300 and 460°C given by [9] is refined to 480°C for this effect; it is characteristic for all carbonates of the RE and is due to the decomposition of the mineral with loss of water and carbon dioxide [3, 4], cf. [10]. The decomposition of ancylite from Sallanlatva in three stages is similar to carbonates of the ankerite group; the thermogravimetric curve of ancylite obtained by heating up to 920°C indicates the main loss of CO_2, H_2O, and OH^- groups to be in the temperature interval 460 to 600°C [11]. From the total weight loss of ancylite from the Mont St. Hilaire, Quebec, 20% have been assigned to water and 80% to carbon dioxide [12].

Products of heating are cubic phases of RE oxides of the fluorite structure type [3].

References for 10.5.4.1.6:

[1] C. Palache, H. Berman, C. Frondel (The System of Mineralogy, 7th Ed., Vol. 2, Wiley, New York 1951, pp. 1/1124, 292). – [2] O. B. Bøggild (Medd. Groenland **149** No. 3 [1953] 1/442, 161). – [3] E. I. Semenov (Mineralogiya Redkikh Zemel', Moscow 1964, pp. 1/412, 110). – [4] E. I. Semenov (in: K. A. Vlasov, Geochemistry and Mineralogy of Rare Elements and Genetic Types of Their Deposits, Vol. 2: Mineralogy of Rare Elements, Israel Program Sci. Trans., Jerusalem 1966, pp. 249/328, 279/80 [Russian Original, Moscow 1964]). – [5] A. F. Efimov, S. M. Kravchenko, E. V. Vlasova (Tr. Inst. Mineral. Kristallokhim. Redk. Elem. Akad. Nauk SSSR No. 16 [1963] 141/75, 173).

[6] F. Emiliani, G. Gandolfi (Mineral. Petrogr. Acta **11** [1965] 123/31, 126). – [7] L. V. Kozyreva, G. A. Il'inskii (Mater. Mineral. Kol'sk. Poluostrova **1** [1959] 69/76, 72). – [8] V. F. Motychko (Tr. Nauchn. Issled. Inst. Geol. Arktiki No. 107 [1959] 23/41, 28). – [9] E. I. Semenov, M. E. Kazakova (Tr. Mineral. Muzeya Akad. Nauk SSSR No. 11 [1961] 196/8). – [10] L. M. Dakhiya (Zap. Vses. Mineral. Obshchestva **98** [1968] 737/9).

[11] M. P. Orlova, Yu. P. Rozhdestvenskii, E. N. Baranova (Tr. Vses. Nauchn. Issled. Geol. Inst. **96** [1963] 3/20, 14). – [12] A. Dal Negro, G. Rossi, V. Tazzoli (Am. Mineralogist **60** [1975] 280/4, 280).

10.5.4.2 Schuilingite $PbCuRE[CO_3]_3(OH)\cdot{\sim}1.5\,H_2O$

The mineral was originally described incompletely and named by Vaes, who did not indicate the locality of its occurrence [1]. Later two samples from a locality designated as Kalompe, Katanga, were studied by Guillemin and Pierrot. Based on an analysis with Ca, erroneously determined instead of the RE, the mineral had been thought to be a carbonate of Pb, Cu, and Ca with the formula $Pb_3Cu_2Ca_6[CO_3]_8(OH)_6 \cdot 6H_2O$ [2]. A new description of this mineral was accomplished by Piret and Deliens on ten crystals from the copper deposits at Menda and Kasompi, southern Shaba, Zaire [3].

Schuilingite at Kalompe was found as crusts of fan-like grouped crystals in weathered vein rocks, consisting predominantly of quartz, clay minerals, sericite, cerussite, pyrite, limonite, calcite, an amorphous black Mn mineral, and rarely also malachite [2]. Schuilingite from Menda and Kasompi occurs in the weathering zone of these copper deposits [3].

A new analysis of the mineral (with determination of Pb, Cu, and RE by microprobe; traces of Tb, Yb, and Lu by neutron activation; CO_2 by chromatography and CO_2+H_2O by thermogravimetric analyses) shows the mineral to be a carbonate of Pb, Cu, and RE, without Ca, having the formula given in the heading. Corrected to 100 wt% it contains 25.81% $(RE,Y)_2O_3$, including 2.06% Y_2O_3. The mean relative amounts of 7 RE and Y in schuilingite from analyses of three crystals for $(RE,Y)_2O_3=100$ are as follows: La 4.0, Pr 4.5, Nd 32.9, Sm 12.8, Eu 9.1, Gd 14.2, Dy 10.7, Y 11.8. The analyses of the same eight elements in ten crystals, recalculated to a total of the oxides = 100%, show, however, a considerably broader range in the relative amounts of the oxides of these eight elements in schuilingite. The variance is for La_2O_3 1.28 to 5.20, Pr_2O_3 2.43 to 6.37, Nd_2O_3 20.72 to 43.97, Sm_2O_3 11.56 to 17.86, Eu_2O_3 8.28 to 11.70, Gd_2O_3 11.96 to 21.01, Dy_2O_3 4.31 to 20.06, Y_2O_3 1.95 to 15.63; it is characterized by a positive correlation of Y_2O_3 with the heavy oxides Gd_2O_3 and Dy_2O_3 and an opposite one between Y_2O_3 and the oxides of the light RE (La to Eu, without Ce). The extreme values of the relation $Y_2O_3+Gd_2O_3+Dy_2O_3/La_2O_3+Pr_2O_3+Nd_2O_3+Sm_2O_3+Eu_2O_3$ are 1.19 and 0.22; further a correlation between the pairs Nd_2O_3-Pr_2O_3 and Sm_2O_3-Eu_2O_3 is confirmed by the values of the ten crystals mentioned, see **Fig. 12** [3].

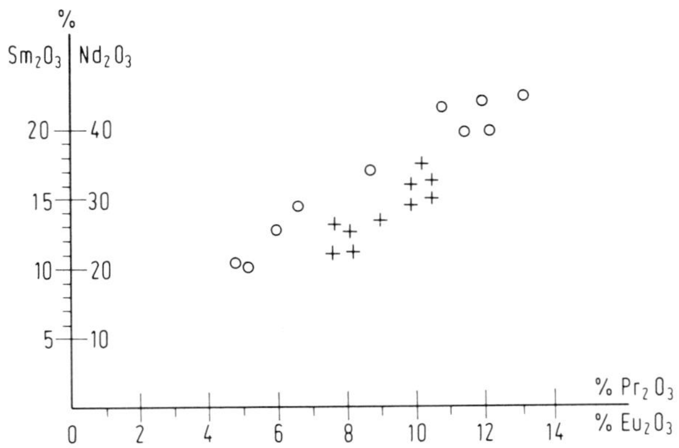

Fig. 12. Relations between pairs of lanthanides in ten schuilingite crystals of southern Shaba, Zaire, from [3].
○ $Nd_2O_3-Pr_2O_3$
\+ $Sm_2O_3-Eu_2O_3$

The mineral is orthorhombic [3], not monoclinic as formerly supposed by [2]. The habit of the crystals is acicular, elongated [001], terminated by prisms with an angle between m and m' = 60° to 70° and by terminal domes; the crystals are generally agglomerated to sheafs or bushels [1]. Elongated monocrystals without well developed faces and flattened to {100} are observed by [3], polysynthetic intergrown crystals by [2]. The cleavage is perfect {110}, poor {100} [1]; it is perfect in the direction of elongation [2].

The unit cell has the parameters a = 7.416(11), b = 18.83(3), c = 6.388(9) Å, determined by powder pattern, and a = 7.420(13), b = 18.91(3), c = 6.381(11) Å, determined by single crystal photograph; the mean values are a = 7.418(9), b = 18.87(2), c = 6.385(7) Å; cell volume V = 893.8 $Å^3$. The space group is $P2_1cn\text{-}C_{2v}^9$ (No. 33, standard setting $Pna2_1$); Z = 4. The positions of the heavy atoms in the structure, obtained by Patterson analyses, show disorder, due to the presence of different RE. Atom coordinates are given in the paper [3].

Schuilingite has adamantine luster, it is transparent and has azure color [1], or is azure to turquoise in color; it has a pale blue streak, light blue color in transmitted light, and no pleochroism [2]. A weak pleochroism in azure-blue and darker blue, orientation X = a, Y = c, and Z = b, a biaxial negative optical character with 2V about 60° and parallel extinction are observed by Vaes [1]; 2V = 66°, refractive indices $n_x = 1.710$ (calculated), $n_y = 1.755 \pm 0.005$, $n_z = 1.775 \pm 0.005$, most crystals with parallel extinction or slightly inclined extinction (8° to 10°) are observed by Guillemin and Pierrot [2].

The hardness $H_M = 3$ to 4; the mineral is brittle. The measured density is 5.2 ± 0.1 g/cm³ [2], the density, calculated for the observed RE variances, D = 4.71 to 4.77, mean value 4.74 g/cm³; theoretically the density might fluctuate between 4.34 g/cm³ (pure Y member) and 4.88 g/cm³ (pure Dy member) [3].

The mineral is easily soluble in dilute acids with strong effervescence [2].

References for 10.5.4.2:

[1] J. F. Vaes (Ann. Soc. Geol. Belg. Bull. **70** [1946/47] B 233/B 236). – [2] C. Guillemin, R. Pierrot (Bull. Soc. Franc. Mineral. Crist. **80** [1957] 549/51). – [3] P. Piret, M. Deliens (Bull. Mineral. **105** [1982] 225/8).

10.5.4.3 Bijvoetite $(RE)_2(UO_2 \cdot CO_3)_4(OH)_6 \cdot {\sim}11\,H_2O$

Bijvoetite, a hydrated carbonate of uranyl and rare earths, has been found together with the new mineral lepersonnite (see p. 232) near primary uraninite, impregnated in beige dolomite in the lower part of the oxidation zone of the uranium deposit at Shinkolobwe, Southern Shaba, Zaire. The new minerals are associated with oxide hydrates of uranium (orange "gummite", consisting mainly of curite and becquerelite), with rutherfordine, sklodowskite, and uranophane.

The chemical composition of this mineral (determination of U and RE by microprobe and of CO_2 by chromatography, water calculated by difference) is characterized by 9.37% RE_2O_3 and 7.52% Y_2O_3 besides 60.66% UO_3; this leads to the formula $(Y_{1.26}Dy_{0.57}Gd_{0.29}Tb_{0.10})U_{4.01}C_{3.82}O_{23} \cdot 14.26\,H_2O$, or, simplified, to the formula given above. According to five analyses the determinable RE (Gd, Dy, and Tb) and Y, recalculated to 100, range for Gd_2O_3 from 13.59 to 19.17, Dy_2O_3 from 27.31 to 39.45, Tb_2O_3 from 2.57 to 9.03, and for Y_2O_3 from 32.91 to 54.19. The RE composition is characterized by an inversely proportional relation between the content of Y on the one hand, and between Dy and Gd + Tb on the other hand.

The mineral is orthorhombic; under the microscope the tabular crystals are flattened to (001) and generally elongated to [110]; polygonal plates are characterized by the faces $(\bar{1}10)$, $(\bar{1}30)$, (010), (130), and $(1\bar{1}0)$; they are of monoclinic appearance. The cleavage is distinct in the plane of the tables.

The unit cell data are a = 21.22(3), b = 45.30(7), and c = 13.38(2) Å; the possible space groups are $Cmma\text{-}D_{2h}^{21}$ (No. 67) with Z = 16 and $C2ma\text{-}C_{2v}^{15}$ (No. 39, standard setting Abm2) with Z = 8. For the existing pseudocell (subcell) with a′ = a/2, b′ = b/2, and c′ = c the space group is $C222_1\text{-}D_2^5$

(No. 20) with Z = 8. In the X-ray powder pattern the reflexes only of this pseudo-lattice are present.

Optical properties: The luster is vitreous and the mineral is transparent and sulfur-yellow or translucent and pale yellow. The refractive indices and pleochroism (in parentheses) are as follows: $n_x = 1.600$ (colorless), $n_y = 1.650$ (pale yellow), and $n_z = 1.722$ (intensive yellow). The optical character is biaxial positive, the axial angle 2V is 83° (calculated) and 84° (measured). The extinction $Z \wedge [110] = 25.1°$, but some crystals elongated to [100] show no angle. The orientation is X = c, Y = a, and Z = b. Fluorescence is not observed for short- and long-wave UV light.

The microhardness (Vickers) is 36 kg/mm². The measured density is 3.95 ± 0.15 g/cm³, the calculated density is 3.907 g/cm³ (for Z = 16).

Reference for 10.5.4.3:

M. Deliens, P. Piret (Can. Mineralogist **20** [1982] 231/8).

10.5.4.4 Lepersonnite $CaO \cdot (RE)_2O_3 \cdot 24\,UO_3 \cdot 8\,CO_2 \cdot 4\,SiO_2 \cdot 60\,H_2O$

The mineral occurs in the same association of minerals as described for bijvoetite (see p. 231) in the lower part of the oxidation zone of the Shinkolobwe uranium deposit, Southern Shaba, Zaire. It is, however, a more common mineral than bijvoetite. As lepersonnite is very similar to studtite, it is possible that some samples in mineralogical collections designated as "studtite" from Shinkolobwe and other localities may be really lepersonnite. The two minerals can be distinguished by the solubility of lepersonnite in cold hydrochloric acid and by its higher refractive indices.

The chemistry of lepersonnite, determined by microprobe (U, RE, Ca, and Si), chromatography (CO_2), and thermogravimetry (water) shows the mineral to be a hydrated carbonate of U, RE, Ca, and Si with 3.28 wt% RE_2O_3 and 0.41 wt% Y_2O_3 (corrected to 100 wt% of oxides). This leads to the formula $Ca_{0.71}(Gd_{1.04}Dy_{0.52}Y_{0.32}Tb_{0.06})U_{23.95}C_{8.10}Si_{4.18}O_{100} \cdot 59.90\,H_2O$, or, simplified, to the formula given above. The relative amounts of the RE in lepersonnite, determined by five analyses, recalculated to 100, range for Gd_2O_3 from 49.60 to 62.34, for Dy_2O_3 from 18.07 to 33.33, and for Y_2O_3 from 10.13 to 26.79; Tb_2O_3 1.07 and 1.87 was determined in two of these analyses. The RE are of the same type as in bijvoetite, the amount, however, seems to be too small to ascribe much significance to the observed variations.

The mineral is orthorhombic and occurs as flexible needles, often terminated by an oblique face and some of them elongated to fine fibers, which are sometimes screw-shaped. The needles are often agglomerated to radiated single spherulites or to mammilary crusts and nodules.

The unit cell data are a = 16.23(3), b = 38.74(9), c = 11.73(3) Å; the space group is Pnnm-D_{2h}^{12} (No. 58) or Pnn2-C_{2v}^{10} (No. 34), Z = 2 (for the simplified formula).

The needles are transparent to translucent, intensely yellow; the optical character is biaxial negative with 2V = 73°. The refractive indices and pleochroism (in parentheses) are $n_x = 1.638$ (pale yellow), $n_y = 1.666$ (intense yellow), $n_z = 1.682$ (intense yellow). The extinction is straight and the elongation (according to Y = c) is positive or negative. The sections XY are strongly pleochroic, the sections YZ are not pleochroic. Screw-shaped fibers show two orientations in one section. The mineral shows no fluorescence in long- and short-wave UV light.

The mineral is brittle; the measured density is 3.97 ± 0.05 g/cm³, the calculated density is 4.01 g/cm³ for Z = 2.

The TG curve of lepersonnite shows a weight loss of about 16% at 500°C and the DTA curve shows endothermal effects at about 40, 85, and 125°C.

Reference for 10.5.4.4:

M. Deliens, P. Piret (Can. Mineralogist **20** [1982] 231/8).

10.6 Sulfates

"Texasite", $Pr_2[O_2|SO_4]$, was described as a supergene alteration product of primary RE minerals from pegmatites of the Llano region, Texas, and from the Luster pegmatite, South Platte group, Jefferson County, Colorado, by Crook III [1, 2]. It has been proved by Peacor et al. that "texasite" only exists as a synthetic phase and all data, given by Crook III, are those of nearly pure $Pr_2O_2SO_4$. The discredition of "texasite" has been approved by the Commission on New Minerals and Mineral Names of the I.M.A. [3]. For crystallographic and physical properties of $Pr_2O_2SO_4$, see "Seltenerdelemente" C 8, 1981, pp. 328/30.

References:

[1] W. W. Crook III (Am. Mineralogist **62** [1977] 1006/8). – [2] W. W. Crook III (Mineral. Rec. **9** [1978] 251/2). – [3] D. R. Peacor, W. B. Simmons Jr., E. J. Essene, E. W. Heinrich (Am. Mineralogist **67** [1982] 156/69, 156/8).

10.7 Phosphates

Most of the RE phosphates have been already described in the Gmelin volume "Phosphor" A, 1965, pp. 251/83, 312/7, 354/5, 396/7, and 485/9. In the following, therefore, only new RE phosphates will be described. For the occurrences of RE phosphates, especially monazite, xenotime, rhabdophane, and churchite, in mineral deposits, see "Rare Earth Elements" A 8, 1984, Chapters 11.7.2 to 11.7.8.

10.7.1 Phosinaite $H_2Na_3(Ca,Ce)[(Si,P)O_4]$

Phosinaite occurs in ussingite veinlets, cutting malignite and foyaite, at Mt. Karnasurt, Lovozero massif, together with nordite, belovite, neptunite, and vuonnemite, and intergrown with aegirine as fillings between anorthoclase crystals in pegmatitic veins, cutting rischorrite, at Mt. Koasha, Khibina, both Kola Peninsula. Phosinaite from Khibina is altered to a rose-colored, nontransparent, X-ray amorphous, isotropic material with RE [1], see also [2].

Analyses with 13.44 and 13.00% RE_2O_3 for two Khibina samples, and with 13.80% RE_2O_3 for one Lovozero sample indicate the general formula, given above, or the crystal chemical formulas

$$(Na_{2.98}K_{0.06})(Ca_{0.72}Ce_{0.26})Si_{1.00}P_{0.99}O_{7.10}\cdot 1.10\,H_2O$$

or

$$(Na_{2.96}K_{0.05})(Ca_{0.70}Ce_{0.24})Si_{1.07}P_{0.97}O_{7.14}\cdot 1.38\,H_2O$$

and

$$(Na_{3.02}K_{0.08})(Ca_{0.41}Ce_{0.27})Si_{1.28}P_{0.94}O_{7.27}\cdot 0.8\,H_2O,$$

respectively. Besides the main elements given by the formula the sample of Lovozero also contains Fe, Mn, and Li; Pb, Sn, Ca, Al, and Cu are spectrochemically determined [1]. The analysis of a sample from the Khibina massif yields 11.67% RE_2O_3 and (besides the elements given in the general formula) in %: 0.45 Fe_2O_3, 0.30 SrO, 0.16 MnO, 0.28 MgO, 0.50 K_2O. Clinophosinaite, the monoclinic analogue of phosinaite, is practically RE free [2]. The composition of RE and Y in the samples of Khibina (1) and Lovozero (2) with $\Sigma RE + Y = 100$ is given as follows [1]:

No.	Y	La	Ce	Pr	Nd	Sm	Eu	Gd	Tb	Dy	Ho	Er	Yb
1	0.6	29	50	4	13	1	0.1	0.8	—	0.2	0.1	0.2	0.1
2	0.6	16.8	53.9	4.4	19.5	3.0	0.5	0.7	0.3	—	—	0.3	—

The mineral is orthorhombic. Lovozero phosinaite forms columnar crystals, radiated concretions, and stretched grains. Crystal forms are {110}, {100}, {010}, and {001}. The cleavage {100} is perfect, those parallel to {010} and {110} are imperfect. Khibina phosinaite occurs as irregular aggregates. Unit cell constants a = 12.23, b = 14.62, c = 7.21 Å, V = 1290 Å^3 (Khibina), a = 12.24, b = 14.59, c = 7.20 Å, V = 1286 Å^3 (Lovozero). Space groups are $P22_12$-D_2^2 (No. 17, standard setting $P222_1$) or $P22_12_1$-D_2^3 (No. 18, standard setting $P2_12_12$) [1]. For a phosinaite of Khibina, Krutik et al. give a = 7.234(3), b = 14.670(4), c = 12.231(4) Å, space group $P2_12_12$-D_2^3 (No. 18), and Z = 2($Na_{11}(Na,Ca)_2Ca_2Ce_{0.67}[Si_4O_{12}][PO_4]_4$). The structure is built up of sheets with chains of eight-coordinated Ce polyhedra (mean distance Ce-O = 2.468 Å), Ca octahedra, and (Ca, Na) seven-pointed polyhedra with Ca : Na = 1 : 1. These polyhedra are braced by two pairs of Na polyhedra. The cationic framework interpenetrates the anionic framework that consists of four-membered Si-O rings [Si_4O_{12}] and mixed rings of P octahedra and Na polyhedra [2].

The luster is vitreous. The mineral is transparent. Colors are brownish rose (Lovozero), colorless or pale rose (Khibina). Columnar crystals have direct extinction. Under the microscope phosinaite is rose or white. Refractive indices $n_x = 1.570$, $n_y = 1.572$, $n_z = 1.573$, $2V_x = 70°$ (Lovozero), $n_x = 1.567$, $n_y = 1.569$, $n_z = 1.570$, $2V_x = 68°$ (Khibina). Orientation Z||a, X||b, Y||c. An IR spectrum of Lovozero sample is given [1].

Hardness, $H_M \approx 3.5$, microhardness (load 20 g) = 207 to 340 kg/mm^2. The mineral is brittle. D = 2.62 g/cm^3 (Khibina), 3.00 g/cm^3 (Lovozero) [1].

Phosinaite slowly dissolves in concentrated acids. DTA and TG curves of samples from both localities show the loss of water at 250°C, the Lovozero material in addition shows an endothermic effect at ~600°C. Melting point 860°C [1].

References for 10.7.1:

[1] Yu. L. Kapustin, A. P. Khomyakov, E. I. Semenov, E. M. Es'kova, A. V. Bykova, Z. V. Pudovkina (Zap. Vses. Mineral. Obshchestva **103** [1974] 567/70). – [2] V. M. Krutik, D. Yu. Pushcharovskii, A. P. Khomyakov, E. A. Pobedimskaya, N. V. Belov (Kristallografiya **26** [1981] 1197/203; Soviet Phys.-Cryst. **26** [1981] 679/82).

10.7.2 Unnamed Ce Phosphate (Ce, Ca, H)[PO_4]·H_2O

The mineral has been found in hydrothermal natrolite-analcime rocks, Narsak River, Ilímaussaq alcaline massif, southern Greenland, where it forms dense microgranular deposits, cut by veinlets of rhabdophane. An analysis gives 56.50% RE_2O_3 besides (in %) 29.63 P_2O_5, 1.34 SiO_2, 4.18 CaO, 1.10 Fe_2O_3, and 7.24 H_2O (by difference).

The mineral is white, optically biaxial, elongation negative, $n_x = 1.62$, $n_z = 1.64$.

Reference for 10.7.2:

E. I. Semenov (Mineralogy of the Ilímaussaq Alkaline Massif, Southern Greenland, Nauka, Moscow 1969, pp. 1/164 from Am. Mineralogist **55** [1970] 2136/7).

10.7.3 **Monazite** $Ce[PO_4]$

The mineral is described in "Phosphor" A, 1965, pp. 251/74. For a compilation of analyses and $(RE,Y)_2O_3$ and Y_2O_3 content, variance and mean of 361 analyses, see "Seltenerdelemente" A 4, 1979, pp. 24/6, 5. For the distribution of RE and Y and relation of $(RE,Y)_2O_3$ content in monazite to its genesis, see "Seltenerdelemente" A 4, 1979, pp. 53/9 and 72ff.

10.7.4 **Xenotime** $Y[PO_4]$

A mineral description is given in "Phosphor" A, 1965, pp. 275/83. For a compilation of analyses and $(RE,Y)_2O_3$ and Y_2O_3 content, variance and mean of 54 analyses, see "Seltenerdelemente" A 4, 1979, pp. 34, 5. For the distribution of Y and RE in xenotime and relation of $(RE,Y)_2O_3$ and Y_2O_3 content of xenotime to its parent rocks and genesis, see "Seltenerdelemente" A 4, 1979, pp. 52/3 and 72ff.

10.7.5 **Cheralithe** (Ca, Ce, La, Th) $[PO_4]$

A mineral description can be found in "Phosphor" A, 1965, pp. 274/5. For a compilation of analyses and $(RE,Y)_2O_3$ and Y_2O_3 content, variance and mean of 3 analyses, see "Seltenerdelemente" A 4, 1979, pp. 17, 7.

10.7.6 **Vitusite** $Na_3RE[PO_4]_2$

Vitusite has been discovered in the Lovozero alkaline massif, Kola, in the pegmatite vein Yubileinaya, Mt. Karnasurt, and at Mt. Sengistchorr. At the first locality it occurs as irregular grains and as fine-grained aggregates disseminated in the natrolite zone of the pegmatite together with belovite, neptunite, leucosphenite, and sazhinite, and in and around steenstrupine grains. A second occurrence is the Ilímaussaq alkaline intrusion, southern Greenland, where the mineral occurs as equidimensional subhedral grains in an arfvedsonite lujavrite together with other accessory minerals such as acmite, analcime, lovozerite, sphalerite, villiaumite, and a mineral of the lomonosovite-murmanite group. It seems to have been formed simultaneously with isotropic steenstrupine [1, p.43].

Microprobe analyses of vitusite from the three localities and in addition a microchemical analysis of a sample from Ilímaussaq lead to the general formula given above. Vitusite from Ilímaussaq has $(La,Ce,Pr,Nd)_2O_3 = 41.1\%$, with La:Ce:Nd:Pr = 37.8:48.5:10.7:3.0. Vitusites from Mt. Karnasurt and from Mt. Sengistchorr have $(La,Ce,Pr,Nd)_2O_3 = 40.6$ and 34.2% with La:Ce:Nd:Pr = 22.8:53.5:19.0:4.7 and 35.7:48.3:12.7:3.3, respectively. Besides the elements noticed in the formula, Ca and Si are determined; the last one being partly due to inclusions of arfvedsonite. The enrichment of Ca that is especially rich in vitusite from Mt. Sengistchorr, Lovozero massif, is compensated by lower contents of Na and RE due to the substitution $2Ca^{2+} \rightleftharpoons Na^+ + RE^{3+}$. This indicates a limited solid solution between vitusite and the meteoritic mineral buchwaldite, $NaCa[PO_4]$ [1, pp.47/8]. An empirical formula of vitusite from Mt. Sengistchorr is calculated as $Na_{2.84}Ca_{0.33}Ce_{0.39}La_{0.29}Nd_{0.10}Pr_{0.03}[P_{2.01}Si_{0.01}]O_8$ [2]. A spectral microanalysis proved the presence of a few tenths of percent of Y and W, and traces of Th and the absence of Ba, B, F, and Cl [1, p. 48].

The mineral is orthorhombic, pseudohexagonal. Cleavages are distinct on {100}, {010}, and {001} (Lovozero material). In thin section twinning after {160} and {120} can be seen. With the pseudo-sixfold axis chosen as c axis the unit cell parameters are:

	Lovozero	Ilímaussaq
a in Å	5.36(2)	5.342(1)
b in Å	18.68(8)	18.680(2)
c in Å	13.96(6)	14.062(2)
V in $Å^3$	1398	1404

Space group is Pcmb-D_{2h}^{11} (No. 57, standard setting Pbcm) or $Pc2_1b$-C_{2v}^{5} (No. 29, standard setting $Pca2_1$), Z = 8 [1, pp.44/5, 47/8]. Unit parameters of the orthorhombic cell given by Karpov et al. for Lovozero sample: a = 14.091(4), b = 5.357(1), c = 18.740(3) Å, with a distinct pseudoperiod 7.05 Å along the b axis. Space group is $Pca2_1$-C_{2v}^{5}*) (No. 29). These authors distinguish two possible variants of heavy atom distributions within this space group. Both models show six-membered units of P tetrahedra surrounding Na and Ce cations. The Ce(1) and Ce(2) polyhedra are octahedra with two additional vertices above the expanded one of the triangular faces; the coordination number for Na is six for all six Na atoms in model 1, but six for only five Na atoms and four for the last Na atom in model 2. This permits the incorporation of Ca (see p. 235) [2].

The luster is vitreous; the mineral is transparent. Colors are pale pink (Lovozero), white or pale green (Ilímaussaq). Under the microscope vitusite is colorless and transparent; biaxial negative. The refractive indices and 2V of vitusite from both localities are:

	Lovozero	Ilímaussaq
n_x	1.602(2)	1.6038(5)
n_y	1.650(2)	1.6465(5)
n_z	1.654(2)	1.6490(5)
$2V_x$	30.0(10)°	28.5(5)°

Orientation X||a, Y||c, Z||b (Lovozero), X||a, Y||b, Z||c (Ilímaussaq) [1, pp. 44/5].

Hardness, $H_M = 4.5$. D(meas.) = 3.60 g/cm³ (Lovozero), 3.70 g/cm³ (Ilímaussaq) [1, p. 44]; D(calc.) = 3.76 g/cm³ [2]. Vitusite easily decomposes in cold dilute hydrochloric and sulfuric acid [1, p. 44].

References for 10.7.6:

[1] J. G. Rønsbo, A. P. Khomyakov, E. I. Semenov, A. A. Voronkov, V. K. Garanin (Neues Jahrb. Mineral. Abhandl. **137** [1979] 42/53). – [2] O. G. Karpov, D. Yu. Pushcharovskii, A. P. Khomyakov, E. A. Pobedimskaya, N. V. Belov (Kristallografiya **25** [1980] 1135/41; Soviet Phys.-Cryst. **25** [1980] 650/3).

10.7.7 Churchite (Weinschenkite) $(Y,Er)[PO_4] \cdot 2H_2O$

A mineral description is given in "Phosphor" A, 1965, pp. 312/3. For a compilation of analyses and $(RE,Y)_2O_3$ and Y_2O_3 content, variance and mean of 8 analyses, see "Seltenerdelemente" A 4, 1979, pp. 17, 6, and for the choice of name p. 33.

*) In the original erroneously C_{2v}^{9}.

10.7.8 **Rhabdophane** $RE[PO_4]\cdot 0$ to 0.5 H_2O

For a mineral description see "Phosphor" A, 1965, pp. 313/4. A compilation of analyses and $(RE,Y)_2O_3$ content, variance and mean of 20 analyses, and distribution of RE in different parent rocks is given in "Seltenerdelemente" A 4, 1979, pp. 28, 6, 59.

10.7.9 **Ningyoite** $(U,Ca,RE)[PO_4]_2\cdot 1.5\,H_2O$

A mineral description can be found in "Phosphor" A, 1965, p. 315.

10.7.10 **Smirnovskite** $(Th,Ce)[PO_4]\cdot H_2O$

A mineral description can be found in "Phosphor" A, 1965, pp. 396/7. For a compilation of analyses and $(RE,Y)_2O_3$ content, variance and mean of 4 analyses, and the new formula, given above, see "Seltenerdelemente" A 4, 1979, pp. 30, 9.

10.7.11 **Lermontovite** $(U,Ca,RE)_3[PO_4]_4\cdot 6\,H_2O$

A mineral description is given in "Phosphor" A, 1965, pp. 316/7. One analysis has $(RE,Y)_2O_3 = 1.67\%$ and yields the formula given above, M. V. Soboleva, I. A. Pudovkina (Mineraly Urana Spravochnik, Gosgeoltekhizdat., Moscow 1957, pp. 1/404, 181/2). The mineral needs verification, M. Fleischer (Am. Mineralogist **43** [1958] 379/80).

10.7.12 **Unnamed Minerals U 2 and U 3**

Both minerals have been found in pegmatites and hydrothermal veins near Bogatynia in the Łuzyce (Lusatia) granitic rocks, Lower Silesia, Poland, where they occur together with feldspars, iron hydroxides, and thorogummite.

U 2 chemically appears to be a member of the series between rhabdophane $RE[PO_4]\cdot 0.5\,H_2O$, brockite $CaTh[PO_4]_2\cdot H_2O$, and hypothetical $FeTh[PO_4]_2\cdot H_2O$ with the formula $(Th,Fe,Ca,RE)[PO_4,SiO_4,CO_3,(OH)?]\cdot 0.5\,H_2O$, while the general formula of U 3 is $(Fe^{2+},Ca)_{1-x}Th_{1-x}(Fe^{3+},RE)_{2x}[PO_4,SiO_4,(OH)]_2\cdot 1$ to $2\,H_2O$.

U 2 is usually found as opal-like microcrystalline cross forms. It is assumed to be hexagonal from the isostructural relation to rhabdophane and brockite. Unit cell parameters are $a = 7.04 \pm 0.06$, $c = 6.36 \pm 0.03$ Å. The color is red-brown to green-brown. In transmitted light the mineral is subtranslucent, white-yellow with a green tinge or red-orange. Reflectance ≈12%. There is no fluorescence in ultraviolet light. U 2 is brittle. Hardness $H_M \approx 4$, microhardness (150 g load) = 192 to 280 kg/mm². After ignition U 2 has generally a monazite-like structure.

U 3 is found as fine-grained aggregates. It is orthorhombic and isostructural with ningyoite. Cell dimension is $a = 6.77 \pm 0.03$, $b = 12.06 \pm 0.07$, $c = 6.41 \pm 0.04$ Å, space group is probably $P222\text{-}D_2^1$ (No. 16), Z = 3. The color is reddish brown. In transmitted light U 4 is subtranslucent, often isotropic due to radioactive decay; brown-red to orange-yellow. Reflectance ≈10%; internal reflections brown-red, red-yellow, and red-orange. No fluorescence has been found in ultraviolet light; electron beam causes emission of soft orange light. DTA, DTG, and TG curves of mixtures, containing U 3, are given.

Reference for 10.7.12:

H. Kucha (Mineral. Polon. **10** No. 1 [1979/80] 3/28, 4, 9/10, 17/23).

10.7.13 Cer- and Yttriumapatite, Saamite, Belovite

Varieties of apatite with considerable amounts of RE replacing Ca(Sr) are described in "Phosphor" A, 1965, pp. 354/5. For $(RE,Y)_2O_3$ content and RE distribution in apatites in relation to their parent rocks and genesis, see "Seltenerdelemente A 4, 1979, pp. 134/56. For refractive indices of RE-containing apatites see "Phosphor" A, 1965, pp. 368/9. Two analyses of belovite with mean of $(RE,Y)_2O_3 = 24.21\%$ are given in "Seltenerdelemente" A 4, 1979, pp. 7, 14.

10.7.14 Florencite $CeAl_3[(OH)_6|(PO_4)_2]$

A mineral description is given in "Phosphor" A, 1965, pp.485/9. For a compilation of analyses and $(RE,Y)_2O_3$ content, variance and mean of 7 analyses, see "Seltenerdelemente" A 4, 1979, pp. 20, 8. For the occurrence of florencite-(La) see p. 160.

10.7.15 Petersite $(Y,RE,Ca)Cu_6[(OH)_6|(PO_4)_3]\cdot 3H_2O$

Petersite has been found as a supergene mineral at the traprock quarry in Secausus, Hudson County, New Jersey; it occurs within the hornfels near the contact of the diabase in association with opal and malachite. A microprobe analysis with 2.5% Y_2O_3 and 5.9% RE_2O_3 yields the crystal chemical formula $(Y_{0.41}Ce_{0.28}Nd_{0.23}Sm_{0.13}La_{0.11}Fe^{2+}_{0.20}Ca_{0.79})Cu_{12.07}(OH)_{11.60}(PO_4)_{6.00}\cdot 6.78H_2O$ or ideally as given above. Petersite thus is a member of the mixite group and the phosphate equivalent of the arsenate agardite, see p. 242.

Petersite occurs as prismatic hexagonal crystals with the prism $\{10\bar{1}0\}$ and the pinacoid $\{0001\}$. The usually euhedral crystals form radiating clusters. The mineral is hexagonal, space group is $P6_3/m\text{-}C^2_{6h}$ (No. 176) or $P6_3\text{-}C^6_6$ (No. 173). Unit cell parameters are a = 13.288(5), c = 5.877(5) Å, V = 898.6(8) Å^3 and Z = 2.

The luster of petersite is dull on the pinacoids and highly vitreous on the prisms. Its color is bright yellowish green. Optically, petersite is uniaxial, positive, with $n_o = 1.666(4)$, $n_e = 1.747(4)$. Pleochroism is strong E green $\gg$ O light yellowish green. There is no discernible fluorescence under ultraviolet radiation.

Petersite is very brittle; its density D(meas.) = 3.41 g/cm^3, D(calc.) = 3.40 g/cm^3.

The mineral is readily soluble in 1 : 1 hydrochloric acid, coloring the solution light greenish yellow.

Reference for 10.7.15:

D. R. Peacor, P. J. Dunn (Am. Mineralogist **67** [1982] 1039/42).

10.8 Arsenates

10.8.1 Chernovite $Y[AsO_4]$

Occurrence, Paragenesis, and Chemistry

Chernovite has been found in a subvolcanic stock-like body of liparite porphyry of a Riphaean dacite-liparite complex at the source of the river Nyarta-syu-yu, a left tributary of the Shchugora, east of Tel'pos-iz in the Polar Urals. It is always associated with molybdoscheelite (seyrigite) in piemontite veinlets, or in quartz-albite zones, close to these veinlets, in the hanging wall of the liparite body and in mineralized zones within the liparite porphyry [1], cf. [2].

In the southern part of Binnatal, Italy, a mineral identical with the synthetic compound $YAsO_4$ had already been found in 1966 in the Cervandone (Cherbadung) massif across the border Binnatal-Val Devero, and has been later described as chernovite [3, 4, 5]. Two further localities in the Binnatal, Switzerland, are Gischihorn and Wannigletscher, where chernovite occurs as a typical product of Alpine fissure-mineral formation [3]. The last found of chernovite described is from China [6]. Chernovite from Binnatal sometimes shows weathering processes at its rims [3].

Three microprobe analyses of chernovite from Binnatal with 36.9 to 38.7% Y_2O_3 and 15.1 to 15.8% Ln_2O_3 [3] and one analysis of chernovite from the Polar Urals [1, 2] show tne mineral to be an Y arsenate. The crystal chemical formula of the Chinese chernovite is $(RE_{0.94}Th_{0.02}Fe^{3+}_{0.03}Mg_{0.03}Ca_{0.02})(As_{0.83}P_{0.06}V_{0.05}Si_{0.05})O_4$ after deduction of U, Zr(Hf), Nb, Ta, Al, Mn as impurities [6]. As can be seen from this formula, As may be substituted by P, V, and Si; this is also the case in the Polar Urals' chernovite [1,3]. There exists an isomorphous series between chernovite and xenotime, at least in the As-rich members; thus chernovite from Binnatal (3 samples) contains 18.0, 26.8, and 36.8 mol% $(Y,Ln)PO_4$ [3]. The distribution of Y and RE shows predominance of Y and Y-earth in the 3 chernovites from Binnatal with 70.4 to 71.4% Y_2O_3 and 28.6 to 29.6% Ln_2O_3 for $(Y,Ln)_2O_3 = 100$ [3] and in chernovite from the Polar Urals with 78% Y and 22% Ln (Ce, Nd through Lu) [1, 2]. In all these minerals the most abundant lanthanides are Gd, Dy, Ho, Er, Yb [1, 2, 3]. RE distribution $(RE,Y)_2O_3 = 100$ in Chinese chernovite is Y 61.5, La 0.1, Ce 1.6, Pr 0.3, Nd 1.2, Sm 2.3, Gd 6.7, Tb 1.4, Dy 10.6, Ho 1.5, Er 5.7, Tm 1.1, Yb 5.3, and Lu 0.8 [6].

Crystal Form and Structure

The mineral is tetragonal with {010}, {011}, and {001}; it has a tetragonal prismatic habit, sometimes it is tabular [1, 2], or pyramidal [6], see **Fig. 13**, p. 240; or it has small bipyramidal shape and forms aggregates along c [3]. The faces are brilliant, only sometimes corroded [1, 2]. The cleavage is perfect on {010} [1, 2], very indistinct on {100} [3].

Unit cell constants are given as follows:

Locality	a in Å	c in Å	Ref.
Binnatal, Switzerland/Italy			
Gischihorn	6.99(3)	6.16(7)	
Wannigletscher	7.09	6.32	
Cervandone massif	7.04(3)	6.26(2)	[3]
Polar Urals	7.039 ± 0.011	6.272 ± 0.022	[1, 2]
China	7.054	6.309	[6]

Space group is $I4_1/amd\text{-}D^{19}_{4h}$ (No. 141), Z = 4 [1, 2]. V = 310.953 $Å^3$ (Polar Urals) [1], 313.433 $Å^3$ (China) [6]. Chernovite is isostructural with xenotime [4], see also above. For a relation between cell parameters and the ratio YPO_4 to $YAsO_4$ in xenotime and chernovite see [3, 6].

Physical Properties

The luster is vitreous. Chernovite is transparent, rarely semitransparent [1, 2]. The color of the inner zones is greenish yellow, changing to bright yellow in the outer zones (Binnatal) [3]; inner zones are colorless, outer zones pale yellow. The streak is colorless (Urals) [1, 2]. The color is lemon-yellow for slightly altered material to bright green for fresh specimens (Cervandone massif) [5]. Pleochroism O pale rose-yellow, E colorless [1, 2], see also [6]. Chernovite is uniaxial, positive [1, 2]. For the refractive indices of chernovite from different localities see the following table:

Locality	n_o	n_e	Ref.
Binnatal, Switzerland/Italy			
Gischihorn	1.753	—	
Wannigletscher	1.772	—	
Cervandone massif	1.764	—	[3]
Polar Urals	1.783 ± 0.001	1.879 ± 0.002	[1, 2]
China	1.775	1.861	[6]

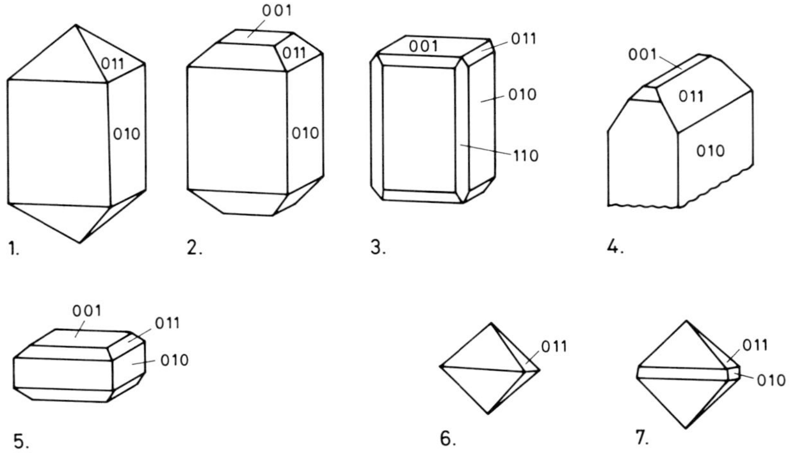

Fig. 13. 1. to 5. chernovite crystals from Polar Urals, natural size <0.65 mm, length mostly 0.24 mm from [1]; 6. and 7. crystals from China, natural size 0.075 to 0.03 mm from [6].

For the relation between refractive indices of xenotime and chernovite and their contents of YPO_4 and $YAsO_4$ see [3, 6].

Microhardness measured on {010} cleavage planes in the direction [010] (load 50 g) = 225 to 429, mean 313 kg/mm², corresponding to $H_M \approx 4.5$ [1, 2]. Microhardness = 197 to 287.5 kg/mm², $H_M = 4$ to 4.4 [6]. Chernovite is brittle [1, 2]. D(meas.) > 4.2 g/cm³, D(calc.) = 4.87 g/cm³ (Urals) [1, 2], D(meas.) = 4.56 g/cm³, D(calc.) = 4.83 g/cm³ (China) [6].

Chernovite entirely concentrates in the electromagnetic fraction [1, 2].

Chemical Behavior

At room temperature chernovite does not dissolve in hydrochloric, nitric, and sulfuric acids, nor in mixtures of them including aqua regia [1, 2].

References for 10.8.1:

[1] B. A. Goldin, N. P. Yushkin, M. V. Fishman (Zap. Vses. Mineral. Obshchestva **96** [1967] 699/704). – [2] B. A. Goldin, N. P. Yushkin, M. V. Fishman (Dokl. Akad. Nauk SSSR **179** [1968] 187/9; Dokl. Earth Sci. Sect. **179** [1968] 120/1). – [3] S. Graeser, H. Schwander, H. A. Stalder

(Mineral. Mag. **39** [1973] 145/51). – [4] C. M. Gramaccioli (Mineral. Rec. **8** No. 4 [1977] 287/93, 291). – [5] S. Graeser, A. G. Roggiani (Rend. Soc. Ital. Mineral. Petrol. **32** [1976] 279/88, 282/3).

[6] D. Chen, C. Li (Acta Geol. Sinica **53** [1979] 125/36 (Chinese with English Abstract p. 136); Mineral. Abstr. **30** [1979] No. 79-4086).

10.8.2 **Retzian** $Mn_2Ce[(OH)_4|AsO_4]$ and **Retzian-(Nd)** $Mn_2Nd[(OH)_4|AsO_4]$

Retzian is only known from the Moss mine, Nordmark, Värmland, Sweden, where it first was described in 1894 by Sjögren [1] as Mn arsenate. Only in 1897 does the same author [2] give an analysis with 10.3% RE_2O_3. The mineral occurs together with jacobsite in small druses in a mixture of braunite and dolomite [1, 2]. From the analysis of 0.08 g from Sjögren [2] the crystal chemical formula $(Ca,Y)_2(Mn,Mg)_4[(OH)_8|(As,Fe,Si)_2O_8]$ was deduced [3, 4]. A new analysis determined that the type material of retzian is Ce-rich and nearly Ca-free; the calculation of a formula, with $\Sigma RE+M^{2+}$ cations = 3, yields

$(Mn_{1.77}Mg_{0.13}Fe_{0.07}Ca_{0.05}Zn_{0.05})(Ce_{0.40}Nd_{0.19}La_{0.08}Y_{0.07}Pr_{0.07}Sm_{0.04}Gd_{0.05}Eu_{0.03})[(OH)_{3.69}(AsO_4)_{1.08}]$

or, generalized as given above [6].

Retzian-(Nd) has been described as a new mineral from the Sterling Hill mine, Ogdensburg, Sussex County, New Jersey, where it occurs together with willemite, sonolite, calcite, barite, and chlorophoenicite in three separate parageneses. An analysis shows that this mineral differs from retzian in having Nd as the dominant RE element. The calculation of a formula with $\Sigma RE+M^{2+}=3$ yields

$(Mn_{1.63}Mg_{0.22}Fe_{0.03}Zn_{0.10})(Nd_{0.29}La_{0.21}Ce_{0.21}Pr_{0.11}Sm_{0.07}Y_{0.06}Gd_{0.05}Eu_{0.02})[(OH)_{3.90}|(AsO_4)_{1.04}]$

or ideally as given above [6].

Retzian always occurs crystallized; its habit is prismatic or thick tabular, elongate parallel to [001]. It has no cleavage [1, 2]. Retzian is orthorhombic; crystal forms, given by Sjögren [1, 2], revised according to the structure cell ratio a:b:c = 0.471:1:0.404, are {001}, {010}, {130}, {110}, and {201} [3]. Retzian-(Nd) in addition to these forms, has {150} and {021}. Its habit is elongate parallel to [100] [6]. Unit cell parameters of retzian are a = 5.670(3), b = 12.03(1), c = 4.863(4) Å, Z = 2, space group Pban-D_{2h}^4 (No. 50), V = 331.7 Å^3 [3], cf. [4]. The unit cell parameters of retzian-(Nd) are a = 5.690(5), b = 12.12(1), c = 4.874(3) Å [6]. The crystal structure of retzian is a sheet structure, ribbons of Mn^{2+}-O edge-linked octahedra are held together in a plane by Y^{3+}-O square antiprisms. A stable three-dimensional structure results by tetrahedrally coordinated As atoms that are fitted between the sheets. The Mn^{2+}-O octahedra are irregular with Mn^{2+} to O distances 2.10, 2.12, and 2.33, mean 2.18 Å, the Y^{3+}-O square antiprism with Y^{3+} to O distances 2.52 and 2.50 Å is only slightly distorted. The As^{5+}-O tetrahedron is nearly regular with As to O distance 1.60 ± 0.06 Å. A substitution of As by Si is assumed [4].

The luster of retzian is vitreous, approaching to greasy. The mineral is subtranslucent. The color is chestnut to chocolate-brown. The streak is light brown. Retzian is strongly pleochroitic with X wine-yellow to orange-yellow, Y dark brown to reddish brown, Z strongly brownish red to crimson with absorption Z > Y > X. The plane of optical axes is parallel to (010) [1, 2]. Retzian is biaxial positive with $n_x=1.777$, $n_y=1.788$, $n_z=1.800$, 2V is large, dispersion is weak $r<v$ [5]. The luster of retzian-(Nd) is vitreous to dull on crystal faces and vitreous on fracture surfaces. Its color is pinkish brown. Retzian-(Nd) has pleochroism X||c yellow, Y||b reddish brown, Z||a brown. It is biaxial positive with $n_x=1.774(2)$, $n_y=1.782(2)$, $n_z=1.798(2)$, and 2V = 69(1)°. The dispersion is weak $r<v$ [6].

Hardness of retzian $H_M = 4$. The fracture is conchoidal to uneven. Density D(meas.) = 4.15 g/cm³ [1, 2], D(calc.) = 4.14 g/cm³ [3]. Retzian-(Nd) has $H_M = 3$ to 4. Its fracture is uneven. D(meas.) > 4.2 g/cm³, D(calc.) = 4.45 g/cm³ [6]. Retzian is easily and completely soluble in hydrochloric acid [2], see also [5].

References for 10.8.2:

[1] H. Sjögren (Bull. Geol. Inst. Univ. Upsala **2** [1894/95] 39/108, 54/9). – [2] H. Sjögren (Geol. Foeren. Stockholm Foerh. **19** [1897] 106/12). – [3] P. B. Moore (Arkiv Mineral. Geol. **4** [1966/69] 425/44, 426, 435/6). – [4] P. B. Moore (Am. Mineralogist **52** [1967] 1603/13, 1604, 1606/7, 1610/1). – [5] E. S. Larsen, H. Berman (U. S. Geol. Surv. Bull. No. 848 [1934] 1/266, 138).

[6] P. J. Dunn, B. D. Sturman (Am. Mineralogist **67** [1982] 841/5).

10.8.3 **Agardite** $(Y,Ca,H)Cu_6[(OH)_6|(AsO_4)_3]\cdot 3H_2O$

Type locality of agardite is the oxidation zone of the Bou-Skour copper deposit in Jbel Sarhro, Morocco, where it occurs associated with azurite, malachite, cuprite, native Cu, quartz, and some other minerals [1, pp. 421, 426/7]. Besides, agardite has been found in the oxidation zone of the vein Saint-Jean, Sainte-Marie-aux-Mines, Haut-Rhin, France, together with tyrolite and pharmacosiderite [2], and in fluorite at the Red Cloud fluorite mine, Gallinas Mountains, New Mexico, associated with bastnaesite [3]. Inspection of several specimens, named chlorotile, mixite or otherwise, by Dietrich et al. had the result of finding further agardite localities. These are: Clara mine at Oberwolfach, middle Black Forest, Western Germany, where the "chlorotile", associated with other secondary Cu minerals and described by Walenta [4, pp. 224/8], is shown to be agardite; Anora, province Córdoba, Spain, with agardite, previously called mixite; Chessy, Rhône, France, where agardite has been identified on azurite; each a sample from Wheal Alfred, Cornwall, Great Britain, and from Tintic, Utah, that were designated as mixite [1, pp. 428/30]. For criteria of distinction see below and p. 243.

An analysis of agardite from Bou-Skour with 7.5% Y_2O_3 and 1.2% RE_2O_3 corresponds with the crystal chemical formula

$$(Y_{1.36}RE_{0.14}CaH_{0.50})(Cu_{11.71}Ca_{0.29})[(OH)_{12}|(AsO_4)_{5.66}(OH)_{1.36}]\cdot 6.04\,H_2O.$$

A spectroscopic analysis of this sample shows, mostly as impurities, in decreasing order of amounts between 10^{-3} to 10^{-4}% Pb, Mn, Si, Al, Zr, and Ti, and 5×10^{-6}% Be. The content of RE and Y, given in ppm with $Y\,n\times10^4$, Nd 3000, Dy, Gd, Sm 1500, Yb, La 1000, Ho, Pr, Tb 400, Lu 250, Tm, Er 100, shows the predominance of Y and absence of Ce [1, pp. 425/6]. The distribution of RE in the newly defined agardites, see above, is as follows: Anora, province Córdoba, Spain, shows predominance of Y, the distribution of RE is $Nd > Gd > Yb > Er \geqslant Dy \geqslant Sm \geqslant Ce \geqslant La > Tb > Pr$; besides there are Ca and traces of Si, Al, Mn in this sample. Chessy, Rhône, France, has high content of Y and $Nd > Gd > Sm > Dy \geqslant Pr > La > Er > ?(Eu, Ho, Lu)$; furthermore Ca and traces of Si. At Wheal Alfred, Cornwall, RE are especially rich in Nd. At Tintic, Utah, the mineral has a high content of Y, only small amounts of RE with $Dy \geqslant Er > Gd > ?$ Ho, some Ca, and traces of Fe, Si, P, S [1, pp. 429/30].

The distinction of agardite from mixite and "chlorotile" is, besides other properties, see p. 243, based on the presence of RE and Y (and absence of Bi) in agardite [1, pp. 427, 433].

Agardite forms very thin needle-like crystals in form of sheafs or tufts, sometimes with (0001) perpendicular to the elongation. It is hexagonal. Unit cell parameters are $a = 13.55 \pm 0.05$, $c = 5.87 \pm 0.02$ Å, Z = 2. The X-ray powder data correspond exactly to that of mixite and show the isotypy of both minerals [1, pp. 421, 426].

The luster is silky; the color green-blue [1, p. 421], yellow-green [2], or bright grass-green [3]. Elongation is positive and extinction of the needles straight. Pleochroism O is yellowish green to nearly colorless, E greenish blue. The mineral is uniaxial positive [1, pp. 421, 424]. The refractive indices may be used to distinguish agardite from mixite: the observed values n_o of agardite lying between 1.700 and 1.725, those of mixite between 1.735 and 1.760. The variations of the refractive indices of agardite are due to the different ratios of Y to RE and the changing content of Ca [1, p. 431].

Refractive indices of agardite from the type locality and from agardites redefined and tabulated, cf. p. 242, are compiled in the following table from [1, pp. 424, 428]:

Locality	n_o	n_e
Bou-Skour, Morocco	1.701 ± 0.005	1.782 ± 0.005
Clara mine, Oberwolfach	1.720 }	> 1.800
(3 samples)*)	1.723 }	
	1.707	1.782
Anora, province Córdoba	1.713	1.788
Chessy, Rhône	1.705	1.777
(3 samples)	1.708	1.778
	1.708	1.780
Wheal Alfred, Cornwall	1.722	1.815
Tintic, Utah	1.715	1.788

*) Data, given for a sample named chlorotile from this locality, are $n_o = 1.725 \pm 0.003$, $n_e = 1.810 \pm 0.005$ [4, p. 227].

Density D(meas.) = 3.72 ± 0.05 g/cm³ (Bou-Skour, Morocco) [1, p. 421], 3.73 g/cm³ ("chlorotile", Clara mine, Oberwolfach) [4, p. 227]. D(calc.) = 3.66 ± 0.04 g/cm³ [1, p. 421]. Agardite is soluble in hydrochloric acid [1, p. 425].

References for 10.8.3:

[1] J.-É. Dietrich, M. Orliac, F. Permingeat (Bull. Soc. Franc. Mineral. Crist. **92** [1969] 420/34). – [2] E. Asselborn, H. Bari (Monde Miner. No. 42 [1981] 56/9). – [3] R. S. DeMark (Mineral Rec. **11** [1980] 69/72). – [4] K. Walenta (Neues Jahrb. Mineral. Monatsh. **1960** 223/36).

10.8.4 Kemmlitzite (Sr, RE, Mg, Ca) $(Al, Fe^{3+})_3$ $[(OH)_{6-n} | AsO_4 | (PO_4, SO_4, SiO_4)] \cdot n H_2O$

Kemmlitzite-(RE) has been found in the kaoline deposit, formed by kaolinization of the quartz porphyry of Kemmlitz near Oschatz, Saxony, German Democratic Republic, together with other heavy minerals, especially zircon, anatase, and apatite. One chemical analysis with total RE_2O_3, computed as $Ce_2O_3 = 12.6\%$, corresponds to the crystal chemical formula $(Sr_{0.42}RE^{3+}_{0.40}Mg_{0.12}Ca_{0.08})(Al_{2.79}Fe^{3+}_{0.05})[(OH)_{5.22}|(AsO_4)_{0.98}|(PO_4)_{0.42}|(SO_4)_{0.39}|(SiO_4)_{0.19}] \cdot 0.875 H_2O$.

Distribution of $RE_2O_3 = 100$ is $Ce_2O_3 = 57.00$, $La_2O_3 = 26.35$, $Nd_2O_3 = 15.77$, $Sm_2O_3 = 0.88$, in addition there are traces of Eu, Pr, Er, Yb, and Th. Further trace elements determined by semiquantitative spectroscopic analyses are with $n \times 10^{-2}\%$ B, Ba, Pb, Ti, Tl, and with $n \times 10^{-3}\%$ Ag, Be, Co, Cu, Ga, Ge, Mn, Mo, Ni, Sb, Sc, Sn, V, Zn, Zr. The substitution of Sr^{2+} by RE^{3+} is compensated by the heterovalent substitution according $Sr^{2+} + [SO_4]^{2-} \rightleftharpoons RE^{3+} + [PO_4]^{3-}$. On account of this RE content the mineral is named kemmlitzite-(RE).

Kemmlitzite-(RE) is trigonal-rhombohedric and forms rhombohedral crystals of pseudocubic type, rarely with {0001}. The cleavage is imperfect {0001}. Unit cell parameters of hexagonal cell $a=7.027\pm0.001$, $c=16.51\pm0.01$ Å, Z=3; rhombohedral cell a=6.837 Å, $\alpha=61°51'$, Z=1.

The color is brownish and cloudy to colorless. Under the microscope the mineral is colorless to brown-yellow, sometimes with zonal structure. Kemmlitzite is uniaxial positive, with $n_o=1.701\pm0.001$, $n_e=1.707\pm0.001$. An infrared spectrum curve is given.

Hardness $H_M=5.5$, D(meas.) = 3.63 g/cm^3, D(calc.) = 3.601 g/cm^3. A low magnetic susceptibility has been determined.

The mineral shows a high resistance toward mineral acids. The DTA curve shows two endothermic reactions, due to the two-step escape of hydroxyl water. An increase in weight, shown by a maximum in a gravimetric thermal curve of kemmlitzite-(RE) heated in air, is caused by oxidation of trivalent RE ions in the lattice.

Reference for 10.8.4:

J. Hak, Z. Johan, M. Kvaček, W. Liebscher (Neues Jahrb. Mineral. Monatsh. **1969** 201/12).

10.8.5 Goudeyite $(Ca,Y,Al)Cu_6[(OH)_6|(AsO_4)_3]\cdot 3H_2O$

Goudeyite occurs as secondary mineral on the Majuba Hill mine, Pershing County, Nevada, a copper-ore body in a subvolcanic complex of rhyolithic porphyries and breccias of Tertiary age. Other secondary minerals of this mine are, besides the other new Ca arsenate parnauite, olivenite, clinoclase, cornwallite, strashimirite, scorodite, pharmacosiderite, arthurite, metazeunerite, chalcanthite, chrysocolla [1]. Goudeyite was also determined in samples from the old copper mine of Chessy, Rhône, France, where it occurs in cavities in montmorillonite [2].

The average of 5 microprobe analyses of goudeyite from Majuba Hill with 3.40% Y_2O_3 leads to the empirical cell content $Ca_{0.24}Al_{0.77}Y_{0.29}Cu_{5.80}(AsO_4)_{2.40}(PO_4)_{0.63}(OH)_{6.17}\cdot 2.95H_2O$ [1].

Goudeyite forms hair-like crystals in groups, encrusting fracture coatings and cross-fiber veinlets. The individual crystals are hexagonal prisms, elongated along the c axis [1, 2]. Unit cell constants of goudeyite from Majuba Hill are a = 13.472(1), c = 5.902(4) Å. Space group (corresponding to mixite) $P6_3/m-C_{6h}^2$ (No. 176) or $P6_3-C_6^6$ (No. 173), Z = 2 [1]. Goudeyite of Chessy has a = 13.50, c = 5.90 Å [2].

The color of goudeyite from Majuba Hill is yellow-green. The mineral is uniaxial positive with $n_o=1.704\pm0.002$, $n_e=1.765\pm0.002$. Pleochroism is O pale yellow-green, E green with E > O [1]. Goudeyite from Chessy is green-blue with $n_o=1.705$, $n_e=1.772$. Its pleochroism is O green-yellow, very clear, E green [2]. Hardness of goudeyite from Majuba Hill $H_M=3$ to 4. Density D(meas.) = 3.50 ± 0.03 g/cm^3 [1]. Density of goudeyite from Chessy D = 3.5 g/cm^3 [2]. The mineral is slowly soluble in dilute hydrochloric acid [1].

References for 10.8.5:

[1] W. S. Wise (Am. Mineralogist **63** [1978] 704/8). – [2] H. Sarp, J. Bertrand, J. Deferne (Schweiz. Mineral. Petrogr. Mitt. **61** [1981] 173/6).

10.8.6 Unnamed Arsenate $3(Y,RE)_2O_3 \cdot Fe_2O_3 \cdot 2As_2O_5 \pm 14H_2O$

The mineral occurs as small euhedral wedge-shaped grains together with zircon in magnetite that is embedded in manganoan calcite and associated with pyrolusite, cerargyrite, and argentian todorokite, at Treasure Hill, Hamilton, Nevada. An electron microprobe analysis results in 21.6% Y and 15.6% RE with (in %): La 4.5, Ce 25.6, Pr 2.6, Nd 19.2, Sm 4.5, Gd 12.8, Dy 14.1, Ho 0.6, Er 12.8, Yb 3.2, and traces of Lu. Besides the elements named in the formula, Ca, Ti, V, Mn, P are detected, furthermore trace amounts of Sc, Si, Al.

The luster is dull. The mineral is anisotropic. In reflected light it is medium brownish gray. Hardness $H_M = 3$ to 3.5.

Reference for 10.8.6:

A. S. Radtke, C. M. Taylor (U. S. Geol. Surv. Profess. Paper No. 575-B [1967] 108/9).

10.9 Vanadates

10.9.1 Wakefieldite $Y[VO_4]$

Wakefieldite has been found on the dumps and in the quarry of the Evans-Lou mine, once a feldspar quarry, about 40 km north of Ottawa, Quebec, Canada. It is intimately associated with quartz and hellandite, and with lesser amounts of kainosite (cenosite) and montmorillonite, and sometimes occurs with montmorillonite in cavities of quartz near yttrian apatite, xenotime, or fergusonite. One spectrographic analysis results in 20.0% Y_2O_3 and 14.0% RE_2O_3; high percentages of Fe_2O_3, CaO, B_2O_3, BeO, and SiO_2, and part of Y and the lanthanides are caused by admixtures of quartz, hellandite, kainosite, and montmorillonite. Some Si may replace V. Substitution between thorite, xenotime, coffinite, and wakefieldite can be expected. The distribution of RE for $RE_2O_3 = 100$ shows the predominance of heavy RE and the maximum of Dy: La 0.6, Ce 3.2, Pr 1.3, Nd 6.4, Sm 5.7, Eu 0.4, Gd 12.1, Tb 2.9, Dy 23.5, Ho 5.0, Er 15.0, Tm 3.6, Yb 17.8, and Lu 2.4.

The mineral forms submicroscopic grains or powder; an electron micrograph shows rectangular, elongated crystals (length ~0.1 μm), prisms with pyramidal or basal terminations, sometimes sawtooth habit. Unit cell constants are $a = 7.105 \pm 0.003$, $c = 6.29 \pm 0.01$ Å, $V = 317.5$ Å^3, space group $I4_1/amd\text{-}D_{4h}^{19}$ (No. 141) (zircon structure type), $Z = 4$. The color is pale tan and canary-yellow. Refractive indices are high. Synthetic YVO_4 is uniaxial positive with $n_o = 2.00$, $n_e = 2.14$ (± 0.01 each). Hardness $H_M \approx 5$. Synthetic YVO_4 has density D(meas.) = 4.21 g/cm^3, D(calc.) = 4.25 g/cm^3.

Reference for 10.9.1:

N. M. Miles, D. D. Hogarth, D. R. Russell (Am. Mineralogist **56** [1971] 395/410, 395, 398/9, 401/6); cf. D. D. Hogarth (Mineral. Rec. **3** [1972] 69/77, 73).

10.9.2 Kusuite $(Ce^{3+}, Pb^{2+}, Pb^{4+})[VO_4]$

Kusuite has been found in silicified limestone in the oxidation zone of the V-Cu deposit Kusu on the river Senge, ~85 km SW Kinshasa, Zaire, together with chalcocite, cuprite, azurite, malachite, descloizite, mottramite, brackebuschite, vanadinite, heyite, chervetite, dioptase, plancheite, and chrysocolla. The kusuite crystals are either inclusions in mottramite, or attached to faces of this mineral, or they are directly associated with the silicified limestone.

Microprobe analyses of two grains with the mean of 34.79% Ce_2O_3 (corrected value) lead to the crystal chemical formula $(Ce^{3+}_{0.61}Pb^{2+}_{0.215}Pb^{4+}_{0.215})[V_{0.98}O_4]$. La is absent.

Kusuite is tetragonal and forms bipyramidal single crystals or twins. The pyramidal faces are weakly bent and mostly striated. The only crystal form is {101}. (001)∧(101) = 41°45′. Unit cell parameters are a = 7.35, c = 6.56 Å (±0.01 Å each), V = 354.4 $Å^3$. Space group $I4_1amd\text{-}D^{19}_{4h}$ (No. 141) (zircon structure type), Z = 4.

The luster is glassy. The color is black, the streak rusty-red. Under the microscope the mineral is transparent, honey-yellow, it has no pleochroism. Kusuite is uniaxial negative, n_o and $n_e > 2.0$, it has strong birefringence. The reflective power maximum = 21%, minimum = 17%. Microhardness (load 100 g) = 440 kg/mm^2 corresponds to hardness $H_M \approx 4.5$. D(calc.) = 5.30 g/cm^3.

Reference for 10.9.2:

M. Deliens, P. Piret (Bull. Soc. Franc. Mineral. Crist. **100** [1977] 39/41).

Table of Conversion Factors

Following the notation in Landolt-Börnstein [7], values which have been fixed by convention are indicated by a bold-face last digit. The conversion factor between calorie and Joule that is given here is based on the thermochemical calorie, $cal_{th\,ch}$, and is defined as 4.184**0** J/cal. However, for the conversion of the "Internationale Tafelkalorie", cal_{IT}, into Joule, the factor 4.1868 J/cal is to be used [1, p. 147]. For the conversion factor for the British thermal unit, the Steam Table Btu, BTU_{ST}, is used [1, p. 95].

Force	N	dyn	kp
1 N (Newton)	1	10^5	0.1019716
1 dyn	10^{-5}	1	1.019716×10^{-6}
1 kp	9.8066**5**	$9.8066\mathbf{5} \times 10^5$	1

Pressure	Pa	bar	kp/m^2	at	atm	Torr	lb/in^2
1 Pa (Pascal) = $1\,N/m^2$	1	10^{-5}	1.019716×10^{-1}	1.019716×10^{-5}	0.986923×10^{-5}	0.750062×10^{-2}	145.0378×10^{-6}
1 bar = $10^6\,dyn/cm^2$	10^5	1	10.19716×10^3	1.019716	0.986923	750.062	14.50378
1 kp/m^2 = 1 mm H_2O	9.8066**5**	$0.98066\mathbf{5} \times 10^{-4}$	1	10^{-4}	0.967841×10^{-4}	0.735559×10^{-1}	1.422335×10^{-3}
1 at = 1 kp/cm^2	$0.98066\mathbf{5} \times 10^5$	0.98066**5**	10^4	1	0.967841	735.559	14.22335
1 atm = 76**0** Torr	$1.0132\mathbf{5} \times 10^5$	1.0132**5**	1.033227×10^4	1.033227	1	760	14.69595
1 Torr = 1 mm Hg	133.3224	1.333224×10^{-3}	13.59510	1.359510×10^{-3}	1.315789×10^{-3}	1	19.33678×10^{-3}
1 lb/in^2 = 1 psi	6.89476×10^3	68.9476×10^{-3}	703.069	70.3069×10^{-3}	68.0460×10^{-3}	51.7149	1

Work, Energy, Heat	J	kWh	kcal	Btu	MeV
1 J (Joule) = 1 Ws = 1 Nm = 10^7 erg	1	2.778×10^{-7}	2.39006×10^{-4}	9.4781×10^{-4}	6.242×10^{12}
1 kWh	$3.\mathbf{6} \times 10^6$	1	860.4	3412.14	2.247×10^{19}
1 kcal	4184.**0**	1.1622×10^{-3}	1	3.96566	2.6117×10^{16}
1 Btu (British thermal unit)	1055.06	2.93071×10^{-4}	0.25164	1	6.5858×10^{15}
1 MeV	1.602×10^{-13}	4.450×10^{-20}	3.8289×10^{-17}	1.51840×10^{-16}	1

1 eV/mol = 23.0578 kcal/mol = 96.473 kJ/mol

Power	kW	PS	kp m/s	kcal/s
1 kW = 10^{10} erg/s	1	1.35962	101.972	0.239006
1 PS	0.73550	1	7**5**	0.17579
1 kp m/s	$9.8066\mathbf{5} \times 10^{-3}$	0.01333	1	2.34384×10^{-3}
1 kcal/s	4.184**0**	5.6886	426.650	1

References:

[1] A. Sacklowski, Die neuen SI-Einheiten, Goldmann, München 1979. (Conversion tables in an appendix.)

[2] International Union of Pure and Applied Chemistry, Manual of Symbols and Terminology for Physicochemical Quantities and Units, Pergamon, London 1979; Pure Appl. Chem. **51** [1979] 1/41.

[3] The International System of Units (SI), National Bureau of Standards Specl. Publ. 330 [1972].

[4] H. Ebert, Physikalisches Taschenbuch, 5th Ed., Vieweg, Wiesbaden 1976.

[5] Kraftwerk Union Information, Technical and Economic Data on Power Engineering, Mülheim/Ruhr 1978.

[6] E. Padelt, H. Laporte, Einheiten und Größenarten der Naturwissenschaften, 3rd Ed., VEB Fachbuchverlag, Leipzig 1976.

[7] Landolt-Börnstein, 6th Ed., Vol. II, Pt. 1, 1971, pp. 1/14.